Strengthening Systems and Realising Human Rights

In Focus – Special Book Series

Strengthening Systems and Realising Human Rights: Strategies to Progress Water, Sanitation and Hygiene (WASH)

Editors

Professor Juliet Willetts, Dr Naomi Carrard and
Dr Mohamad Mova Al'Afghani

IWA
PUBLISHING

Published by

IWA Publishing
Unit 104–105, Export Building
1 Clove Crescent
London E14 2BA, UK
Telephone: +44 (0)20 7654 5500
Fax: +44 (0)20 7654 5555
Email: publications@iwap.co.uk
Web: www.iwapublishing.com

First published 2022

British Library Cataloguing in Publication Data
A CIP catalogue record for this book is available from the British Library

ISBN13: 9781789063691 (paperback)

Contents

doi: 10.2166/h2oj.2022.104

Editorial: Systems strengthening and human rights as entry points for WASH

INTRODUCTION

This special issue explores two frameworks for progressing structural change that supports safe, universal and equitable water, sanitation and hygiene (WASH): systems strengthening and the human rights to water and sanitation. As such, it elaborates and defines the intersection between these two areas. Whilst inherently interconnected, the two also represent different entry points for change, with opportunity for increased intentional integration to improve progress towards Sustainable Development Goal 6: Clean water and sanitation.

Systems strengthening is widely recognised as foundational for ensuring equitable access to sustainable WASH services in low- and middle-income countries, leaving no-one behind. Systems strengthening requires clear sector policy and strategy, effective institutional arrangements, sufficient sector financing, integration of environmental and water resources concerns, regular cycles of planning, monitoring and review and capacity development across the sector.

The human rights to water and sanitation articulate governments' obligation to progress realisation of the right of all people to enjoy access to these basic services. Research and practice in this domain encompass two areas of focus: empowering people to claim their rights, and supporting and advocating for governments at all levels to act.

In this editorial we provide an overview of the ways in which the special issue papers draw on systems strengthening strategies or human rights approaches and principles, including accountability as an interlinking concept across human rights and systems strengthening. We identify opportunity for further integration of the two approaches, including for those applying systems strengthening to engage more deeply with the human rights framework. We note the potential for greater focus in both approaches on critical resource management and climate resilience considerations. We also identify opportunity for both practitioners and researchers to strengthen knowledge and understanding of their relevant disciplinary foundations in governance, management and legal studies, to avoid partial and uncritical use of the terms and approaches.

The special issue encouraged contributions from practice-based and emerging academic writers. It piloted a mentoring scheme to support newer authors to navigate academic writing and peer review processes. We hope that others build on this experience towards facilitating inclusive authorship and incorporation of diverse forms of knowledge to WASH scholarship.

SYSTEMS STRENGTHENING ENTRY POINTS

Two papers provide insights about addressing WASH systems as a whole. Kimbugwe *et al.* (2022), examine monitoring of the overall system as a crucial aspect of systems strengthening, arguing the importance of this process to support the system to evolve. They apply an empowerment evaluation approach with system actors to self-assess the barriers and bottlenecks to WASH sustainability and inclusion in Cambodia and Uganda. This participatory system analysis and monitoring is argued to equip WASH actors to apply systems thinking and course-correct their own work. Huggett *et al.* (2022) focus specifically on gender equality, disability and social inclusion (GEDSI) in the context of strengthening WASH systems, pointing to the need to address both self-belief of disadvantaged groups as well as shift wider norms and attitudes which present structural barriers to empowerment. Their review of field approaches in Cambodia and Timor-Leste describes how to address systemic normative and attitudinal barriers which prevent equality and inclusion.

Policy and strategy is a key dimension of the WASH system, providing a broad entry point for systemic change. Bikram & Mishra (2022) describe a policy-level state-wide initiative in India that leveraged a national policy commitment to improve WASH in government-run elementary schools. Their paper describes key elements of this initiative, including an embedded management information system (MIS) across more than 100,000 schools and champion-led transformation where well-advanced schools served as 'learning labs' for other schools. Ardhianie *et al.* (2022) focus on the city of Jakarta, Indonesia, examining sector strategy, policy and monitoring. They highlight the importance of analysis of demand and supply and suggest ways in which evidence of key drivers, pressures, status, impacts, and responses can inform a system-wide understanding of the problem and solutions. Zisa *et al.* (2022) discuss the use of social art to encourage handwashing in five Latin American countries. Except for one country, this intervention, among other factors, has increased the proportion of the population handwashing with soap. Elements of the Social Art for Behavioural Change (SABC) have also been adopted to encourage monthly tariff payment in Guanajuato, Mexico. VanRiper *et al.* (2022) paper makes the case for container-based sanitation forming part of wider city-wide inclusive sanitation strategy. They provide evidence related to user demand, outcomes and worker experiences in Haiti as the basis to advocate for public sector investment in container-based sanitation as a viable strategy for meeting sanitation needs in densely populated, low-resource environments. Chouhan *et al.* (2022) discuss the factors that contributed to improvements in WASH in schools in Odisha, India, including the importance of multi-layered stakeholder coordination and tracking of key WASH indicators. Using Social Network Analysis (SNA), the authors found that 'community representatives' had the highest overall influence, followed by teachers and headmasters and cluster level coordinators. Community level stakeholders played a vital role in the program's bottom-up approach.

As water, sanitation and hygiene commonly bridge multiple ministries and agencies, institutional arrangements are a critical aspect of WASH systems strengthening. Dhoba (2022) describe national WASH sector coordination in Zimbabwe. Focusing on the National Action Committee mandated with sector coordination, they assert the importance of legal clarity, procedural adherence, and administrative capacity to address ongoing coordination challenges. They emphasise the importance of national leadership to ensure the multiple actors involved in WASH can effectively deliver on their roles. In contrast to this, Coultas *et al.* (2022) have a sole focus at subnational level. Their study explores experiences across three local governments in East Africa towards strengthening sanitation governance, noting the importance of political will, identifying stakeholder motivations, easy-to-use monitoring and reporting systems and documentation of processes and commitments to hold decision-makers to account. Effective regulatory mechanisms are a necessary part of functioning institutional frameworks, and these are explored by Mbogo *et al.* (2022) in relation to faecal sludge management in Kenya. They describe a diagnostic, consensus-based tool to support sector stakeholders to evaluate the quality and adequacy of policy in guiding equity, targeting of resources, quality of service, financial considerations and institutional roles and responsibilities. Saxena *et al.* (2022), also focus on institutional arrangements for faecal sludge management, specifically in relation to how such services can be designed to span urban and surrounding rural areas in Odisha, India. The model included a data-led situational assessment, model development, stakeholder consultation, legal formalization of urban-rural partnership, and capacity building.

The area of finance was covered by Michaels *et al.* (2022), who discuss inequalities in access to water provided by private enterprises in rural areas in Cambodia. Their analysis demonstrates the complexity of reaching all poor households, which includes the need for additional household subsidies and improved marketing and socialisation of services. A wider question is how to incentivise service providers to reach all.

In addition to the papers by Kimbugwe *et al.* (2022), Coultas *et al.* (2022) and Bikram & Mishra (2022), noted above, planning and monitoring was also considered by Ceaser *et al.* (2022). They explore monitoring as an entry point to strengthen WASH systems in Uganda, Cambodia, PNG and Myanmar. Their premise is that government service level monitoring processes and data use are vital for effective decision making and accountability. Their analysis highlights the impact of system strengthening activities on data coordination, timely and relevant data availability and data use to inform decision making in WASH service delivery. They articulate a vision for progressive realisation of 'data-based accountability culture'. Bikram & Mishra (2022) also focus on the collection and use of data, highlighting how ICT-based tools generated evidence and informed planning to improve school-based WASH in Uttar Pradesh, India.

Finally, capacity building is a fundamental tenet of WASH systems strengthening approaches, and was described in four papers. Nhim & McLoughlin (2022) share insights from an innovative leadership program in Cambodia, providing evidence of its effectiveness by comparing with other leaders. They also describe changes in leaders' capacity, skills and perceptions of leadership and factors influencing their effectiveness in promoting sanitation. Yusuf *et al.* (2022) describe processes to support government capacity building and community empowerment at local level in Nigeria. They point out the challenges of implementing partnerships to achieve the desired changes, which included non-discriminatory spaces for participation of communities. Leal *et al.* (2022) evaluate how a Massive Open Online Course (MOOC) can be utilized to increase knowledge and skills to strengthen WASH systems.

They compare different models, demonstrating that a providing structured support and blended-learning arrangement had respectively higher completion rates compared with a self-paced stand-alone course.

While connections between WASH, environment and water resources was not a central focus for any of the papers, we wish to emphasise this critical component of systems strengthening, particularly in the face of climate change and its impact on the water cycle and resources on which WASH services depend. A number of sanitation-focused papers did note the links between sanitation and clean, healthy environments (e.g. VanRiper *et al.* 2022; Saxena *et al* 2022; Mbogo *et al.* 2022). In addition, Ardhianie *et al.* (2022) focus on the importance of supply and demand analysis – including links to available water resources – as critical for water services planning. Hepworth *et al.* (2022) highlight the smaller number of papers discussing accountability issues on water resources or agricultural water management as compared to WASH, arguing that more discussion on accountability across sectors is required given they are highly interlinked and affected by climate change. Given the foundational nature of environment and water resources for WASH service availability and sustainability, we hope to see further work on this area in future systems strengthening initiatives.

HUMAN RIGHTS ENTRY POINTS

Human rights were often mentioned across the papers in this special issue, however mostly with respect to the outcomes sought – realizing the human rights to water and sanitation – rather than as the entry point for change. By human rights outcomes, we refer to water and sanitation services that are available when needed, accessible, affordable, acceptable and ensure quality and safety. By human rights as an entry point, we mean approaches drawing directly on human rights principles (non-discrimination and equality, access to information and transparency, participation, accountability, sustainability), approaches that specifically operationalize the human rights legal framework, and approaches that place citizen-state relations centrally.

There were exceptions however, with two papers specifically focused on human rights-based approaches to drive change. Gosling *et al.* (2022) critically reflect on WaterAid's human rights-based approach in Nepal, Ghana, Mali, and Burkina Faso, finding that a specific focus on the human rights framework created awareness for duty-bearers about their responsibilities, and for rights-holders about their entitlements. The approach was described to contribute to increased WASH budgets and prioritisation of WASH funding at the local level, a stronger base for citizen demand and mobilisations, integration of the most marginalized in policy and strategy, and increased openness of officials to sit together with citizens. An interesting finding was that explicit constitutional recognition of water and sanitation as a human right is more beneficial from an advocacy standpoint when compared with an implicit or interpretive recognition. Masiangoako *et al.* (2022) analyse two strategies for claiming the human right to water in South Africa – protest and litigation. They identify benefits in combining approaches, with litigation best employed alongside complementary strategies such as community mobilisation and protest. They also describe how water rights can be claimed in tandem with other rights such as housing, exploring how case study communities indirectly gained access to water services by employing legal tools of land expropriation and provincial intervention. This article resonates with the situation in low-middle income countries where the human right to water is recognized constitutionally, but is challenging to implement, and highlights the importance of litigation and protest in holding local governance systems accountable.

Other papers explored WASH from the perspective of specific human rights principles. Yusuf *et al.* (2022) focus on the principle of participation, exploring participatory processes employed in WASH programs across two local government areas in Nigeria. They connect participation to collaborative models whereby international non-government organisations work in partnership with local government authorities. Key findings include the importance of engaging with power dynamics to facilitate effective participatory spaces that are non-discriminatory and account for political, socio-economic and cultural differences. As previously mentioned, Huggett *et al.* (2022) highlight the importance of changes at both individual/self-empowerment and enabling environment levels in pursuit of equality and non-discrimination. Hepworth *et al.* (2022), Ceaser *et al.* (2022), Dickin *et al.* (2022) and Coultas *et al.* (2022) all address the principle of accountability, as described further below. The human rights principles of transparency and access to information are given relatively less focus in included papers. The principle of sustainability is addressed as it relates to continuity of services, though without consideration of its environmental aspects as noted above.

ACCOUNTABILITY AS AN INTERLINKING CONCEPT FOR RIGHTS AND SYSTEMS

Accountability was dealt with in four papers. It is a concept that is common to both WASH systems strengthening approaches and human rights. Hepworth *et al.* (2022) place this concept centrally in their review of global evidence on accountability and advocacy interventions in the water sector, with the aim of improving practice, policy and research related to strengthening WASH systems. They provide a detailed theory of change for how such interventions achieve outcomes and the factors that shape performance. Dickin *et al.* (2022) argue for the importance of accountability between multisectoral stakeholders at national level, known as mutual accountability. Findings from across five countries, namely Bangladesh, Indonesia, Kenya, Peru and Somalia, demonstrate the nascent status of mutual accountability in the WASH sector. Multistakeholder platforms existed, primarily supporting coordination and communication, and in the future, these could serve as a foundation mutual accountability. Ceaser *et al.* (2022) also discuss the importance of accountability in the context of service delivery monitoring approaches, as such systems facilitate mechanisms to hold decision-makers to account. Lastly, Coultas *et al.* (2022) describe efforts to support local leadership and note the importance of making commitments publicly to ensure commitments were held and served to strengthened sanitation governance.

INTEGRATION OF HUMAN RIGHTS AND SYSTEMS STRENGTHENING APPROACHES

In concluding this editorial, we present four key points for the WASH sector to consider in relation to the practical application of human rights and systems strengthening.

Firstly, human rights provide a globally recognized normative framework which offers guidance, standards and direction for development. There is significant further opportunity for systems strengthening to engage more deeply with this framework by supporting its operationalization and enhancing the capacity of duty bearers to fulfil their roles.

Secondly, there is opportunity for both systems strengthening and human rights proponents to engage more with critical sustainability considerations related to resource management and climate resilience. The recently declared human right to a clean, healthy and sustainable environment (Resolution 48/13, October 2021) may prompt stronger connections, with potential to both support advocacy for WASH and to drive a stronger environmental sustainability orientation within WASH work. The resolution mentions key considerations related to WASH including environmental degradation, impacts of climate change and pollution of water, asserting the state's role to protect the environment and ensure water resources management.

Thirdly, as noted above, accountability has an important and critical place in both human rights and systems strengthening approaches. However, the area of accountability remains under-researched, and is rarely a central focus of WASH implementation practice. Further efforts to bring it into both thinking and practice are needed, along with the requisite funding to enable this focus.

Fourthly, given the historical dominance of technical and technocratic approaches to improving WASH outcomes, there is more for the WASH sector to learn about the respective disciplinary fields of systems strengthening – namely governance and management – and human rights, which are based on legal studies. Often the language of human rights or systems is used loosely by practitioners and researchers alike, without accompanying depth of knowledge on the nuances of each area. We hope that this special issue can provide a foundation for deeper appreciation of the underlying concepts such that they can inform and underpin more effective research and practice, towards the ultimate end of improving universal access to water, sanitation and hygiene.

Guest Editors
Professor Juliet Willetts
Institute for Sustainable Futures, University of Technology Sydney, Sydney, Australia

Dr. Naomi Carrard
Institute for Sustainable Futures, University of Technology Sydney, Sydney, Australia

Dr. Mohamad Mova Al'Afghani
Center for Regulation, Policy and Governance (CRPG), Universitas Ibn Khaldun Bogor, Bogor, Indonesia

REFERENCES

Ardhianie, N., Daniel, D., Purwanto, P. & Kismartini, K. 2022 Jakarta water supply provision strategy based on supply and demand analysis. *H2Open Journal* **5** (2), 221–233. https://doi.org/10.2166/h2oj.2022.076.

Bikram, K. & Mishra, R. M. 2022 Evidenced transformations: using data to demonstrate improvement in school-based access to water, sanitation and hygiene in Uttar Pradesh, India. *H2Open Journal* **5** (2), 264–274. https://doi.org/10.2166/h2oj.2022.060.

Ceaser, K., Tim, D., Fraser, G., Ellen, G., Sokhadeva, C. & Benjamin, K. 2022 Strengthening country-led water and sanitation services monitoring and data use for decision-making: lessons from WaterAid experience in four countries. *H2Open Journal* **5** (2), 348–364. https://doi.org/10.2166/h2oj.2022.028.

Chouhan, N. S., Nielsen, M. O., Singh, P., Manchikanti, S., Pandey, V., Walters, J. P. & Kadyan, K. 2022 A systems approach to improving access to water, sanitation, and hygiene (WASH) in schools in Odisha, India. *H2Open Journal* **5** (3), 395–411. https://doi.org/10.2166/h2oj.2022.044.

Coultas, M., Chanza, M. M., Iyer, R., Karangwa, L., Kariuki, J. E., Kosugi, H., Meeuwissen, T. T., Mwende, E., Myers, J., Son, P. & Ssemwanga, D. K. 2022 Galvanising and fostering sub-national government leadership for area-wide sanitation programming. *H2Open Journal* **5** (1), 1–10. https://doi.org/10.2166/h2oj.2022.022.

Dhoba, L. 2022 Strengthening water, sanitation and hygiene governance: a critical review of Zimbabwe's WASH sector institutional arrangements. *H2Open Journal* **5** (2), 248–263. https://doi.org/10.2166/h2oj.2022.063.

Dickin, S., Syed, A., Qowamuna, N., Njoroge, G., Liera, C., Al'Afghani, M. M., Chowdhury, S., Sanchez, Z., Salad, A. M., Winterford, K., Uijtewaal, E., Roaf, V., Butterworth, J. & Willetts, J. 2022 Assessing mutual accountability to strengthen national WASH systems and achieve the SDG targets for water and sanitation. *H2Open Journal* **5** (2), 166–179. https://doi.org/10.2166/h2oj.2022.032.

Gosling, L., Rai, T., Obani, P., Traore, M. A., Ouangre, L., Aliu, F. & Shah, S. K. 2022 Analysis of experience using human rights to accelerate WASH access in four countries. *H2Open Journal* **5** (2), 234–247. https://doi.org/10.2166/h2oj.2022.073.

Hepworth, N. D., Brewer, T., Brown, B. D., Atela, M., Katomero, J., Kones, J. & Gashaw, M. 2022 Accountability and advocacy interventions in the water sector: a global evidence review. *H2Open Journal* **5** (2), 307–322. https://doi.org/10.2166/h2oj.2022.062.

Huggett, C., Da Costa Cruz, L., Goff, F., Pheng, P. & Ton, D. 2022 Beyond inclusion: practical lessons on striving for gender and disability transformational changes in WASH systems in Cambodia and Timor-Leste. *H2Open Journal* **5** (1), 26–42. https://doi.org/10.2166/h2oj.2022.039.

Kimbugwe, C., Sou, S., Crichton-Smith, H. & Goff, F. 2022 Practical system approaches to realise the human rights to water and sanitation: results and lessons from Uganda and Cambodia. *H2Open Journal* **5** (1), 69–83. https://doi.org/10.2166/h2oj.2022.040.

Leal, A., Saleh, A. & Verhoeven, J. 2022 E-learning for WASH systems strengthening: lessons from a capacity-building platform. *H2Open Journal* **5** (2), 379–394. https://doi.org/10.2166/h2oj.2022.066.

Masiangoako, T., Khunou, K. & Potter, A. 2022 Fighting for water in South Africa: public participation, water rights claiming and strengthening governance. *H2Open Journal* **5** (1), 98–114. https://doi.org/10.2166/h2oj.2022.023.

Mbogo, H. N., Home, P. G., Raude, J. M. & Mwangi, R. W. 2022 Scorecard diagnostic tool for the enabling regulatory environment of faecal sludge management. *H2Open Journal* **5** (2), 333–347. https://doi.org/10.2166/h2oj.2022.070.

Michaels, T., Suong, C., Chi, C., Morm, L., Mab, P. & Denis, J. 2022 Increasing access to piped water for poor households: an analysis of water connection subsidy projects implemented by the GRET. *H2Open Journal* h2oj2022024. https://doi.org/10.2166/h2oj.2022.024.

Nhim, T. & Mcloughlin, C. 2022 Local leadership development and WASH system strengthening: insights from Cambodia. *H2Open Journal* **5** (3), 469–489. https://doi.org/10.2166/h2oj.2022.129.

Saxena, S., Chouhan, N. S., Mojumdar, S., Nielsen, M. O., Manchikanti, S., Mukherjee, A., Agarwal, N., Rani, M., Dutta, A., Vignesh, R. & Walters, J. P. 2022 Bridging the rural–urban divide in sanitation with a cluster-based approach to faecal sludge management: a case study from Dhenkanal district in Odisha, India. *H2Open Journal* **5** (4), 549–566. https://doi.org/10.2166/h2oj.2022.054.

Sarah Dickin, Abu Syed, Nishrin Qowamuna, George Njoroge, Carla Liera, Mohamad Mova Al'Afghani, Sidratun Chowdhury, Zoraida Sanchez, Abdulwahab Moalin Salad, Keren Winterford, Erma Uijtewaal, Virginia Roaf, John Butterworth, Juliet Willetts; Assessing mutual accountability to strengthen national WASH systems and achieve the SDG targets for water and sanitation. *H2Open Journal* 1 June 2022; **5** (2): 166–179. doi: https://doi.org/10.2166/h2oj.2022.032

VanRiper, F., Russel, K. C., Tillias, D., Tilt, J. & Laporte, J. 2022 Container-based sanitation in urban Haiti: how can it improve human rights as a component of citywide inclusive sanitation? *H2Open Journal* **5** (1), 135–152. https://doi.org/10.2166/h2oj.2022.037.

Yusuf, T. S., Murray, A. & Okereke, C. 2022 Working with local governments to increase access to WASH services: a case of WaterAid's participatory approaches in Nigeria. *H2Open Journal* **5** (3), 424–437. https://doi.org/10.2166/h2oj.2022.061.

Zisa, A., Nilsson, K., Mirza, R. & Vachon, T. 2022 Achieving handwashing with Social Art for Behaviour Change: the experience of the Lazos de Agua programme in Latin America. *H2Open Journal* **5** (2), 323–332. https://doi.org/10.2166/h2oj.2022.029.

doi: 10.2166/h2oj.2022.022

Galvanising and fostering sub-national government leadership for area-wide sanitation programming

Mimi Coultas [a,*], Mable Mideva Chanza[b], Ruhil Iyer [a], Lambert Karangwa[c], Jimmy Eric Kariuki[d], Hodaka Kosugi [d], Talia Timani Meeuwissen [d], Emmah Mwende[e], Jamie Myers [a], Philip Son[f] and David Katwere Ssemwanga[g]

[a] Sanitation Learning Hub, Institute of Development Studies, UK
[b] Siaya County Department of Health, Ministry of Health, Kenya
[c] WaterAid, Rwanda
[d] UNICEF Kenya Country Office, Kenya
[e] Kenya Ministry of Health, Kenya
[f] Water Supply and Sanitation Collaborative Council, Geneva, Switzerland
[g] Environmental Health Department, Ministry of Health, School of Public Health, Makerere University, Uganda
*Corresponding author. E-mail: m.coultas@ids.ac.uk

MC, 0000-0002-4391-158X; RI, 0000-0002-6223-9637; HK, 0000-0002-6388-9498; TTM, 0000-0002-9169-2609; JM, 0000-0002-0973-7565

ABSTRACT

Government leadership at both the national and sub-national levels is an essential step towards ensuring safely managed sanitation services for all. Though the importance of sub-national government leadership for water, sanitation and hygiene is widely acknowledged, to date much of the focus has been on the delivery of water services. This article sets out to start to address this imbalance by focusing on practical ways to galvanise and foster sub-national government leadership for sanitation programming. By focusing on the experiences across three sub-national areas in East Africa where positive changes in the prioritisation of sanitation by local governments have been witnessed, we (a group of researchers, local government representatives and development partner staff) cross-examine and identify lessons learnt. The results presented in this paper and subsequent discussion provide practical recommendations for those wishing to trigger a change in political will at the local level and create the foundation to strengthen sanitation governance and the wider system needed to ensure service delivery for all.

Key words: local government, sanitation, systems strengthening, WASH

HIGHLIGHTS

- Operationalises systems strengthening thinking for sanitation and hygiene (S&H) at local government level.
- Draws lessons on supporting local governments to lead and prioritise S&H from three cases.
- Learning generated by local government and development partner staff from three countries.
- Makes practical recommendations on galvanising and fostering local government leadership and support for sanitation.

INTRODUCTION

The introduction of the sustainable development goals (SDGs) in 2015 brought a new level of ambition to the water, sanitation and hygiene (WASH) sector. The SDGs commit signatories to reaching everyone with sustainable WASH services by 2030, placing a specific emphasis on women, girls and those in vulnerable situations. To meet this target, new approaches and ways of working are required.

Shifting away from projects that deliver WASH-related outputs (such as through the provision of infrastructure or behaviour change interventions), development partners are increasingly seeking to understand and strengthen the wider social, political, environmental, institutional, financial and technical context – or 'system' – in which services are delivered (Casey & Crichton-Smith 2020; Valcourt *et al.* 2020).

Following widespread decentralisation reforms, including across Africa (Cabral 2011), responsibility for sanitation and hygiene (S&H) in many countries sits with sub-national governments. Sub-national or local government is often the only structure with the mandate and infrastructure required to lead sanitation programmes at the scale needed (Perez *et al.* 2012). Strengthening systems at this level is, therefore, critical to reach everyone. This paper explores influences on local government leadership and their prioritisation of sanitation as a critical component of sub-national systems strengthening. We draw on practical experiences in three rural sub-national areas in East Africa where an increase in local government action had been identified: Moyo, Uganda, Nyamagabe, District, Rwanda, and Siaya County, Kenya. We then examine how political will and commitments were galvanised and fostered.

Context

Uganda, Rwanda and Kenya all have decentralised structures, with responsibility for sanitation provision sitting with sub-national governments.

Kenya is the largest and most populous of the three countries, with a population of 52.6 m people (World Bank Group n.d.). Constitutional reform approved in 2010 and implemented in 2013 devolved roles and responsibilities from the national government to 47 new elected county governments (Cheeseman *et al.* 2016). Under this structure, the national government retains decision making powers – creating policies, guidelines and strategies – while the county governments are responsible for implementation. Overall responsibility for sanitation provision rests with county governments. Counties have relatively high amounts of power and central government funding compared to many comparable units in other countries, making 'the Kenyan system of decentralisation closer to Nigerian federalism than the sort of limited decentralisation practiced in Malawi and Uganda' (Cheeseman *et al.* 2016).

Uganda is smaller in size and population with 44.3 m people (World Bank Group n.d.) and divided into 135 Districts and the capital Kampala. Decentralisation was introduced in the 1980s as a means of improving service delivery, devolving extensive central government powers and responsibilities to local governments (Muriisa 2008; Green 2015). However, this level of decentralisation may not have been achieved consistently in practice, with Muriisa (2008) finding that 'decisions at local government are frequently influenced and/or affected by the financial and political dependence of local governments on the central government'.

Rwanda is the smallest of the three countries, with a population of 12.6 m people (World Bank Group n.d.). Rwanda's four provinces are made up of 30 districts, which are further divided into sectors, cells and villages. Following a history of highly centralised government, decentralisation reforms were introduced in 2000 (Hasselskog & Schierenbeck 2015). However, according to Tsinda *et al.* (2021), here also 'districts have not gained the powers and resources to carry out their mandate'. Individual, institutional and collective sanitation and hygiene targets are set centrally by the Rwanda government in a five-year plan but implemented through annual plans led by district councils.

Although decentralisation has been implemented differently in the three contexts and in practice the governments have varying degrees of power to shape programmes, their responsibilities with respect to sanitation are similar allowing for meaningful comparison. Table 1 shows the different levels in the administrative system in Kenya, Rwanda and Uganda. In Kenya, the Country is mandated to provide sanitation services, in Uganda and Rwanda, it is the district level. It is this level we refer to as sub-national or local government throughout the article.

BACKGROUND LITERATURE

WASH systems strengthening approaches aim to examine the multiple, interrelated and dynamic actors and factors which influence access to and uptake and management of improved WASH services. Proponents recognise that change is often messy, unpredictable and non-linear and seek to address the complexity of reaching everyone with sustainable services by understanding where the system's strengths and weaknesses occur so that programming can be designed accordingly (Huston & Moriarty 2018; Hollander *et al.* 2020).

Different frameworks have been developed to describe and strengthen the WASH system. These include those used by Agenda for Change (Huston & Moriarty 2018) and Sanitation & Water for All (2020). These frameworks use 'building blocks' identified as key elements necessary for positive WASH outcomes in different contexts. Though not referred to explicitly as a WASH systems framework, the Africa Ministers' Council on Water's also use building blocks, describing the Ngor Commitments as 'areas of the enabling environment that need to be addressed… [and] the building blocks of an effective sanitation and hygiene sector' (AMCOW n.d.). These frameworks build on previous work linking the WASH enabling environment through a set of core governance functions (building blocks) (Jiménez *et al.* 2016) and thinking around water systems (Harvey & Reed 2004).

Table 1 | Administrative set-up of Kenya, Rwanda and Uganda with the bold indicating where the mandate for sanitation services falls

Government of Kenya	Government of Rwanda	Government of Uganda
County	Province	***District***
Sub-county	***District***	Sub-county
Ward	Sector	Parish
Village	Cell	Village
	Village	

While minor variations exist, these 'building blocks' – can be summarised as (1) policy, strategy and planning; (2) institutional arrangements; (3) infrastructure and service delivery; (4) monitoring, evaluation and learning; (4) financing; (5) accountability and regulation; (6) capacity development (Huston & Moriarty 2018; Jiménez *et al.* 2020; SWA n.d.). Each of these blocks requires specific, context-dependent actions to be strengthened. For example, to improve financing, it is likely necessary to develop financing streams, whereas for institutional arrangements, clarifying ambiguities related to roles and responsibilities of different actors may be required (Huston & Moriarty 2018; SWA n.d.).

A key factor that serves as an enabler to the other building blocks is government leadership or political will. The World Bank's Water and Sanitation Program (WSP) outlined the importance of national, state and local governments, alongside the local private sector, communities and households for at scale approaches for sustainable service delivery. Ensuring political will was identified as the 'first step' of this process (Perez *et al.* 2012). Other guidance documents also note how government leadership influences the strength and shape of the building blocks, from creating and arranging institutions related to sector policy and service provision (Agenda for Change n.d.) to ensuring prioritisation of WASH for investment (Casey & Crichton-Smith 2020). In short, it is government leadership that directs and coordinates internal and external resources – ensuring alignment to government policy – to reach scale and ensure no one is left behind (Casey & Crichton-Smith 2020; SWA n.d.).

The importance of government leadership and political will in reaching sanitation goals is also emphasised by academic literature as well. O'Reilly and & Louis (2014), for example, establish a 'toilet tripod' model that puts political will as one of the three pillars behind successful toilet adoption by drawing on their work in rural India. Building on this, McMichael (2018) shows how political will at multiple levels were key to toilet adoption in Nepal, and Ekane *et al.* (2014), after comparing sanitation outcomes between Rwanda and Uganda, conclude that political leadership and commitment in the former contributed to improved sector performance and effective service delivery.

While political will is widely acknowledged to be a key enabler to success, research and guidance on *how* it can be fostered is limited. Though Perez *et al.* (2012) acknowledge the importance of political will and the roles, responsibilities, capacity development needs and resources of local governments they do not detail how subnational political will could be fostered. The Making Rights Real approach is an example of an emerging approach that attempted to do this. Rooted in human rights, the approach fosters interaction between local civil society organisations and local government officials to empower the latter to become 'heroes' of human rights to water and sanitation. While traditional rights-based approaches focus on advocacy, placing emphasis on demanding rights and holding duty bearers to account without acknowledging of the duty bearers' (often lack of) capacity, the Making Rights Real approach aims to support government staff to understand the human rights to water and sanitation and their responsibility on a personal level so that they can begin to make realistic, context-adapted changes (Carrard *et al.* 2020). Institutional triggering is another approach that aims to increase political will. A participatory method based on the Community Led Total Sanitation (CLTS) approach aims to trigger administrative, elected, traditional and community leaders to take action to eliminate open defecation. The approach is dependent on understanding different leaders' motivations and designing activities to tap into these. However, there is limited published guidance on how to do this or the effectiveness of it.

To our knowledge, there is limited research on the ways to foster political will in practice for improved sanitation. Though studies have highlighted there is a sense that local government efficacy was influenced in part by external multi-level pressure from NGOs and donors – forming a 'multi-scalar political will' (O'Reilly & Louis 2014) – details of *how* the local authorities became 'proactive' is not included.

This paper aims to fill this gap by providing practical examples and recommendations. Doing so is particularly important given that political motivation has been low for sanitation for many years (Moe & Rheingans 2006; Rosenqvist *et al.* 2016) and that sanitation governance has been little studied in comparison to water governance (Rosenqvist *et al.* 2016). As Valcourt *et al.* (2020) argue, 'the strong preference of systems approaches to focus on rural water services (48% of all studies) also points to another significant knowledge gap in the existing literature.'

This gap is particularly evident at the sub-national government level. Given that responsibility for implementing sanitation and hygiene improvements often sits at the local rather than national government level (Cabral 2011; Ekane *et al.* 2014; Jiménez *et al.* 2014; Carrard *et al.* 2020), it is important that the local government is motivated and have the capabilities and capacity to lead sanitation improvements. According to Huston & Moriarty (2018), 'in many cases a particular building block is well developed at the national level and in sector frameworks but remains poorly applied at the district level'. Jiménez *et al.* (2014) meanwhile note that 'the combination of these processes (new approach, change of governmental responsibilities and decentralised service delivery) poses a formidable challenge to local governments. Despite the importance of this aspect from the practitioner's point of view, the role of local government in this new service delivery approach has received very little research attention.'

This work, therefore, focuses on practical ways to support sub-national government leadership to prioritise sanitation as an integral element of systems strengthening for improved service delivery.

METHODS

In late 2020–early 2021, the Sanitation Learning Hub (SLH) worked with UNICEF and Siaya County in Kenya, WaterAid and Nyamagabe District in Rwanda and WSSCC/Uganda Sanitation Fund and Moyo District in Uganda, to explore how to increase sub-national government commitments and leadership for sanitation. Specific local government areas were selected in consultation with WaterAid, WSSCC/Ugandan Sanitation Fund and UNICEF. Selection criteria were that the chosen sub-national government must have demonstrated prioritisation of sanitation and that they were willing and able to participate actively in the case study development process and online analysis workshops. Stakeholders involved in the research process – both in terms of collecting and analysing data – included researchers, sub-national and national government officials and development partners. The process involved two stages:

1. Case study development
2. Participatory online analysis workshops

The process was facilitated by the SLH who took an iterative approach to both stages of the research. This included regular reflection and adaption points designed to ensure the approach remained emergent to the needs of participants and evolution of discussions.

Case study development

Firstly, local government and development partners formed a case study development team to document their experiences relating to the overarching research question: *what influenced the local government* [county and districts] *to increase their leadership and/or prioritisation of sanitation and hygiene?* The case studies were later analysed collectively in three online workshops (see the following section). In Uganda, the team was made up of Moyo District representatives, staff from the Uganda Sanitation Fund and an independent consultant, who reviewed quarterly and annual progress reports and interviewed District leaders and staff. In Rwanda, the team included WaterAid Rwanda and staff from Nyamagabe District who used national, sub-national and organisational documents and undertook 18 key informant interviews. In Kenya, the team included UNICEF and government staff from Nairobi and Siaya County who reviewed project documents and undertook 13 key informant semi-structured interviews.

To support case study development teams, SLH developed a case study development guide. This included a suggested process as well as a case study template with examples of questions to consider in each section. This aimed at ensuring a comparable process as well as quality and level of detail. Teams were also provided with a cases study template to ensure they followed a common format to aid subsequent analysis across the cases. The case study development guide also included an annex on data management and confidentiality, a draft participant information sheet, consent form and information about obtaining permission to use third-party images. The guide, annexes and template have been published in the Supplementary Material.

The case study development guide drew on two participatory monitoring and evaluation methods: Most significant change (MSC) (Davies & Dart 2005) and outcome harvesting (Simister & Napier 2017). MSC engages stakeholders involved in a particular process under investigation to identify and agree on a change to record as well as analyse the relevant data (Davies & Dart 2005). Outcome harvesting is a process to identify, describe, verify and analyse change – evidence of a change is collected with harvesters working backwards to assess what contributed to the change (Simister & Napier 2017). Drawing on the MSC method case study teams were first tasked with identifying a change in the prioritisation and leadership around sanitation by answering the following questions: *what change had been seen, who has been influenced, what had they changed, when and where did the change take place and what was being done differently*. After the change had been identified, outcome harvesting was used to work back from the MSCs – or outcome(s) – and identify what may have contributed to them. Teams were required to identify evidence they already had available as well as list and address questions which would require further investigation through reviewing additional documents and/or interviewing key informants. Appreciative inquiry principles were promoted throughout to encourage participants to be open about their experiences and focus on what had been successful that could be built on (AI Commons n.d.).

Prior to teams undertaking literature reviews, interviews and focus group discussions, they were asked to produce a research plan. SLH provided tailored support to each case study development team throughout this phase based on demand. For most, this involved reviewing research plans, interview guides and draft case studies. Support conducting interviews was also provided for the Kenya team.

Participatory online analysis workshops

Following the development of the case studies, all participants came together to analyse and draw out learning from their experiences in a series of three 3-h online participatory analysis workshops facilitated by SLH. Each workshop was recorded and notes of groupwork taken to ensure discussions were captured to enable subsequent documentation.

The objectives and main topics of discussion for each of these collective analysis workshops are shown in Table 2. These formed the basis within which participants collectively analysed their experiences to draw out common findings, nuances, lessons and recommendations. The collective analysis approach added further rigour to the research, ensuring timely analysis by local experts-by-experience with a range of perspectives (Burns 2018). An inductive approach was used to guide the data analysis with themes emerging from across the case studies.

To inform this adaptive process, facilitators reflected regularly on the process and content of the workshops: debriefing sessions following each workshop were held with facilitators to capture immediate reflections, more in-depth reflection sessions were held periodically to reflect on the overall process to date and a learning diary was kept to record ongoing reflections as they emerged. Participants were also invited to share feedback on each workshop, with space to do so provided in the workshops, via email or through an anonymous online Padlet.

Table 2 | Workshop overview

Workshop	Objectives	Topics of discussion
1: Introduction	• introduction to participants and the different cases • understand case study contexts • begin case study comparison	• participant introductions • presentation of each case study • group discussions on similarities and differences between case studies
2: Exploration	• explore 'how local governments have achieved success' in more detail across common themes emerging from Workshop 1	• group discussions exploring the questions: 1. How was ownership created within local government? 2. What role did plans and targets play? 3. How was political will/leadership fostered? 4. What did effective MIS/data systems look like?
3: Consolidation	• draw out lessons learned and recommendations	• group discussions to identify: 1. What have we learned from each other? 2. What do we still need to learn? 3. What recommendations do we have?

Workshop one took place on 26 November 2020 and workshop two on 14 December 2020. Workshop three was delayed due to the shutting down of internet access in Uganda during the presidential and parliamentary elections, it was rescheduled and took place on 4 February 2021. Workshop participants included SLH researchers, government representatives from Siaya County, Nyamagabe District and Moyo District, the Ministry of Health and Ministry of Water and Sanitation and Irrigation in Kenya, UNICEF Kenya, WaterAid Rwanda and WaterAid UK. The Uganda team was unable to join the first workshop; however, a separate meeting was organised with them and SLH staff to share with them the discussion they missed, and space was provided in workshop two for them to present their case study.

Limitations

There are some limitations to the methodology described above. Firstly, regarding case study selection, there may be more representative sub-national governments or local administrations that respond to different incentives. However, this is offset by working with three different cases and then exploring commonalities and differences. Secondly, there is no way of knowing if local administrations will successfully sustain progress towards sanitation and hygiene outcomes over the long term. Thirdly, the MSC and outcome harvesting methods include those directly involved in the process under investigation. They are the selectors of the change(s) and primary researchers. Consequently, there is a potential bias in selecting a change and highlighting contributions that portray researchers positively. It is important to note that the change makers (government actors) were directly engaged in this process; however, the findings, discussion and conclusion should be considered as one way and not the only way to galvanise and foster leadership for sanitation.

FINDINGS AND DISCUSSION

Through the case study development process, the following significant change(s) were identified.

Changes in Nyamagabe, Rwanda, included improved coordination of activities with the establishment in 2016 of the Human Security Issue Taskforce, who met monthly, to work on sanitation and other human security issues. The district established a Situation Room which compiled monthly achievements and sets new targets. Additional district stakeholders were also involved in S&H programming – for example, the District Business Unit who allocated staff time to support the administration of a revolving fund for S&H for the first time. Furthermore, there was increased planning with the launch in the 2018/2019 fiscal year of the district's first WASH investment plan with targets and strategies for achieving basic services. This plan was incorporated into the District Development Strategic Plan which had previously not included WASH-related targets. These actions have been cited as changes which enabled Nyamagabe to become ODF in 2020.

A baseline survey conducted in 2013 had found that sanitation indicators in Moyo District, Uganda, were lower than the national average with latrine coverage at 64.5% compared to a national average of 72% and no village being ODF. In April 2014, Moyo district's political leadership convened a special council meeting – the highest political making forum in a district – and decided to put sanitation as a priority on their district's development agenda. Sanitation was added as an agenda item to every council session agenda. In addition, the dormant District Water and Sanitation Coordination Committee was re-activated and tasked to give a quarterly update to the council on progress made. Sanitation has remained as an agenda item ever since. The district also made it mandatory for those who receive benefits from livelihood programmes to have and use a toilet. As of June 2020, latrine coverage was 95%, handwashing with soap 65.3% and 130 villages out of 165 declared ODF.

Latrine coverage in Siaya county, Kenya, was only 20% in 2010. Though there had been steady progression, a critical change occurred in 2016 with the county allocating budget to sanitation-related activities. Siaya county invested 5 million shillings, with UNICEF investing 10 million shillings, for CLTS activities in the country for the 2016/2017 financial year. Additionally, the county agreed to continually finance CLTS activities with the governor signing a financial commitment letter to ensure the funds were not redirected to other programmes. The county WASH budget the previous year had been 3 million shillings, however, had been used to tackle a cholera outbreak with minimal funds used for CLTS implementation. Siaya county also implemented performance contracting for Community Public Health Offices and Community Health Volunteers, who were also provided with stipends to motivate them to undertake regular household sanitation-related visits. Siaya county became ODF in the 2017/2018 financial year with latrine coverage at 87%.

Learnings from across the case studies

The case studies and collective analysis workshops identified advocacy efforts at different administrative levels, data presented in concise and visual ways, engaging multiple actors at the sub-national level, learning exchanges and long-term partnerships all contributed to increasing sub-national government's commitments to sanitation. While documenting and publicising commitments, continuing to display data and regular reporting contributed to continued attention on sanitation.

Advocacy efforts are likely needed at multiple government levels. In all three case studies, a strong enabling environment with clear policies and strategies for sanitation at a national level was key to increasing the prioritisation of sanitation at the county and district levels. In Uganda, advocacy by a range of elected officials and technical staff working with the national WASH parliamentary forum was identified as helping to increase political prioritisation of sanitation at the national level. In Kenya, a series of sanitation conferences helped both national and county governments recommit to their ODF targets. While in Rwanda, the national government declared sanitation a human security issue impacting priorities at the sub-national level.

These advocacy efforts relied on data to achieve political buy-in and leadership at both national and sub-national levels. Data presented included health prevalence statistics, economic impact and data on sanitation access. Our research indicates that presenting compelling sanitation and hygiene-related data that speak to different leaders' motivations in concise, visual ways is a particularly effective way to engage decision makers. Sharing select information concisely was also found to be important when working with busy leaders who have multiple competing demands on their time. The research identified a range of practical ways that this can be done. In both Kenya and Uganda, health data were shared with leaders to emphasise the health benefits of improved S&H, and the long-term cost-effectiveness of disease prevention, helping to increase leaders' prioritisation of sanitation. In Kenya, UNICEF developed short, easily accessible county sanitation profiles using census data. These four-page profiles gave county leaders a snapshot of the county's sanitation situation, their county's position in the sanitation 'league table' and an estimated budget for reaching ODF status. In Uganda, institutional triggering of district staff was undertaken by the Ministry of Health. Firstly, staff were presented with photographs displaying good aspects of the district such as tourist attractions and cultural leaders from the district, this was then followed by pictures of open defecation. In an interview the District Chairperson cited, this activity really challenged them to take action on sanitation, leading them to reactivate the District Water and Sanitation Coordination Committee and make sanitation a standing agenda item at quarterly meetings. Similar triggering events were then held with government staff working at lower levels conducted by technical staff from the district.

Learning or exchange visits were also highlighted as valuable to support programme advocacy efforts and increase motivation. In Moyo, Uganda, politicians were taken to different districts on learning journeys to see what others were achieving. While community champions were engaged in learning exchange visits across sub-counties and villages to motivate them. In Kenya, County Public Health Officers were sent to different counties to learn from other's success.

The case studies indicate the need to engage a diverse range of government representatives at the sub-national government level. It was clear that sanitation decision making and programming involves both elected officials and technical staff, sometimes spread across multiple departments, with different motivations and interests. These differences impacted the entry points used to engage sub-national governments. For instance, in Siaya, Kenya, the Ministry of Health targeted mid-level county public health staff first who then worked to increase focus on sanitation at both higher and lower levels. Conversely, in Moyo, Uganda, political and technical leaders at the district (highest local government) level were targeted first to get their buy in and to encourage them to be a role model by improving district sanitation and hygiene facilities, before then leading similar processes at more local levels.

Strong partnerships between sub-national governments and development partners over long periods were found to be key to gaining and maintaining government leadership and increasing political commitments. In the case studies, partnerships had been in place for approximately 10 years.

A number of different activities have been credited with maintaining commitments. In Siaya County, Kenya, for example, to counter-act changes in personnel, leaders were encouraged to document decisions and commitments. This was found to lead to improved continuity across political transitions. Furthermore, a letter from the governor was issued to prevent the redirection of funds committed for sanitation and hygiene. In Nyamagabe, Rwanda, the use of the Situation Room with data displayed visually on the walls and was updated regularly was reported to contribute to continued attention on sanitation with district leaders and the Mayor visiting the room to engage with the data. In Moyo, Uganda, the case study reported ranking political leaders based on their sub-counties progress on sanitation and hygiene, with those not doing well being put on a 'shame list' and leaders given warnings from the county leadership, as an effective mechanism.

Each sub-national area still faces challenges. Though attempts have been made to maintain commitments across political leadership transitions, this is still a major stumbling block. It was also noted that maintaining commitments beyond the declaration of ODF status, especially when targets are theoretically met is challenging. Though data is highlighted as key, there are still no adequate resources in place to develop, maintain and update monitoring systems. Furthermore, development partners' still play an active role through providing financial support, leveraging county resources, technical support for county and district sanitation and hygiene plans and monitoring progress.

Wider implications for WASH systems thinking

The aims of this study were to identify practical ideas for how development partners, and national governments, can foster local government prioritisation for sanitation. Reflecting on the findings, there are a number of areas where actions are recommended which can help support the continuous strengthening of the WASH system at a sub-national level.

The results highlight that the sub-national system operates within a larger national system. Sub-national advocacy efforts alone are unlikely to make a substantial impact without a supportive national enabling environment. Therefore, the results suggest advocacy efforts at a sub-national level should not be undertaken without work towards developing strong policies, frameworks and commitments at the national level – this should be a priority for both the national government and development partners operating at this level. National level actors can also support sub-national political will through fostering healthy competition between sub-national levels, by ranking commitment levels and actions or through performance contracts.

Furthermore, our findings highlighted that advocacy efforts are not a one-off activity. Plans and budgets change, new priorities emerge and both political and technical staff move on. Examples from the case studies found that follow-up with leaders is needed to hold them to account and public declarations and the public documentation of decisions and commitments are effective mechanisms to ensure continued support through political transitions. As was the identification and support of champions who will remain in place, and who can lobby new staff (both political and non-political). Linked to this point is the need to engage and work with both technical staff and political appointees – both will need to be engaged in different ways and at different points in time. Different stakeholders are involved in resource allocation and planning activities. An initial stakeholder analysis to determine where advocacy efforts are needed, entry points (mid-level, political leadership) and how to best trigger action could help achieve maximum impact.

In addition, our findings support the idea that longer timelines are needed to see results from this kind of work (Casey & Crichton-Smith 2020). Development partners and national governments, therefore, should be prepared to work with sub-national governments over a longer period of time. However, this comes with challenges as replicating this long-term partnership model at scale is challenging and more research is needed to find ways to strengthen systems with less intensive engagement from non-governmental partners. Investigating the role national-level actors can play in supporting this and how commitments can influence surrounding sub-national areas should be a priority.

A strong message has emerged about how best to use data to support these efforts. It has been recognised as effective across all cases to galvanise support from local officials and maintain this commitment over time. This study identified creative ways for displaying data in a concise and interactive way.

All three cases have provided examples of local governments taking a more of a leading role in service delivery. However, it is important to note that all cases are works in progress, and challenges do remain both in terms of maintaining commitments, especially across political leadership transitions and beyond declarations of ODF. Furthermore, though data are noted as powerful and influencing, decision makers' resources for the development and maintenance of monitoring systems remain low. Challenges also persist with reaching universal coverage with vulnerable communities and households still underserved. This is especially important when considering the link between a systems approach and a human rights approach and ensuring everyone, everywhere has access to improved sanitation services.

CONCLUSION

This work was conceived to provide practical examples of ways sub-national governments can be supported to provide both leadership and resources for sanitation programming. By first identifying positive changes in the prioritisation of sanitation by local government, we used a participatory process working with both local government representatives and development partners to analyse what triggered this change. We then brought case study author teams and additional stakeholders involved to cross-examine and identify lessons learnt and practical recommendations.

Our findings indicate that to galvanise commitments, an understanding of the existing sub-national landscape and stakeholders' motivations is key. This should then be followed by ongoing engagement and strong documentation of processes and commitments to hold decision makers to account. Collection and use of timely and relevant sanitation and associated economic health data are also important and serve multiple purposes for sanitation system strengthening, including forming a basis for evidence-based planning and adaptation; informing and strengthen advocacy for resource allocation; and motivating stakeholders to continue (or improve) sanitation efforts. Investment in easy-to-use monitoring and reporting systems that generate ongoing data that can be rapidly analysed and simply presented is essential. This process can help provide an important foundation of which other 'building blocks' can then be strengthened.

DATA AVAILABILITY STATEMENT

Data cannot be made publicly available; readers should contact the corresponding author for details.

REFERENCES

Agenda for Change, n.d. Strong WASH Systems [WWW Document]. Agenda for Change. URL https://washagendaforchange.org/strong-wash-systems/ (accessed 24 March 2021).

AI Commons n.d. *Introduction to Appreciative Inquiry*. The Appreciative Inquiry Commons. Available from: https://appreciativeinquiry.champlain.edu/learn/appreciative-inquiry-introduction/ (accessed 18 May 2021).

AMCOW n.d. *Addressing Africa's Sanitation Crisis: Monitoring the Sharm el-Sheikh Ngor Commitment on Sanitation and Hygiene*. Available from: https://knowledgehub.amcow-online.org/blogs/addressing-africa-s-sanitation-crisis-monitoring-the-sharm-el-sheikh-ngor-commitment-on-sanitation-and-hygiene (accessed 12 January 2021)

Burns, D. 2018 Deepening and scaling participatory research with the poorest and most marginalised. *European Journal of Operational Research* **268**, 865–874. https://doi.org/10.1016/j.ejor.2017.11.025.

Cabral, L. 2011 *Decentralisation in Africa: Scope, Motivations and Impact on Service Delivery and Poverty*. FAC Working Paper. 20 14.

Carrard, N., Neumeyer, H., Pati, B. K., Siddique, S., Choden, T., Abraham, T., Gosling, L., Roaf, V., Alvarez-Sala Torreano, J. & Bruhn, S. 2020 Designing human rights for duty bearers: making the human rights to water and sanitation part of everyday practice at the local government level. *Water* **12** (2), 378. https://doi.org/10.3390/w12020378.

Casey, V. & Crichton-Smith, H. 2020 *System Strengthening for Inclusive, Lasting WASH That Transforms People's Lives: Practical Experiences From the SusWASH Programme*. WaterAid, UK.

Cheeseman, N., Lynch, G. & Willis, J. 2016 Decentralisation in Kenya: the governance of governors. *The Journal of Modern African Studies* **54**, 1–35. https://doi.org/10.1017/S0022278X1500097X.

Davies, R. & Dart, J. 2005 *The 'Most Significant Change' (MSC) Technique: A Guide to Its Use*. Clear Horizons, Melbourne, Australia.

Ekane, N., Nykvist, B., Kjellén, M., Noel, S. & Weitz, N. 2014 Multi-level sanitation governance: understanding and overcoming challenges in the sanitation sector in sub-Saharan Africa. *Waterlines* **33**, 242–256.

Green, E. 2015 Decentralization and development in contemporary Uganda. *Regional & Federal Studies* **25**, 491–508. https://doi.org/10.1080/13597566.2015.1114925.

Harvey, P. A. & Reed, R. A. 2004 *Rural Water Supply in Africa: Building Blocks for Handpump Sustainability*. WEDC, Loughborough, UK.

Hasselskog, M. & Schierenbeck, I. 2015 National policy in local practice: the case of Rwanda. *Third World Quarterly* **36**, 950–966. https://doi.org/10.1080/01436597.2015.1030386.

Hollander, D., Ajroud, B., Thomas, E., Peabody, S., Jordan, E., Javernick-Will, A. & Linden, K. 2020 Monitoring methods for systems-strengthening activities toward sustainable water and sanitation services in low-income settings. *Sustainability* **12**, 7044.

Huston, A. & Moriarty, P. 2018 *Understanding the WASH System and its Building Blocks*. IRC, The Hague. Available from: https://www.ircwash.org/sites/default/files/084-201813wp_buildingblocksdef_newweb.pdf (accessed 19 July 2021).

Jiménez, A., Mtango, F. F. & Cairncross, S. 2014 What role for local government in sanitation promotion? Lessons from Tanzania. *Water Policy* **16**, 1104–1120. https://doi.org/10.2166/wp.2014.203.

Jiménez, A., Deunff, H. L., Avello, P. & Scharp, C. 2016 *Accountability of Sustainability*. UNICEF, New York. Available from: https://www.watergovernance.org/resources/enabling-environment-and-water-governance-a-conceptual-framework/ (accessed 01 December 2021).

Jiménez, A., Saikia, P., Gine, R., Avello, P., Leten, J., Lymer, B. L., Schneider, K. & Ward, R. 2020 Unpacking water governance: a framework for practitioners. *Water* **12** (3), 827. https://doi.org/10.3390/w12030827.

McMichael, C. 2018 Toilet talk: eliminating open defecation and improved sanitation in Nepal. *Medical Anthropology* **37**, 294–310. https://doi.org/10.1080/01459740.2017.1371150.

Moe, S. & Rheingans, R. 2006 Global challenges in water, sanitation and health. *Journal of Water and Health* **4**, 41-57. https://doi.org/10.2166/wh.2006.004 (accessed 24 March 2021).

Muriisa, R. 2008 Decentralisation in Uganda: prospects for improved service delivery. *African Development.* **33**. https://doi.org/10.4314/ad.v33i4.57344.

O'Reilly, K. & Louis, E. 2014 The toilet tripod: understanding successful sanitation in rural India. *Health & Place* **29**, 43–51. https://doi.org/10.1016/j.healthplace.2014.05.007.

Perez, E., Cardosi, J., Coombes, Y., Devine, J., Grossman, A., Kullmann, C., Kumar, C. A., Mukherjee, N., Prakash, M., Robiarto, A., Setiawan, D., Singh, U. & Wartono, D. 2012 *What Does It Take to Scale Up Rural Sanitation?* Water and Sanitation Program, Washington, DC. Available from: https://www.wsp.org/sites/wsp/files/publications/WSP-What-does-it-take-to-scale-up-rural-sanitation.pdf (accessed 01 December 2021).

Rosenqvist, T., Mitchell, C. & Willetts, J. 2016 A short history of how we think and talk about sanitation services and why it matters. *Journal of Water, Sanitation and Hygiene for Development* **6**, 298–312. https://doi.org/10.2166/washdev.2016.118.

Simister, N. & Napier, A. 2017 *Outcome Harvesting*. Available from: https://www.intrac.org/wpcms/wp-content/uploads/2017/01/Outcome-harvesting.pdf (accessed 19 July 2021).

SWA n.d. *Five Building Blocks*. Available from: https://www.sanitationandwaterforall.org/sites/default/files/2020-02/Five%20building%20blocks.pdf (accessed 19 July 2021).

Tsinda, A., Abbott, P., Chenoweth, J. & Mucyo, S. 2021 Understanding the political economy dynamics of the water, sanitation and hygiene (WaSH) sector in Rwanda. *International Journal of Urban Sustainable Development* **0**, 1–14. https://doi.org/10.1080/19463138.2021.1881787.

Valcourt, N., Javernick-Will, A., Walters, J. & Linden, K. 2020 System approaches to water, sanitation, and hygiene: a systematic literature review. *Int. J. Environ. Res. Public Health* **18**. https://doi.org/10.3390/ijerph17030702.

World Bank Group n.d. Population, total | Data [WWW Document]. Available from: https://data.worldbank.org/indicator/SP.POP.TOTL (accessed 14 May 2021).

First received 29 January 2021; accepted in revised form 19 December 2021. Available online 6 January 2022

doi: 10.2166/h2oj.2022.039

Beyond inclusion: practical lessons on striving for gender and disability transformational changes in WASH systems in Cambodia and Timor-Leste

Chelsea Huggett [a,*], Livia Da Costa Cruz [b], Fraser Goff [a], Pharozin Pheng [c] and Doeurn Ton [c]
[a] WaterAid Australia, Melbourne Victoria, Australia
[b] WaterAid Timor-Leste, Dili, Timor-Leste
[c] WaterAid, Phnom Penh, Cambodia
*Corresponding author. E-mail: chelsea.huggett@wateraid.org.au

CH, 0000-0001-9063-8259

ABSTRACT

Women and girls, and people with disabilities are often excluded from governance systems and decision-making on water, sanitation and hygiene (WASH) at all levels – from household to the national level. This paper analyses inclusion and empowerment outcomes for these two groups in three WASH system-strengthening programmes in Cambodia and Timor-Leste. The analysis used WaterAid's 'Equality and Inclusion Framework' to assess the quality of inclusion and empowerment outcomes at different levels of WASH systems. The review found that household gender equality changes in Timor-Leste were empowering because they shifted women's and men's roles, relationships and attitudes. It found that women's leadership and disability-inclusive approaches in Cambodia achieved inclusive outcomes because they enhanced women's critical consciousness and increased WASH leaders' awareness of the importance of disability inclusion. There was evidence of either individual/self-empowerment changes or changes to the enabling environment; however, no intervention achieved a comprehensive combination of both. The authors conclude that both individual and environmental outcomes contribute to empowerment outcomes in WASH systems at different levels that can shift gender norms, redistribute power and benefit women and people with disabilities. These findings will enable WASH practitioners to better understand empowerment outcomes and how to integrate them with system-strengthening approaches.

Key words: disability, empowerment, gender, norms, system strengthening, WASH

HIGHLIGHTS

- Gender equality and disability reduce inequalities in WASH system strengthening.
- Partnerships with women's rights groups and disabled people organisations.
- Practitioner perspectives from Cambodia and Timor-Leste.
- Testing field approaches and leveraging this learning to inform the influence of WASH systems nationally.

INTRODUCTION

Globally, 3.6 billion people lack access to safely managed sanitation services at home, and 2 billion people lack access to safely managed drinking water at home (WHO/UNICEF 2021). Despite progress towards increased access to water, sanitation and hygiene (WASH), WHO/UNICEF's Joint Monitoring Programme (2021) found inequalities in access to WASH based on geography, wealth quintiles and other factors. As Cambodia and Timor-Leste drive progress towards the Sustainable Development Goal 6 by strengthening WASH systems and services (WHO/UNICEF 2021), it is critical that they include, benefit and empower women and girls in all their diversity, and that they do the same for people with disabilities (Overseas Development Institute/WaterAid 2017).

There is a deep understanding of the importance of gender equality in efforts to strengthen WASH systems (Dery *et al.* 2020; MacArthur *et al.* 2020) and an increasing awareness of the importance of disability inclusion (Mactaggart *et al.* 2018). However, the literature on these intersecting issues is limited, and there is a shortage of studies that focus on the importance of integrating them. To support women's empowerment and gender equality, WASH system interventions must contribute to the transformation of political, economic and social systems.

This paper examines the changes to gender equality, disability and social inclusion (GEDSI) which occurred in three programmes designed to strengthen WASH systems. These findings are presented in the following three programmatic cross-country case studies:

- Shifting gender norms in rural community and households (Timor-Leste).
- Building women's WASH leadership (Cambodia).
- Government-led disability-inclusive WASH systems (Cambodia).

The authors analyse GEDSI outcomes across these programmes by drawing on the following: WaterAid's outcome and performance data obtained from its monitoring of GEDSI changes; qualitative studies of WASH system interventions; key informant interviews with participants; face-to-face questionnaires and field observations by WaterAid staff. The analysis tracks changes in behaviours among WaterAid's partners including the water user committees called *Grupu Maneja Fasilidade* (GMFs) and communities in Timor-Leste, and provincial and district government staff in Cambodia.

What emerges is a set of practical solutions for improving gender equality and disability rights through WASH system efforts in Cambodia and Timor-Leste. These findings are useful to WASH practitioners wishing to better understand empowerment outcomes and what works within a system-strengthening approach.

Exploring concepts of gender equality and disability inclusion in WASH system-strengthening programmes

Evidence shows that poor access to WASH disproportionately affects marginalised groups, poorer households and women. Women and girls experience challenges in relation to WASH due to their practical needs such as the need for safe access to WASH services and the need to manage menstrual bleeding (Fisher *et al.* 2017). In addition, their strategic gender interests are affected by shifts in power, social structures and status (Moser 1989; Kabeer 2002; MacArthur 2020). Despite this, women are often left out of governance systems and decision-making processes on WASH at all levels – from households to national policy and agenda setting (Dery *et al.* 2020).

People living with disabilities make up 15% of the global population (WHO 2010) and 80% live in low- and middle-income countries (WHO 2010). People with disabilities experience disproportionate challenges in accessing WASH (Mactaggart *et al.* 2018). Barriers range from physical ones such as inaccessible toilet facilities in homes and schools to social issues such as the risk of violence, discrimination or harassment when utilising services (Groce *et al.* 2011). People with disabilities are often left out of decision-making processes and leadership positions due to stigma and discrimination, or because they have not had opportunities to develop technical knowledge or skills (UNICEF 2016).

Systems thinking and practice for WASH has evolved significantly in the past decade, largely in response to concerns about the sustainability of WASH services. The literature on WASH systems describes programmatic approaches that aim to bring together diverse perspectives in order to address 'wicked problems' – problems for which there is no easy answer, and often different start and end points based on the context – encountered in WASH service provision (Neely 2019). The literature also presents systems thinking as a means of understanding and addressing the ways in which a multitude of factors affect the sustainability of WASH services (Valcourt *et al.* 2020).

The broader academic literature on influencing systems refers to the need to bound a system, understand it, assess how its component parts interact and then find 'levers of change' (Foster-Fishman *et al.* 2007) or leverage points (Meadows 1999), which can be used to influence the system. Meadows (1999) describes leverage points that include system goals; the structure of information flows; the distribution of power and decision-making; norms, attitudes, values and beliefs; and the mindset or paradigm out of which the system arises. There is limited literature on GEDSI in WASH systems, despite many of these leverage points being very relevant to transformational gender and disability programming. Power, norms and attitudes, decision-making and mindsets are highly gendered in any context and are likely to reinforce inequalities (Moser 1989; Kabeer 2005; Cornwall 2016). Gender or disability, if they are referenced in the WASH system literature at all, tend to be treated as two among many 'factors' (Valcourt *et al.* 2020) affecting how a system functions, or as one potential 'leverage point' for change, such as improved school WASH leading to improved educational outcomes for girls (Grant & Willetts 2019). Grant & Willetts (2019) identify gender equality and inclusion as one aspect of the mindset and paradigm out of which a system arises, and give an example:

> Historically the 'technocratic' paradigms for WASH interventions were found to be inadequate for creating change in the system, and so other paradigms (such as those around the human rights to water and sanitation and gender equality and inclusion) emerged, but they continue to be marginalised as discourses due to the power of dominant paradigms (Grant & Willetts 2019, 126).

The empowerment literature, particularly in relation to gender equality, explores empowerment as an ongoing process, which enhances a person's ability to make decisions or strengthen their agency (Kabeer 2005). It also examines notions of power and decision-making (ISF-UTS 2019; Dery *et al.* 2020). Empowerment is seen as arising from a combination of a person having the capacity to make choices and a shifting of structural opportunities around them, so that those choices can be actioned (Cornwall 2016). As Rowlands (1997) describes it, empowerment is an iterative process, which is also about collective action and shifts in power relations away from 'power over' towards 'power within' (self-representation), 'power with' (social capital) and 'power to' (agency). It is a process in which all four power shifts are required. Carrard *et al.* (2013) drew on Moser's framework (Moser 1989) to describe how shifting the social dynamics around gender equality can occur within the personal dimension and/or the relationship dimension. Changes in the personal level include changes related to the roles of women and men, and changes to attitudes or self-perceptions. Changes in the relationship level include women and men changing the ways in which they relate with one another (Moser 1989; Carrard *et al.* 2013). Dery *et al.*'s (2020) review of the WASH research on empowerment found the following five common interrelated dimensions of empowerment: access to information; participation; capacity building; leadership and decision-making. The authors noted that empowerment is both a cause and an outcome of successful gender-sensitive WASH programmes. It has also been recognised that gender equality work in WASH can result in backlash, resistance and negative outcomes, which may cause harm if the gender work is not done sensitively and does not employ strong monitoring and mitigation strategies (Willetts *et al.* 2014). As there is limited literature on the empowerment of people with disabilities in WASH, for the purposes of this paper, empowerment literature related to gender is also applied to analyse empowerment outcomes for people with disabilities.

STUDY DESIGN

In this paper, we analyse findings related to gender equality and disability outcomes across three WASH system-strengthening programmes (WaterAid 2017, 2020). An internal review team was established. It consisted of three field-level equality and inclusion practitioners, who are based in Timor-Leste and Cambodia, and lead programming with partners in the community, and of two technical leads (system strengthening and equity and inclusion – E&I). The team established the following criteria to select three programme examples from Cambodia and/or Timor-Leste:

- *Technical focus*: Gender and/or disability as a core focus of programme outcomes.
- *Strong evidence*: Qualitative monitoring evidence which had been collected throughout or following implementation which demonstrates changes to inclusion or empowerment.
- *WASH system levels*: Represent a cross-section of the different 'levels' at which we seek to strengthen the WASH system and influence transformative/empowering GEDSI outcomes: household; community; sub-national and national.

Of the eight programmes reviewed, the following three were considered to have met the core criteria through a review team workshop (see Table 1 summary):

- Gender equality changes in households and communities (Timor-Leste).
- Women's leadership in the sub-national WASH (Cambodia).
- Disability-inclusive sub-national WASH system building (Cambodia).

The evidence used in this paper has been taken from WaterAid's programme-monitoring systems and processes for each programme. These evidence sources are designed on a programme level to respond to specific programme outcomes and assumptions, and are intended to respond to the contextual GEDSI issues and desired influence of the programmes. This means that while the monitoring evidence and methods for evidence gathering are based on a common framework, the *Equality, Inclusion and Rights Framework* (Figure 1), the methods and evidence type vary between the three programmes. See the evidence type and source in Table 1 (WaterAid, 2021).

Table 1 | Programmes selected against criteria

Programme, country	Programme aims/outcomes from Theory of Change	Evidence type and source
Gender equality changes in households and communities Timor-Leste	• Men and women in target areas have increased knowledge and they practise equitable roles and responsibilities related to WASH at household level. • Women and people with disabilities in target communities have increased opportunities to engage on gender equality and social inclusion issues, especially in technical and leadership roles.	Qualitative evidence, generated externally by the Institute for Sustainable Futures at the University of Technology Sydney: • Qualitative evidence: A 2019 qualitative study which examined changes in gender norms and moves towards gender equality at the community and household levels (Grant & Megaw 2019). • The study engaged 172 people in total. Focus group discussions and participatory research methods were held across nine communities with 146 participants; and three semi-structured interviews were conducted with government and NGO stakeholders.
Women's leadership in the sub-national WASH Cambodia	• Increased meaningful participation of women in WASH decision-making	Qualitative and quantitative evidence, generated internally by WaterAid: • Qualitative evidence: A Gender Power Analysis (2019) drawing on perspectives of men and women from national and provincial levels through focus group discussions (50 participants) key informant interviews (17 participants). • Qualitative evidence: Real-time monitoring data from facilitators and participants (2020). • Qualitative interviews: 11 participants 1 year after leadership course (2021) using the Kirkpatrick model for evaluating training (Kirkpatrick & Kirkpatrick 2016) to assess changes in behaviour and desired results targeted by the training programme.
The disability-inclusive sub-national WASH system strengthening Cambodia	• Increased inclusive WASH evidence to inform sector • The most marginalised communities have sustainable and equitable models to provide and maintain clean water supply.	Qualitative evidence, generated internally by WaterAid: • Qualitative evidence: Monitoring and real-time learning evidence collected by WaterAid staff across 4 years from field visits, meetings and workshops. • Data were collected from WaterAid staff reflective discussions from two facilitators following a national inclusive WASH 2-day training (68 participants, 2017) and two regional follow-up workshops (2018); seven members of the Rolear District-Inclusive WASH Committee during regular committee meetings (2018–2021); and a CLTS field visit (six villages, 2019). Data were collected from four WaterAid field staff through reflective group discussions (2020) and the Inclusive WASH Annual Learning Forum (70 participants, 2019).

Once these programmes were selected, the review team applied WaterAid's gender equality framework to reflect on and analyse the programmes using the following learning questions:

- What inclusion or empowerment outcomes were achieved in each programme? What is the strength of the evidence?

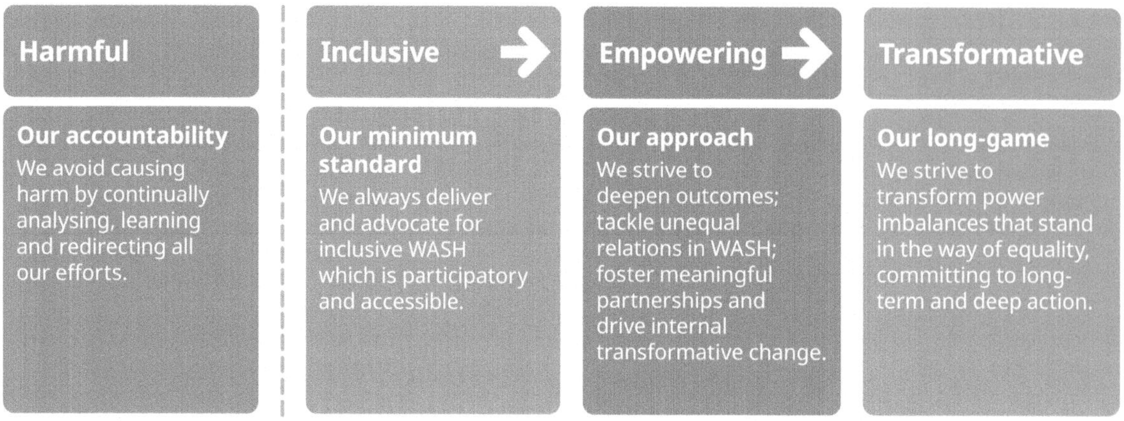

Figure 1 | Four outcome levels of WaterAid's *Equality, Inclusion and Rights Framework* WaterAid (2021).

- What system-strengthening approaches may have contributed to those outcomes?
- What were the common and divergent themes and what do the findings mean for practice?

WaterAid's method for monitoring gender equality and disability inclusion changes in WASH systems

The results presented in this review come from WaterAid's ongoing performance monitoring of GEDSI changes within WASH system programmes. The results track changes in behaviours among WaterAid's partners including the GMFs and communities in Timor-Leste, and provincial and district government staff in Cambodia. To monitor, evaluate and learn from GEDSI changes in programming, WaterAid applies outcome mapping methods. Using outcome mapping, a WASH system programme can target a behavioural change, and it can aim to measure the contribution the change has made to an intended outcome, while recognising that other actors in the local system may also have contributed to the observed change (Earl *et al.* 2001). This is of particular relevance to gender and disability changes, where specific behavioural changes related to challenging norms or power can be mapped. Changed behaviours can indicate changes to the underlying mindsets, understandings and norms of individuals and groups.

To map empowerment outcomes, WaterAid developed and later revised an *Equality, Inclusion and Rights Framework* (2021). WaterAid's framework drew together gender empowerment theory (Moser 1989; Carrard *et al.* 2013; Inter-agency Gender Working Group (IGWG) 2017), to create four 'levels' of equality and inclusion outcomes – harmful, inclusive, empowering and transformative (see Figure 1). It aims to guide how we work and what outcomes we want to achieve in WASH system-strengthening efforts at levels ranging from the individual and the household to policy and government actors (see Figure 2) (WaterAid 2017, 2020).

RESULTS

In the results section, we analyse the evidence and show what inclusion or empowerment outcomes were achieved in each programme, and what system-strengthening approaches may have contributed to those changes/outcomes. These are summarised in Table 2.

Gender equality changes in rural communities and households participating in WASH system-strengthening (Timor-Leste)

Since 2016, WaterAid has led participatory gender dialogue sessions within the rural Community Action Planning for Sanitation and Hygiene (PAKSI in Tetun language), a localised version of the Community-Led Total Sanitation (CLTS) which uses disgust rather than shame, to trigger community behaviour change and structured community action planning. The approach aims to motivate households to share WASH labour more equally between women and men. The sessions facilitate dialogue about gender norms, roles and rights in support of achieving WASH outcomes. A range of facilitator tools make up five modules, which local partners and women's rights organisations implement. The tools are implemented at strategic moments alongside WASH activities that establish community water user groups – GMF. While the dialogue sessions do not explicitly encourage greater women's involvement in the GMFs, they seek to influence the societal norms that would otherwise limit gender equity in the community-level system for the management of WASH services.

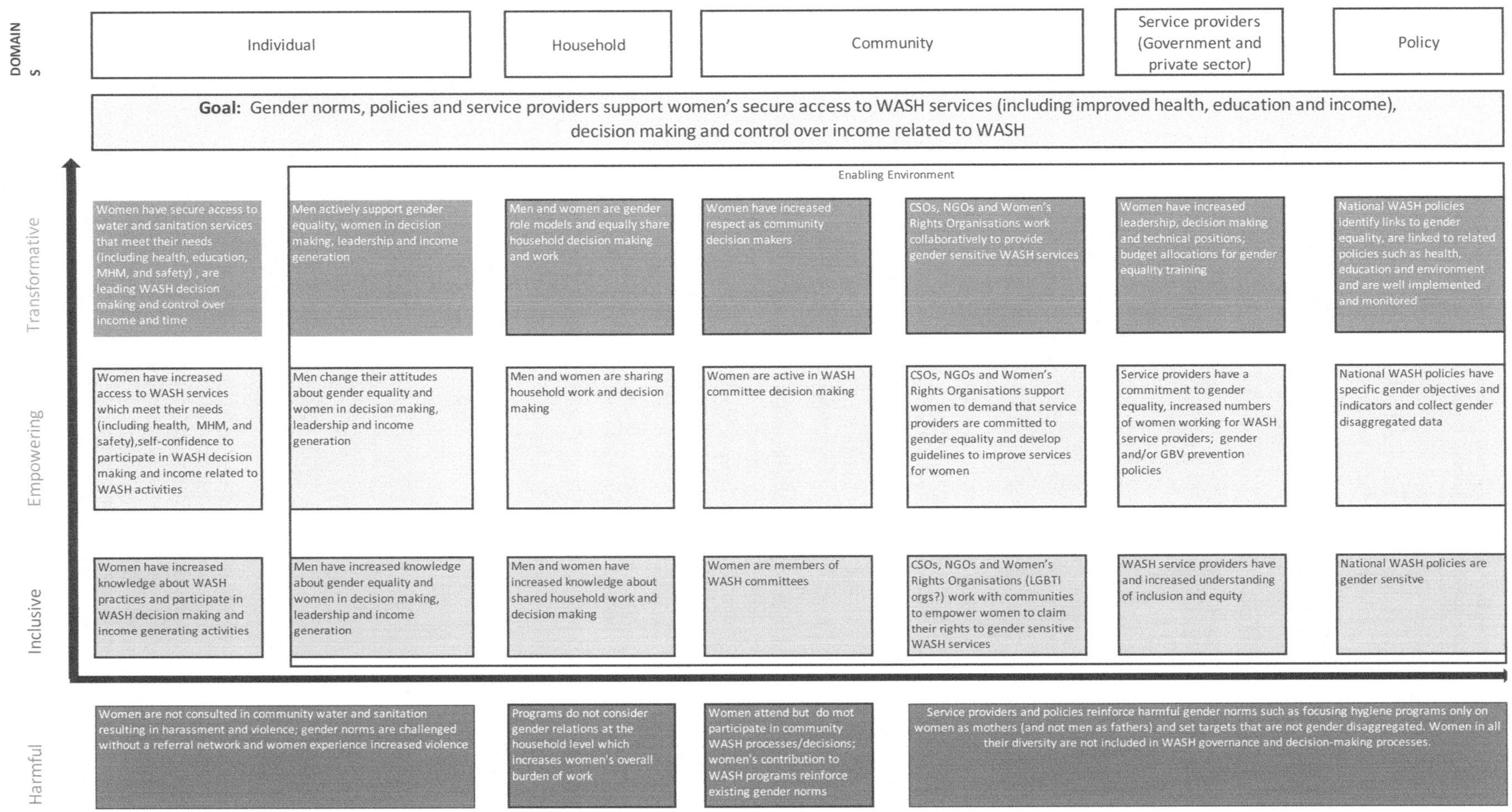

Figure 2 | Gender in WASH system outcome framework WaterAid.

Table 2 | Results of programming evidence and findings of the review team

Outcome/design of programme	Results of real-time Monitoring Evaluation and Learning and programme evidence	Review using WaterAid framework and the empowerment literature
Timor-Leste		
Men and women in target areas have increased knowledge and adopt equitable roles and responsibilities related to WASH at household level.	a. Changes in the household WASH role division between men and women b. Women were more involved in household-level decision-making.	WaterAid's '**Empowering**' outcome level, demonstrated by evidence of: • Attitudinal change within individuals (both women and men) and a shift in the of roles and relationships between women and men (Carrard *et al.* 2013) • Practical needs of women (WASH) necessary to improve gender equality were met (Moser 1989) • Access to resources as one core dimension of empowerment was met (Kabeer 2005) • Type of power shifts: 'power to' and 'power with' (Rowlands 1997; Cornwall 2016)
Women and people with disabilities in target communities have increased opportunities to engage on gender equality and social inclusion issues, especially in technical and leadership roles.	a. Increase in the status of women – greater opportunity to fill leadership roles in GMF structure. b. Women have increased opportunities for decision-making c. Men support women to participate more in village and sub-village meetings d. Follow-up modules need to go deeper to identify changes occurring at the household and community levels, including indicators to measure decision-making. e. The approach needs to include ways to measure negative impacts.	WaterAid's 'Empowering' outcome level, demonstrated by evidence of: • Strategic interests of women were realised through leadership and decision-making (Grant & Megaw 2019) • Gender norms were effectively challenged and there were changes to the power structures that prevent women's active participation beyond the home, and changes to relationships with others (Carrard *et al.* 2013) • Service providers and community began to accept and support women's leadership roles (Rowlands 1997) • Changed notions of what women and men 'should' do (Cornwall 2016) • Type of power shifts: power within; power to and power with (Rowlands 1997; Cornwall 2016) WaterAid's 'harmful' outcome level at risk here, demonstrated by a lack of evidence of: • Lack of monitoring of backlash or resistance (Willetts *et al.* 2014)
Cambodia		
Increased meaningful participation of women in WASH decision-making.	a. Participants in the women's leadership training reported increased confidence and capacity to express opinions and speak in meetings. b. Women felt other training participants had been strong role models. c. Women reported greater support from their supervisors.	WaterAid's 'Inclusive' outcome level, demonstrated by evidence of: • Increased confidence and contribution to decision-making shows self-belief, increase in critical consciousness and practical skill-building (Carrard *et al.* 2013) • Women's increased engagement in decision-making and WASH

(*Continued.*)

Table 2 | Continued

Outcome/design of programme	Results of real-time Monitoring Evaluation and Learning and programme evidence	Review using WaterAid framework and the empowerment literature
	d. There are ongoing limitations to women's ability to contribute such as not being listened to or having limited decision-making power.	management are some dimensions of empowerment (Dery *et al.* 2020) • Limited challenging of the gender norms that prevent women's active participation (Sen 1997; Batliwala 2007; Cornwall 2016) • Type of power shifts: power within; power with; however, power to and power over were not shifted (Rowlands 1997; Cornwall 2016)
Increased inclusive WASH evidence to inform sector. The most marginalised communities have sustainable and equitable models to provide and maintain clean water supply.	a. District WASH committees took ownership of inclusive WASH standards and independently supported other actors from government and private sector to adopt inclusive approaches. b. Communes, pagodas and healthcare centres invested their own money to construct inclusive public WASH facilities. c. The district WASH committee deliberately included a greater diversity of participants in their activities but ongoing engagement of marginalised groups remained limited.	WaterAid's 'Inclusive' outcome level, demonstrated by evidence of: • Attitudinal change among WASH decision-makers (Rowlands 1997; Sen 1997) • However, it did not achieve active participation, engagement or leadership by people with disability, which meant there was limited evidence of changes in self, critical consciousness or stronger self-belief (Batliwala 2007; Carrard *et al.* 2013) • Types of power shifts: power with, to some degree although limited (Rowlands 1997; Cornwall 2016)

To ascertain what gender equality changes had occurred in communities, a qualitative study (Grant & Megaw 2019) examined the strategic and practical gender equality changes experienced by men and women at the community and household levels. One hundred and seventy-two people across nine communities participated. As documented in Table 2, the review found that the qualitative study's findings aligned with WaterAid's 'Empowering' outcomes across household and community change domains of the rural Timor-Leste's WASH system.

Changes in WASH responsibilities were reported at the household level, as men and women were taking on new roles. Overall, both women and men reported 'a lot of change' with respect to WASH-related roles. Eight communities reported men and women being more willing to share household tasks. Men, in particular, reported that the modules gave them a newfound understanding of the work that women did in the home.

Participants also reported changes in gender roles at the household and community levels. Eight communities reported that women had become more involved in household-level decision-making. Nine communities reported a rise in the status of women, with more opportunities to fill leadership roles. Nine communities also reported that women had increased access to opportunities for decision-making.

Women in six communities reported that men supported them to participate more in village and sub-village meetings. A women's group discussion noted that because of the gender sessions 'we are confident to participate in trainings and meetings' and that this improved equality in access to the WASH and health information. Also, it was reported that women were taking up new roles in the GMF structure, and in technical and construction roles in WASH projects.

The research also indicated that more women now shared responsibility with men for controlling resources (such as financial and property resources) and/or had control of household finances and livestock. As stated in one women's group discussion: 'Through recent training and activities, women can also make decisions about when to build pipes, and we can also become managers of technical finance in the community'.

The review found a lack of evidence of the avoidance of 'harmful' outcomes as articulated in WaterAid's framework. The qualitative study reported some limitations: the need to go deeper into identifying gender equality changes; unintended consequences were not measured such as backlash or increasing women's work burdens and prevailing social norms made change slow and difficult.

Women's leadership in the sub-national WASH system building (Cambodia)

In Cambodia, WaterAid's gender framework informed a Gender Power Analysis and a participatory process to identify barriers and outcomes for equality in the WASH system. Both processes identified limited women's participation and representation in WASH sector decision-making and limited support to women from their male colleagues and supervisors as key barriers to equity in the WASH system. In response, WaterAid developed a Women's Leadership Training Program in Kampong Chhnang to strengthen women WASH professionals' networks and leadership confidence. Twenty women from different government WASH-related departments participated in the programme which aimed to improve leadership skills and confidence on WASH topics and brought women leaders together in a peer support model. The programme facilitated a series of six sessions over a period of 6 months (December 2019–July 2020).

The programme's contribution to outcomes and changes were recorded from:

- Participants' self-reflections during sessions (recorded by facilitators).
- Formal feedback at the end of each session to inform training rounds.
- Facilitators' reflections back to the WaterAid team.
- Interviews with participants led by WaterAid at the end of the programme.
- Follow-up interviews with 11 participants 1 year after the training.

As documented in Table 2, the review found that data from the programme's real-time monitoring and follow-up interviews aligned with WaterAid's 'inclusive' outcome. A range of inclusive outcomes were documented across individual people connected to the sub-national level of Kampong Chhnang's WASH system.

During the course, participating women self-reported that they had increased capacity and more confidence to express opinions around male colleagues. They felt they had increased their leadership skills and improved their public-speaking abilities. Some women reported at the later stages of the training that they had greater support from their supervisors and some even reported promotions (although this may not be attributable to the programme).

In an interview conducted immediately after she attended the leadership course (June 2020), one participant, who is a Deputy District Governor, reported having the confidence to take on a bigger role in the effective functioning of an inclusive WASH Committee. She reported that since taking part in WaterAid's leadership programme, she confidently leads meetings with relevant people to make collective decisions on WASH. She now led the quarterly and monthly district WASH Committee meetings with members and commune councillors to check and review the progress of the WASH plan implementation:

> 'Before I didn't know much about WASH, but after I joined a lot of activities with WaterAid such as women leadership and inclusive WASH training, I have gained much knowledge on WASH and I have confidence to push and speak with community people and senior government leaders in WASH. Now I am committed to improving my district to be Open Defecation Free and push local authorities to take more actions on WASH improvement' (Interview with training participant, 29 June 2020).

Real-time learning data from facilitators' reflection notes documented changes in women's confidence, knowledge and skills. One facilitator reported that at the end of the programme, a female participant said: 'My confidence increased after the leadership training'. Another facilitator reflected following the training that rather than the theoretical study, there needed to be practical learning which met the needs of women and their own critical consciousness 'especially building more on Power with, Power Within, and Power To' (WaterAid real-time monitoring, November 2019).

Some participants reported that improvements to their knowledge and skills helped empower other women they worked with in the community. One facilitator noted that the sessions were 'building a group of supportive women'. However, another facilitator reflected that there had been limited progress in building a strong collaborative platform.

One end-of-training reflection noted that the course was a pathway that was 'important in [enabling] transformation [for] women to move up [to] the next level of their leadership and capacity at workplaces'. However, other real-time data revealed challenges to the ways in which women applied leadership skills within their organisations. One facilitator reflected that an overall learning from delivering the programme was that there is a lack of pathways for career progression and role models in workplaces and that this 'really restricts women's efforts to develop facilitative leadership style and use of their leadership potential no matter how much they try to break this glass [ceiling]'.

Another facilitator's reflection highlighted the barriers created by existing leadership styles. The facilitator reported that half of the female participants rated their organisation's leadership style as 'dictatorial' and the majority of the participants reported that their preference was for 'facilitative' leadership.

One year after the training, WaterAid undertook follow-up qualitative interviews using the Kirkpatrick Model (Kirkpatrick & Kirkpatrick 2016) to assess how the programme had contributed to gender equity in WASH decision-making. The Kirkpatrick Model was used to design interview questions based on the desired changes to behaviours and outcomes identified in the training programme plan. All training participants were invited to be interviewed with 11 of the 16 participants agreeing to an interview. All participants reported that there had been at least some change in their knowledge or their capacity, or that they had put their learning into practice. Many women reported using what they learnt to advocate for women's participation in the community level and formal meetings. One woman reported advocating for men to become champions for women.

The most common change reported was an improved confidence to speak up and share ideas. Most (8/11) participants reported that their participation in decision-making had changed. This included being more involved in decision-making, raising their voices and being heard in meetings, and positive improvements in participation generally.

Most women (9/11) reported they had observed other participants being strong role models, for example, by contributing ideas in meetings and fulfilling their roles. However, fewer (4/11) participants reported that other women had changed their participation in decision-making. Where it was reported, observable changes included an increased confidence to speak and/or contribute to decision-making. Few participants also reported making progress against the personal goals they had set in the training. Only one respondent reported their goal had been achieved and four reported making *some* progress.

Participants reported a wide range of unexpected results from the training, including changes in WASH knowledge and mobilisation of resources for WASH. Several women (4/11) observed that their increased confidence had noticeably improved their community WASH promotion work which was helping to strengthen community WASH knowledge. Two of these respondents especially noted that conveying WASH knowledge to women in the community was fundamentally about supporting women to understand their rights and to improve their decision-making role in their family. One participant noted their improved confidence to collaborate with commune leaders was helping to mobilise both commune budget for WASH and also increase philanthropic contributions from better-off community members.

The review of these findings also found a lack of evidence of avoiding 'harmful' outcomes as articulated in WaterAid's framework. Three women reported that there are ongoing barriers to women's ability to contribute, such as women not being promoted or listened to, and an inability for women to decide things in meetings.

Disability-inclusive sub-national WASH services (Cambodia)

To address the limited operationalisation of Cambodia's National Inclusive WASH Guidelines, since 2018, WaterAid has partnered with the Rolea Bier District Administration and Organisations for People with Disabilities (OPDs) to create an inclusive WASH learning hub to showcase practical implementation of the guidelines.

The inclusive learning hub adopted a range of approaches including:

- OPD-led orientations on the guidelines.
- Participatory trainings.
- Accessibility audits at homes and public institutions.

These efforts aimed to build the Rolea Bier District WASH Committee's understanding of the experiences and barriers encountered by people with disabilities in their WASH access.

The review analysed monitoring data gathered through real-time documentation of changes referred to as 'Inclusive' WASH outcomes under WaterAid's Framework, as captured in Table 2. The data that were analysed came from WaterAid staff field-monitoring reports, workshop notes, partners' reports and anecdotes shared by district government partners in online learning forums. Collectively, the evidence from these sources shows what action government counterparts began taking to increase the inclusion of people with a disability.

Evidence from the monitoring reports shows the following:

- In late 2018, WaterAid and the District WASH Committee used field trips to showcase existing accessible infrastructure as a method to inspire and trigger change. The monitoring report states, 'After training the participant created their own plan to build toilet for all in their own district. They will target three communes, three pagodas, and their district office'. (WaterAid real-time monitoring, October 2018)
- Three months later (February 2019), a (non-disability focused) system-strengthening workshop found 'evidence that the recent E&I training has influenced people's perceptions in Kampong Chhnang. When the district WASH committees were asked to draw their vision for their district in 2020 most of the toilets drawn were accessible [to people with disabilities]'. (WaterAid real-time monitoring, February 2019)
- In November 2019, the District WASH Committee learning on accessibility standards had led them to conduct an inclusive CLTS process. There was greater diversity of participants in the process, including people with disabilities, older people, pregnant women and women from women-led households; the field note reflected that this was a change from the past when only a few older people and no people with disabilities joined.
- Throughout 2020, there was evidence of a growing commitment on the part of the District WASH Committee to address disability-inclusive WASH. The Committee organised training on inclusive WASH standards, inclusive principles and construction monitoring to other government staff, religious leaders and toilet construction businesses (February–March 2020). One real-time monitoring report (November 2020) captures how the influencing components worked: 'They also visit each commune to follow up and encourage all commune offices to have accessible toilets at their community especially at public space.…This is a role model, and hope [it will] have a positive influence on other communes in Kampong Chhnang' (WaterAid real-time monitoring, November 2020). As a result, three commune councillors (Kork Banteay, Krang Lvea and Pong Ro) used their own budgets to build accessible toilets with technical support from the District WASH Committee.
- In March 2021, the Rolea Bier District WASH Committee had led two inclusive WASH activities by themselves, demonstrating that the Committee members had taken on the responsibility of implementing their own inclusive WASH plans based on their willingness and capacity. However, one challenge reported by this note was that the Committee had limited engagement of marginalised groups in its activities.

Observed changes in the mindsets and priorities of the district administration as they became the inclusive WASH champions appear to have led to public investment in inclusive WASH facilities at public offices, healthcare centres and pagodas, and the mainstreaming of inclusive principles in WASH planning and community activities. Other districts have approached Rolea Bier or the Rolea Bier District Administration to learn from their example.

DISCUSSION

The review found that two of the three GEDSI-related WASH system initiatives might have contributed to inclusive outcomes; one might have contributed to empowering outcomes and as yet none demonstrated evidence suggesting transformative outcomes. The review also found limited evidence of initiatives, ensuring that there were no harmful outcomes. The key themes and their implications for practice that emerged from the review were as follows.

Supporting individual and collective empowerment

Both the Timor-Leste gender dialogue sessions and the women's leadership development programme in Cambodia sought to empower women within the WASH system by building two dimensions of empowerment: 'power within' (whereby women recognise their own power) and 'power with' (whereby collective action is built to achieve social change) (Cornwall 2016).

The analysis of gender module sessions in Timor-Leste found evidence of changes to women's strategic interests (Moser 1989) including confidence to participate in meetings; to contribute more to community decisions about WASH and to participate more actively in the GMFs that were formed to manage the water systems. At the household level, the evidence demonstrating a shift towards men taking on more WASH responsibilities in households represents a change in the demands placed on women (Moser 1989). This reflects progress towards WaterAid's programming outcomes targeted through the approach of dialogues between men and women. Since women have traditionally been expected to take on the role of WASH work in the home, the evidence from eight communities reporting men and women being more willing to share household tasks is a positive change in social dynamics (Carrard *et al.* 2013).

The qualitative data showed that broader community members wanted to support women to speak up and find practical ways to enable women to participate more. This is an example of collective power (Batliwala 2007) and shows evidence suggesting that the gender dialogue approach led communities to adopt a 'power with' approach (Rowlands 1997).

A review of the data of Cambodia's leadership programme gathered from women and facilitators during and immediately after the course, as well as 1 year later, revealed that women felt more confident in themselves. For example, the leadership courses focused on public speaking and self-esteem, and women participants and the facilitators reported that their confidence to speak out and raise their voices in the decision-making processes had increased. This reflects a positive change of 'power within' (Rowlands 1997). It also aligns with the notion that empowerment is a process of building self-esteem (Cornwall 2016). The programme's empowerment changes demonstrate 'Capacity Building', one key component of Dery *et al.*'s (2020) empowerment outcomes in WASH.

Tackling systemic inequality barriers in the enabling environment

Both the Timor-Leste gender dialogue sessions and the inclusive WASH learning hub in Cambodia sought to address systemic barriers within the WASH system's enabling environment – norms, attitudes and entrenched decision-making structures – which limit empowerment.

The evidence from Timor-Leste demonstrates a contribution to changing men's attitudes and actions, which is a critical part of empowerment work. As Sen (1997) describes it, the prevailing social norms also have to shift in order for women to be able to act on their critical consciousness, which requires space and an enabling environment for this change to be supported. The review found an attitudinal change reflected empowerment outcomes (Cornwall 2016) as evidenced by the qualitative study's documented changes of men's attitudes and actions, such as supporting women in village meetings, sharing household decision-making and having a better understanding of the gendered vision of labour.

The review found in Cambodia changes in disability-inclusive practice were not empowering of people with disabilities themselves. Rather, the changes were to the political commitment of district authorities and to the actions taken by existing leaders to make WASH more inclusive. So, the approach did not shift power relationships; however, it could be seen as an example of 'power with' (Rowlands 1997) if these efforts are viewed as part of collective action (Batliwala 2007). The evidence of the District WASH Committee independently training others on inclusive WASH standards and advocating for greater prioritisation of inclusive WASH services by other WASH actors may indicate the beginning of collective support among WASH service providers for people with disabilities.

The review also found evidence in Rolea Bier of 'mindset changes' among WASH actors (Grant & Willetts 2019; Neely 2019) including changing attitudes towards people with disabilities and responding to their specific WASH needs. This was evidenced by the provincial and district governments regularly meeting with the ODP and building the capacity of marginalised groups, especially persons with disabilities. They did this by listening to their voices, conducting public forums, and visiting places where marginalised groups would otherwise not have been listened to at all.

The Timor-Leste and Cambodia case studies demonstrate that even where empowerment interventions intended to change gender attitudes at a community level (Timor-Leste) or among one cohort (women leaders in Cambodia), such change is slow due to prevailing social norms. Persistent cultural and societal norms (such as attitudes towards women) must be addressed in order to achieve more empowering outcomes for women and people with disabilities in decision-making.

Moving from inclusive to empowering outcomes

The review team identified that the key to moving from *inclusive* to *empowering* was a holistic approach which focused on strengthening the confidence and self-belief of women and/or people with disabilities at an individual level while simultaneously tackling the systemic barriers preventing their empowerment (Cornwall 2016). Among the three case studies, the Timor-Leste gender dialogue sessions were found to be contributing to *empowering* outcomes, while the women's leadership and inclusive learning hub in Cambodia were contributing to *inclusive* outcomes.

The Timor-Leste gender dialogue sessions shifted women's understandings of their ability to participate in decision-making and also challenged wider community attitudes to gender norms. This combination led to changes in gender relations and roles.

The two interventions in Cambodia did not attempt to achieve empowerment outcomes with such a holistic approach and consequently did not achieve meaningful empowerment outcomes for women or people with disabilities.

Women who participated in the women's leadership training in Cambodia reported feeling more confident, and having improved knowledge within themselves, but have not quite reached empowerment achievements such as exercising leadership or greater power (Kabeer 2005). Some of the participants even named ongoing barriers in their environment which meant that their confidence did not translate to more meaningful participation in WASH decision-making. Barriers such as women not being promoted, or listened to, or having access to decision-making authority, are all examples of limiting paradigmatic factors (Foster-Fishman *et al.* 2007; Cornwall 2016) and relational issues that are part of the women's local WASH system. This shows that in order to create meaningful change and personal transformation, women's leadership efforts in WASH systems need to maximise power to and power with (Rowlands 1997). The leadership programme addressed only 'power within'. Batliwala (2007) describes collective action as being essential to the process of empowerment. The leadership programme promoted the creation of role models of women who shared ideas. However, without the engagement of male champions this did not enable women's role to extend to decision-making or collective action between men and women.

Conversely, the Rolea Bier inclusive WASH learning hub changed attitudes towards people with a disability among WASH decision-makers. However, while local ODPs were involved in building knowledge and awareness, the limited leadership and ownership by people living with disabilities of the ongoing roll-out of the inclusive WASH guidelines in the district missed opportunities for meaningful participation (Dery *et al.* 2020) and true collective action with a shift in power relations from 'power over' to 'power with' (Rowlands 1997). There was little evidence throughout the real-time learning data to demonstrate a shift in power within and power to (Rowlands 1997). The four requirements from Dery *et al.* (2020) – participation, capacity building, leadership and decision-making – were not met. The review found there was little evidence of self-empowerment of people with disabilities due to the focus of the activities being on the WASH systems and actors rather than on individuals or groups themselves.

These case studies reinforce the framing in the literature of empowerment as a process which requires both shifts in consciousness and changes in culturally embedded normative beliefs and ideas (Cornwall 2016).

Building a case for a systems approach to gender, disability and social inclusion in WASH

The review of outcomes across the three case studies found changes at different levels of WASH systems. It highlights that integrating gender, disability and social inclusion outcomes within system-strengthening programming provides useful opportunities and leverage points for addressing systemic barriers to inclusion and empowerment.

Building commitment for gender equity and disability inclusion among those in positions of power and authority within the WASH system can contribute to shifting norms and behaviours, as demonstrated by the Rolea Bier district decision-makers regarding disability-inclusive WASH. Changing fundamental aspects of the system such as norms, mindsets and paradigms, especially at the societal level, is known to be difficult (Meadows 1999; Foster-Fishman *et al.* 2007). Working with active change agents (Meadows 1999) to change how system stakeholders see the world, in this case by encouraging visions of more inclusive services, leads to changed practices and behaviours and eventually alters the system context (Foster-Fishman *et al.* 2007), both for WASH and for other public services. Early evidence that this is occurring in Rolea Bier includes the District WASH Committee influencing communes, pagodas and other public facilities to the extent that they utilised their own budgets to build accessible toilets, which is itself a positive outcome for inclusive services. Notwithstanding the limited leadership by people with disabilities themselves, the District WASH Committee's increasing engagement of diverse stakeholders and perspectives to solve the multiple problems (financing, accessibility and inclusive participation) facing WASH services in their local system shows the district authorities are beginning to demonstrate more participatory leadership (Dery *et al.* 2020) and use their 'power to' and 'power over' to advocate for more inclusive practice among other actors in the system.

While prevailing norms may change slowly, community gender dialogues in Timor-Leste demonstrate that there can be key moments of WASH system changes when introducing equity, and inclusion initiatives can have rapid impacts on the paradigms within which WASH systems function (Meadows 1999). In the Timor-Leste example, the creation of GMFs for managing water infrastructure provided opportunities for community collective action (Rowlands 1997) and new power relations within the new decision-making structures (Meadows

1999). An initial shift in social norms occurred through facilitating community dialogues. This process contributed to the emergence of women role models and demonstrated women's proficiency in decision-making and resource management. It is an example of 'insert[ing] people with the new paradigm in places of public visibility and power' (Meadows 1999, p. 18). It is also an example of gender empowerment as both a cause and an outcome of successful gender-sensitive programming (Dery *et al.* 2020) and of the creation of positive feedback loops that enable women's more visible leadership in WASH to reinforce new social norms (Meadows 1999).

This review did not find sufficient evidence to assess whether greater gender equity or disability inclusion can create a stronger WASH system. In both Timor-Leste and Cambodia, positions of formal power and authority in WASH systems continue to be dominated at all levels (household, community, district and national) by able-bodied men. It would be useful for future research to consider whether WASH decision-making, planning and service provision that is more equitable, participatory and representative of service users, performs more strongly on system metrics such as service access, accountability, equality, transparency and sustainability.

Limitations

As well as the programmatic limitations mentioned in sections 'Results' and 'Discussions' (such as limited monitoring of potentially harmful consequences of gender programming and the need to deepen the monitoring of gender equality changes), the authors note two key limitations of the study design and methodology:

1. The sources of data in all three case studies did not include baseline or data collected from control groups. Therefore, changes observed cannot be fully attributed to WaterAid's programming, but as a contribution towards social changes. These findings need to be considered as one potential explanation of how the change has happened and is not representative of all empowerment outcomes in WASH systems.
2. Only the qualitative evidence of the gender equality changes in households and communities in Timor-Leste was collected by independent reviewers. The evidence from the programmes in Cambodia may be affected by social desirability bias by participants in their responses to the women's leadership interviews. The real-time observations made by the WaterAid staff in documenting the inclusive WASH learning hub may have been biased towards particular programme targets and objectives. However, this bias was mitigated through triangulation. In Cambodia, the data were collected on a broader set of outcome questions. The monitoring strategies aimed to measure broader outcomes and it was a participant-led approach therefore some potential self-reported bias was mitigated. For example, the follow-up surveys encouraged participants to reflect generally on what changes they had observed and we later analysed based on what outcomes we sought to harvest evidence of.

These limitations mean the study's analysis and results are not representative of all contexts and WASH systems' outcomes. There is a limited ability to apply these findings to other settings, programmes and monitoring systems. The difficulties in attributing change, as well as the analysis and interpretation of findings, being internally led means the findings need to be interpreted from a practitioner lens. The authors are not suggesting that these results represent a complete narrative of how our work has shifted norms in WASH systems. The authors sought to understand the extent to which our programming has been effective against our own objectives/targets. In future, we recommend practitioner-led research to measure impact having quality baseline data to improve rigour and improve triangulation to limit social desirability bias.

RECOMMENDATIONS

The following recommendations are for practitioners addressing GEDSI in WASH system-strengthening and based on the review team's findings of limitations and areas for improvement in the above case studies:

- WASH system-strengthening interventions can leverage strategic entry points to address systemic inequalities

While WASH systems' thinking evolved in response to the sustainability challenges of WASH services, it has great potential to reorientate prevailing power structures and harmful attitudes. The leverage points and factors which system-strengthening interventions seek to influence – power, norms and attitudes, decision-making and mindsets – are highly gendered in any context and are likely to reinforce inequalities (Cornwall 2016). Applying a GEDSI lens such as WaterAid's 'Equality and Inclusion Framework' when bounding, understanding and assessing the WASH system, and designing interventions, can reduce inequalities and move from inclusive outcomes towards transformative GEDSI outcomes.

- To achieve transformative change, GEDSI interventions need to shift individuals' self-perceptions and also address attitudes and norms within the broader WASH system.

Among the three case studies, the most empowering outcomes were demonstrated when interventions raised critical consciousness and engagement (Batliwala 2007) and normative beliefs (Cornwall 2016). Increasing women's confidence and skill-building are not sufficient for empowerment if women still face cultural barriers and negative attitudes. Similarly, changing attitudes among WASH system leaders is not sufficient for empowerment if it is not complemented by empowering participation and collective action. Changing four dimensions of power – 'power within', 'power with', 'power to' and 'power over' – is necessary for meaningful empowerment outcomes.

- Employ a variety of methods to generate collective evidence to monitor deeper empowerment changes.

While the evidence in the qualitative study of Timor-Leste's gender dialogues was strong, one qualitative study finding was that programmatic monitoring needed to go deeper to identify empowerment changes. For example, there was a lack of indicators in WaterAid's programme monitoring to measure decision-making changes. The review team noted that, had the qualitative study not been done, regular programme monitoring would not have documented changes to decision-making. In the Cambodia examples, systematically monitoring indicators could have improved tracking of changing attitudes and marginalised groups participation and thus informed programme adaptations.

Overall, the review found that examining empowerment outcomes through analysis of real-time monitoring and learning data from Cambodia related to both disability and gender programming was an effective method to understand small changes over time. In contrast, as a 'point in time' study, the Timor qualitative study provided deeper insights which can be used to reorientate programming. The review team recommends that a mixture of both methods would be best to analyse and understand empowerment changes across the levels of WaterAid's E&I framework.

- Actively monitor and respond to harmful outcomes in GEDSI and system-strengthening programming.

While inclusive or empowering outcomes were reported in both Cambodia and Timor-Leste case studies, a common gap was that the 'harmful' outcome articulated in WaterAid's framework was not strongly mapped and tracked; therefore, the 'harmful' outcomes of WaterAid's framework may not have been fully avoided. The literature shows that resistance and backlash are common in gender equality work (Willetts *et al.* 2014), and it is strongly recommended that all the gender and WASH programming in system strengthening has practical tools to map, monitor and mitigate any unintended harm.

CONCLUSION

A review of three field approaches in Timor-Leste and Cambodia demonstrates the evidence of WASH system-strengthening initiatives providing an opportunity to move beyond inclusive outcomes towards achieving empowering outcomes. The review found that a system approach to WASH which integrates GEDSI can help to address the systemic normative and attitudinal barriers which prevent equality and inclusion. To achieve empowering outcomes, WASH system-strengthening actors must seek to shift the self-belief of women and people with disabilities while at the same time addressing the wider norms and attitudes to GEDSI which may present structural barriers to empowerment.

ACKNOWLEDGEMENTS

Funding for WaterAid to undertake the programmes reviewed in this paper was provided by the Australian Government's Department of Foreign Affairs and Trade/Australian Aid through the Civil Society WASH Fund and Water for Women Fund (Timor-Leste) and the H&M Foundation through the SusWASH Program (Cambodia). The authors wish to acknowledge the community members and participants who contributed their knowledge and information to this study, as well as the contribution of the ISF-UTS who led the gender dialogue review in Timor-Leste (Grant & Megaw 2019).

DATA AVAILABILITY STATEMENT

All relevant data are included in the paper or its Supplementary Information.

REFERENCES

Batliwala, S. 2007 Taking the power out of empowerment: an experiential account. *Development in Practice* **17** (4/5), 557–565.

Carrard, N., Crawford, J., Halcrow, G., Rowland, C. & Willetts, J. 2013 A framework for exploring gender equality outcomes from WASH programmes. *Waterlines* **32** (4), 315–333. https://doi.org/10.3362/1756-3488.2013.033.

Cornwall, A. 2016 Women's empowerment: what works? *Journal of International Development* **28** (3), 342–359. https://doi.org/10.1002/jid.3210.

Dery, F., Bisung, E., Dickin, S. & Dyer, M. 2020 Understanding empowerment in water sanitation and hygiene: a scoping review. *Journal of Water, Sanitation and Hygiene for Development* **10** (1), 5–15. https://doi.org/10.2166/washdev.2019.077.

Earl, S., Carden, F. & Smutylo, T. 2001 *Outcome Mapping: Building Learning and Reflection Into Development Programs*. International Development Research Centre, Ottawa.

Fisher, J., Cavill, S. & Reed, B. 2017 Mainstreaming gender in the WASH sector: dilution or distillation? *Gender and Development* **25** (2), 185–204.

Foster-Fishman, P., Nowell, B. & Yang, H. 2007 Putting the system back into systems change: a framework for understanding and changing organizational and community systems. *American Journal of Community Psychology* **39**, 197–215.

Grant, M. & Megaw, T. 2019 *Review of the Implementation of WaterAid's Gender Manual and Facilitated Sessions*. Institute for Sustainable Futures, University of Technology Sydney, Sydney.

Grant, M. & Willetts, J. 2019 Learning for adaptive management: using systems thinking tools to inform knowledge and learning approaches. In: *Systems Thinking and WASH Tools and Case Studies for A Sustainable Water Supply* (Neely, K., eds). Practical Action Publishing, Rugby, UK, pp. 107–132.

Groce, N., Bailey, N., Lang, R., Trani, J. F. & Kett, M. 2011 Water and sanitation issues for persons with disabilities in low- and middle- income countries: a literature review and discussion of implications for global health and international development. *Journal of Water Health* **9**, 617–627.

Institute for Sustainable Futures, University Technology Sydney 2019 *Womens Empowerment: Sharpening Our Focus*. Available from: https://multisitestaticcontent.uts.edu.au/wp-content/uploads/sites/57/2019/03/05164744/ISF-UTS_2019_WomensEmpowermentArticle1_GenderinWASH_WaterforWomen.pdf (accessed 17 November 2021).

Kabeer, N. 2002 Resources, agency, achievements: reflections on the measurement of women's empowerment. *Development and Change* **30**, 435–464. https://doi.org/10.1111/1467-7660.00125.

Kabeer, N. 2005 Gender equality and women's empowerment: a critical analysis of the third millennium development goal 1. *Gender and Development* **13** (1), 13–24. doi:10.1080/13552070512331332273.

Kirkpatrick, J. & Kirkpatrick, W. 2016 *Kirkpatrick's Four Levels of Training Evaluation*. ATD Press, Alexandria, VA.

Macarthur, J., Carrard, N. & Willetts, J. 2020 WASH and gender: a critical review of the literature and implications for gender-transformative WASH research. *Journal of Water Sanitation and Hygiene* **10** (4), 818–817.

Mactaggart, I., Schmidt, W. P., Bostoen, K., Chunga, J., Danquah, L., Halder, A. K., Jolly, S. P., Polack, S., Rahman, M., Snel, M., Kuper, H. & Biran, A. 2018 Access to water and sanitation among people with disabilities: results from cross-sectional surveys in Bangladesh, Cameroon, India and Malawi. *BMJ Open* **8**, e020077. https://doi.org/10.1136/bmjopen-2017-020077.

Meadows, D. 1999 *Leverage Points: Places to Intervene in a System*. The Sustainability Institute, Hartland, VT.

Moser, C. 1989 Gender planning in the third world: meeting practical and strategic gender needs. *World Development* **17** (11), 1799–1825. https://doi.org/10.1016/0305-750X(89)90201-5.

Neely, K. 2019 Systems thinking and trans-disciplinarity in WASH. In: *Systems Thinking and WASH Tools and Case Studies for A Sustainable Water Supply* (Neely, K., eds). Practical Action Publishing, Rugby, UK, pp 17–28.

Overseas Development Institute/WaterAid 2017 *How to Reduce Inequalities in Access to WASH Synthesis Report*. Available from: odi.org/sites/odi.org.uk/files/resource-documents/11604.pdf.

Rowlands, J. 1997 *Questioning Empowerment: Working with Women in Honduras*. Oxfam Publishing, Oxford, UK.

Sen, G. 1997 *Empowerment as an Approach to Poverty*. Working Paper Series 97.07, Background Paper for the UNDP Human Development Report. UNDP, New York.

United Nations Children's Fund 2016 *WASH Disability Inclusion Practices*. United Nations Children's Fund, New York, NY, USA.

USAID Interagency Gender Working Group 2017 *Gender Integration Continuum*. Available from: https://www.igwg.org/wp-content/uploads/2017/05/FG_GendrIntegrContinuum.pdf.

Valcourt, N., Javernick-Will, A., Walters, J. P. & Linden, K. 2020 System approaches to water, sanitation, and hygiene: a systematic literature review. *Faculty Publications – Biomedical, Mechanical, and Civil Engineering* **93**. https://digitalcommons.georgefox.edu/mece_fac/93.

WaterAid 2017 *Gender in WASH Systems Strengthening Guidance Note*. Available from: https://washmatters.wateraid.org/publications/practical-guidance-gender-equality-strengthening-water-sanitation-hygiene-systems.

WaterAid 2020 *System Strengthening for Inclusive Lasting Water, Sanitation and Hygiene (WASH) That Transforms People's Lives: Practical Experiences From the SusWASH Programme*. Available from: washmatters.wateraid.org/publications/suswash-system-strengthening-for-inclusive-lasting-water-sanitation-hygiene (accessed 22 January 2020).

WaterAid 2021 *Equality, Inclusion and Rights Framework*. Available from: https://washmatters.wateraid.org/publications/equality-inclusion-and-rights-framework.

Willetts, J., Carrard, N., Crawford, J., Rowland, C. & Halcrow, G. 2014 Working from strengths to assess changes in gender equality. *Development in Practice* **23** (8), 991–1006. doi:10.1080/09614524.2013.840564.
World Health Organisation 2010 *Monitoring the Building Blocks of Health Systems: A Handbook of Indicators and Their Measurement Strategies*. Available from: https://www.who.int/healthinfo/systems/WHO_MBHSS_2010_full_web.pdf (accessed 14 April 2021).
World Health Organisation (WHO) & United Nations Children's Fund (UNICEF) 2021 *Progress on Household Drinking Water, Sanitation and Hygiene 2000–2020: Five Years Into the SDGs*. Available from: https://washdata.org/sites/default/files/2021-06/jmp-2021-wash-households-LAUNCH-VERSION.pdf (accessed 15 July 2021).

First received 31 January 2021; accepted in revised form 21 December 2021. Available online 17 January 2022

doi: 10.2166/h2oj.2022.040

Practical system approaches to realise the human rights to water and sanitation: results and lessons from Uganda and Cambodia

C. Kimbugwe[a], S. Sou[b], H. Crichton-Smith[c] and F. Goff[d,*]

[a] WaterAid Uganda, Kampala, Uganda
[b] WaterAid Cambodia, Phnom Penh, Cambodia
[c] WaterAid UK, London, UK
[d] WaterAid Australia, Melbourne, Australia
*Corresponding author. E-mail: fraser.goff@wateraid.org.au

FG, 0000-0002-1091-6536

ABSTRACT

The United Nations General Assembly's 2010 legal recognition of the human rights to water and sanitation shaped approaches of many actors working to improve access to water, sanitation and hygiene (WASH). Persistent challenges of poor WASH sustainability, scale and inclusion are increasingly being tackled through system thinking and system strengthening. However, little has been written about how participatory system analysis and monitoring can equip and empower WASH actors to apply system thinking, self-assess and course-correct in their own work to improve sustainable WASH for all. WaterAid's Sustainable WASH Services at Scale (SusWASH) programme applies a system approach, underpinned by human rights principles. In this paper, we share perceptions of local stakeholders, engaged in empowerment evaluation in Cambodia and Uganda, and lessons learned for future initiatives that seek to catalyse WASH system improvements for the realisation of the human rights to water and sanitation. We argue that a system approach, underpinned by human rights principles, can help advance progress towards inclusive and sustainable WASH for all. Working in this way fosters inclusive, locally led decision-making about how system blockages can be overcome, strengthening local ownership of a shared vision for change and the capacities and skills required to achieve it.

Key words: empowerment evaluation, human rights, participatory development, system approaches, WASH

HIGHLIGHTS

- WaterAid's approach seeks to tackle systemic blockages to WASH sustainability, inclusion and scale.
- This paper illustrates how empowerment evaluation, system concepts and human rights principles can coalesce into a practical, accessible, participatory approach that helps to strengthen WASH systems.

INTRODUCTION

Calls for 'system change' are often heard in discussions about the realisation of the human rights to water and sanitation. But how is system change – change to how a system works and what happens as a result (Posthumus *et al.* 2020) – brought about in practice? 'System change initiatives are rooted in the assumption that significant improvement in the outcomes of a targeted population … will not occur unless the surrounding system … adjusts to accommodate the desired goals' (Foster-Fishman *et al.* 2007, p. 197). System thinking, originating from Complex Adaptive Systems theory, enables a more networked understanding of the environment in which actions are taken (Neely 2019a). WaterAid's definition of the water, sanitation and hygiene (WASH) system aligns with relevant literature and summarises the WASH system as all of the actors (people and institutions), factors (social, economic, political, environmental and technological) and the interlinkages between them that affect the provision of sustainable WASH for all (Neely 2019b; Valcourt *et al.* 2020). The word 'system' can sometimes conjure up thoughts of 'the establishment' or government bureaucracy. However, system thinking is much broader than that. In this context, the WASH system includes rights groups and civil society, marginalised people excluded from WASH access, community WASH users, private sector operators, politicians and local government, regulators, and many more, as well as the political economy, and gender and cultural norms which influence how services are provided and progressively improved to achieve the human rights to water and sanitation (Figure 1).

There is extensive literature on system thinking and approaches, including a growing body of both academic and grey literature on WASH systems, which has emerged largely in response to the sustainability challenges facing WASH services (Lockwood *et al.* 2010; Huston & Moriarty 2018; Neely 2019b; Valcourt *et al.* 2020). Much of the academic literature on WASH systems explores the 'application of complex analytical methods' to identify and document system factors affecting WASH sustainability and, less frequently, the interactions between factors or the ultimate impact of system approaches on WASH service sustainability (Valcourt *et al.* 2020). While these persistent sustainability challenges have increasingly informed practitioners' programming interventions to strengthen systems through which WASH services and behaviours are delivered (WaterAid 2020), there remains a need within the literature of WASH systems for 'a diversification in the methods, scopes, and applications of system approaches for WASH' (Valcourt *et al.* 2020, p. 14), especially methods 'that are likely accessible to those who are tasked with making critical decisions at the local level around WASH service provision' (Valcourt *et al.* 2020, p. 13). While practical, locally accessible tools and approaches are an identified gap in academic system thinking, a number of implementing agencies, including WaterAid, have been exploring ways to fill this gap through their interventions.

The rights to water and sanitation not only ensure accessibility, availability, quality, affordability and acceptability of water and sanitation services, but also promote principles of participation, non-discrimination, accountability, transparency and sustainability in how those services are provided and progressively realised by governments (de Albuquerque 2014). WaterAid (2020) and others have written about the complementarity of system approaches and human rights–based approaches that seek to strengthen the accountability of local government institutions for the realisation of the rights to water and sanitation. Carrard *et al.* (2020, p. 7) write how 'a systems-strengthening approach explicitly recognises the key role of local government institutions in water and sanitation service delivery and acknowledges the challenges they face in contexts where the decentralisation process is incomplete. Described as a "philosophy of action" rather than intervention type, a systems-strengthening approach seeks to engage with complexity and work towards meaningful, sustainable solutions'. Grant & Willetts

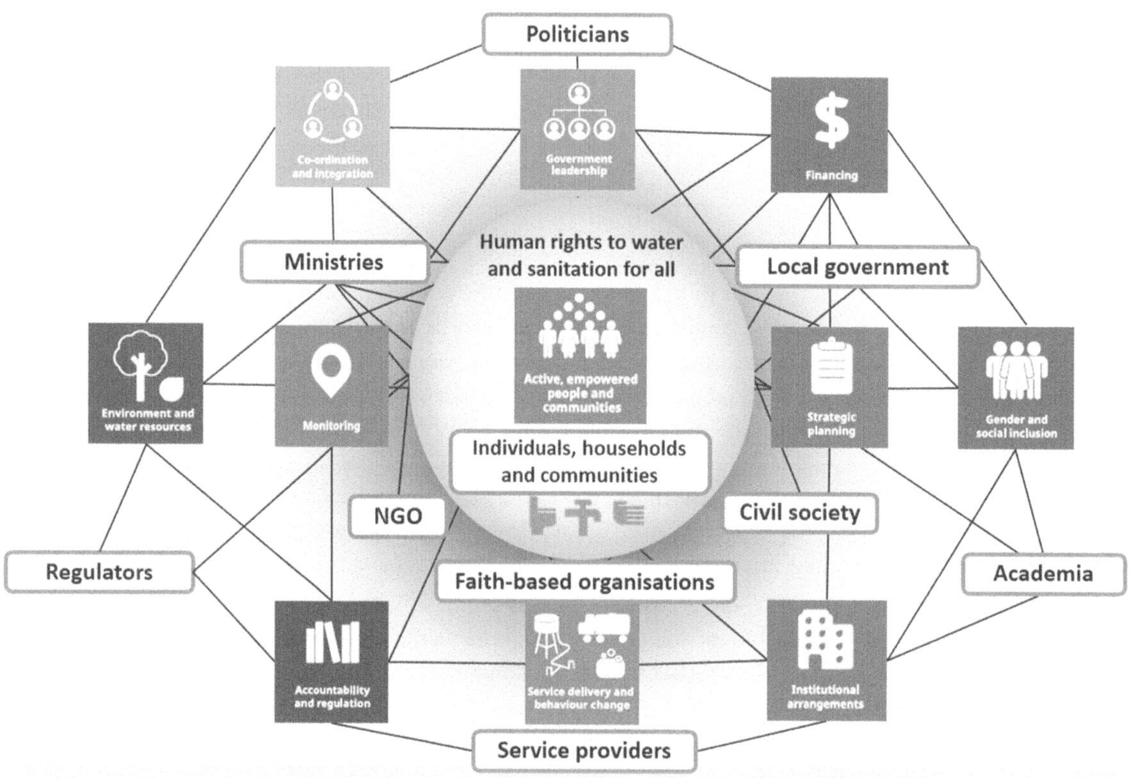

Figure 1 | Conceptualisation of the WASH system showing factors, actors and interlinkages (WaterAid).

(2019) identify the human rights to water and sanitation as one of the goals of WASH systems and also note that attention to human rights influences whose knowledge is 'legitimised and drawn upon' (p. 126), affecting the flow of information and its use in decision-making; key elements that determine how a system functions (Meadows 1999). Despite the recognised complementarity of system approaches and human rights–based approaches, a gap in the literature remains about how to effectively apply human rights principles in practical WASH system-strengthening interventions.

This paper seeks to contribute to the literature by showing how a practical systems approach, underpinned by human rights principles, can help advance progress towards inclusive and sustainable WASH for all. Strengthening WASH systems with a focus on human rights principles and with an emphasis on locally accessible WASH system assessments and regular reflections can help to foster inclusive, locally led decision-making about how system blockages can be overcome, cultivate local ownership of a shared vision for change and build the capacities, skills and mindsets required to achieve it.

WaterAid's approach to strengthening WASH systems: empowerment evaluation

As with many other organisations, the tools, frameworks and approaches which WaterAid uses to define and monitor systems involve participatory exercises that bring together diverse perspectives (Foster-Fishman *et al.* 2007; Ramalingam *et al.* 2008; Neely 2019b). WaterAid's repeated use of exercises to engage experts, decision-makers and stakeholders from across the system to monitor system change provides evidence for programme monitoring, but also provides a platform for system change through structured reflection on the state of the system and on the paradigms and mindsets out of which the system arises (Meadows 1999). Such participatory exercises have potential, as 'social deliberative processes' to be transformative in the way participating system actors perceive the complex WASH system and their role within it (König 2018). By supporting learning, adaptation and reflection at the individual and collective level (Brown 2019; Grant & Willetts 2019), WaterAid's approach aims to equip and empower WASH actors to apply system thinking, self-assess and course-correct in their own work to provide and improve lasting WASH access for all.

WaterAid employs a version of 'empowerment evaluation' to design, implement, monitor and adapt system-strengthening interventions. In this approach, WaterAid plays the role of a 'critical friend' (Fetterman 1994) and uses 'evaluation concepts, techniques and findings to foster improvement and self-determination' (Fetterman & Wandersman 2007, p. 186). Empowerment evaluation aims to 'increase the probability of achieving program success by: (a) providing program stakeholders with tools for assessing the planning, implementation, and self-evaluation of their program, and (b) mainstreaming evaluation as part of the planning and management of the program/organization' (Wandersman & Snell-Johns 2005, p. 422). By applying an empowerment evaluation approach to facilitate reflection and learning and to ask difficult questions, we hope to guide system actors and decision-makers to self-assess the barriers and bottlenecks to WASH sustainability and inclusion within their system and to identify resources and support needs that can help progressively realise their intended change.

Designing system interventions typically involves 'bounding' the system, understanding and documenting it, and assessing how its component parts interact in order to identify 'levers of change' (Foster-Fishman *et al.* 2007). Beyond the initial programme design, documenting and measuring system change occurring throughout the programme intervention is necessarily complex, reflecting the dynamic, complex and emergent nature of systems (Brown 2019; Hollander *et al.* 2020). To capture this dynamism and complexity within WASH system monitoring, Hollander *et al.* (2020, pp. 6–7) recommend applying a combination of 'qualitative monitoring methods focused primarily on the process of system change, analytical monitoring and analysis of the overall structure of the system and actor networks and the use of indicators or proxy indicators that adequately capture impacts and results while also considering the complex nature of such impacts'.

WaterAid's empowerment evaluation approach to WASH system strengthening, exemplified in the SusWASH programme, places local decision-makers and stakeholders at the centre of WASH system analysis from the programme design phase, addressing the issue of poor locally accessible system tools identified in the literature. This aims to keep decision-making power about what interventions the programme prioritises in the hands of those who live the reality of local WASH systems rather than external WASH system practitioners (Chambers 1997).

A core component of WaterAid's empowerment evaluation approach is the use of tools and frameworks based on 'system building blocks' to facilitate system-wide factor assessments (Hollander *et al.* 2020). Numerous frameworks for 'system building blocks' exist, which have been developed by multilateral institutions and international non-government organisations (NGO)s since they were first documented in 2007 by the World Health

Organization (2007). While acknowledging the widely accepted limitations of building block frameworks as a monitoring tool and the need to complement them with other monitoring approaches (World Health Organization 2007; WaterAid 2019a; Hollander *et al.* 2020), we have found them to be an accessible and practical framing to introduce system thinking to local WASH actors and to facilitate the analysis of various WASH system functions (Figure 2).

METHODOLOGY

In this paper, we present the results of a process review of WaterAid's empowerment evaluation approach to design, support, monitor, adapt and learn about WASH system-strengthening interventions in the SusWASH programme.

In practice, empowerment evaluation involves facilitators guiding participants through three steps to (a) establish their mission or purpose, (b) assess their current state of affairs using an evaluative framework and (c) plan for the future (Fetterman & Wandersman 2007). Empowerment evaluation emphasises the following 10 principles (Fetterman & Wandersman 2005):

- Improvement: improves programme performance, builds on successes and re-evaluates areas requiring attention.
- Community ownership: values and facilitates community control.
- Inclusion: invites involvement, participation and diversity from all.
- Democratic participation: ensures that decision-making is open and fair.
- Social justice: addresses social inequities in society.
- Community knowledge: respects and values community knowledge.
- Evidence-based strategies: respect and use the knowledge base of scholars.
- Capacity building: enhances stakeholders' ability to conduct an evaluation and to improve programme planning and implementation.
- Organisational learning: uses data to inform decisions and supports organisations to learn from experience.

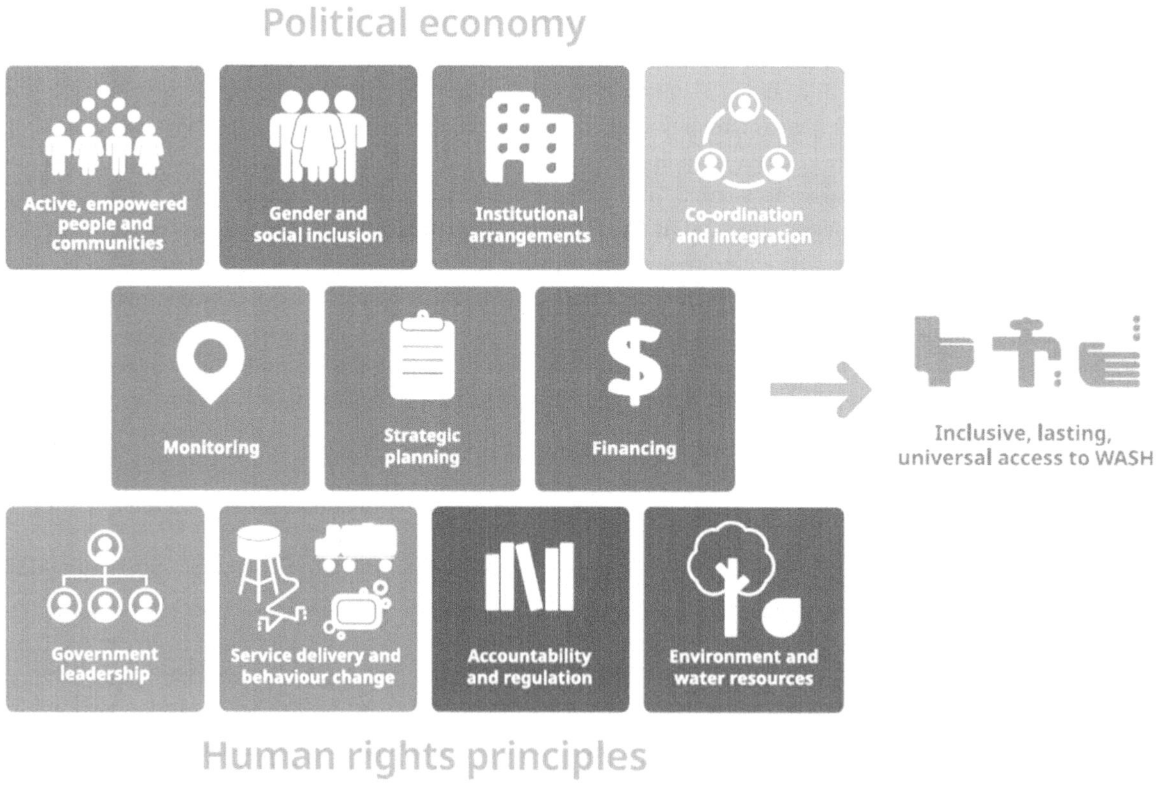

Figure 2 | WASH system building blocks, contextualised within the political economy and supported by human rights principles (WaterAid 2020).

- Accountability: supports accountability towards outcomes within a context of existing policies, standards and measures of accountability.

There is a strong alignment between empowerment evaluation principles and the human rights principles outlined above.

During the programme design and periodically throughout implementation, WaterAid used building blocks as an evaluation framework to facilitate and guide empowerment evaluation in participatory workshops with diverse stakeholders to (re)assess and (re)prioritise programming areas and to build collective ownership of issues and a vision for change. To facilitate this, we created outcome mapping tools based on the building block framework (Figure 3) (Earl *et al.* 2001; Hollander *et al.* 2020).

Methods used in this study

For this paper, we applied a process review method involving questionnaires administered online to understand the extent to which our engagement with system actors in the SusWASH programme upheld the empowerment evaluation principles and led to improvements in the WASH system strength. The SusWASH programme has been implemented in Cambodia, Ethiopia, Pakistan and Uganda since 2017. Two of the four country contexts, Cambodia and Uganda, were selected based on the consistency of the empowerment evaluation approach used in the programme in those contexts to engage actors within the system and the countries' application of participatory building block frameworks to drive cycles of assessment, action and self-reflection. Review participants were selected based on their involvement in the SusWASH programme in Kampong Chhnang province, Cambodia and Kampala city, Uganda, and included representatives from sub-national government, WASH line departments, and community and civil society representatives. A total of 15 participants from Cambodia and 12 participants from Uganda took part in the review. Participants were invited to respond to a series of Likert-scale statements and open questions. The Likert-scale questions sought to assess participants' knowledge and understanding of applying building block frameworks to assess their local WASH system, their perception of changes in the strength of the WASH system building blocks during the course of the programme and the extent to which they felt WaterAid's programming approach embodied the 10 principles of empowerment evaluation. Question responses were rated on a scale from −2 to 2; whereby −2 indicated a strongly negative result (e.g. deterioration), while on the other end of the scale, 2 indicated a strongly positive result (e.g. improvement). Open questions provided opportunities for further qualitative descriptions of the changes they had perceived in their organisation and WASH sector during the 4 years of SusWASH programme interventions.

The process review results are complemented by a snapshot of stakeholder building block assessment results in which WASH stakeholders assessed the state of building blocks of their local WASH system and identified actions to progressively strengthen the system's functionality.

In the following section of this paper, we discuss the findings while highlighting the complementary nature of empowerment evaluation approaches, human rights principles and system concepts outlined in the aforementioned literature.

Building Block	Weak	Medium	Strengthening	Desired – all actors using sub-national sector systems
Sector Coordination and Integration	PTWG exists on paper but inactive.	PTWG meet irregularly but limited ongoing commitment and department representatives are not consistent.	PTWG meet regularly and is skilfully led by secretariat. Information sharing is good between actors within PTWG and District WASH committee but the two groups don't share.	Agencies aligned behind comprehensive government-led strategy and there is strong communication flows between governmental levels.
	District WASH committee do not exist.	District WASH committee exists on paper but limited activity and engagement by commune and district authorities.	District WASH committee meets regularly.	
	CCWC exists but does not prioritize WASH.	CCWC understand their role in WASH service provision but are not actively coordinating WASH work	CCWC includes updates about WASH in their regular meetings and actively coordinates local WASH actors	
	Lack of coordination with women's groups lack of knowledge of service providers and government structures capacity for gender (equality, inclusion, empowerment, transformation)	Limited coordination of agencies, women's rights organisations or knowledge of agencies and their capacity for gender (equality, inclusion, empowerment, transformation)	Geographical coordination with Women's Rights groups and services within the district. Knowledge and understanding of partnering service providers' capacity for gender (equality, inclusion, empowerment, transformation)	Full coordination with women's rights groups and / or services. Knowledge and understanding of partnering service providers' capacity for gender (equality, inclusion, empowerment, transformation) and plans are in place to support partnering service providers' in maintaining and / or achieving desired gender transformation
Strategic planning	PAP exists but is not well-known or understood.	PAP is known by sector actors. PAP focuses on extending services but not clear plan to sustain services.	Government-led PAP and district plans are in place and all relevant departments, NGOs and private sector know about it.	Province and district have a realistic plan to deliver sustained universal access, which is supported by non-government actors and by stated actors and integrated from province to district and commune planning
	Districts do not have a WASH plan or targets	Districts have a plan for WASH but it is not well-integrated into district and community planning processes.	The WASH plans are partly integrated with district and commune investment and development planning.	
	Lon-term development plans do not include gender inclusion targets.	No gender analysis plan.	National WASH policies are gender sensitive.	National WASH policies are gender sensitive & plans to sustain gender equality and inclusion.
	No consultations involving women representatives prior to implementation No gender analysis plan	General planning is inclusive of women & girls physically, but not from the perspective of donors, rather than through consultations with women and girls	Gender analysis plan is in place. Consultations with women & girls are occurring at planning stages.	Gender analysis in place, with appropriate and supportive follow up

Figure 3 | Example of a building block assessment framework used in participatory workshops in the SusWASH programme in Cambodia (WaterAid).

FINDINGS AND DISCUSSION

This section presents the results of the process review and selected supporting evidence gathered through empowerment evaluation activities held throughout the SusWASH programme. In Tables 1 and 2, we present the results of the process review, showing local stakeholders' perceptions of the empowerment evaluation approach to system strengthening used in the SusWASH programme and the relative change in each WASH system building block over the life of the programme. In Table 3, we present selected results from participatory WASH system building block assessments as examples of stakeholders' analysis and assessment of their local WASH systems.

In Table 1, we present the extent to which local stakeholders in Cambodia and Uganda feel the approach used in the SusWASH programme upheld the 10 empowerment evaluation principles. In general, scores in Uganda were more positive than the scores in Cambodia. On average, stakeholders in Uganda perceived all 10 principles to have been demonstrated to some extent, with 'Evidence-based strategies' scoring the highest (2.0). The principle the least well demonstrated in Uganda was 'Improvement' in the system goal of providing WASH

Table 1 | Strength of the application of empowerment evaluation principles in the SusWASH programme

Principle	Number of indicators	Cambodia $n=15$	Uganda $n=12$
1. Improvement	1	1.67	1.33
2. Community ownership	2	0.91	1.67
3. Inclusion	1	1.20	1.83
4. Democratic participation	1	1.27	1.92
5. Social justice	3	0.94	1.75
6. Community knowledge	1	1.43[a]	1.92
7. Evidence-based strategies	1	1.33	2.00
8. Capacity building	4	0.95[a]	1.67
9. Organisational learning	2	0.90[b]	1.50
10. Accountability	1	1.40	1.42

The ratings in the table are based on the following scale: 2, principle strongly demonstrated; 1, principle weakly demonstrated; 0, principle neither demonstrated nor lacking; −1, principle weakly lacking; −2, principle strongly lacking.
[a]Only 14 responses to one or more indicators.
[b]Only 13 responses to one indicator.

Table 2 | Participants' perception of change in each WASH system building block over the life of the SusWASH programme

Building block	Number of indicators	Cambodia $n=15$	Uganda $n=12$
1. Active empowered people and communities	1	1.20	1.83
2. Gender and social inclusion	2	0.84	1.71
3. Institutional arrangements	2	0.90[a]	1.58
4. Coordination and integration	1	1.13	1.92
5. Monitoring	2	1.21[ba]	1.38
6. Strategic planning	2	1.22	1.79
7. Financing	2	0.94	1.25
8. Government leadership	2	0.91	1.67
9. Service delivery and behaviour change	1	1.67	1.33
10. Accountability and regulation	2	1.22	1.63
11. Environment and water resources	0	n/a	n/a

The ratings in the table are based on the following scale: 2, strong improvement perceived in this building block; 1, weak improvement perceived in this building block; 0, no change perceived in this building block; −1, weak deterioration perceived in this building block; −2, strong deterioration perceived in this building block.
[a]Only 14 responses to one indicator.
[b]Only 13 responses to one indicator.

Table 3 | Selected results of participatory building block assessments

Building block	Strengths, Weaknesses, Opportunities, Threats (SWOT) analysis		Action points	Status rating
	Uganda (2021)			
Monitoring	**Strengths:** □ Work plans □ Technical staff □ Organisational structure	**Weaknesses:** □ Weak reporting system □ Limited facilitation for monitoring □ Limited acceptance of ICT innovations	□ More innovations to collect evidence-based data □ Build a Knowledge Management Unit	Medium
	Opportunities: □ WASH monitoring forums i.e. KWSF and NSWG □ Already using built M&E systems i.e. HMIS □ Use of ICT to get evidence-based information	**Threats:** □ Political environment/leadership □ IT systems and technology change quickly which can require continuous and costly investment		
WASH service delivery, equity and inclusion	**Strengths:** □ Technical resources □ Established governance structures (VHT) □ Clear partnership guidelines.	**Weaknesses:** □ Inadequate consultation with key players □ Inadequate monitoring and evaluation □ Design technologies not meeting needs of the end user	□ Total inclusion of the vulnerable groups at inception stage (women, youth, disabled, children) □ Establish city sanitation fund for pooling funds from donors, partners and private sector actors □ Enhance KCCA's capacity in areas of supervision, M&E, behaviour change & WASH operation and maintenance □ Document best practices to increase awareness and learning	Medium
	Opportunities: □ Enabling environment □ Community willingness to receive and pay for services □ Innovative approaches to WASH activities	**Threats:** □ Low capacity to satisfy the ever-growing population □ Poor physical planning (high water table, floods, poor soils) □ Absentee landlords limiting WASH infrastructure development □ Prohibitive water tariff structure		
	Cambodia (October 2020)			
Government leadership	**Strengths:** □ Well-established government-led civic champion program to develop local leadership □ Good intra-governmental coordination for WASH planning	**Weaknesses:** □ Women in leadership roles have limited skills/experience □ Male sector leaders don't provide support and space to promote women and hear their inputs	□ Support institutionalisation of civic champion leadership development program □ Sector leaders to show leadership commitment –	Strengthening
Building block	Strengths, Weaknesses, Opportunities, Threats (SWOT) analysis		Action points	Status rating
	□ Growing local government capacity in WASH data management □ Districts support most marginalised households to build latrines □ Women occupy 25-30% of leadership and decision-making roles in sub-national WASH coordination		e.g. budget allocation for inclusive and sustainable WASH services □ Sub-national authorities to allocate budget for institutional capacity development □ Promote women's meaningful participation, leadership and decision-making.	
	Opportunities: □ Empower women to actively fulfil leadership and decision-making roles □ Promote leadership roles for women □ Promote peer-to-peer learning and competition between leaders	**Threats:** □ Limited willingness and commitment from high-level political leaders □ Limited space for freedom of expression		
Gender and social inclusion	**Strengths:** □ Some districts promote and demonstrate inclusive WASH to other institutions and public forums □ Increasing women's participation in sector coordination mechanisms and in local government planning processes □ Gender-related indicators included in WASH data collection □ Some districts and communes mobilise funds to support people who are marginalised □ Private water operators provide discount for low-income household connections	**Weaknesses:** □ Limited meaningful participation of women and marginalised group representatives in decision making □ Limited understanding and space to discuss human rights to water and sanitation	□ Build openness and willingness among sub-national authorities to prioritise meaningful participation of women and marginalised groups' representatives □ Build awareness and capacity among sub-national authorities on human rights to water and sanitation □ Strengthen sub-national authorities' ability to train on and promote human rights to water and sanitation to others □ Create safe and effective spaces for women and marginalised people to share their WASH situation and needs □ Increase confidence and capacity of women and Civil	Medium-Strengthening
	Opportunities: □ Create space for people who are marginalised to raise their voices and advocate for their rights to water and sanitation □ Promote championing of inclusive WASH among sector leadership	**Threats:** □ Negative social norms and attitudes towards gender and disability □ Geographical issues with marginalised groups often living in challenging environments		
Building block	Strengths, Weaknesses, Opportunities, Threats (SWOT) analysis		Action points	Status rating
	□ Promote peer-to-peer learning and competition	□ Limited willingness and commitment from high-level political leaders □ Limited space for freedom of expression	Society Organisations representing marginalised people □ Promote government-led women's leadership development program	

services (1.33). Conversely, in Cambodia, the highest scored principle was 'Improvement' (1.67) followed by 'Community knowledge' (1.43). The lowest scored principle in Cambodia was 'Organisational learning' (0.94).

In Table 2, we present the results of participants' perception of the change in each WASH system building block over the life of the SusWASH programme. We see that stakeholders in both Cambodia and Uganda consider that there have been improvements, of some kind, across all 10 rated building blocks. The lowest scored building block in Cambodia was 'Gender and social inclusion' (0.84). While in Uganda, the lowest scored was 'Financing' (1.25), although this was still rated as having had a 'weak improvement'. The highest scored building block in Cambodia was 'Service delivery and behaviour change' (1.67), while in Uganda, the highest scored building block was 'Coordination and integration' (1.92).

Table 3 presents selected results of participatory WASH system building block assessments conducted by subnational WASH stakeholders during the SusWASH programme. These results reflect the second and third step of the empowerment evaluation process (Fetterman & Wandersman 2007); stakeholders who have already set their overall mission and goal for WASH in their context assessed the current state of the system through the building block framework (Figure 3) and identified actions that they could collectively take to respond to weaknesses.

Following building block assessments and regular reflection sessions, WaterAid worked with others to support the operationalisation of the action points identified by local stakeholders.

In Uganda, WaterAid supported the Ministry of Health and the Kampala Capital City Authority (KCCA) to align their WASH monitoring indicators with the global Joint Monitoring Programme service level indicators. WaterAid also built the capacity of government staff and community-level Village Health Teams (VHTs) to collect service level data using electronic data capture tools. The transition from paper-based data collection to electronic has helped ensure the accuracy and reliability of the data collected. This new electronic data collection approach is being tested by the KCCA for learning and potential scale-up.

Faced with the high cost of water, public schools often struggled to pay their water bills. This challenge grew as the government began promoting the installation of water-borne toilets in all public schools. To address this, WaterAid, in collaboration with the KCCA, developed the *Single Flush Female-Friendly Toilet*. The toilet design was built with water-saving technology, integrating the pit latrine model with pour-flush technology thus significantly reducing water usage by over 10 times compared to a normal flush toilet. The five target schools that previously struggled to pay their bills can now afford their water bills and have access to water for flushing, handwashing, cleaning and drinking at all times. The toilet design also provides for a safe disposal of menstrual hygiene materials for girls.

To build government leadership in Cambodia, WaterAid worked in partnership with a local NGO to strengthen local authority leadership through the Civic Champions Programme[1]; helped to facilitate sector coordination and multi-stakeholder collaboration; promoted sector learning and peer-to-peer learning exchange, including the sharing of best practices through mass media and media engagement; promoted good sanitation and hygiene behaviours through the family champion initiative which identified and publicly recognised families practicing good WASH behaviours and gender equality; and encouraged high-level sector and political leaders to discuss and share their strategic vision with the public through social media.

To strengthen gender, equity and social inclusion (GESI) in Cambodia, WaterAid engaged with and influenced sector and political leaders to allocate and mobilise resources to support most marginalised households gain access to clean water and sanitation; empowered vulnerable groups by building their capacity and confidence to share their WASH experiences with local authorities through storytelling on social media and in public forums; challenged gender norms by promoting women's leadership, facilitating women's participation in WASH networks and working with male leaders to change their mindsets towards gender relations.

In the subsequent paragraphs, we discuss the outcomes and lessons learned from the combination and practical application of empowerment evaluation principles, human rights principles and system concepts outlined above.

Strengthening organisational learning and capacity to improve the system by identifying and responding to blockages

The routine participatory evaluation of the WASH system has helped to strengthen the awareness and capacity of system actors, and the processes they use, so that they may continue to practically assess their system and respond to emerging issues beyond the life of WaterAid's intervention (Table 4). On average, survey respondents from both Cambodia and Uganda felt that their engagement with the SusWASH programme had improved their personal and organisational capacity to evaluate WASH sector needs, rating the capacity building principle as 0.95 and 1.67, respectively. They felt better equipped to use data for evaluating their practice and improve their

---

[1] More information on the Civic Champions Programme can be found in WaterAid's website: https://washmatters.wateraid.org/publications/civic-champions-leadership-development-programme-for-strengthening-sub-national-wash and on local partner WaterSHED's website: https://watershedasia.org/civic-champions/.

Table 4 | Key lessons learned and relevant empowerment evaluation principles regarding strengthening organisational learning and capacity

Key lessons learned	Relevant empowerment evaluation principles
• Building blocks can provide an accessible framework through which local decision-makers can begin to analyse the WASH system and progressively strengthen it. • Monitoring frameworks can and should evolve with the context. • Efforts should be made to clarify the role played by any implementing agency in bringing about a system change for accountability and transparency purposes. Often system changes are the result of collective action. • Institutionalising system thinking and scaling systems assessments beyond NGO programming may be possible, if an adequately resourced institution can play a convening role and tools are adequately contextualised.	1. Improvement 7. Evidence-based strategies 8. Capacity building 9. Organisation learning

decision-making, rating the organisational learning principle as 0.9 and 1.5, respectively. Respondents also felt that WASH services had improved in their area over the corresponding period, rating the improvement principle as 1.67 and 1.33 in Cambodia and Uganda, respectively. These perceptions are supported by government monitoring data which show that between 2017 and 2020, 15,667 households (13% of total households) gained access to basic sanitation in SusWASH programme areas in Cambodia and sanitation access rates have increased in SusWASH programme areas of Uganda by 2% from 21% in 2018 to 23% in 2020.

Regular reflection and system assessments have supported local stakeholders to build on their successes and re-evaluate areas requiring further effort. As explained in the process review by a government stakeholder in Kampong Chhnang, Cambodia, the results from WASH sector evaluation workshops are used to 'compare what has been done, compare between the work plan and the real practice, to check if it responds to the requirement or not. Then we consider what areas we need to improve and how to do it better for the next steps'. The collective identification of blockages within the system each year informed tangible actions for strengthening the system (examples in Table 3) as per the three steps of the empowerment evaluation approach (Fetterman & Wandersman 2007). The following are two examples of actions taken as a result of collective system blockage identification. The idea to create a knowledge management unit within the Kampala City Council Authority in Uganda was proposed to strengthen the facilitation of WASH monitoring and reporting systems. In Cambodia, the identification of limited meaningful participation of women and marginalised people in WASH decision-making (participation being one of the human rights principles) led to collective agreement to create safe and effective spaces for those groups to raise their voices and share their concerns.

Alongside these strengthened capacities have been tangible improvements occurring in WASH system functions, many of which have arisen from the cycles of system assessment and action planning. These include:

- Formation of district WASH committees and open defaecation plans under the leadership of district administrations in Cambodia.
- Roll-out of community WASH monitoring systems in Uganda.
- A joint assessment of WASH accessibility and sustainability and life-cycle costing of WASH services in health-care facilities across Kampala, Uganda.
- Creation of cross-department district WASH committees in Kampong Chhnang, Cambodia.
- A piloting of public–private partnerships for toilet pit emptying and partial subsidies for marginalised households in Kampala, Uganda.
- Peer-to-peer sharing between communes who had used their own budget to promote sanitation and hygiene to inspire other communes in Kampong Chhnang, Cambodia.
- Development of local leaders' leadership skills and engagement in WASH motivated increases in sanitation coverage in Kampong Chhnang, Cambodia.

Regular reflections in Cambodia and Uganda provided a platform for raising and problem-solving issues among the participating stakeholders. Collectively assessing system problems from a range of different perspectives leads to participants re-assessing their own ways of thinking (König 2018) and reinforces the self-organising nature of complex systems (Ramalingam *et al.* 2008). One civil society stakeholder in Cambodia highlighted that the programme's assessments contributed to clarifying unclear sector roles: 'Private Water Operators and sub-national authorities get better understanding on their roles and responsibilities on WASH'. In some instances, involving and engaging non-traditional WASH stakeholders in the discussions to assess the WASH system has created feedback loops that have shaped how information is shared within the WASH system (Meadows 1999). One government survey respondent from Uganda noted that the discussions encouraged different actors to engage in the system: 'the political leadership has more ownership of WASH planning, delivery and monitoring'. In Uganda, the generation and use of data (empowerment evaluation principle 9), which is accessible to all, helps inform decisions and investments in WASH, while also strengthening transparency. Building the capacity of Kampala authorities to collect, store, visualise and use JMP service level data (the 'Monitoring' building block scored 1.38 in Table 2) as part of sanitation monitoring has improved decision-making and transparency. These examples further demonstrate how the human rights principles of transparency and accountability have been operationalised within system-strengthening approaches.

Through regular cycles of reflection and action, the approach tries to model a culture of learning, reflection and adaptation among WASH actors (Grant & Willetts 2019). WaterAid deliberately did this through facilitating regular participatory 'pause and reflect' sessions, and the use of online knowledge hubs and real-time learning and sharing networks through platforms such as Facebook, Telegram and WhatsApp. Learning activities and cycles of action, reflection and learning draw on WaterAid's own adaptive programming which encourages planned learning activities to intentionally deepen the understanding of the context and system dynamics.

Our reflection and learning approaches during the SusWASH programme prompted review and updates to the building block framework itself to ensure it remained relevant (WaterAid 2019b) to the changing programme contexts over time. For example, at the beginning of the interventions, we did not have a specific building block articulated for government leadership, active and empowered people or GESI. Adding these helped to focus the discussions and awareness among stakeholders of the importance of these aspects of the system. The definitions used in frameworks and outcome mapping also evolved according to changes in context – such as when policy or regulation changes created new mechanisms for coordination, financing or monitoring. While such changes over time complicate the use of building block frameworks for programme monitoring, they reflect the dynamic nature of the WASH systems in which they were being applied and improve their value as tools for empowerment evaluation.

WaterAid and several of the stakeholders responding to the process review survey have noted the value of the building block framework for categorising, assessing and planning to respond to problems in WASH systems. We have begun to see small examples of how these practical, locally accessible tools have potential for take-up by those making critical decisions around WASH service provision and scaling beyond WaterAid's direct intervention areas. In Cambodia, several district administrations have fixed their latest building block assessments to the office wall and refer to it in their monthly WASH progress meetings and action planning. In Uganda, the KCCA have used their own budget to mentor two municipality authorities through a municipal WASH investment planning process that draws on participatory system assessment processes. At the national level, we also see potential for institutionalisation of participatory building block assessments within government and sector efforts to strengthen WASH systems. Sector working groups in Cambodia engaging with Sanitation and Water for All and Agenda for Change have drawn on the SusWASH experience in Kampong Chhnang and are discussing and framing sector priorities using participatory building block assessments, and in Uganda, elements of the work in Kampala such as training and WASH assessment and planning processes in healthcare facilities are being scaled nationally by the Ministry of Health through their own investment. While we have not transitioned to sector stakeholders using the framework to facilitate participatory discussions independently of WaterAid, the above examples indicate that actors, or groups of actors, other than large NGOs like WaterAid could potentially scale this way of working at either sub-national or national level. Based on our experience in these two countries, scaling and institutionalising system approaches could be possible if there is an adequately resourced organisation or institution that can play a convening and facilitating role, and that the tools and frameworks used are appropriately contextualised and locally owned, though we note that this takes time even when led by an organisation like WaterAid. Achieving the outcome of participatory WASH system building block assessments being led and institutionalised by either civil society or government actors would provide further evidence of the usefulness of the approach and the confidence and capacity of users to self-assess and course-correct (Fetterman & Wandersman 2007). It would also contribute to the human rights principle of sustainability in both service sustainability and sustained embedment of human rights principles in the WASH system.

Critiques and limitations of these lessons

WaterAid is cognisant that building blocks do not represent the WASH system in its entirety and should be complemented by other monitoring approaches (WaterAid 2019a; Hollander *et al.* 2020). In one of the formative documents which first described the building blocks of the health system, the World Health Organization (2007, p. v) notes that 'while the building blocks provide a useful way of clarifying essential functions, the challenges facing countries rarely manifest themselves in this way. Rather, they require a more integrated response that recognises the inter-dependence of each part of the health system'. Hollander *et al.* (2020), building on previous analysis by WaterAid (2019b), highlight that building blocks are incomplete as a full programme monitoring tool because they 'neglect the full complexity of systems engagement and change' (p. 4) and therefore capture only a simplified and static view of predefined categories. Monitoring indicators and corresponding interventions may become siloed and deprioritise the interlinkages and emergent properties of the system (WaterAid 2019a; Hollander *et al.* 2020). For these reasons, and the inherently subjective nature of the participatory assessment process, WaterAid has separate internal processes and tools for monitoring programme impact which draw on the participatory assessments alongside other evidence sources. The analysis must also be considered within the context's political economy and influencing factors such as power structures, relationships and norms (Foster-Fishman *et al.* 2007). In practice, during the programme start-up phase, for the purposes of 'bounding the system' (Foster-Fishman *et al.* 2007), WaterAid complements the building block assessments with simple stakeholder mapping tools for local sector stakeholders to identify which actors have willingness, ability and influence to advance improvements in WASH, while simple power and gender analysis exercises help to prompt reflection on issues of marginalisation and power exclusion.

While we accept the above critiques of building block frameworks, we see their value in the way in which they can introduce, and make accessible, the concept of the WASH system to local stakeholders. Similarly, we see the value in the way in which they can draw diverse stakeholders together to share perceptions of persistent problems and discuss strategies for tackling them. Indeed, it is this process which helps to uncover issues of power, attitudes and behaviours, even if the building block framework itself does not explicitly focus on these components; it is nearly impossible to discuss aspects of leadership, community empowerment, GESI and accountability without considering power, agency and the dynamic nature of rights holder–duty bearer relationships. We also recognise that identified system changes cannot always be attributed to WaterAid, particularly in areas crowded with other WASH system actors (Ramalingam *et al.* 2008; Hollander *et al.* 2020). For this reason, we are mindful of clearly stating whether or not WaterAid has played a leading role, or more of a supportive role, in bringing about any identified system change.

Inclusive participation to re-shape WASH sector power relations

Through an empowerment evaluation approach, regular reflections among sector stakeholders about changes in their system over time went beyond system monitoring to become a tool for facilitating dialogue about system issues in which diverse voices and perspectives could be heard and engaged (Table 5) (Neely 2019b). As one government stakeholder from Uganda stated, 'the WASH agenda needs everyone's involvement and at my level I can contribute to this change'. Respondents from Cambodia and Uganda felt that WaterAid's programme approach demonstrated the principles of 'Inclusion' (1.20 and 1.83, respectively), 'Democratic participation' (1.27 and 1.92, respectively) and 'Community knowledge' (1.43 and 1.92, respectively). Stakeholders from both countries attributed the engagement of diverse stakeholders in WASH activities to the system assessment process.

Each stakeholder's understanding of the systems they are part of will differ; such subjectivity reinforces the importance of involving multiple and diverse stakeholders in exploring and assessing systems (Foster-Fishman *et al.* 2007). WaterAid does this through facilitating participatory multi-stakeholder discussions because we see

Table 5 | Key lessons learned and relevant empowerment evaluation principles regarding inclusive participation

Key lessons learned	Relevant empowerment evaluation principles
• Taking time to build and maintain rapport and trust is critical to locally owned and accessible system analysis. • Participatory multi-stakeholder assessments of the system can contribute to building a collective understanding about where power in the system does, and should, reside.	2. Community ownership 3. Inclusion 4. Democratic participation 6. Community knowledge

the value in the process of stakeholders exchanging and challenging each other's perspectives. Using participatory approaches inevitably invites questions of power and perceptions of power relations. In the words of Chambers (2005, p. 290), 'participation is about power relations. It is about much else as well; but power relations are pervasive: they are always there, and they affect the quality of process and experience'. Gathering political leaders, technical experts, community representatives and social and business leaders to discuss the state of the WASH system and create shared plans to improve WASH services is a process for encouraging participants to acknowledge their own power and build awareness about who is often absent from decision-making within their local WASH system (thus linking to the human rights principle of participation and non-discrimination). The process review identified several examples of changing power relations that are helping to resolve system blockages. One government bureaucrat in Uganda reported that 'the political leaders have become the champions of WASH interventions which makes the work of us, the technical people, easy'. Another government stakeholder from Uganda noted 'some of the research has helped us engage different groups in our intervention', while one civil society stakeholder from Cambodia observed that the participatory process enabled 'multiple partners to work on WASH, includ[ing] private sector and political group'. We have observed additional early signs of shifting power dynamics as a result of SusWASH programme interventions:

- In Uganda, WaterAid mapped power dynamics and incentives of local actors to identify leverage points for system change. In Kampala, power is highly centralised. Little community participation in government planning processes can mean community WASH needs are not met or prioritised. The SusWASH programme has created spaces for citizen engagement (scoring 1.83 in Table 2) through the WASH Mayors Forum and community-level Participatory Budget Clubs. Community hygiene promotion groups locally called 'Weyonje' volunteers have become more empowered to engage service providers including public hospitals in their area about expenditure of public funds in support of WASH, something they had never done before and thought was not their responsibility.
- In Cambodia, discussions with men and women about their roles and capabilities helped build a safe and open dialogue on gender equality and non-discrimination among district government staff, enabling fulfilment of WASH roles and responsibilities while beginning to address barriers to women's participation. Participatory barrier analysis and public storytelling also helped marginalised people to identify their own power in driving WASH improvements. Despite these efforts, it is interesting to note that stakeholders from Cambodia participating in the process review rated a change in the 'Gender and social inclusion' building block the weakest (0.84). This could be linked to the level of attention paid by WaterAid to increase local stakeholder understanding of gender and social inclusion issues and local stakeholder recognition of the efforts required to address them.

These examples support the observations of Foster-Fishman *et al.* (2007) that simply 'the act of having stakeholders explore and accommodate differences among competing worldviews can serve to create transformative shifts in stakeholders' understandings of the problems' (p. 201).

As an NGO temporarily participating in the WASH systems we seek to strengthen, and with no mandate or responsibility for ongoing WASH delivery, WaterAid has learned that it takes time to build trust and acceptance of this way of working among local stakeholders. It took years for local stakeholders from government, civil society and the private sector to value the system-focused contributions of WaterAid teams in Cambodia and Uganda in contexts where NGOs typically fill a role in delivering services outside of government processes. It took at least one annual cycle for a shared vocabulary and understanding of the WASH system to begin to emerge and even longer before stakeholders expressed the appreciation of the value of honest and open reflection in order to identify and address weaknesses in the system. Such openness is a necessary step towards developing a collective vision and a space in which diverse actors feel able to meaningfully direct and own the process; as Chambers (1997, p. 133) puts it 'relaxed rapport between outsiders and local people, and some measure of trust, are minimum conditions for [participatory appraisals]'. This reflection raises questions about the viability of this approach in more closed contexts where suspicion of civil society and international NGO activity is particularly high and trust is low.

Critiques and limitations of these lessons

Although small examples of positive progress have been observed, complex and inequitable power dynamics persist in both of the systems described and will continue to persist for a long time. Despite WaterAid's efforts to engage and facilitate meaningful participation of rights groups, representatives of marginalised communities, actors who possess formal power and authority, and other actors in between, not all groups have been represented, and group discussions may continue to favour the voices of some actors over others (Chambers 1997). We have observed that deep-seated distrust, prejudices and suspicion (mentioned above) can perpetuate harmful attitudes that prevent the safe spaces needed for meaningful participation or shifts of power within the system (Foster-Fishman *et al.* 2007). WaterAid is not immune from blindness to power; only four of the 27 process review responses were from representatives of civil society and rights groups. More can and must always be done in system-strengthening programmes to continuously and consciously create space for empowerment of those who experience marginalisation and ensure that where momentum for change builds among actors who possess formal power that it is accompanied by accountability, non-discrimination and transparency (de Albuquerque 2014).

Shifting mindsets to focus on equitable WASH services for all

Paradigms, which Meadows (1999, p. 18) describes as 'the sources of systems', are the shared ideas and beliefs about how the world works. 'Normative elements' such as people's attitudes, values, beliefs and expectations maintain the system's existence, and explain why and how the system and its members operate, and 'can often be identified as the root causes of systems problems' (Foster-Fishman *et al.* 2007, p. 205). To an extent, all system-strengthening interventions, therefore, should aim to identify and leverage change in such paradigms, and this lies at the core of WaterAid's participatory approach based on empowerment evaluation (Table 6). Process review respondents from Cambodia and Uganda felt on average that WaterAid's approach emphasised social justice principles (0.94 and 1.75, respectively) and encouraged accountability within roles and responsibilities between rights holders and duty-bearers (accountability principle rated 1.4 and 1.42, respectively). Notably, civil society representatives from Uganda felt that the social justice and accountability principles were less well demonstrated than other respondents (rated 1.0 and 1.5, respectively), and civil society representatives from Cambodia also rated the accountability principle lower than the average (1.0). This finding could be explained by the role and focus of civil society organisations in holding government to account and tackling issues of social injustice; they may have higher expectations and a higher level of understanding about what is required to meet them.

The process review identified examples of how participants' beliefs had changed through their involvement in the SusWASH programme, especially regarding the rights to water and sanitation services. One Government stakeholder from Uganda reported, 'at first I thought it was someone else's task to demand for better wash services but now I know and I have learnt that policies are influenced rightly from the grass roots from the lowest citizens as a right'. Though changing individual mindsets, attitudes and beliefs may happen quickly, changing them across an entire system takes continuous engagement, champions and a variety of tactics (Meadows 1999). WaterAid's partnerships and regular activities bringing together Mayors, government bureaucrats, community groups and disabled people's organisations have sought to contribute to this process (in line with the human rights principles of participation and transparency). A district government respondent from Cambodia shared their revelation that 'everyone has the right[s] to water and sanitation. To meet the basic needs of water supply, [we] collaborate with institutions, NGOs and the private sector to expand WASH services and encourage participation from all relevant stakeholders'. Changes in mindset such as this may have come from storytelling and photography exhibitions we arranged for groups who experience marginalisation to share their experience and WASH needs with decision-makers, from visits by government officials to households of those groups to help them better understand people's situation and their access to WASH, and from 'early-adopter' decision-makers championing rights and equality among their peers (Meadows 1999).

Alongside experiential and relationship-building activities, we also facilitated targeted training and awareness raising sessions with disabled people's organisations on the rights to water and sanitation and on issues of inclusion and non-discrimination. Working to strengthen the skills and processes of government, private

Table 6 | Key lessons learned and relevant empowerment evaluation principles regarding shifting mindsets

Key lessons learned	Relevant empowerment evaluation principles
Empowerment evaluation's emphasis on participation and community knowledge creates opportunities for individual and collective self-reflection and self-improvement. In time, this can help to shift deeply held attitudes and beliefs, helping to strengthen accountability and tackle issues of social injustice.	5. Social justice 10. Accountability

sector and civil society actors in areas of planning, budgeting, monitoring and coordination helped to open opportunities for discussing equality and sustainability in WASH service planning and delivery and make rights part of duty-bearers' everyday practice (Carrard *et al.* 2020). Indeed, following participatory life-cycle costing assessments, we observed a shift in some duty-bearers' mindsets around the use of resources for WASH and who should be involved in decisions regarding resource use. One government stakeholder in Uganda shared that they previously 'had beliefs that women had no option in making decisions concerning sanitation and hygiene', while another shared that they had 'learnt that all engagements need to be sustainable, hence community ownership is key'. A third Ugandan government stakeholder had come to realise that 'schools can do a lot more and create a positive WASH impact within their available resources'. These findings illustrate that participatory system-strengthening approaches, designed based on a detailed understanding of the incentives and motivations of the target system actor(s), can shift and alter deeply embedded attitudes, beliefs and norms.

Critiques and limitations of these lessons

WaterAid's WASH system monitoring approach has not been designed to systematically document changes in mindsets over time; the above lessons are based on programme participants' reflections at a single point in time. We continue to evolve our monitoring framework and its accompanying guidance to better capture these changes and to understand how these changes manifest in improvements in WASH system strength and service levels. One means of facilitating this could be to have more regular personal reflections and use of semi-structured interviews to draw out the changes in mindsets, attitudes and norms, as has been done for gender power analysis and power mapping. Complementary tools already in use include simple stakeholder mapping and political economy analysis.

CONCLUSION

This paper illustrates how empowerment evaluation, system concepts and human rights principles can coalesce into a practical, participatory approach that helps to strengthen WASH systems for the realisation of sustainable WASH for all. System thinking, framed through simple frameworks like system building blocks, can help local actors to understand the interconnected nature of how the people, institutions and resources required for the realisation of the human rights to water and sanitation interact, behave, flow and function. Approaching system-strengthening interventions with an emphasis on human rights principles can help to guide efforts that seek to strengthen how a WASH system functions in a way which empowers and builds capacity among local actors, shifts power relations and challenges people's attitudes towards equity and social justice. System perspectives do not presuppose a pathway to change but rather develop pathways from initial analysis and continuously adapt over time. Systems analyses, when undertaken in an empowerment evaluation approach with the participation of diverse voices, can help to identify potential leverage points to target and effect change in the system depending on the context. What is unique to this approach, and which we think is critical to the realisation of the human rights to water and sanitation, is taking system actors on a journey to deepen understanding of which parts of the system, which principles, in which context and when, it makes most sense to target and prioritise. Other WASH agencies working to realise the human rights to water and sanitation should continue to test, share lessons, and where appropriate, work with others to scale practical approaches that combine these concepts together.

ACKNOWLEDGEMENT

The funding for WaterAid to undertake the SusWASH programme was provided by the H&M Foundation.

DATA AVAILABILITY STATEMENT

All relevant data are included in the paper or its Supplementary Information.

REFERENCES

Brown, C. 2019 Measuring impact in WASH from a complex adaptive systems perspective. In: *Systems Thinking and WASH Tools and Case Studies for a Sustainable Water Supply* (Neely, K., ed). Practical Action Publishing, Rugby, pp 17–28. http://dx.doi.org/10.3362/9781780447483.008.

Carrard, N., Neumeyer, H., Pati, B. K., Siddique, S., Choden, T., Abraham, T., Gosling, L., Roaf, V., Alvarez-Sala Torreano, J. & Bruhn, S. 2020 Designing human rights for duty bearers: making the human rights to water and sanitation part of everyday practice at the local government level. *Water* **12**, 378–388. https://doi.org/10.3390/w12020378.

Chambers, R. 1997 *Whose Reality Counts?* Practical Action Publishing, Rugby.

Chambers, R. 2005 *Ideas for Development.* Taylor & Francis Group, London.

de Albuquerque, C. 2014 *Realising the Human Rights to Water and Sanitation: a Handbook by the UN Special Rapporteur.* United Nations Office of the High Commissioner for Human Rights, Lisbon.

Earl, S., Carden, F. & Smutylo, T. 2001 *Outcome Mapping: Building Learning and Reflection Into Development Programs.* International Development Research Centre, Ottawa.

Fetterman, D. M. 1994 Empowerment evaluation. *American Journal of Evaluation* **15** (1), 1–15. doi:10.1177/109821409401500101.

Fetterman, D. M. & Wandersman, A. 2005 *Empowerment Evaluation: Principles in Practice.* Guildford Press, New York.

Fetterman, D. M. & Wandersman, A. 2007 Empowerment evaluation: yesterday, today and tomorrow. *American Journal of Evaluation* **28**, 179–198. doi:10.1177/1098214007301350.

Foster-Fishman, P., Nowell, B. & Yang, H. 2007 Putting the system back into systems change: a framework for understanding and changing organizational and community systems. *American Journal of Community Psychology* **39**, 197–215. doi:10.1007/s10464-007-9109-0.

Grant, M. & Willetts, J. 2019 Learning for adaptive management: using systems thinking tools to inform knowledge and learning approaches. In: *Systems Thinking and WASH Tools and Case Studies for a Sustainable Water Supply* (Neely, K., ed.). Practical Action Publishing, Rugby, pp. 107–132. http://dx.doi.org/10.3362/9781780447483.007.

Hollander, D., Ajroud, B., Thomas, E., Peabody, S., Jordan, E., Javernick-Will, A. & Linden, K. 2020 Monitoring methods for systems-strengthening activities toward sustainable water and sanitation services in low-income settings. *Sustainability* **12**, 7044–7060. doi:10.3390/su12177044.

Huston, A. & Moriarty, P. 2018 *Understanding the WASH System and its Building Blocks.* IRC WASH, Delft, The Netherlands. Available from: https://www.ircwash.org/resources/understanding-wash-system-and-its-building-blocks.

König, A. 2018 Sustainability science as a transformative social learning process. In: *Sustainability Science: Key Issues* (König, A. & Ravetz, J., eds). Earthscan/Routledge, London.

Lockwood, H., Smits, S., Schouten, T. & Moriarty, P. 2010 Providing Sustainable WASH Services at Scale, International Symposium on Rural Water Services – Background Paper, Kampala. Available from: https://www.ircwash.org/sites/default/files/Lockwood-2010-Providing.pdf.

Meadows, D. 1999 *Leverage Points: Places to Intervene in a System.* The Sustainability Institute, Hartland, VT.

Neely, K. 2019a Introduction: WASH projects and complex adaptive systems. In: *Systems Thinking and WASH Tools and Case Studies for a Sustainable Water Supply* (Neely, K., ed.). Practical Action Publishing, Rugby, pp. 1–16. http://dx.doi.org/10.3362/9781780447483.001.

Neely, K. 2019b Systems thinking and transdisciplinarity in WASH. In: *Systems Thinking and WASH Tools and Case Studies for a Sustainable Water Supply* (Neely, K., ed.). Practical Action Publishing, Rugby, pp. 17–28. http://dx.doi.org/10.3362/9781780447483.002.

Posthumus, H., Shah, R., Miehlbradt, A. & Kessler, A. 2020 *A Pragmatic Approach to Assessing System Change.* Available from: https://beamexchange.org/resources/1334/ (accessed 10 December 2021).

Ramalingam, B., Jones, H., Reba, T. & Young, J. 2008 *Exploring the Science of Complexity: Ideas and Implications for Development and Humanitarian Efforts*, 2nd edn. Overseas Development Institute, London.

Valcourt, N., Javernick-Will, A., Walters, J. & Linden, K. 2020 System approaches to water, sanitation, and hygiene: a systematic literature review. *International Journal of Environmental Research and Public Health* **17**, 702–720. doi:10.3390/ijerph17030702.

Wandersman, A. & Snell-Johns, J. 2005 Empowerment evaluation: clarity, dialogue, and growth. *American Journal of Evaluation* **26** (3), 421–428. doi:10.1177/1098214005278774.

WaterAid 2019a *Beyond Building Blocks? Identifying and Monitoring Dynamic Drivers of System Performance.* Available from: https://washmatters.wateraid.org/publications/beyond-building-blocks-identifying-and-monitoring-dynamic-drivers-of-sector (accessed 23 July 2021).

WaterAid 2019b *Glossary of System Strengthening and Empowerment Terms.* Available from: https://washmatters.wateraid.org/sites/g/files/jkxoof256/files/wateraids-system-strengthening-and-empowerment-glossary_1.pdf.

WaterAid 2020 *System Strengthening for Inclusive Lasting Water, Sanitation and Hygiene (WASH) That Transforms People's Lives: Practical Experiences From the SusWASH Programme.* Available from: https://washmatters.wateraid.org/publications/suswash-system-strengthening-for-inclusive-lasting-water-sanitation-hygiene.

World Health Organization 2007 *Everybody's Business: Strengthening Health Systems to Improve Health Outcomes: WHO's Framework for Action.* World Health Organization, Geneva, Switzerland. Available from: https://www.who.int/healthsystems/strategy/everybodys_business.pdf.

First received 31 January 2021; accepted in revised form 7 February 2022. Available online 21 February 2022

IWA PUBLISHING

doi: 10.2166/h2oj.2022.023

Fighting for water in South Africa: public participation, water rights claiming and strengthening governance

Thato Masiangoako [a], Kelebogile Khunou [a,*] and Alana Potter [b]

[a] Research and Advocacy, Socio-Economic Rights Institute of South Africa, 6th Floor Aspern House, 54 De Korte Street, Braamfontein, Johannesburg, Gauteng 2001, South Africa

[b] Socio-Economic Rights Institute of South Africa, 6th Floor Aspern House, 54 De Korte Street, Braamfontein, Johannesburg, Gauteng 2001, South Africa

*Corresponding author. E-mail: kelebogile@seri-sa.org

TM, 0000-0002-6423-5509; KK, 0000-0002-3301-0673; AP, 0000-0001-7746-0679

ABSTRACT

Communities in South Africa employ a diverse range of strategies to actively claim their right to water. This paper examines two rights claiming strategies – protest and litigation – through the lens of two case studies. The first case study examines the struggles of the residents of Makhanda, a small town in the predominantly rural Eastern Cape Province, who formed a coalition to advocate for the dissolution of the municipal council for failing to fulfil its constitutional mandate to provide basic services. The second case examines the struggles of the residents of Marikana, an informal settlement in the City of Cape Town, where residents are forced to live in unlit, unhygienic and undignified conditions as a result of inadequate services provision. Although access to water is a justiciable right in South Africa, there is a curious paucity of legal rights mobilisation, with only one court case reaching the Constitutional Court. The article presents the following findings. First, communities employ different methods to claim their water rights, including engagement with government through formal channels, mobilisation, protest, litigation and self-supply. Water is a justiciable human right. Second, litigation is a valuable component of rights-claiming but is most effective if used alongside other strategies like community organisation and protest. Third, water rights can be achieved alongside struggles to secure other rights, like the right to housing, or holding local government accountable. In the two case studies, the legal tools of land expropriation and provincial intervention were employed to indirectly gain access to water services. Finally, community mobilisation, combined with strategic partnerships, plays a critical role in sustaining long-term efforts to claim water rights.

Key words: claiming water rights, governance, legal mobilisation, protest, public interest litigation, public participation

HIGHLIGHTS

- Communities employ different methods to claim water rights including engagement with government through formal channels, community mobilisation, protest, litigation and self-supply.
- Litigation is a vital component of rights-claiming but is most effective used alongside other strategies like protest and community mobilisation.

INTRODUCTION

While the number of households in South Africa with access to a basic water supply is higher now than at the end of apartheid in 1994, today, fewer households have access as a percentage of all households. Despite an investment of more than R1.3 billion in water resources and water services infrastructure over the last 27 years (Department of Water & Sanitation 2019), hundreds of thousands of households, primarily located in rural and informal settlements, lack access to safely managed water services. According to Statistics South Africa, by 2016, approximately 89.8% of households had access to piped water and of that figure, only 44% had access to water in their dwellings (Lehohla 2017). A higher percentage of households without access to basic water supply is prevalent in provinces with higher rural populations, for example, 31% in the Eastern Cape and 27.2% in Limpopo (South African Human Rights Commission 2014).

Communities in South Africa employ a diverse range of strategies to actively claim their right to water. This paper examines two rights claiming strategies – protest and litigation – through the lens of two case studies. Although water is a justiciable right in South Africa, there is a curious paucity of legal rights mobilisation, with only one water-related court case reaching the Constitutional Court, namely, *Mazibuko & Others v City of Johannesburg & Others*. In both case studies in this article, litigation and protest were used to secure tenure, water and other municipal services as alternative strategies to accessing water services.

The first case study examines the struggles of the residents of Makhanda, in the jurisdiction of Makana Local Municipality in the Eastern Cape, who formed a coalition to advocate for the dissolution of the municipal council because it failed to fulfil its Constitutional mandate to provide basic services. The case examines the effectiveness of a legal intervention in terms of Section 139 of the Constitution, in response to municipal failure. Twenty-one percent of municipalities in the country are currently either under provincial administration or facing potential Section 139 interventions (Griffiths 2020).

The second case study examines the struggles of the Marikana informal settlement community, in the jurisdiction of the City of Cape Town in the Western Cape, where residents are forced to live in unlit, unhygienic, undignified and often dangerous conditions as a result of inadequate services provision. The case illustrates the use of expropriation in terms of the Housing Act 107 of 1997 as a tool to widen access to well-located urban land and ultimately, to permanent municipal services like water.

The rest of the article is organised into five sections: The next section is a literature review on central concepts explored in the article largely focused on access to water and methods of rights claiming in South Africa, namely, the water rights legal framework and mandates of local government in delivering services; protest action and the use of public interest litigation in social transformation. The literature review is followed by an overview of the methodology. The article then includes discussions of the Makana and Marikana informal settlement case studies in Sections four and five, respectively, followed by concluding remarks.

ACCESS TO WATER IN SOUTH AFRICA: RIGHTS AND RIGHTS-CLAIMING

Rights and mandates

The right to water is enshrined in Section 27(1)(b) of the Constitution which states that everyone has the right to have access to sufficient water. The legal framework that gives effect to the Constitution includes the National Water Act 36 of 1998, the Water Services Act 108 of 1997 (Services Act), the National Norms and Standards for Domestic Water and Sanitation Services of 2017 and the Municipal Systems Act 32 of 2000 (Systems Act).

The legal framework jointly places the responsibility for water resources management and water services provision on all spheres of government – national, provincial and local. The national Department of Water and Sanitation is the custodian of the country's water resources, 'ensuring that water is protected, used, developed, managed, conserved and controlled in a sustainable and equitable manner for the benefit of all persons and for setting national policy frameworks and standards for the delivery of water services' (Socio-Economic Rights Institute 2019a: 24). The delivery of domestic water and sanitation services is placed squarely within the mandate of district, metropolitan and authorised local municipalities (Socio-Economic Rights Institute 2019a: 24). The national and provincial governments have the authority to ensure that local government effectively executes its functions. The Executive Council of a province (MEC) may choose to take over the obligations of the municipality where a municipality fails to deliver water services.

The democratic developmental state envisioned local government as central to poverty reduction through the delivery of essential services, with a focus on implementing policies of free or subsidised basic services, such as electricity, water and sanitation to poor households (Ledger & Rampedi 2019). As the sphere of government 'closest to the people', local government has a constitutional obligation to ensure that services are delivered to communities. Twenty-seven years later, most municipalities have failed to meet their constitutional obligations. In 2015, the Presidency revealed that 80% of municipalities failed to meet their obligations including 'to ensure the provision of services to communities in a sustainable manner' among others (Khambule *et al.* 2018).

Ledger & Rampedi (2019) summarise the debate on the reasons for local government's failure to include poor financial management, widespread corruption and a lack of skills and capacity to effectively manage the delivery of services. The authors, however, focus their attention on the systemic factors contributing to municipal failure. They argue that the overarching fiscal framework within which municipalities function poses the most significant challenge to municipalities because it assumes that municipalities are able to raise sufficient revenue through property taxes and service charges. However, according to Ledger and Rampedi, in aggregate, local government is able to fund less than 50% of its operating expenditure from property taxes and service charges, with the funding gaps being greatest outside of metropolitan areas. Dugard *et al.* (2014) argue that this inability to raise sufficient revenue leads municipalities to view basic services more as a revenue stream than as a public health or development issue. This has led to inadequate extension of infrastructure coupled with a technocratic approach to the delivery of services, widespread limiting of services and even disconnections, ultimately reducing the initial gains in connecting poor communities to basic services grids. Communities across South Africa are stepping into the gap left by local government, by mobilising to claim their rights, employing a range of strategies including engagement with different levels government, self-supply and protest and litigation (Socio-Economic Rights Institute 2020a).

Protest

The right to protest is enshrined in Section 17 of the Constitution, which states, 'Everyone has the right, peacefully and unarmed, to assemble, to demonstrate, to picket and to present petitions' (The Constitution of the Republic of South Africa 1996). A lack of basic services like electricity, water and sanitation are the most common issues driving dissent in township and informal settlement communities (Jain 2010 as cited by Dugard *et al.* (2014)). South Africa has come to be known as the protest capital of the world (Alexander 2012). Between 2005 and 2017, South Africa has seen significant increases in the number of community protests per year (Alexander *et al.* 2018). This increase in protest action has been referred to as a 'rebellion of the poor', largely concentrated in poor urban centres and characterised by disruption and sometimes violence, reaching 'insurrectionary proportions' (Alexander 2010). Drawing from a range of indicators and sources, Alexander *et al.* also show that the period coincides with an increase in community protests that were disruptive and/or violent (Alexander *et al.* 2018).

Communities and groups have increasingly relied on the right to protest as a vital means of political participation and expression of dissent (Botes 2018). The surge in protest action is attributed to exclusion from formal avenues of participation in local governance, also referred to as 'invited' spaces of participation (Miraftab 2004; Mkhize 2015; Socio-Economic Rights Institute 2019b). For many poor and marginalised communities, this exclusion forces them to rely on 'invented' spaces and participation mechanisms which fall outside the formal channels and can include protest action as a method of engagement (Socio-Economic Rights Institute 2019b). Scholars from the Centre for Social Change at the University of Johannesburg argue that groups, communities and social movements resort to protest action usually after having exhausted both formal and informal channels of participation which typically yield minimal outcomes (Alexander *et al.* 2014).

The main obstacles to the right to protest in South Africa stem from a skewed perception of protestors and protest action, where protestors are viewed as a nuisance and inconvenience at best, and criminal at worst. The policing of protests is criticised for excessive use of force and abuses by police, often carried out with impunity due to a lack of effective remedies and accountability (Madlingozi 2014; Mkhize 2015; Lancaster 2016; Socio-Economic Rights Institute 2017).

Water is often at the centre of 'service delivery' protests, as part of a broader set of socio-economic demands for decent roads, sanitation, refuse collection, electricity and housing. The term 'service delivery protests' has been critiqued by the Centre of Social Change which argues that the term masks wider concerns including South Africa's democracy as well as other issues like corruption, crime, unemployment, education, housing and repression. The emerging field of hydropolitics, defined by Jankielsohn as 'attempts to mobilise support in order to consolidate a power base which can secure the equitable and sustainable supply, management, and distribution of water resources to specific areas, communities and activities', notes additional factors affecting access to water which have socio-political and economic implications, potentially giving rise to unrest at the local, regional and international levels (Jankielsohn 2012). These factors include water scarcity, management of available water resources, pollution of water sources, sewerage management as well as broader issues affecting water services infrastructure. According to the National Climate Change Response White Paper, 'based on current projections, South Africa will exceed the limits of economically viable land-based water resources by 2050' (Department of Environmental Affairs 2011). As surface water sources are reduced and water services become more expensive to access, we can expect water-related unrest to continue and possibly increase (Jankielsohn 2012).

In South Africa, according to data on protest, of 2,880 incidents between 2013 and 2015, 46 (2%) were water-related while 262 (9%) were unspecified 'service delivery' protests (Lancaster 2016). The Civil Society Coalition

Campaign for South Africa's Ratification of the International Covenant on Economic, Social and Cultural Rights (ICESCR) and its Optional Protocol stated in its shadow report (2018) to the United Nations Committee on Economic, Social and Cultural Rights (CESCR) that 'the poorer and more rural the municipality a person lives in, the more likely he or she is to receive sub-standard water and sanitation services'.

Protests related to shortages in vital resources like water are not unique to South Africa. Water-related protests have taken place in Venezuela, where two people were allegedly killed during water and electricity protests in 2019; in India, police allegedly killed five farmers during water protests in 2017; in the United States, protests resisting the construction of an oil pipeline that threatened water supplies erupted in 2018; in Morocco, protests over water shortages turned violent in 2017; in Somaliland, protestors were allegedly killed during protests over water shortages in 2013; and in Mexico, protests to prevent the diversion of local spring water to a new development led to more than 100 people getting injured (The Pacific Institute 2019).

More recently, Iranian authorities responded violently to water protests caused by chronic water shortages in the province of Khuzestan. In July 2021, the United Nations High Commissioner for Human Rights called on the Iranian government to urgently address the water crisis and refrain from using excessive force against protestors (United Nations Human Rights Office of the High Commissioner (OHCHR) 2021). As water sources around the world are threatened, more groups and communities will be forced to turn to protest in an effort to defend their human right to water.

Public interest litigation

After years of struggle to engage government authorities to improve services delivery, communities are increasingly turning to the courts as one of their strategies to reclaim their power and resist state repression of their constitutional rights. In fact, law and legal services have been assigned a prominent role in the transformation of South African society because one of the primary vehicles to achieve social transformation has been the Constitution, which is considered one of the most progressive in the world because it contains socio-economic rights like the right to have access to health care, sufficient food and water, a basic education and access to housing among others (Socio-Economic Rights Institute 2015). The socio-economic rights literature presents significant examples where courts have been used to correct government failure through public interest litigation, for example government's failure to respond to the HIV/AIDS crisis in *Minister of Health v Treatment Action Campaign (No.2)(2002) 5 SA 721 (CC)* and its failure to protect people facing homelessness through eviction or disaster in *Government of the Republic of South Africa v Grootboom 2001 (1) SA 46 (CC)* (Socio-Economic Rights Institute 2015). Black's Law Dictionary (1990) defines public interest litigation as 'a legal action initiated in a court of law for the enforcement of public interest or general interest in which the public or class of the community have pecuniary interest or some interest by which their legal rights or liabilities are affected'. It seeks to achieve changes to laws and policy that will benefit individuals and communities beyond those directly involved in a court case. Public interest litigation is of particular value to marginalised groups, without the resources to be directly influential in policy issues that affect their communities (Durbach *et al.* 2013).

Socio-economic rights cases which have been heard at the Constitutional Court include *Mazibuko v City of Johannesburg 2010 (4) SA 1 (CC)* which is a seminal case on the provision of adequate water services and *Joseph v City of Johannesburg 2010 (4) SA 55 (CC)*, a case about unlawful electricity disconnections. Another notable case that reached the High Court is *Concerned Residents of Flag Boshielo West and Others v Sekhukhune District Municipality and Others,* Case No. 15437/2015 (11 August 2020), a case about five villages with approximately 5,000 households in Limpopo who have been living without a basic water supply since 2009 (Centre for Applied Legal Studies 2017).

A wealth of literature exists on the contribution of public interest legal services, in particular litigation, to social transformation in the United States (Scheingold 1974; Rosenberg 1991; McCann & Sliverstein 1998; Cummings & Rhode 2009). The most prominent debate on the impact and value of public interest litigation is between Rosenberg and McCann. Rosenberg posits that judicial strategies are usually ineffective in bringing about meaningful social change, and can even have negative effects for change; while McCann, on the other hand, holds an optimistic perspective, arguing that litigation can have a positive impact on social movements and provide leverage for political mobilisation (Durbach *et al.* 2013). While Rosenberg focuses on measurable direct outcomes, McCann focuses on the mobilising effect that litigation can have (Durbach *et al.* 2013).

Over the years, the discourse has evolved to the more nuanced and contextual approach of the *legal mobilisation* framework, which, led by McCann, recognises that public interest law is 'imperfect yet indispensable…'

(Cummings & Rhode 2009) and has the potential to contribute meaningfully to social transformation when viewed as a political tool forming part of a larger arsenal of strategies (McCann & Sliverstein 2009). Scheingold (2004) discusses how rights can both empower and constrain social movements and warns against the 'myth of rights' – the assumption that 'litigation can evoke a declaration of rights from courts; that it can be used to assure the realisation of these rights; and finally, that realisation is tantamount to meaningful change'. In analysing the usefulness of litigation, he advocates for a political approach that places the redistribution of power at its centre (Scheingold 2004).

While much of the literature emanates from the context of the United States, the impact of public interest legal services is also being debated in South Africa, one of the countries advancing thought and practice in this discourse. According to the Public Interest Legal Services Project Report, some scholars hold that the purpose of public interest legal services is to win 'concrete positive outcomes' for clients, and then implementing the strategies and tactics which will ensure the enforcement of a favourable judgment (Socio-Economic Rights Institute 2015). A 2008 report (and subsequent 2015 follow-up report) commissioned by Atlantic Philanthropies by Budlender *et al.* (2015) discusses strategies to be used in combination with litigation to ensure success as well as several factors like proper organisation of clients; coordination and information sharing; and proper follow-up after litigation, which they hold are essential to ensuring that litigation succeeds and achieves maximum social change. Other factors include having an overall long-term strategy, timing, research and characterisation (Budlender *et al.* 2015).

Dugard & Langford (2011) critique Budlender, Marcus and Ferreira for taking a 'materialist' position and being 'too narrow'; and that the value of public interest legal services cannot be reduced to favourable outcomes in court. They argue that public interest litigation is 'generally too unpredictable and diffuse for it to be adequately assessed through a formulaic or scientific approach' by using the example of *Mazibuko*, as a case which 'ticked all the boxes' prescribed by Budlender, Marcus and Ferreira and yet did not result in a favourable judicial outcome, and *Joseph*, a case which did not tick all of boxes and resulted in a win in court. Moreover, the authors argue that, while litigation is unpredictable, when combined with *legal mobilisation*, defined by Zemans (1983) as the point at which 'a desire or a want is translated into a demand as an assertion of one's rights', it has the power to shape the politics around an issue in less obvious but profound ways. On the other hand, Brickhill (2014) argues that Dugard and Langford's critique of Budlender, Marcus and Ferreira 'speaks past' the theory proposed in the report, which is primarily concerned with 'how' to best conduct public interest litigation, while Dugard and Langford's critique is about 'why' public interest litigation should be used. Brickhill (2014) insists that both theories are important for public interest lawyers.

In order to avoid overestimating or underestimating the impact of legal strategies, another legal mobilisation theorist, Madlingozi (in Langford *et al.* 2013), argues that the role and impact of socio-economic rights strategies on post-apartheid social movements should be located in a mix of variables that shape social movement success or failure, namely, political opportunities and threats (for example, the existence of state repression), resources, action repertoire (the range of tactics and tools, like protest, available to social movements) and a movement's collective action frame. With the background of a lack of participatory democracy and governance at the local level, coupled with the absence of any significant change to the lives of poor people in South Africa, Tissington (2012) argues that collaborative efforts between individuals, communities and social movements, although complex and messy, should be encouraged. While acknowledging the critiques levelled against the impact of public interest litigation, she argues that public interest legal organisations have an important role to play, and can offer social movements and communities political currency and platforms, in their efforts to advocate for socio-economic transformation. The gains from litigation, according to Tissington, should be examined through a broader lens, focused on the less material role of rights and law as a politicising agent, and for the potential of politicisation of this nature to counterbalance party and electoral politics which dominate the present South African landscape.

METHODOLOGY

The article draws on research conducted by the Socio-Economic Rights Institute (SERI) in its 'Claiming Water Rights in South Africa' research project which is part of the #ClaimYourWaterRights campaign by End Water Poverty. All authors were employed at SERI and contributed to the project. The cases were selected to present a diversity of ways in which communities, representing a variety of settings, claim their water rights. The cases included rural and urban (towns and informal settlements) communities, spread across different provinces in South Africa, who adopted a wide range of strategies. The two case studies in this article were selected to compare, contrast and draw learnings from two disparate communities which utilised protest and litigation as strategies to claim water rights. The Marikana case study also draws from a 2019 SERI research series on existing norms, practices and agency across three informal settlements in South Africa, one of which was the Marikana informal settlement. SERI also litigates on behalf of the Marikana informal settlement and has a long-standing relationship with the community. SERI's interest in conducting research in Marikana, beyond the litigation relationship, is the sheer size of the settlement (consisting of approximately 60,000 people), the local reality in terms of land management practices, access to basic services, livelihood strategies and local leadership arrangements. The case studies were compiled using desktop research and semi-structured interviews with individuals from the communities discussed in the article and others involved in the litigation. The theoretical framework was compiled using academic articles and media reports.

The article examines two cases in the Claiming Water Rights series. While there are surprisingly few water rights legal cases in South Africa, there are numerous socio-economic rights cases, which we used to examine the role of litigation in rights-based struggles. The Claiming Water Rights cases did not include interviews with government officials serving the communities discussed in the article because the research was designed to be a resource for other communities facing similar challenges and therefore focused on the experiences and perspectives of the communities in the case studies. As local government is legally mandated with the delivery of water services, this presents a limitation.

CASE STUDY 1: MAKANA LOCAL MUNICIPALITY

The Makana Local Municipality, in the Eastern Cape Province, is home to approximately 86,000 people, many of whom live in and around Makhanda, its largest town and economic hub (Makana Local Municipality 2021). Makhanda (formerly known as Grahamstown) is the seat of the High Court and home to Rhodes University, the municipality's largest ratepayer (Makana Local Municipality 2021). In 2005, the Makana Local Municipality had financial reserves of over R50 million and approximately 350 employees, however, by 2014, the municipality had accrued over R150 million in debt and over 1,500 employees (Unemployed Peoples' Movement v Premier Province of the Eastern Cape & Others 2020). Makana's 2020–2021 Integrated Development Plan (IDP) acknowledges poor service delivery, lack of good corporate governance and inadequate revenue collection as some of the challenges that have affected the municipality since 2014 (Makana Local Municipality 2021).

Access to water varies: some communities have taps in their homes or yards, while in other areas like informal settlements and township, residents rely on communal standpipes, municipal tankers and rain water (Hamer *et al.* 2018). Informal settlement residents have also relied on relationships with people living in formal houses nearby to share water (A. Kota, interview, 5 May 2020). Those relying on communal taps have reported needing to travel very early for up to 30 minutes on foot to access water before it is depleted (Hamer *et al.* 2018). In May 2018, independent laboratories discovered unsafe levels of *E. coli* in Makana's drinking water, sparking outrage among residents who had already been battling with access to clean and safe water (Nowicki 2018). Water supply has been hampered by aging water infrastructure and a severe drought that lasted 4 years, almost depleting water levels in the main dams (T. Bull, interview, 5 May 2020). This resulted in water supply disruptions across the municipal area, reportedly ranging from between four and six times a week. In extreme cases, water is inaccessible for weeks at a time (Hamer *et al.* 2018; Kota, radio interview, 2019). Currently, water in Makhanda is rationed on a 'two-days on and one-day off basis' by the municipality, however, the urban townships of Fingo, Tantyi and Joza report water cuts for 5 days at time and in Alicedale, a rural town, there is no water at all (Majavu 2020; Makana Local Municipality 2020; Dyongman 2021).

Different actors in Makana place emphasis on different causes of the water crisis. Municipal officials emphasise drought-induced dam levels and aging infrastructure, while the Grahamstown Residents' Association (GRA) highlights a lack of municipal planning for population growth and associated pressure on existing infrastructure (T. Bull, interview, 5 May 2020). A Rhodes University research unit, the Public Service Accountability Monitor (PSAM), blames persistent instability in municipal governance (J. Kruuse, interview, 12 May 2020), while the Unemployed Peoples' Movement (UPM), a South African social movement active primarily in the Eastern Cape and KwaZulu-Natal, with over 2,000 members, believes that corruption and incompetence have resulted in poorly maintained infrastructure which has gutted the municipality's ability to provide water (A. Kota, interview, 5 May 2020).

Community mobilisation

The water crisis and poor service delivery in Makana have brought a wide range of groups together over the years. The residents of Makana have collaborated and mobilised for improved services delivery since 2008 (A. Kota, interview, 5 May 2020). UPM, PSAM, GRA and the Makana Civil Society Coalition (MCSC) have led various efforts to engage government at all levels to hold the Municipality accountable and to agitate for provincial intervention in terms of Section 139 of the Constitution which is discussed further below. This collaboration culminated in the litigation brought by the UPM in which the Eastern Cape High Court ordered the dissolution of the Makana Municipal Council.

The situation in Makana is unique in that the town is relatively small, with a water crisis that affects both suburban residents who are predominantly white and township and informal settlement residents who are largely black and poor. Because the water crisis in Makana affects all residents and key entities including local businesses and Rhodes University, it became an issue around which groups could collaborate and mobilise their own constituents despite differences in their backgrounds, socio-economic statuses or political leanings. When asked about the relationship their organisations had with each other, Tim Bull of the GRA (now known as the Makana Residents Association) responded as follows:

> 'There exists a very close relationship and we are in communication regularly because there is a shared sense that we all in this together'.

Similarly, Ayanda Kota of UPM responded as follows:

> 'The relationship is honest and cordial and informed by the shared concerns that affect everyone – rich or poor, black or white. The issues affect everyone and so it would not be strategic for UPM to isolate itself. It would be foolish to do so. There is a need to work well with everyone'.

Litigation

Section 151 of the Constitution provides that local government functions autonomously, but that provincial and national government have oversight powers to ensure that municipalities fulfil their obligation to deliver basic services (The Constitution of the Republic of South Africa 1996; Ledger & Rampedi 2019). This oversight system is set out in Section 139 of the Constitution and elaborated in the Municipal Finance Management Act 56 of 2003. Section 139 provides options for provincial intervention which include pressurising the municipality to improve through a directive; assuming the responsibilities of the municipality for the unfulfilled obligations; and in exceptional circumstances, dissolving the municipal council and appointing an administrator (The Constitution of the Republic of South Africa 1996; Chamberlain & Masiangoako 2021).

Prior to the January 2020 court ruling ordering the dissolution of Makana's municipal council, the municipality had already been subject to two previous provincial interventions beginning in September 2014 in which two consecutive administrators had been appointed (Nowicki 2020; Unemployed Peoples' Movement v Premier Province of the Eastern Cape & Others 2020). By October 2016, at the end of the second provincial intervention, there was no noticeable improvement in services delivery or in the management of the municipality's finances. This confirms Ledger and Rampedi's finding that since 1998, of a total of 140 provincial interventions involving 143 municipalities, 48 had more than one intervention initiated previously and that very few of these provincial interventions resulted in sustained improvements in municipalities' functioning and financial circumstances (Ledger 2019; Ledger & Rampedi 2019).

Between 2016 and 2019, different residents' groups in Makana agitated for urgent intervention by reaching out to a range of bodies and officials, from Makana's municipal manager and mayor, to the South African Human Rights Commission (SAHRC), different ministers of the Department of Cooperative Governance and Traditional Affairs (CoGTA), and various Parliamentary Portfolio Committees (A. Kota, interview, 5 May 2020).

When these efforts failed to yield any favourable responses, UPM, with the support of the GRA, approached the courts seeking yet another intervention but this time with the dissolution of the municipal council. For UPM, previous interventions had not succeeded partly because appointed administrators were forced to work with 'the same corrupt [municipal] councillors' (A. Kota, interview, 5 May 2020). In preparation for the litigation, UPM and the GRA went further by consulting the community and collected about 22,000 signatures in person from Makhanda residents who all felt that council needed to be dissolved (A. Kota, interview, 5 May 2020). This is an important example of how community mobilisation intersected with the litigation to reinforce it.

In *the Unemployed People's Movement v Makana Municipality and Others*, the Court was asked to declare the Makana Municipality in breach of its constitutional duty to 'ensure the provision of services to its community in a sustainable manner and to promote a safe and healthy environment' (Section 152) as well as its duty to 'structure and manage its administration, budgeting and planning processes in order to give priority to basic needs and promote the social and economic development of its community' (Section 153(a)) (The Constitution of the Republic of South Africa 1996). The Court also had to determine whether *discretionary* or *mandatory* provincial intervention was warranted and whether the municipal council should be dissolved (Chamberlain & Masiangoako 2021). Discretionary provincial intervention takes place 'in cases of executive failure and, if the province decides to intervene, and if exceptional circumstances exist, [Province] may dissolve the municipal council'; mandatory provincial intervention takes place 'in instances of a finance driven failure to deliver services or honour financial commitments, and when [Province] intervenes, [it] must dissolve the municipal council if the municipality fails to pass legislative measures to implement the recovery plan' (Chamberlain & Masiangoako 2021: 437).

Ultimately, the Court found that a mandatory intervention and the dissolution of council was indeed warranted. The Court also noted that the residents had exhausted every avenue available to them to address the failure of the municipality. The Makana Local Municipality and the provincial government have since sought to appeal the judgment arguing partly that the High Court's decision is a violation of the separation of powers and that it amounts to judicial overreach. At the time of writing, the matter was before the Supreme Court of Appeal (Chamberlain & Masiangoako 2021).

In the meantime, services delivery remains poor, sparking continued unrest. The January 2020 judgment has been hailed as a landmark judgment with significant implications for other failing municipalities across the country. Other litigation against municipalities failing to deliver basic services has since been pursued in other provinces including Enoch Mgijima Municipality also in the Eastern Cape and Kgetlengrivier Municipality in the North West Province (Heese & Allen 2021; Potter 2021).

Protest

UPM has used protest as a tool for mobilisation and consciousness-raising among Makhanda residents (A. Kota, interview, 5 May 2020). Over the years, residents have staged numerous protests, ranging from sit-in demonstrations, peaceful marches and submitted memoranda to more disruptive protest action in the form of road blockades to which police have often responded violently.

For example, in 2011, UPM alongside the Women's Social Movement staged a sit-in at the municipal offices in protest against rape, recurring water outages and poor services delivery in Makhanda's townships and informal settlements (Unemployed Peoples Movement 2011). Following a violent removal from the offices, protestors staged road blockades in a number of townships to which the police reportedly responded by beating protestors, firing rubber bullets, using pepper spray and setting dogs on the protestors (Tabensky 2011).

Various UPM leadership were arrested and released the following day with restrictive bail conditions prohibiting them from inciting any protest or participating in any further protest action and keeping at least 100 m distance from the Makana Municipality and Magistrates Court (Unemployed Peoples Movement 2011; Duncan 2016). The response to the protests was disproportionate and according to Jane Duncan, the bail conditions were unconstitutional (Tabensky 2011).

In 2018, following the discovery of *E. coli* in the community's drinking water, Ayanda Kota, a prominent community leader in Makhanda and founding chairperson and UPM spokesperson, and 25 members of the UPM, delivered buckets of dirty water collected from local taps to a public consultation meeting. The protest was widely reported and is said to have led to the resignation of the Water and Sanitation manager of the municipality.

During the COVID-19 lockdown in 2020 and 2021, protests erupted due to continued water outages and other services delivery concerns. In May 2021, Makhanda residents staged a series of road blockades to protest poor services delivery including 'potholed roads, unreliable water supplies, illegal dumping, demands for the electrification of informal settlements and sewage spills' (Maclennan 2021). The protest also sought a direct response from the Premier of the Eastern Cape Province as to why the province is opposing the dissolution of Makana's municipal council (Maclennan 2021). These protests were halted to allow authorities to develop a plan to address the grievances, however in June 2021, these protests resumed (Dayimani 2021).

Confronting failing local governance and water services

The struggle to secure sustainable access to water in Makana is located within a larger struggle to restore an effective local government. Over the years, residents in Makana have made persistent efforts to participate in local government structures, including social accountability efforts led by the university-based PSAM, to no avail. In fact, a lot of these efforts were well documented, important pieces of evidence for the litigation. Litigation and protest have had far more impact on holding the system of local governance accountable than the local government public participation processes designed precisely for that purpose. However, it is yet to be seen what the overall impact of this litigation and protest action will have on governance and, more specifically, the restoration of reliable water services to households in Makana.

According to Zemans (1983), law is mobilised when a desire or a want is translated into a demand as an assertion of one's right or lawful claim. Legal mobilisation in Makana, by the collaborating organisations and pro-bono legal assistance, has been invoked to seek a legal solution to a failure in an administrative and political arena of local governance. This strategic use of the law in Makana's case was also aimed at mobilising around the restoration of socio-economic rights, and in particular, the right to water.

The use of litigation and protest action serve as reinforcing strategies employed by a coalition of actors in Makana, all of whom are negatively affected by unreliable water provision and uncertain water quality. UPM turned to the courts as a last resort to resolve the crisis in Makana, as noted by Kota, 'When UPM sat and met with the lawyers… it became clear that all channels were exhausted'. However, in light of the ongoing appeal process and the resultant delays to enacting the January 2020 judgment, the latest wave of protests in Makana are a direct challenge to what is being seen as an effort by the executive to delay the enactment of a decision by a court of law. At the 2021 protest, Kota explained the purpose of the protest: 'We have no memorandum. We just want [the Premier] to explain why he decided to oppose the court ruling' (Maclennan 2021). Given the upcoming local government elections scheduled for late 2021, such delays could render the judgment moot.

Both strategies of litigation and protest have depended heavily on sustained relationships and coalition-building across the various organisations active in Makana. What has been especially remarkable about Makana's experience has been the energetic use of collaboration among the various actors and organisations, including social movements, informal settlement residents, residents of the middle and upper-class suburbs, unions and the university. The collaboration and coalition-building has taken place over several years as the water crisis and service delivery failures in Makana worsened. These organisations have had to sustain their efforts over a long period of time which culminated in a successful court challenge. As noted by one of the interviewees, Jay Kruuse from PSAM, the litigation depended on an established paper trail that could prove persistent municipal non-compliance, which required sustained collaboration between all the affected groups (Socio-Economic Rights Institute 2020d). These alliances span class, race and geographic divides and have partly been made possible because Makhanda is a relatively small town with services delivery and water crises that affect all its residents. The extent to which such successful coalition-building sustained over a period is possible in more populous areas is yet to be seen.

Ultimately, Makana's experience highlights the critical role that communities themselves play in strengthening governance by actively, fearlessly and persistently holding the state accountable through the use of protest and litigation. Their reliance on protest and litigation, both measures of last resort, speaks to a failure in systems of local governance, which in Makana excluded the community and until litigation has been unresponsive to their concerns. For years, the Makana Local Municipality has limped along from crisis to crisis despite two provincial interventions (Chamberlain & Masiangoako 2021). The UPM judgment, the first of its kind in South Africa, is the result of more than a decade of persistent struggle, in which protest played a prominent role. It was litigation, however, that got the attention of Makana's Local Municipality and Provincial Executive. The judgment, which orders the dissolution of a municipal council, is significant because of the implications it could have for more than 80% of the other struggling municipalities in South Africa.

CASE STUDY 2: MARIKANA INFORMAL SETTLEMENT

The Marikana informal settlement, located in Philippi East, falls in the jurisdiction of the City of Cape Town Metropolitan Municipality. The settlement is home to an estimated 60,000 people living in over 12,000 households (Ground Up 2017; Socio-Economic Rights Institute 2019a). Named after the site of the massacre of mineworkers on the Platinum Belt in 2012, the residents of the settlement compared the state's violent response to their land occupation to the state's violent response to the protesting mineworkers (Socio-Economic Rights Institute 2019a: 12). Settlement, in what became known as Marikana, began slowly in the early 2000s. In 2013, however, an influx of people occupied the area and by August 2014 more than 1,500 structures had been built on 11 adjoining properties. With each wave of occupation came violent demolition of homes and unlawful evictions by the Anti-Land Invasion Unit, the South African Police Service and/or the Cape Town Metropolitan Police Department in an effort to quell the occupation (Socio-Economic Rights Institute 2019a: 21–24).

Due to the lack of affordable accommodation in well-located areas, poor households continue to 'self-allocate' land where systems of private and public allocation of land are failing (Socio-Economic Rights Institute 2019b: 5). Reports indicate that there has been a steady increase in informal settlement in metropolitan areas over the last 10 years (Cirolia 2014: 399). Where a person resides determines their ability to access basic services. The majority of black and poor people living in South Africa including those residing in informal settlements have disproportionally poor access to basic services including water, sanitation, electrification and health (Socio-Economic Rights Institute 2016: 3).

South Africa has a progressive legislative and policy framework and case law governing the right to housing and a comprehensive state-subsidised housing programme that seeks to redress the apartheid legacy (Socio-Economic Rights Institute 2020a: 48). Despite the gains over the last two decades, systemic challenges persist mainly due to challenges in implementation, including poor planning, a lack of coordination, insufficient capacity and a lack of political will among other issues (Pithouse 2009; Huchzermeyer 2010; Socio-Economic Rights Institute 2020a:48). The Upgrading of Informal Settlements Programme (UISP) sets out a system that prioritises the provision of basic services, sanitation, bulk infrastructure and tenure security, and an extensive community participatory process facilitated by municipalities. However, municipal implementation of the UISP is poor; municipalities are still choosing to relocate informal settlements to vacant land, far from social and economic networks (Omenya & Huchzermeyer 2006, Huchzermeyer 2011).

There is little available disaggregated information about basic services in informal settlements because informal dwellings in informal settlements are not considered households, and therefore, data on residents' access to services is rarely collected (ICESCR CSO Coalition South Africa 2018: 8). According to the Housing Development Agency (2013), in 2001, 26% of households in informal settlements had piped water; 33% could obtain piped water within 200 m of their homes; 32% had access to piped water in excess of 200 m from their homes (there is no clear indication of how far away the water source is); and 9% had no access to water at all.

Government provision of basic services like electricity, water and sanitation, and refuse collection in Marikana has been dismal. Communal taps, located on the periphery of the settlement, are the only government-provided source of water in the settlement (Socio-Economic Rights Institute 2019a). This makes it difficult for people with disabilities, the elderly or sick who may not be able to travel long distances and carry jerry cans from the communal standpipes to their homes (ICESCR CSO Coalition South Africa 2018: 8). Furthermore, the responsibility to collect water often falls on women and girls (World Health Organization & UNICEF 2010: 29) and as such the position of the standpipes, especially where there is a lack of electrification and adequate policing, increases their risk of violence and sexual assault (Social Justice Coalition 2009). The City claims to have installed 100 communal tap stands, however, when SERI conducted its Informal Settlement Action Research in 2018, it counted between 50 and 60 taps, and noted that water did not flow from at least one in every five of the taps (Socio-Economic Rights Institute 2019a: 46). Assuming that 100 communal taps are available to Marikana residents, each tap is shared by 600 people. Additionally, residents often had to wait in queues for hours to collect water; the taps had low pressure and sometimes ran dry due to high demand over weekends (Danti 2017: 32; Socio-Economic Rights Institute 2019a: 46).

Where water services were not provided or were unaffordable, inadequate or insufficient, Marikana households drew on their own resources to fill in the gap through self-supply (Socio-Economic Rights Institute 2019a: 43). Some bought water from neighbouring areas (at a cost of up to R50 for a 25-litre bucket of water); others paid for water connections to be installed, connecting to main lines on the periphery of the settlement or to municipal infrastructure which was discontinued; and a smaller number of residents bought water from shops or accessed water from natural springs (Socio-Economic Rights Institute 2019a: 46–47).

Community mobilisation

Community representation in Marikana was initially comprised of a number of committees and task teams responding to evictions and demolitions and vying for leadership status. This was the case particularly in Rolihlahla and Marikana I, and less so in Marikana II, which is a far larger part of the settlement (T. Ebrahim, Interview, 17 April 2020). In the early stages of the occupation, leadership was a 'free for all' and leaders lacked legitimacy in the view of local residents and there were sporadic conflicts between rival committees (Rayner 2016).

This shifted with legal representation in early 2014. A clear leadership structure needed to be identified and the residents of Marikana organised themselves into three committees representing different settlement zones elected through mass meetings (Socio-Economic Rights Institute 2019a: 25). The process was, however, fraught with tension and conflict, and posed the question of whether this external influence served to validate and legitimise claims to leadership, interfering with what should have been an organic process (T. Ebrahim, Interview, 17 April 2020).

Nevertheless, residents used these structures to advocate for their interests, which included the right to water, through a variety of means including ward participation, taking part in spaces convened by the local councillor, staging protests, submitting memoranda and in the end through litigation (Socio-Economic Rights Institute 2019a: 80–83). Initially, when the Marikana community attempted to engage government about service delivery issues, the City dismissed the residents as 'land invaders' (Socio-Economic Rights Institute 2019a: 47–48). In paragraph 4 of the *Occupiers of Portion R25 of the Farm Mooiplaats 355 JR v Golden Thread Ltd and Others* judgment, the Constitutional Court disapproved of the term 'invaders' to occupiers of vacant, unused land as it detracts from the humanity of the occupiers and is 'emotive and judgmental and comes close to criminalising the occupiers' and held that the word 'occupiers' might be more appropriate.

Litigation

Though the initial legal strategies focused on defending the residents against the demolitions and evictions, the strategy switched to proactively calling for expropriation of the occupied land. The residents launched a counter-application against eviction proceedings launched by the landowners in 2014, in which they argued that the state should purchase or expropriate the land on which the settlement is located, as this was the only way of avoiding homelessness (Socio-Economic Rights Institute 2020a: 47). The eviction of 60,000 people was not feasible. They relied on Section 9(3) of the Housing Act which provides a legislative framework within which the state can purchase or expropriate land for the purposes of developing state-subsidised housing (Housing Act 107 of 1997).

After 6 years of legal battles, on 4 March 2020, the residents of Marikana informal settlement, the property owners, the City of Cape Town and the Western Cape Provincial Minister of Human Settlements reached a landmark agreement, which was made an order of the Supreme Court of Appeal, in *Fischer v Unlawful Occupiers, Erf 150, Philippi, 2018 (2) SA 228 (WCC) 99* (Socio-Economic Rights Institute 2020c). The order requires the City of Cape Town to 'purchase the properties that are the subject of the High Court's order at a price to be determined in an arbitration between the City of Cape Town and the property owners' (Socio-Economic Rights Institute 2020c).

Protest

> 'Protest is a tool to bring about change… [but] there's a risk involved. Even if you are shot by a rubber bullet. When the shot takes place, when you hear that noise, it affects your emotions. Remember, there are those who are brave, who can go forward with gunshots taking place, and there are those that are scared. But everybody needs services' – Resident of Marikana informal settlement (in Socio-Economic Rights Institute 2019a).

Beginning in June 2013, after the first wave of demolitions when residents organised a protest and handed over a memorandum of grievances to the City, and throughout the litigation process, protest was used as a tool for mobilisation among the Marikana residents (Socio-Economic Rights Institute 2019a: 23). Given their unproductive engagements with a City which regarded them as 'land invaders', and fruitless attempts to participate in the 'invited' spaces of ward committees, the residents used protest as a means of political participation and expression of their grievances, when all formal channels of participation did not yield results. Protest for some of the residents constituted collective resistance against evictions, for others it was a way of making a claim to land and advocating for the provision of basic services, not only for themselves, but for future generations (Socio-Economic Rights Institute 2019a: 84–85).

Residents undertook protest action in Philippi East and Cape Town's Central Business District to communicate to government at all levels about their challenging living circumstances (Socio-Economic Rights Institute 2019a: 87). Notable protest action by the Marikana community took place in 2013 when residents blocked the N2, a national road, disrupting the entry and exit point for trucks transporting goods in and out of Cape Town; as well as the protests outside the Cape High Court in 2017, during the Marikana eviction case court hearing (Socio-Economic Rights Institute 2019a: 86). Integral to the protests was the attention the residents got from the media. One resident explained that they incorporated singing into protests as a way to draw the attention of outsiders and journalists (Socio-Economic Rights Institute 2019a: 85). Singing during protests is a way to demand recognition and is significant for informal settlement residents who have been largely ignored and rendered invisible by the City.

Demanding recognition for access to water services in informal settlements

The Marikana case study showcases how expropriation can be used as a tool to widen access to urban land and to provide marginalised communities with services where they already live. Section 25 of the Constitution contains the right to property and provides that property may be expropriated in terms the of law of general application, for a public purpose or in the public interest, and subject to compensation (The Constitution of the Republic of South Africa 1996). This subject has dominated the public discourse in recent years when in February 2018, the motion to expropriate land without compensation was passed in the National Assembly, starting a debate on how this could be implemented. SERI argues that expropriation can be used as one of the tools to help widen access to urban land to people, like informal settlement communities, who would not be able to access the urban housing and property market (Socio-Economic Rights Institute 2019c: 5). This can be done by the state through taking control of buildings and land currently occupied informally by poor people as a first step towards securing their tenure.

Section 9(3) of the Housing Act provides for the expropriation of land for the purposes of housing development in terms of any housing programme, including the UISP (Socio-Economic Rights Institute 2020b: 22). The UISP, in fact, specifically provides that funding to municipalities can be allocated for the acquisition of land either through negotiation or expropriation, where the land to be developed is privately owned. Moreover, the UISP requires the provision of services like water, therefore there is a clear route from expropriation to the realisation of water rights and other basic services (Socio-Economic Rights Institute 2020b: 22).

It should be noted that even though the litigation has been settled, the City of Cape Town is yet to purchase the properties as per the order of the Supreme Court of Appeal, and as such the community of Marikana has not yet received secure tenure and access to basic services. Nevertheless, as the Marikana story is unfolding, many lessons can be used by communities elsewhere in their struggles to obtain access to basic services.

Legal mobilisation in Marikana began with residents organising themselves into committees and seeking legal assistance to resist unlawful evictions and demolitions. This initiation of a juridical process moved the politics of the settlement from the ground to the courts, where legal protections in Section 26 of the Constitution and the Prevention of Illegal Eviction and Unlawful Occupation of Land Act (PIE Act) were invoked when residents' access to land was threatened. Throughout the litigation process, the community utilised protest to further draw attention to their cause.

In preparation for the litigation process, SERI collaborated with the committees to obtain information about the number of households, the names of deponents and other information. This was used by community leaders to establish a level of legitimacy in the community. This can be understood in two ways: litigation used as a rallying point or litigation influencing what should be an organic process (Socio-Economic Rights Institute 2020b: 31). With large numbers and significant levels of organisation in the settlement, a defence against eviction turned into positive action for claiming land, and thus provided the footing from which water rights can be claimed (Socio-Economic Rights Institute 2020b: 31). Having a united settlement involved in one case rather than separate evictions cases put the argument for expropriation on the table, as evicting the combined numbers of the sections of the settlement was not a reasonable course of action. Furthermore, the high levels of organisation in the settlement meant that the steps after expropriation and the roll-out of services as the UISP is being implemented were plausible (Socio-Economic Rights Institute 2020b: 31). It is important to note, however, that while litigation is often afforded prominence, side-lining the daily organising of communities, it is often used as a means of last resort, which was the case with *Fischer* (Socio-Economic Rights Institute 2019a: 33).

Although it is too soon to tell whether the litigation will in fact lead to the delivery of water to Marikana, through the lens of legal mobilisation theory, there are several ways to understand the role and value of litigation. Most importantly, the litigation shifted the perception of the Marikana residents from 'land invaders' to 'rights-bearers', people who are deserving of a home, with access to services like water (Socio-Economic Rights Institute 2020b: 32). Furthermore, the litigation process bridged the differences between the different settlement sections and led to a 'single case', uniting leadership structures that previously functioned independently of each other (Socio-Economic Rights Institute 2020b: 31).

The litigation fostered collaboration among the different NGOs working to support the residents of Marikana. While SERI was drafting papers and making court appearances, organisations like Social Justice Coalition (SJC) and Ndifuna Ukwazi (NU) provided the community with logistical assistance in the form of transportation to court, food for the day in court and letters to employers for leave to attend court (Socio-Economic Rights Institute 2020b: 31). Most importantly, the long-term relationships which SJC and NU established with the community of Marikana will outlast the litigation process. Furthermore, the community's relationship with these NGOs, which have significant access to mainstream media, meant that the story of Marikana and their advocacy messages made their way into the public discourse.

The Marikana residents utilised various interweaving strategies to claim their water rights including critical community organising, engagement with government, protest, self-supply and litigation. At the foundation of this was community mobilisation through the election of the three committees. Without the committees, other strategies would not yield the desired results or may not have been possible at all. Initially, the committees functioned independently of each other, differed in structure and responded to separate eviction cases individually. Eventually, the residents recognised that there is power in large numbers and a sense of collective identity based on their shared need to secure a home (Socio-Economic Rights Institute 2019a: 92). It should be noted, however, that proper organisation of clients, although valuable, is not a necessary prerequisite to successful litigation as argued by Dugard & Langford (2011), and that basic services cases consisting of 'disparate individuals' can be successful.

The self-supply demonstrated extraordinary resilience and agency; however, these mechanisms should not be romanticised because they come at high cost and risk, rendering already vulnerable people even more so (Socio-Economic Rights Institute 2020b: 31). Self-supply mechanisms can be dangerous and pose a threat to public health of residents, for example, drinking polluted water from streams, injury and other health risks from building and maintaining pit latrines, and electrocution or fires from unlawful electricity connections.

CONCLUSION

While there has been an improvement in access to water and other basic services in South Africa, access to safe water, alongside other socio-economic rights like housing and sanitation, remains geographically and racially determined (Civil Society Coalition Campaign for South Africa's Ratification of the ICESCR & its Optional Protocol 2018). While most people living in South Africa have access to improved water sources, and at first glance, the human right to water is being realised, the poor and marginalised have the least adequate water services, and as systems of local government become increasingly dysfunctional, water poverty and injustice expands.

Most of the people who are 'left behind' are deliberately excluded from formal participation processes. This exclusion forces groups and communities to resort to protest as a measure of last resort. Such communities in South Africa are largely black and poor and are often marginalised and criminalised. Under these circumstances, communities like the residents of Makana and the Marikana informal settlement are forced to confront the state at immense personal cost and risk. They often have to confront formal forms of state violence including excessive use of police force during protests, misuse of the criminal justice system to stifle dissent, as well as clandestine and more informal forms of state violence like intimidation, threats of assault and assassination. Ayanda Kota, for example, recently had to go into hiding because of threats made against his life for the work that he does in Makana (Majavu 2020).

Including these communities in participation processes means that the state (and other duty-bearers) must change its view and treatment of them. To view undocumented migrants, homeless people, so-called slum dwellers, land occupiers, as lawful and legitimate participants, with rights, to whom the state and other duty-bearers, are accountable, means making a paradigm shift. It means legitimising people who do not have a formal presence and whom governments characterise as undocumented, unlawful, illegal, invading and so on. These kinds of shifts are seldom brought about by tidy rights claiming methods such as social accountability and formal participatory processes. Systems of local governance in the cases of Makana and Marikana only shifted when disruptive rights claiming methods such as litigation and protest were employed. Indeed, these were the only strategies that had any effect.

In addition to using protest, communities are increasingly employing the tools of legal activism to realise their rights. Justiciable human rights are only the beginning. Communities play a critical role in strengthening governance by persistently holding the state accountable and employing the tools of legal activism to realise their rights. A rights-based legal framework is by no means sufficient and in many cases, not even necessary. It is the agency and actions of communities that activate legal and policy frameworks and political promises. In many instances, procedural law rather than the Constitutional right to water is utilised in water-related litigation, and the right to health in sanitation-related litigation.

As these case studies show, communities need to actively claim their legal rights in order to realise them. The two case studies juxtapose how the use of legal tools of land expropriation on the one hand and provincial intervention, on the other, are used to further struggles for access to water services. In both cases, litigation is turned to as a last resort. The residents of Makana confronted their local municipality's breach of its constitutional obligation of services delivery. In the Marikana informal settlement, residents were forced to defend their rights to adequate shelter by resisting eviction and compelling their local government to expropriate the land they live on to ensure their security of tenure and ultimately gain access to basic services like water and sanitation.

The case studies also show that while the law and litigation are valuable components of rights-claiming, they may not be sufficient. The case studies highlight the need for litigation to be used as a complementary strategy, alongside other rights claiming strategies like community mobilisation, engagement with government, relationship-building, media advocacy, self-supply and sometimes protest, to derive maximum benefit to communities and apply significant pressure to local government structures. Many of these non-litigious strategies continue after litigation in Makana and Marikana, and become especially necessary where the court victories depend on the very local governments litigated against to implement the court orders.

Finally, as both cases show, community mobilisation underlies both protest and litigation. Without community mobilisation, effective protest and litigation would not be possible. Furthermore, both case studies demonstrate the vital role of building and sustaining strategic partnerships within communities. The article also shows how important this is for sustaining long-term efforts to claim water rights, often spanning many years. These connections are forged through struggle and in the face of high risks, costs and ongoing challenges, residents displayed great resilience, courage, and a deep commitment to demanding an improvement in services delivery and claiming their rights to water. Ultimately, communities play a critical role in strengthening governance by actively, fearlessly and persistently holding the state accountable.

DATA AVAILABILITY STATEMENT

All relevant data are included in the paper or its Supplementary Information.

REFERENCES

Alexander, P. 2010 Rebellion of the poor: South Africa's service delivery protests – a preliminary analysis. *Review of African Political Economy* **37** (123), 25–40. https://doi.org/10.1080/03056241003637870.

Alexander, P. 2012 A massive rebellion of the poor. *The Mail & Guardian*. Available from: https://mg.co.za/article/2012-04-13-a-massive-rebellion-of-the-poor/

Alexander, P., Runciman, C. & Ngwane, T. 2014 *Community Protests 2004–2013*. Available from: https://www.youtube.com/watch?v=XqwBSNBMPCU

Alexander, P., Runciman, C., Ngwane, T., Moloto, B., Mokgele, K. & van Staden, N. 2018 Frequency and turmoil: South Africa's community protests 2005–2017. *South African Crime Quarterly* **63**, 27–42.

Black's Law Dictionary 1990 *Public Interest Litigation*.

Botes, L. 2018 South Africa's landscape of social protests: a way forward for developmental local government? *African Journal of Public Affairs* **10** (4), 241–256.

Brickhill, J. 2014 Public interest alchemy: combining art and science to litigate for social change. *SSRN*. doi:10.2139/ssrn.3172855.

Budlender, A., Marcus, G. & Ferriera, N. 2015 *Public Interest Litigation and Social Change in South Africa: Strategies, Tactics and Lessons.* Atlantic Philanthropies, Hamilton, Bermuda.

Bull, T. 2020 *Interview with Timothy Bull of the Makana Residents' Association (Formerly Grahamstown Residents' Association) [Telephonic].*

Centre for Applied Legal Studies 2017 *Case Study: Limpopo Water.* Centre for Applied Legal Studies. Available from: https://www.wits.ac.za/cals/our-programmes/basic-services/water-and-sanitation/

Chamberlain, L. & Masiangoako, T. 2021 Third time lucky? Provincial intervention in the Makana Local Municipality. *South African Law Journal* **138** (2), 423–459. https://doi.org/10.47348/SALJ/v138/i2a7.

Cirolia, L. 2014 South Africa's emerging housing programme: a prism of urban contest. *Development Southern Africa* **31** (3), 397–411. https://doi.org/10.1080/0376835x.2014.887998.

Civil Society Coalition Campaign for South Africa's Ratification of the ICESCR and its Optional Protocol which comprises 2018 *Joint Submission to the United Nations Committee on Economic, Social and Cultural Rights*, 64th Session, 24 September–12 October 2018 [Joint Submission to the United Nations CESCR], pp. 1–42. Available from: https://tbinternet.ohchr.org/Treaties/CESCR/Shared%20Documents/ZAF/INT_CESCR_CSS_ZAF_32156_E.pdf

Cummings, S. & Rhode, D. 2009 Public interest litigation: insights from theory and practice. *Fordham Urban Journal* **36** (4), 603–651.

Danti, N. 2017 *Critical Assessment of Right to Safe Water and Sanitation in a South African Informal Settlement: A Case Study of Marikana, Cape Town. Master of Philosophy,* Faculty of Engineering and the Built Environment at University of Cape Town.

Dayimani, M. 2021 Makhanda shutdown protest ends after municipality obtains court order. *News24.* Available from: https://www.news24.com/news24/southafrica/news/makhanda-shutdown-protest-ends-after-municipality-obtains-court-order-20210618

Department of Environmental Affairs 2011 *National Climate Change Response White Paper.* Department of Environmental Affairs, Pretoria.

Department of Water and Sanitation 2019 *National Water and Sanitation Master Plan.* Department of Water and Sanitation, Pretoria.

Dugard, J. & Langford, M. 2011 Art of science? Synthesising lessons from public interest litigation and the dangers of legal determinism. *South African Journal on Human Rights* **27** (3), 39–64. https://doi.org/10.1080/19962126.2011.11865004.

Dugard, J., Madlingozi, T. & Tissington, K. 2015 Rights-compromised or rights-savvy? The use of rights-based strategies to advance socio-economic struggles by Abahlali baseMjondolo, the South African shack-dwellers' movement. In: *Social and Economic Rights in Theory and Practice: Critical Inquiries* (Alviar García, H., Klare, K. & Williams, L. eds). Routledge, Oxford.

Duncan, J. 2016 The right to freedom of expression in South Africa. In: *Socio-Economic Rights: Progressive Realisation?* (J. Dugard, M. Clark, K. Tissington & S. Wilson eds.) Foundation for Human Rights, Johannesburg.

Durbach, A., McNamara, L., Rice, S. & Rix, M. 2013 Public interest litigation: making the case in Australia. *Alternative Law Journal* **38** (4), 219–223.

Dyongman, L. 2021 'We have been rationing water for almost a year': Makana crisis undented by recent rains. *Sunday Times.*

Ebrahim, T. 2020 *Interview with Tiffany Ebrahim, Former SERI Researcher [Zoom].*

Griffiths, J. 2020 *The Decline and Fall: One in Five Municipalities in a State of Collapse.* Daily Maverick. Available from: https://www.dailymaverick.co.za/opinionista/2020-10-06-the-decline-and-fall-one-in-five-municipalities-in-a-state-of-collapse/

GroundUp 2017 *Does Marikana Really Have 60,000 People?* GroundUp. Available from: https://www.groundup.org.za/article/does-marikanareally-have-60000-people/

Hamer, N. G., Lipile, L., Lipile, M., Molony, L., Nzwana, X., O'Keeffe, J., Shackleton, S. E., Weaver, M. & Palmer, C. G. 2018 Coping with water supply interruptions: can citizen voice in transdisciplinary research make a difference? *Water International* **43** (5), 603–619. https://doi.org/10.1080/02508060.2018.1497863.

Heese, K. & Allen, K. 2021 *The Floodgates of Local Government Dysfunction Have Opened.* Business LIVE. Available from: https://www.businesslive.co.za/bd/opinion/2021-03-30-the-floodgates-of-local-government-dysfunction-have-opened/

Housing Act 107 of 1997

Housing Development Agency 2013 *Informal Settlement Status in South Africa.* Housing Development Agency.

Huchzermeyer, M. 2010 Pounding at the tip of the iceberg: the dominant politics of informal settlement eradication in South Africa. *Politikon* **37** (1), 129–148. https://doi.org/10.1080/02589346.2010.492153.

Huchzermeyer, M. 2011 *Cities with 'Slums': From Informal Settlement Eradication to a Right to the City in Africa.* University of Cape Town Press, Cape Town.

Jankielsohn, R. 2012 Defining hydropolitics: the politics of water in South Africa. *Journal for Contemporary History* **37** (1), 123–141.

Khambule, I., Nomdo, A. & Siswana, B. 2018 *The Social Cost Violent and Destructive Service Delivery Protests in South Africa.* Human Sciences Research Council.

Kota, A. 2019 *Water Crisis in Makhanda: A Microcosm of SA's Looming Health Catastrophe – Expert [Radio].* Available from: http://www.702.co.za/articles/342480/water-crisis-in-makhanda-a-microcosm-of-sa-s-looming-health-catastrophe-expert

Kota, A. 2020 *Interview with Ayanda Kota of the Unemployed Peoples Movement (UPM) [Telephonic].*

Kruuse, J. 2020 *Interview with Jay Kruuse of the Public Service Accountability Monitor (PSAM)*. Rhodes University [Telephonic].

Lancaster, L. 2016 At the heart of discontent – measuring public violence in South Africa. *Institute for Security Studies Papers* **2016** (292), 1–20.

Langford, M., Cousins, B., Dugard, J. & Madlingozi, T. (eds) 2013 Socio-Economic Rights in South Africa: Symbols or Substance? Cambridge University Press, Cambridge.

Ledger, T. 2019 *We Already Have a Solution to Failing Municipalities: It's in the Fine Print*. Daily Maverick. Available from: https://www.dailymaverick.co.za/opinionista/2019-10-06-we-already-have-a-solution-to-failing-municipalities-its-in-the-fine-print/

Ledger, T. & Rampedi, M. 2019 *Mind the gap: Section 139 Interventions in Theory and in Practise*. Public Affairs Research Institute (PARI).

Lehohla, P. 2017 *The State of Basic Service Delivery in South Africa: In-Depth Analysis of the Community Survey 2016 Data*. Statistics South Africa. Available from: http://www.statssa.gov.za/publications/Report%2003-01-22/Report%2003-01-222016.pdf

Maclennan, S. 2021 *Makhanda Shutdown Set to Continue Tuesday*. Grocott's Mail. Available from: https://www.grocotts.co.za/2021/05/24/makhanda-shutdown-set-to-continue-tuesday/

Madlingozi, T. 2014 Post-apartheid social movements and legal mobilisation. In: *Socio-Economic Rights in South Africa: Symbols or Substance?* Cambridge University Press, pp. 92–130.

Majavu, A. 2020 Activist Ayanda Kota fears for his life. *New Frame*. Available from: https://www.newframe.com/activist-ayanda-kota-fears-for-his-life/

Makana Local Municipality 2020 *Update on Water Situation in Makhanda Issued by Makana Communications Office*.

Makana Local Municipality 2021 *Makana Municipality Final Integrated Development Plan 2021–2022*.

McCann, M. & Sliverstein, H. 1998 Rethinking law's 'allurements': a relational analysis of social movement lawyers in the United States. In: *Cause Lawyering: Political Commitments and Professional Responsibilities* (Sarat, A. & Scheingold, S. A. eds). Oxford University Press, New York.

Miraftab, F. 2004 Invited and invented spaces of participation: neoliberal citizenship and feminists' expanded notion of politics. *Wagadu* **1** (1), 1–7.

Mkhize, M. C. 2015 Is South Africa's 20 years of democracy in crisis? Examining the impact of unrest incidents in local protests in the post-apartheid South Africa. *African Security Review* **24** (2), 190–206. https://doi.org/10.1080/10246029.2015.1038285.

Nowicki, L. 2018 *Grahamstown's Dirty Water Blamed on Municipality's Inefficiency*. GroundUp News. Available from: https://www.groundup.org.za/article/grahamstowns-dirty-water-blamed-municipalitys-inefficiency/

Nowicki, L. 2020 *Court Orders Dissolution of Makana Municipality*. GroundUp News. Available from: https://www.groundup.org.za/article/court-orders-dissolution-makana-municipality/

Omenya, A. & Huchzermeyer, M. 2006 Slum upgrading in the complex context of policy change: the case of Nairobi. In: *Informal Settlements: A Perpetual Challenge?* (Huchzermeyer, M. & Karam, A. eds). University of Cape Town Press, Cape Town, pp. 290–312.

Pacific Institute 2019 *Water Conflict*. The World's Water. Available from: https://www.worldwater.org/water-conflict/

Pithouse, R. 2009 A policy without progressive politics: lessons from the failure to implement breaking new ground. *Town and Regional Planning* **54**, 1–14.

Potter, A. 2021 *Beyond Kgetlengrivier: citizen groups taking over collapsed municipal services is only a short-term solution*. Daily Maverick. Available from: https://www.dailymaverick.co.za/article/2021-02-28-beyond-kgetlengrivier-citizen-groups-taking-over-collapsed-municipal-services-is-only-a-short-term-solution/

Rayner, T. 2016 *The Organisation of Land Occupation: A Case Study of Marikana, Cape Town. Master of Social Science in Sociology*, University of Cape Town.

Rosenberg, G. N. 1991 *The Hollow Hope: Can Courts Bring About Social Change?* University of Chicago Press, Chicago.

Scheingold, S. A. 1974 *Politics of Rights: Lawyers, Public Policy, and Political Change*. Yale University Press, New Haven.

Scheingold, S. A. 2004 *The Politics of Rights: Lawyers, Public Policy and Political Change*, 2nd edn. University of Michigan Press, Ann Arbor.

Social Justice Coalition 2009 https://sjc.org.za/campaigns/sanitation (accessed 27 August 2021).

Socio-Economic Rights Institute 2015 *Public Interest Legal Services in South Africa*. Socio-Economic Rights Institute.

Socio-Economic Rights Institute 2016 *Evictions and Alternative Accommodation in South Africa 2000–2016: An Analysis of the Jurisprudence and Implications for Local Government*. [Written by Clark, M.]. Socio-Economic Rights Institute.

Socio-Economic Rights Institute 2017 *A Double Harm: Police Misuse of Force and Barriers to Necessary Health Care Services*. [Written by Rayner, M., Baldwin-Ragaven, L., & Naidoo, S.]. Socio-Economic Rights Institute.

Socio-Economic Rights Institute 2019a *Our Place to Belong: Marikana Informal Settlement*. [Written by Royston, L., Potter, A., Webster, D., Ebrahim, T. and Molopi, E.]. Socio-Economic Rights Institute.

Socio-Economic Rights Institute 2019b *Here to Stay: A Synthesis of Findings and Implications From Ratanang, Marikana and Siyanda*. [Written by Royston, L., Potter, A., Khunou, K., Ebrahim, T. and Molopi, E.]. Socio-Economic Rights Institute.

Socio-Economic Rights Institute 2019c *Submission on Draft Expropriation Bill*. Socio-Economic Rights Institute.

Socio-Economic Rights Institute 2020a *Claiming Water Rights in South Africa*. [Written by Chamberlain, L., Khunou, K., Masiangoako, T., & Potter, A.]. Socio-Economic Rights Institute.

Socio-Economic Rights Institute 2020b *Residents of Marikana Use Expropriation as a Tool*. [Written by Chamberlain, L., Khunou, K., Masiangoako, T., & Potter, A.]. Socio-Economic Rights Institute.

Socio-Economic Rights Institute 2020c *Fischer v Unlawful Occupiers*. Available from: https://www.seri-sa.org/index.php/19-litigation/case-entries/491-fischer-v-unlawful-occupiers-erf-149-philippi

Socio-Economic Rights Institute 2020d *Makana Local Municipality – Provincial Intervention in a Municipal Crisis*. [Written by Chamberlain, L., Khunou, K., Masiangoako, T., & Potter, A.]. Socio-Economic Rights Institute.

South African Human Rights Commission 2014 *Report on the Right to Access Sufficient Water and Decent Sanitation in South Africa*.

Tabensky, P. A. 2011 *The flames of Phaphamani. Pambazuka News*. Available from: https://www.pambazuka.org/governance/flames-phaphamani

The Constitution of the Republic of South Africa of 1996. 1996.

Tissington, K. 2012 Tacticians in the struggle for change? Exploring the dynamics between legal organisations and social movements engaged in rights-based struggles in South Africa. In: *Contesting Transformation: Popular Resistance in 21st Century South Africa* (Dawson, M. & Sinwell, L. eds). Pluto Press, London.

Unemployed Peoples Movement 2011 *Press Statement by the Unemployed People's Movement*. Anti-Eviction Campaign. Available from: https://westerncapeantieviction.wordpress.com/2011/02/14/solidarity-the-rebellion-of-the-poor-comes-to-grahamstown/.

Unemployed Peoples' Movement v Premier, Province of the Eastern Cape and Others 2020 *Case No. 553/2019*. Eastern Cape High Court, Grahamstown.

United Nations Human Rights Office of the High Commissioner (OHCHR) 2021 *Bachelet Urges Iran to Focus on Addressing Water Crisis in Khuzestan Rather than Crushing Protests*. United Nations Human Rights Office of the High Commissioner (OHCHR). Available from: https://www.ohchr.org/EN/NewsEvents/Pages/DisplayNews.aspx?NewsID=27335&LangID=E

World Health Organization and UNICEF 2010 *Progress on Sanitation and Drinking Water: Joint Monitoring Programme*.

Zemans, F. K. 1983 Legal mobilization: the neglected role of the Law in the political system. *American Political Science Review* **77** (3), 690–703. https://doi.org/10.2307/1957268.

First received 29 January 2021; accepted in revised form 7 February 2022. Available online 24 February 2022

doi: 10.2166/h2oj.2022.037

Container-based sanitation in urban Haiti: how can it improve human rights as a component of citywide inclusive sanitation?

Froggi VanRiper [a,*], Kory C. Russel IWA [b], Daniel Tillias[c], Jenna Tilt[d] and Jessica Laporte[e]

[a] Oregon State University, Corvallis, OR, USA
[b] University of Oregon, School of Architecture and Allied Arts, Eugene, OR, USA
[c] FatraKa, Cite Soleil, Haiti
[d] College of Earth, Ocean and Atmospheric Sciences, Oregon State University, Corvallis, OR, USA
[e] Sustainable Organic Integrated Livelihoods (SOIL), Cap Haïtien, Haiti
*Corresponding author. E-mail: vanriper@oregonstate.edu

FV, 0000-0002-2564-5177; KCR, 0000-0002-0496-0856

ABSTRACT

Health, sanitation, and livelihoods are interrelated human rights and essential components of community wellbeing. Despite the United Nations recognizing the human right to sanitation, one-third of the world's population lack access to improved sanitation. Furthermore, in many low- and middle-income countries, urban latrines are excavated manually without physical or regulatory protections. Container-based sanitation (CBS) has promise as a component of Citywide Inclusive Sanitation (CWIS) in densely populated, low-resource environments. Before investing in CBS, however, governments and funders require evidence of user acceptance, public demand, and labor aspects of service delivery. To provide this evidence base, we completed 633 interviews with active and former users of EkoLakay, a CBS service in Cap Haïtien, Haiti, creating a profile of the user base and their motivations for subscribing. We also compiled and analyzed secondary data to determine the impacts of widespread CBS expansion in northern Haiti. Results reveal that CBS presently serves a geographical subset of Haiti's most resource-insecure residents, and that CBS significantly reduces handling of untreated excreta in Haiti's Nord Department, while doubling livelihood opportunities through safe and dignified jobs. Given its cost-effectiveness, this provides an opportunity to achieve widespread sanitation coverage by integrating and cross-subsidizing CBS within a CWIS strategy.

Key words: Citywide Inclusive Sanitation, container-based sanitation, human rights, safely managed sanitation, sanitation systems, SDG 6

HIGHLIGHTS

- First systematic study of CBS's impact on vulnerable households.
- Haiti's sanitation system leaves 19% of urban households without improved sanitation access.
- Haitian CBS users had above-average rates of open defecation and limited sanitation pre-CBS.
- CBS could help urban sanitation systems in LMICs achieve SDG 6.
- CBS can replace 1 M hours of unsafe excreta-handling labor with 500+ safer jobs in northern Haiti.

GRAPHICAL ABSTRACT

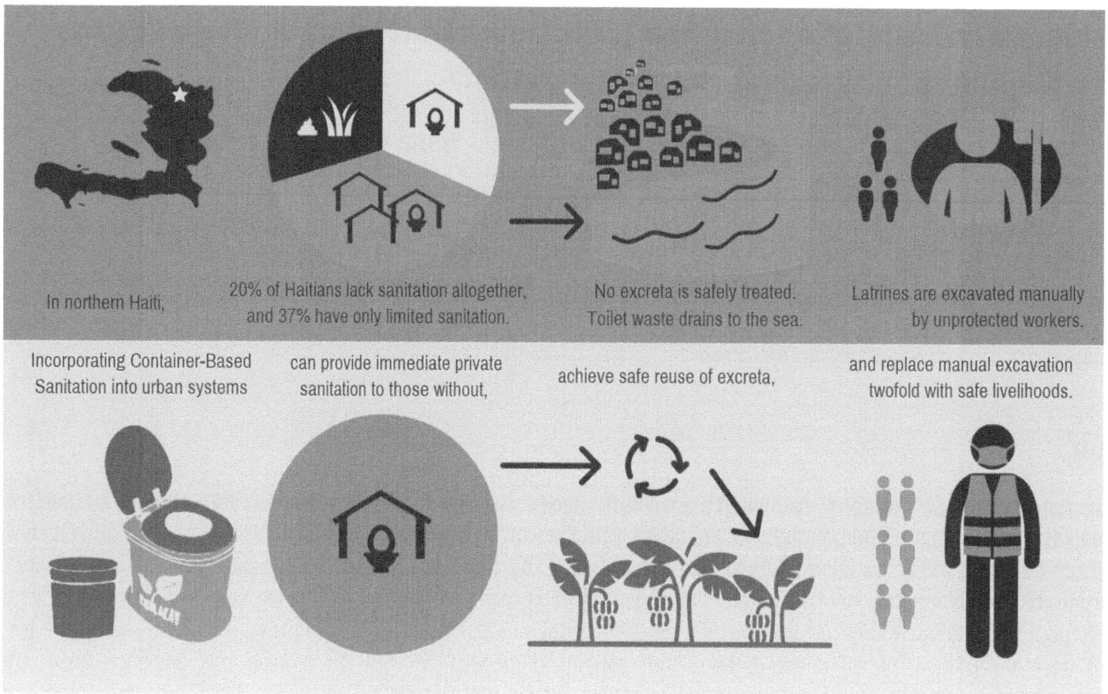

INTRODUCTION

Sanitation systems and human rights in the global South

Access to safe sanitation is an essential underpinning of human rights, recognized by the United Nations (UN) Committee on Economic Social & Cultural Rights (2010) and the UN General Assembly (2016). However, as of 2020 only a little over half of the global population has access to safely managed sanitation (WHO 2021). This includes six hundred million people in urban areas, where the negative health implications of open defecation are intensified by population density and lack of environmental 'sinks' for excreta. UN Sustainable Development Goal (SDG) 6.2 challenges the world to 'achieve access to adequate and equitable sanitation and hygiene for all and end open defecation, paying special attention to the needs of women and girls and those in vulnerable situations' by 2030 (UN General Assembly 2015). At the present pace of progress, we will fall far short of this target; the number of households without sanitation is increasing, and the gap between rich and poor is exacerbating, as urbanization outpaces the development of public services (UNICEF 2019). While many technologies meet the definition of improved sanitation, they cannot achieve the conditions described in SDG 6.2 without connections to processes for safe excreta management. Such is the case in many urban environments in fragile contexts (WHO 2021). Sewers that discharge excreta into the environment without treatment threaten the rights to health and safe drinking water. Latrines that must be manually excavated undermine the right to safe working conditions protected by the Universal Declaration of Human Rights (UN 1948). Where the proliferation of latrines creates a class of citizens who are stigmatized and subject to public abuse (Neiburg & Nicaise 2011), it violates the right to be free from inhuman or degrading treatment, 'laying the groundwork for violations of human rights' (de Albuquerque 2012).

Some scholars suggest that the global sanitation deficiency is partly a product of fixation on centralized, water-based technologies (Gambrill *et al.* 2020). Homogeneous systems certainly benefit from physical and organizational simplicity, making them attractive to public service coordinators. Centralized sewers, however, are inappropriate or unaffordable in many contexts, leaving them unable to meet SDG 6.2 alone (Andersson *et al.* 2017).

In recent years, the emerging Citywide Inclusive Sanitation (CWIS) framework has begun to guide the pursuit of equitable and sustainable urban sanitation access (Schrecongost *et al.* 2020). The CWIS framework views sanitation provision from a system-wide approach, recognizing the value of integrating varied technologies, service models, and economic strategies to achieve universal sanitation coverage. Rather than focusing on specific technologies, CWIS framework measures success by access outcomes, considering the full sanitation service chain.

Container-based sanitation (CBS) may be an important tool within a CWIS portfolio; CBS, recognized by the World Health Organization (WHO) as a form of safely managed sanitation (WHO/UNICEF 2017) is a non-sewered sanitation strategy through which excreta is captured in sealable containers, and transported to facilities for treatment. CBS has low life-cycle costs (EY 2020; Sainati *et al.* 2020; Delaire *et al.* 2021), provides immediate sanitation coverage in locations without preexisting infrastructure, operates without water, and is resilient to environmental disruption (Russel *et al.* 2019). Where a CBS service is available, households can transition directly to private improved sanitation from any step on the sanitation ladder,[1] without substantial upfront investment in infrastructure. In flood-prone environments, CBS provides further advantages over pit latrines; the technology allows excreta to be stored above-ground and sealed against environmental release. In areas vulnerable to flooding, latrines (the most affordable traditional sanitation technology) are not capable of maintaining a safe barrier between a community and their excreta. A single flooding event can spread suspended excreta from latrines across the surrounding environment. After floodwaters have receded, pathogens in fecal solids that remain on the earth's surface can continue to pose a threat to human health for up to a year (Carr 2001). In addition, the portable nature of CBS technology enables it to provide sanitation coverage even where land tenure is disputed or unclear (such communities are home to the most vulnerable households). Finally, CBS separates both user and processor from direct contact with excreta (SOIL 2019), unlike commonly practiced manual latrine excavation (World Bank 2019).

Thus, expanding CBS as a component of municipal sanitation portfolios has the potential to alleviate the human rights crisis associated with manual excavation, in addition to meeting SDG 6.2. Before funders and public entities will confidently invest in CBS as a component of citywide sanitation delivery, however, they require real-world research for evidence-based decision making. Previous research has explored willingness to pay for CBS (Russel *et al.* 2015), practicalities of scaling CBS services, and cost-effectiveness of CBS (Tilmans *et al.* 2015; Sklar & Faustin 2017; EY 2020; Delaire *et al.* 2021). There remains a need for research into user acceptance, public demand, and labor aspects of service delivery.

In this study, we approach the challenges and opportunities of an equitable sanitation future by considering the role of CBS in the context of a complex system of interrelated economic, social, ecological, and technical subsystems (Clayton & Radcliffe 1996). For example, implementing innovations to the sanitation system, such as CBS, face many barriers which require new approaches to strategic planning (Spuhler & Lüthi 2020). However, the investment in financial and human capital to overcome these barriers is needed because 100% public sanitation coverage is likely unachievable without incorporating innovative sanitation solutions like CBS that can be implemented in areas with limited infrastructure or disputed lands. Such factors and their interactions are also explored by Hyun *et al.* (2019), in their Augmented Sanitation Service Chain framework. The authors define a sanitation system as more than its physical components; it is the full range of public and private actors engaged in sanitation provision, as well as the various practices, technologies, transactions, regulatory conditions, and social values driving local excreta flows. Through this lens, we see that the flow of excreta is accompanied by flows of power and capital, which can exacerbate inequalities or contribute to safe livelihoods. The technological aspects of excreta collection, conveyance, and treatment interrelate with human rights, dignity, health, and social wellbeing.

CWIS and the Augmented Sanitation Service Chain approach community sanitation from a system-wide perspective. Both approaches recognize the relevance of interacting systems at varying scales, to the efficacy of public sanitation endeavors. Applying these frameworks to low- and middle-income countries (LMICs), we can see that the novel characteristics of CBS may strengthen urban sanitation systems, improving their ability to yield equitable sanitation, health, and dignity outcomes. This research sets out to evaluate the potential system benefits of expanding the role of CBS in the urban sanitation system of an LMIC. Using Northern Haiti as a case study, we will characterize the users of an existing CBS service to determine the user base in need of such services, identify changes to excreta-management livelihoods that would likely occur with the

[1] The World Health Organization Joint Monitoring Programme's sanitation ladder ranks sanitation technologies based on safety and privacy characteristics.

expansion of CBS, identify changes to excreta flows and sanitation coverage that could result, and describe barriers that must be removed for these benefits to be realized.

Research objectives

One of the longest-running providers of household CBS services, EkoLakay provides an ideal case for exploring CBS and its role in urban sanitation systems of low-income cities. Based on household interviews with EkoLakay users in Cap Haïtien, this study addresses the following research questions:

- What are the characteristics of households that subscribe to the EkoLakay CBS system?
- What factors (including household, environment, social, and governance-related factors) affect households' decision to subscribe to EkoLakay?
- How do users of the EkoLakay CBS system perceive it in comparison to other available sanitation options?
- What are the implications of a CBS-integrated urban sanitation system on the exposure, labor hours, and livelihoods of northern Haiti's bayakou (professionals who manually excavate latrines and cesspits)?

Study context

The sanitation crisis is particularly acute in the country of Haiti. In 2014, a quarter of the Urban Haitian population lived in impoverished informal communities (World Bank World Development Indicators 2018b), and as of the most recent assessment by the WHO, nearly 20% of the population have no access to sanitation facilities (World Bank World Development Indicators 2018a). Although an estimated 81% of the Haitian population has access to a toilet or latrine, these technologies are rarely connected to a sanitation system for treating the excreta before release into the environment. Haiti ranks among the lowest 20 countries in the world for urban sanitation coverage, with less than 1% of the country's excreta and wastewater safely treated (WHO Joint Monitoring Programme 2017). The only safely managed sanitation available in the Cap Haïtien metropolitan area, where this study takes place, is a CBS service, EkoLakay (Table 1).

In communities with adequate road infrastructure, cesstanks can be emptied mechanically by pump vehicles. However, the density of pit latrine solids and inaccessibility of many dense, low-income environments makes mechanical excavation impractical, so these latrines must be excavated manually (Neiburg & Nicaise 2011; World Bank 2018). While there has been significant work to improve latrine pumping technology, solutions have still not adapted to the context of most dense urban informal settlements (Kabange & Nkansah 2019). As of 2017, 55.7% of urban Haitians relied on latrines (WHO Joint Monitoring Programme 2017).

While the work of the bayakou – those that manually excavate latrines in Haiti – is essential to community health, the job itself is stigmatized. These professionals work under cover of night, brokering their services through a middle party to protect their identity (Neiburg & Nicaise 2011; World Bank 2018). There are no regulations governing the working conditions of manual excavators, who often perform their work submerged in human excreta and chemical solvents, without equipment to protect them against disease or injury (Lozano-Gracia & Garcia Lozano 2018; World Bank 2019). In addition, Northern Haiti has no sewage treatment facilities, and the nation lacks regulations to ensure that fecal sludge is disposed of properly; leaving manual excavators to dispose of pit sludge in open land, waterways, and/or the sea (Sklar & Faustin 2017; Mazars & Earwaker 2019).

In 2017, diarrhoeal disease was the fifth leading contributor to loss of life in Haiti, at a rate of 20.5 years of life lost (YLL) for every 1,000 people in the country (Fene *et al.* 2020). About half of the diarrhoeal disease burden in Haiti is attributable to inadequate excreta management (WHO 2018). The structure of Haiti's sanitation landscape creates risk for households, the public, and essential workers who manage sanitary wastes. Haiti is unlikely to achieve the UN's SDG 6, 'to end open defecation and ensure safe and equitable sanitation for all' by 2030 (UN General Assembly 2015), and significant changes would be necessary to realize a sanitation system sufficient for the protection of public health.

One of the only safely managed sanitation options in Haiti is a CBS service, EkoLakay.[2] Operating for 7 years at the time of this study, EkoLakay serves neighborhoods that are flood-prone, densely populated, urban or peri-urban, and have low sanitation coverage and income levels. EkoLakay's CBS service provides a urine-diverting container-based toilet to households on a monthly subscription basis. Subscribers pay a monthly service fee of

---

[2] In Port au Prince metropolitan area, a small number of flush toilets are connected to a treatment plant.

Table 1 | Reported sanitation options available in metro Cap Haïtien, Haiti

<table>
<tr><th colspan="3">WHO JMP sanitation ladder categories</th><th>Options available in Cap Haïtien, Haiti</th></tr>
<tr><td rowspan="3">Improved sanitation</td><td rowspan="2">Private improved sanitation</td><td>Safely managed facilities</td><td>• EkoLakay CBS service</td></tr>
<tr><td>Basic facilities</td><td>• Private latrine, flush, or dry toilet (no treatment after discharge)</td></tr>
<tr><td>Non-private improved sanitation</td><td>Limited facilities</td><td>• Shared or public latrine, flush, or dry toilet (no treatment after discharge)</td></tr>
<tr><td colspan="2">Unimproved sanitation</td><td>Unimproved facilities</td><td>• Shallow pit
• Pit with no slab</td></tr>
<tr><td colspan="2">Open defecation</td><td>No facilities</td><td>• In a bag
• In a field
• By the river</td></tr>
</table>

200 to 300[3] Haitian gourdes (2.11 to 3.15 USD in August 2019) for rental of the portable toilet structure and weekly container exchange. The EkoLakay toilet houses a sealable five-gallon container that receives solid excreta, as well as a one-gallon jug that receives liquid waste through a urine-diverting funnel in the seat structure. After users defecate, they cover the fecal matter with a dry carbon-based cover material. This cover material is made from agricultural byproducts (currently ground sugarcane bagasse) and is provided by EkoLakay during the weekly exchange of full excreta containers for sanitized containers containing cover material. Users dispose of urine (which has a comparably negligible pathogen load) locally, reducing the volume and weight of excreta to be transported and treated. EkoLakay transports the sealed, containerized excreta to a compost treatment site where trained staff, equipped with personal protective equipment (PPE), transform the material through thermophilic composting. Studies of possible points of contamination along the CBS value chain have found that exposure risk is far below that of manual pit emptying (Mackinnon *et al.* 2018; Bischel *et al.* 2019). As a formal business, SOIL is able to implement training and strict safety protocols (SOIL 2019), providing all employees with PPE, health checks, and vaccinations. The final compost product is laboratory-tested for quality and safety before being sold to local consumers.

Globally, CBS services are largely delivered by non-governmental actors, and most operate at a small scale relative to municipal sanitation services. This arrangement is affected by multiple system characteristics: division of cost burden, scale of operation (EY 2020), and policy/regulatory barriers (Mara 2018). Recent research indicates that the true cost of CBS service delivery is lower than that of centralized sewerage (EY 2020; Delaire *et al.* 2021). The cost of sewerage is largely borne by the municipality, even if only through historical investment in infrastructure (Bhagwan *et al.* 2019). Private entities providing sanitation services like CBS must pass the full cost on to the consumer, or seek supplemental funding from other parties. Considering that users of CBS services are often resource-insecure households, this asymmetrical cost structure exacerbates existing inequities. Among strategies for implementing the CWIS framework, targeted subsidies and public-private partnerships can mitigate such inequities. Funding structures based on sanitation coverage outcomes will increase accessibility of CBS services and strengthen urban sanitation systems.

Another key barrier to the implementation of CBS services are regulatory structures in place. In many areas, sanitary regulations do not include safe alternatives to traditional sanitation infrastructure, and sometimes, the areas regulations expressly forbid excreta reuse or manual collection of waste flows. Existing CBS services must either adapt their operational model to work within existing regulatory frameworks, or work with governing agencies to modify regulations (World Bank 2019).

Community sanitation coverage is affected by interacting complex systems at multiple scales. To gain insight into the potential for CBS to achieve long-term sustainability, we seek to understand the users of this sanitation system. Understanding CBS at the user scale will allow insight into how larger scales of system management,

[3] Payment structure details in Supplementary Material, Table S1.

including governance, financing, and physical infrastructure, influence and are influenced by characteristics of the user base.

As a component of a CWIS approach to sanitation in LMICs, CBS may facilitate progress toward multiple SDGs. By characterizing the user base of a long-running CBS service, and understanding what household and broader-system characteristics drive users to subscribe, we can contextualize the role that CBS plays, and can play, in the ability of low-income urban residents to maintain access to improved sanitation. By calculating selected impacts of expanding CBS service, we hope to understand the potential that this innovative sanitation approach holds for the health and wellbeing of the sanitation workforce. Considering the potential of CBS alongside barriers to widespread expansion, we can conceptualize a path toward optimizing urban sanitation systems in LMICs and toward a goal of safety, health, and equity.

METHODS

Research team and tools

This study used a mixed methodology approach (Greene 2007), deriving qualitative and quantitative data from EkoLakay subscriber records and structured interviews with active and former subscribers of the EkoLakay CBS service. Primary data was collected by a Haitian team involved in the development of the research instrument. To understand user relationships with CBS and the broader sanitation system, we developed a structured household survey that included open- and closed-ended questions. Interview questions and prompts were structured around self-reporting of all parameters, for consistency of results regardless of telephone or in-person interview format. All research staff training, tool-development, interviews, and other respondent interactions were conducted in Haitian Kreyol.

To answer questions about the safety, dignity, and employment implications of widespread CBS adoption, we compiled secondary data on Northern Haiti's urban population and their sanitation coverage; technical conditions of manual pit latrine excavation, excreta-handling labor and livelihoods; annual fecal generation values (see Supplementary Material, Table S3 for data sources and analysis); EkoLakay subscriber-to-excreta handler ratios; and service scale economies. From these variables, we calculated the volume of excreta produced by latrines in the urban areas of Haiti's Nord Department; the number of exposure events and cumulative hours of unsafe manual excreta handling that could be replaced by safely managed handling of containerized excreta; and the replacement ratios of working hours and livelihoods that would result from incorporation of CBS into urban sanitation systems in the Cap Haïtien area.

Household sample selection

At the start of data collection in August 2019, there were 1,008 households actively enrolled in the EkoLakay service, and 1,323 former-subscriber households. SOIL provided lists of former and active EkoLakay subscribers for whom telephone contact information was available. To ensure that active subscribers would have sufficient exposure to EkoLakay to answer questions about their experience and satisfaction, we filtered out subscribers who had enrolled after April 30, 2019. We further eliminated households of SOIL employees. The total number of potential participants meeting these criteria was 1,199 former and 733 active subscriber households.

We randomized the lists of active and former subscribers and assigned each household a confidential identifier. Research staff recruited participants by phone, and scheduled interviews with a single household decision maker over 18 years.[4] Respondents who had moved away from the Cap Haïtien area, or were unwilling to meet in-person, were offered the opportunity to be interviewed by phone.

DATA COLLECTION AND ANALYSIS

Data collection

Members of the team that developed the survey instrument also served as interviewers. Interviewers collected survey responses using Qualtrics® (Qualtrics, Provo, UT) on a handheld smartphone. Responses were accessible only to investigators, after uploading to secure storage at the end of each day. From August 24 to October 6, 2019, the team conducted 633 structured household interviews. Interviewers obtained and documented verbal

[4] Haitian households rarely recognize a single household head; they are more likely to have multiple adult members engage in financial decision making.

acknowledgement of informed consent. Interviewers identified themselves, in person and by telephone, as representatives of an independent research group with no affiliation to EkoLakay. Due to significant disruption to statewide transportation and energy infrastructure, in mid-September our field team ceased household visits and completed the remaining 286 interviews by telephone. Recruitment methods did not change during this period; interviewers continued to schedule willing respondents for an interview date and time, but did not offer the option of a home visit.

Initial survey prompts identified household population by age group and sex, the number of rooms in the home, and how long the respondent had lived in the home (see Supplementary Material, Table S4 for complete questionnaire). Closed-ended questions addressed property and household characteristics, as well as satisfaction questions based on a Likert scale and household expense questions (questions fully described in Supplementary material). Open-ended questions addressed subjective perceptions and preferences. These questions were left open-ended to capture unanticipated variability in the respondent population. Interviewers categorized these responses in the field using predefined codes, summarizing responses that did not fit the available codes for later categorization by investigators.

Data transformation

After all data had been collected, investigators assigned uncategorized responses to appropriate codes. In some cases, new categories were necessary to house these responses. Investigators further aggregated categorical responses through an iterative process of team-based coding (Tolley *et al.* 2016). To compare pre-subscription household sanitation of EkoLakay users to the general Haitian population, responses were aggregated into categories associated with the WHO sanitation ladder (Table 1).

Household density was calculated as the quotient of reported household members per reported room in the home. Reported expenses and frequencies, along with the currency exchange rate average for the month of their interview, were used to calculate estimated monthly electricity, phone, and transportation expenses in USD. From EkoLakay's full list of active and former subscribers, we calculated the average cumulative time that households remain subscribed to the EkoLakay service. In the case of active subscribers, cumulative time was calculated as the difference between the contract open date and August 24, 2019. We used the respondent's reported residence time and their start-of-service date to calculate the approximate duration of residence at the time they opened their EkoLakay subscription.

Data analysis

While this study collected responses from both active and former subscribers to the EkoLakay CBS service, some analyses in this report involve only active subscribers. When characterizing subscriber households, we use only active respondent data. When analyzing reported reasons for subscribing to the service and pre-subscription household sanitation, we use data from both active and former-subscriber respondents.[5]

Our respondent sample included only households and individuals who had subscribed to EkoLakay at some point in time, and as a result, there is no comparison group of general Haitian households in this study. Thus, household demographic and property data were sourced from publicly available national and international datasets (UN Department of Economic and Social Affairs, WHO/UNICEF Joint Monitoring Programme, USAID Demographic and Health Surveys, and World Bank) for comparison purposes. To compare proportions (e.g., sanitation access types, residence status, and utility access) between active EkoLakay subscriber and urban Haitian households, we used a one-sample proportions test. To identify a difference in household size between EkoLakay subscribers and the average urban Haitian household, we conducted a two-tailed *t*-test.

To determine whether selected household characteristics correlated with household sanitation status prior to opening an EkoLakay subscription, and to determine the likelihood of reported motivations falling into each code category, we performed odds ratio analysis. Where analysis involved large counts, we used Pearson's Chi-square test. For analyses involving cell counts below ten, we used Fisher's exact test. To compare continuous household variables to reported motivations, we applied logistic regression. All statistical analyses were performed with R 22.0 (R Core Team 2020).

[5] Analysis of data showed that this transition did not yield significant differences between groups (full details in Supplementary material).

RESULTS AND DISCUSSION

Secondary data analysis: human rights and work in the sanitation sector

A 2018 report estimated that the sanitation system of the Port au Prince, Haiti metropolitan area supports 270 manual excavation livelihoods (World Bank 2018). Based on the population (IHSI 2015) and household density (Institut Haïtien de l'Enfance – IHE/Haiti & ICF 2017) of this metropolitan area, proportion of the regional urban population using latrines (WHO Joint Monitoring Programme 2017), and a conservative latrine-emptying frequency of 4.5 years (Sklar & Faustin 2017), we estimate a ratio of 0.001 manual emptying professionals per latrine-using household. Furthermore, reports of manual excavators' team structure and working hours indicate an average team of three to four professionals working for two to three nights per latrine emptied (Reed 2010; Neiburg & Nicaise 2011; Lozano-Gracia & Garcia Lozano 2018; World Bank 2019). Assuming 2.5 professionals per team and the population of urban Nord Department using latrines, we estimate 42,000 manual pit-emptying exposure events per year; and with 16 h per job, this translates to over 800,000 annual hours of direct exposure by manual emptying professionals in the urban areas of the Nord Department.

In contrast, EkoLakay currently operates at a rate of 0.016 CBS excreta-handling professionals (collection and processing) per household. Assuming an 86% increase in efficiency brought about by increases in scale and customer density (Boyer *et al.* 2009), if latrine-using households instead adopt CBS, this would yield a replacement ratio of two CBS excreta handlers per manual excavator (Table 2; detailed calculations in Supplementary Material, Table S3). Furthermore, based on urban Nord Department latrine users, the equivalent sanitation coverage, provided by CBS services, would support over 500 safe excreta-handling livelihoods. An additional 100 plus livelihoods would be created if CBS was extended to the 20% of Haitian households who currently lack sanitation altogether (Table 2; detailed calculations in Supplementary Material, Table S3). Thus, a transition to CBS would replace unsafe labor conditions with safe and reliable livelihoods and potentially increase employment opportunities while dramatically reducing exposure to sanitation workers. These changes would lift multiple community members from circumstances undermining the human right to sanitation, as well as safe and dignified employment. Many previous international efforts to improve the working conditions of manual excavators have been deemed unsuccessful and abandoned (World Bank 2018). While political systems may lack the enforcement resources to protect independent workers, CBS services provide a structure compatible with safety monitoring and worker protection, and confer dignity to employees (Tilmans *et al.* 2015). CBS is gaining recognition around the world, and recent research has taken on the question of scaling these operations and integrating them with inclusive sanitation planning (Shepard *et al.* 2017; Russel *et al.* 2019; EY 2020). From a systems perspective, the social benefit of a stable and increased employment base in the community is created by CBS expansion.

Table 2 | Potential impact of expanding CBS in Cap Haïtien metropolitan area

Extending EkoLakay CBS service to homes without sanitation	
Persons transitioning from open defecation to safely managed sanitation	53,000
Annual excreta safely managed (metric tonnes)	600
Safe excreta-handling livelihoods created	100
Replacing latrines with EkoLakay CBS service	
Persons transitioning from unsafe to safely managed sanitation	288,000
Annual excreta safely managed (metric tonnes)	3,300
Safe excreta-handling livelihoods created	500
Person-exposure-events per year avoided in urban Nord	41,700
Hours direct exposure avoided per year	833,000
CBS/Bayakou livelihoods ratio	2.0
CBS/Bayakou working hours ratio	1.2

All figures are the result of calculations based on secondary data sources. Full calculations and source citations are available in Supplementary Material Table S3.

Primary household data collection: response rate, confidence, and representativeness

Due to a high rate of inactive phone numbers, it became necessary to attempt contact with all active and former subscribers to achieve an adequate sample size. Of 733 active subscribers, we made contact with 383 subscribers, 9 of whom declined to participate, and 49 were unable to be scheduled; thus resulting in an active subscriber completion rate of 84.5% for a sample $n = 325$, with a 4% margin of error at the 95% confidence level. Of the 1,199 potential former subscribers, we successfully contacted 406 former subscribers, 30 of whom declined to participate, and 68 were unable to schedule; thus resulting in a former-subscriber completion rate of 75.9% for a sample of $n = 308$, with a 5% margin of error at the 95% confidence level. Sample sizes vary by analysis, as respondents could decline to answer any prompt.

Where are EkoLakay users?

The EkoLakay service area spans four communes in the Cap Haïtien metropolitan area, spanning a range of urban and peri-urban sites (map and further details in Supplementary material). All neighborhoods serviced by EkoLakay are characterized by low infrastructure and high poverty (Russel *et al.* 2015). The flood-prone nature of many of these zones is an important factor in their inclusion in the EkoLakay service area. The EkoLakay toilet container sits above the ground, protecting it from water intrusion in minor flood events, and can be sealed to secure against excreta release in the case of extreme flooding or other disruption.

Who are EkoLakay users?

Household demographics

The sex ratio of active EkoLakay subscriber households matches the Haitian distribution ($p = 1.000$), as does the distribution of age groups across households ($p = 0.987$) (Table 3). Active EkoLakay subscriber households are larger than the national urban average, at 5.1 persons per household, compared to the 2017 urban Haitian average of 4.4 ($p < 0.001$). Household density varies from as few as one person to as many as eight persons per reported room in the house, with an average density of 2.2 persons per room (Table 3). This provides insight into the resource strain of the EkoLakay subscriber base. In addition to general resource strain, highly populated households may also experience pressure on existing sanitation infrastructure, requiring an alternative or auxiliary sanitation option.

Resource access

Nearly 10% of active subscriber respondents report living in an unauthorized informal dwelling, such as a temporary shelter constructed on public or privately owned land. This number is 48 times higher than the 2003 urban Haitian population ($p < 0.001$). Active subscribers are less likely to report renting their home ($p < 0.001$) and more likely to report owning their home than the 2003 urban Haitian population ($p < 0.001$) (Table 3).[6] Homeownership statistics should be contextualized; as of 2010, only 40% of Haitian landowners possessed a legal title to their property, or formal documentation of land ownership (USAID 2010). With this in mind, and given the vulnerable state of the neighborhoods that comprise EkoLakay's service area, respondents reporting homeownership may only own the structure itself, or have otherwise insecure tenure on their land. Furthermore, this vulnerability is not temporary. Active subscriber respondents reported residence times of up to 70 years, with a mean residence time of 12.2 years. One EkoLakay service area, an informal community of over 50 years, was leveled without warning by the municipal government since the time of this research.

Active subscribers describing their household sanitation situation circumstances immediately prior to accessing EkoLakay reported higher rates of open defecation ($p < 0.05$), higher rates of non-private improved sanitation use ($p < 0.05$), and lower rates of private improved sanitation ($p < 0.001$) than the national urban average (Table 3; Supplementary Material, Figure S3). Upon subscribing to EkoLakay, 29% of households gained sanitation where they had previously lacked any toilet access; 39% of households gained privacy, and all subscribers gained safely managed sanitation status.

Active EkoLakay subscribers are only two-thirds as likely to have access to electricity as the average urban Haitian household ($p < 0.001$).[7] Active subscribers did not differ from the urban national average in access to on-site

---

[6] Subsequent to 2003, national demographic surveys did not include residence status statistics.

[7] It is not clear whether the SE4ALL dataset considers households with generators among those considered to have access to electricity. However, regardless of whether generators are included in this grouping, the statistical significance of this finding remains the same.

Table 3 | Household characteristics of EkoLakay subscribers vs. Urban Haitian population

	EkoLakay subscriber households (2019)	Urban Haiti (2017)	*p*-value
Respondent sex ratio (M:F) (household spokesperson, active and former-subscriber respondents; $n = 633$)			
Individual respondents	46%:54%	49%:51%	$p = 0.086$
Household sex ratio (M:F) (active subscriber respondents; $n = 322$)			
Whole household	49%:51%	49%:51%	$p = 1.000$
Household size (active subscriber respondents; $n = 322$)			
Household members	5.1	4.4	$p = 5.74 \times 10^{-06}$***
Household composition (active subscriber respondents; $n = 322$)			
Adult women	1.7	1.5	$p = 0.987$
Adult men	1.4	1.1	
Youth ages 5–17	1.5	1.4	
Children under 5 years	0.4	0.5	
Households with 1 or more members with a disability or chronic illness	34.2%	–	
Households with youth ages 5–17	72.6%	65.1%	$p = 5.02 \times 10^{-131}$***
Households with children <5 years	31.4%	35.4%	$p = 8.78 \times 10^{-40}$***
Household density (active subscriber respondents; $n = 322$)			
Persons per room in house	2.2	–	
Household resources (active subscriber respondents)			
Has access to electricity ($n = 314$)	grid-connected electricity: 52.2%	(2018) 78.2%	$p = 3.95 \times 10^{-31}$***
	grid and/or generator: 59.9%		$p = 6.32 \times 10^{-15}$***
Has on-site water ($n = 314$)	(includes cistern for delivered water) 24.8%	20.5%	$p = 0.065$
Has bank account ($n = 321$)	41.1%	37.1%	$p = 0.147$
Has no formal money-management system	12.8%	–	
Residence status (active subscriber respondents; $n = 306$)			
Own	71.7%	(2003) 55.9%	$p = 1.38 \times 10^{-08}$***
Rent	15.5%	(2003) 40.6%	$p = 7.81 \times 10^{-20}$***
Informal (authorized)	3.1%	(2003) 3.3%	$p = 1.000$
Informal (unauthorized)	9.6%	(2003) 0.2%	$p = 1.40 \times 10^{-252}$***
Mean residence time (years)	12.2	–	
Household sanitation prior to subscribing (active and former-subscriber respondents; $n = 619$)			
Open defecation or unimproved sanitation	29.4%	19.4%	$p = 0.012$*
Non-private improved sanitation	39.2%	36.8%	$p = 0.013$*
Private improved sanitation	31.2%	43.8%	$p = 2.01 \times 10^{-05}$***

Dark gray field indicates a higher value than the urban Haitian average
Light gray field indicates a lower value than the urban Haitian average
*Significant at a 0.05 level.
**Significant at a 0.01 level.
***Significant at a 0.001 level.
Haitian household and demographic data from Haiti Demographic and Health Survey 2016–17 Dataset.
Haitian electricity access data from World Bank, Sustainable Energy for All (SE4ALL) database.
p-value for percentages is the result of a 1-sample proportions test with continuity correction (1 df).
p-value for means is the result of a 1-sample *t*-test against the Haitian average (n–1 df).
p-value for household composition is a result of Pearson's Chi-square test against the Haitian distribution (3 df).
Sample sizes vary by analysis due to nonresponse.

water in the form of a well, cistern, or piped water (Table 3). The majority of active EkoLakay subscriber respondents (61%) reported having none of the selected amenities used as a proxy for wealth in this study (generator, indoor shower, inverter, solar panels, water cistern on house, well), further illustration of the marginal economic status of EkoLakay subscriber households.

Household financial practices

The EkoLakay service operates on a subscription basis, with fees collected monthly. This payment structure is somewhat unique in the Haitian context and may play a role in the accessibility of the service to some users. To better understand the EkoLakay subscriber population, we gathered information on financial practices, including frequency of payments for various necessities.

Household expenses varied broadly within each reported category. It is possible that the most recent week or payment interval did not represent a respondent's average expenses (the country experienced a gas shortage and political unrest during the data collection period), and/or that respondents' memory or estimations were flawed. Furthermore, respondents may have interpreted this inquiry differently, reporting family or personal transportation expenses depending on their individual role in family finances. Comparing these monthly expenses to the cost of EkoLakay, however, can contextualize its role in household finance.

The average subscriber was more likely to pay for sanitation as opposed to electricity; two-thirds of subscribers report no electricity expenses, though this finding does not parse out whether electricity services were readily available (note: these figures do not include fuel costs for the 8% of active subscribers who have generator-powered electricity at home). The relative burden and/or priority of communication expenses, however, averages 1.5 to 3 times the cost of an EkoLakay subscription (Figure 1(a) and 1(b); note that the *y*-axis on visual comparisons is log scale). Transportation is by far the highest monthly service expense incurred by subscribers during the study period, averaging 10 to 17 times the cost of EkoLakay service (this figure may have been heavily impacted by the fuel shortage, as it exceeds the ratio observed by researchers in 2015 (Russel, personal communication February 13, 2021). The majority of Cap Haïtien residents do not own a personal vehicle, so transportation expenses largely describe fees for transit by multi-passenger vehicle or motorcycle taxi.

Why do households subscribe to EkoLakay?

Themes and specific reasons

There were 324 active subscribers and 306 former subscribers to the EkoLakay CBS service who responded to the open-ended question. Do you remember why you chose to join the EkoLakay service? Responses to this question were coded into 14 categories at the time of interview. Responses that did not fit one of the existing codes were described in an open-text 'other' field. From these descriptions, six new categories were created by investigators. On review of all responses, six major themes emerged to describe factors affecting the decision to open an EkoLakay subscription. Themes indicated relationships to economic, social, governance, and technical systems.

The majority (60%) of all respondents cited some form of cost and resource savings as their major motivation for subscribing to EkoLakay. This was followed by quality of life themes (40%), health, cleanliness, and security themes (27%), favorable comparison to existing or other options (15%), external influence (6%), and targeted or auxiliary use of the technology (4%) (Table 4).

Within the cost and resource savings theme, well over half of respondents (57%) stated that they subscribed to EkoLakay because they lacked the resources to invest in permanent sanitation infrastructure (Table 4). This is an indication that EkoLakay fills an existing sanitation gap; it meets the needs of those who seek private improved household sanitation but do not have the economic means of attaining other options on the 'market'. As cost and resource savings was mentioned by 60% of all respondents, former and active, it is evident that EkoLakay is serving an important role in household economy for low-income subscriber households in Cap Haïtien. In many cases, it appears to be the only accessible form of private sanitation. In others, it is a valuable tool that allows the household to free up needed financial resources for other important expenses.

Within the quality of life theme, 30% of respondents described having a private toilet as a key reason for subscribing to EkoLakay. A further 15% simply mentioned improvements to quality of life in general (Table 4). Many responses categorized in this way included the simple phrase, to have a toilet. Such a statement implies that the household, prior to their EkoLakay subscription, lacked fundamental resources for meeting their basic human needs. Rights pertaining specifically to sanitation have been reiterated and clarified twice by the United Nations (UN Committee on Economic Social & Cultural Rights 2010; UN General Assembly 2016). Based on our

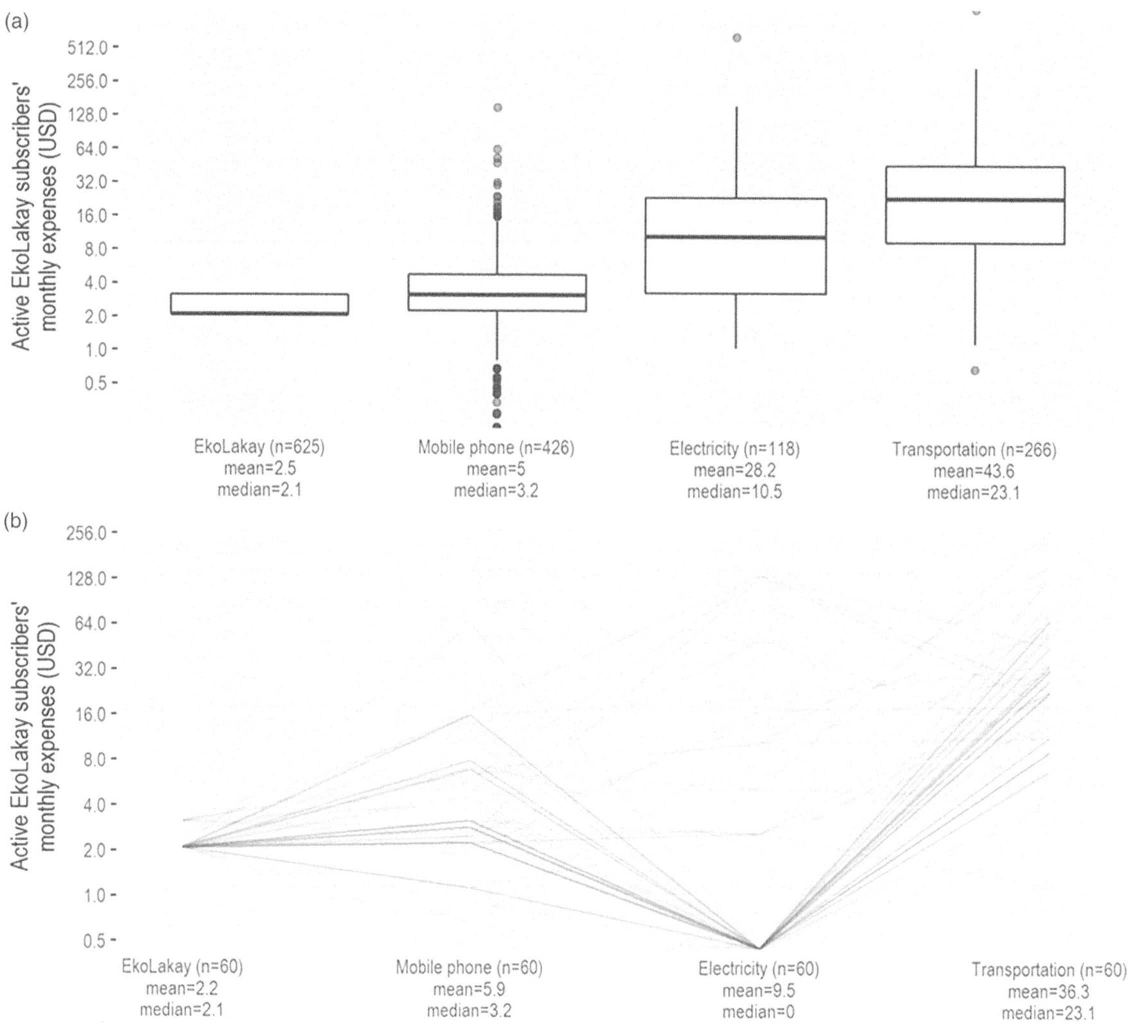

Figure 1 | (a) Active EkoLakay subscribers' monthly expenses in USD (not including respondents without Electricity); (b) Relative priority of EkoLakay compared to other monthly service expenses in USD (including respondents without access to Electricity). Each line represents an individual household. Only those households reporting all four categories are included in this visualization.

findings, over a third of EkoLakay subscribers have no technological or infrastructure alternative for meeting their sanitation needs. For citizens whose sanitation needs remain unmet, CBS may be an accessible tool for use by the Haitian government as they work to secure the basic rights of their citizens.

Among reasons categorized under health, cleanliness, and security, the most-cited motivation for subscribing to EkoLakay was cleanliness, cited by 15% of all respondents (Table 4). This continues to illustrate the vulnerable condition of EkoLakay subscriber households, as well as the important role that CBS is serving in protecting health and meeting basic needs. The specific mention of community environment by 10% of all respondents indicates recognition of individuals' role in community health, and the value of CBS access in advancing the wellbeing of their community. This indicates the influence of social factors in sanitation decision making, and suggests a potential for community-led initiatives to drive demand for CBS in the region.

The dominant response categorized as favorable comparison to existing/other options was that the household was unable to resolve an issue with their existing sanitation technology[8] (14% of all respondents) (Table 4). This challenge may have been technical or financial in nature, or the result of a landlord's failure to maintain the sanitation infrastructure. Sanitation scholars have documented the prevalence of unusable latrines in urban Haiti, subject to disputes over stewardship responsibility between tenants and landlords (Neiburg & Nicaise 2011), and our respondents provide further evidence of the extent of this challenge. Responses within this theme further

[8] Such issues may also include unreliable water and electricity access, necessary to gravity cistern function.

Table 4 | Cited factors affecting decision to subscribe to EkoLakay ($n = 630$)

Themes; specific reasons	respondent count	% within theme	% of all respondents
Cost and resource savings	**378**		**60%**
To save water	4	1.1%	0.6%
To save money	9	2.4%	1.4%
Cheaper than other options	25	6.6%	4.0%
Don't have the money to build another san option	360	95.2%	57.1%
Quality of life	**251**		**39.8%**
To improve quality of life	92	36.7%	14.6%
To have a private toilet	189	75.3%	30.0%
Health, cleanliness, and security	**171**		**27.1%**
Security	32	18.7%	5.1%
Health	58	33.9%	9.2%
To improve/protect community environment	63	36.8%	10.0%
For cleanliness	93	54.4%	14.8%
Favorable comparison to existing / other options	**97**		**15.4%**
Better system/service than other options	9	9.3%	1.4%
Problem with management of existing sanitation	89	91.8%	14.1%
External influence	**38**		**6.0%**
Was offered a subsidy	1	2.6%	0.2%
Felt under pressure by neighbors/others	8	21.1%	1.3%
Positive influence of others	29	76.3%	4.6%
Targeted or auxiliary use	**23**		**3.7%**
Visitors	2	8.7%	0.3%
For semi-public use (church, school, or business)	3	13.0%	0.5%
For renters	4	17.4%	0.6%
Temp solution awaiting permanent sanitation infrastructure	4	17.4%	0.6%
Household member with illness or limited mobility	10	43.5%	1.6%
All respondents	**630**		

Cumulative column sums may exceed total respondent count, as respondents could offer up to three 'reasons' for subscribing.

build on the indications that the technological landscape of the Cap Haïtien metro area offers few options for households to access safe, reliable sanitation. Thus, EkoLakay appears to be filling this gap, and if CBS services were subsidized or otherwise made more accessible to low-infrastructure and resource-strained households, the system may be able to alleviate a portion of exigent community needs.

External influence was cited by 6% of respondents ($n = 38$) as a factor in their decision to subscribe. Most of these responses referred to the positive influence of others, while some also expressed that they felt pressure from community members to subscribe. These responses provide evidence of the role of social systems in sanitation decision making.

Of the subscribers who reported having obtained an EkoLakay toilet for a targeted or auxiliary use, nearly half (1.6% of all respondents) did so to provide accessible sanitation to a household member with illness or limited mobility. While this targeted purpose was mentioned as a primary driver for a smaller segment of the user population, the portability and small footprint of EkoLakay's CBS technology may be a meaningful advantage for the 34% of respondents who reported having a household member living with a disability or chronic illness (Table 3). Furthermore, this theme illustrates that CBS has the potential to fill another unique market gap – that of increasing accessibility, privacy, and convenience for households whose basic needs are already met. While EkoLakay has not historically targeted such households (Russel *et al.* 2015), such a clientele could be a valuable subscriber segment in CBS service delivery models.

Relationships between household characteristics and motivations for subscribing

Subscribers who previously practiced open defecation were more likely than the average respondent to mention quality of life ($p < 0.001$) and health, cleanliness, and security ($p < 0.001$) among reasons for subscribing to EkoLakay, while those who previously had access to private improved sanitation were less likely to cite these themes ($p < 0.01$; $p < 0.001$) (all calculations in Supplementary Material, Table S7). These responses display a continuous-style relationship along the sanitation ladder; evidence that the severity of quality of life and health, cleanliness, and security-related concerns decreases as a household accesses increasingly safe, reliable, and private sanitation. A subscriber coming from a situation without sanitation and/or privacy is motivated strongly by these concerns, while other factors are more likely to affect decision making for households who have maintained sanitation access. Furthermore, the probability of mentioning quality of life among reasons for subscribing to EkoLakay is positively correlated with household density (persons per room in home) (coef = 0.199, $p = 0.002$) (Supplementary Material, Figure S4). This may be a response to the difficulty of securing privacy in a dense living environment. It is also possible that densely populated households are characterized by heightened resource scarcity, leading to priorities associated with more basic human needs.

When comparing EkoLakay favorably to other sanitation options, we found that preference for EkoLakay is positively correlated with residence time (coef = 0.003, $p = 0.004$) (Supplementary Material, Figure S5). It is likely that this relationship reflects the condition of existing infrastructure in long-term residences and completed homes. Aging sanitation infrastructure may malfunction or require expensive maintenance or emptying. Over half of respondents comparing EkoLakay favorably to alternatives already had some form of private sanitation in the home.

CBS user satisfaction and future sanitation aspirations

While EcoLakay subscribers expressed a general preference for flush toilets (VanRiper 2021), the majority of former and active subscribers (87%) also report satisfaction with their household EkoLakay CBS experience (Table 5).

This high rate of satisfaction illustrates that CBS is not only filling a critical sanitation gap, but is an agreeable experience for subscriber households. Inadequate sanitation is more than an inconvenience; disease caused by poor sanitation interrupts livelihoods, earning opportunities, and education (Santiago Ortiz-Correa *et al.* 2016; WHO 2018). The ability of the EkoLakay CBS service to meet immediate household sanitation needs without major expenditures may allow resource-insecure households to maintain a basic quality of life while pursuing economic advancement that would enable them to invest in their preferred sanitation.

These CBS users demonstrate that CBS is a priority for many households, most of whom choose to pay for the service. These subscribers are part of a social system of extreme poverty in which they are trying to improve their quality of life with minimal investment and careful financial prioritization. They understand the importance of improved sanitation in terms of safety, convenience, and cleanliness, but despite the obvious preference and benefits to the social system, this is linked to other systems of governance and infrastructure that impede the full realization of the social benefits.

CONCLUSION

This study provides insights into the potential for CBS as a tool for achieving multiple Sustainable Development Goals. CBS has the potential to strengthen sanitation systems in LMICs, extending the human right of safe sanitation to the most vulnerable households while also protecting the health and dignity of sanitation professionals.

Table 5 | Reported satisfaction with EkoLakay CBS experience, by subscriber status ($n = 625$)

	Respondent count (% of column)		
	Former subscribers	Active subscribers	**Grand total**
Absolutely satisfied	56 (18.7%)	60 (18.5%)	**116** (18.6%)
Satisfied	205 (68.3%)	225 (69.2%)	**430** (68.8%)
Neither satisfied nor dissatisfied	33 (11%)	35 (10.8%)	**68** (10.9%)
Not satisfied	5 (1.7%)	5 (1.5%)	**10** (1.6%)
Not at all satisfied	1 (0.3%)	0 (0%)	**1** (0.2%)
Grand total	**300**	**325**	**625**

The human rights implications of these findings are promising; internationally, many communities share similar challenges and opportunities to those observed in northern Haiti. In urban areas where permanent infrastructure is not feasible, accessible, or functional, CBS could meet the need for safe sanitation, a human right recognized by the UN. In addition, CBS expansion has implications for the wellbeing and human rights of the community workforce, the bayakou, that manually empty pit latrines exposing themselves to physical danger and stigma.

Despite the many benefits of CBS found in this study, large-scale implementation of CBS faces a multitude of systems barriers. A system-based approach to public sanitation, including integrating CBS into community-wide sanitation delivery models will enable aspects of social, governance, and technological systems to interact effectively, thus allowing the benefits of this innovation to reach those who need it most. First and foremost, CBS will need investment from the public sector. As shown in this study, users noted cost and resource savings as a key motivator for subscribing to CBS services. CBS fills an important role in providing safe sanitation for those that do not have the resources for permanent forms of improved sanitation, but without subsidy those who desire CBS access may be unable to afford or remain subscribed to the service.

Just as CWIS calls for multiple technological approaches, the framework also recognizes that varied economic models are required to correct the market failures inherent to services that meet basic human needs (McGranahan 2015; Schrecongost *et al.* 2020). To achieve equitable sanitation coverage for the entire population, urban sanitation programs often need to be managed as a public service. Public investment in CBS as part of a CWIS portfolio could take many forms, from strategic public-private partnerships and blended financing models to full provision of services. Laguna Water, in the Philippines, uses water service fees from across the customer base to subsidize CBS service provision for low-income households (GHD 2018). As seen in this example, leveraging the potential of private sector collaboration will allow public entities to fund sanitation coverage for fragile context residents, including those living in areas of unclear land tenure.

Funding structures based on sanitation coverage outcomes could increase accessibility of CBS services and strengthen urban sanitation systems. With expansion, per-capita service delivery costs are likely to decrease, creating a positive feedback loop (EY 2020). Furthermore, public service providers investing in broad-scale CBS application may be able to avail themselves of atypical funding resources, such as carbon mitigation credits (Shaw *et al.* 2021) or other environmentally linked financial incentives, due to the unique characteristics of CBS technologies. Alternatively, with growing recognition of the ability for CBS to meet SDG 6.2 and human rights or alleviate some environmental pollution crises, emerging regulations or international agreements could create positive pressure to adopt CBS.

Support from the public sector is also necessary to dismantle policy-related obstacles to expansion of this appropriate technology. From clarifying recognition of CBS as a safely managed sanitation solution as has been done by the Kenya Ministry of Health (2016), to defining conditions and markets for safe reuse of excreta as seen in Ghana and multiple European countries (Jayathilake *et al.* 2019), coordination among governing entities can facilitate the ability of CBS to fill gaps in urban sanitation systems.

Other system-wide concerns include regulatory constraints and labor rights. Haiti lacks regulatory structures, and the capacity to enforce such regulation necessary to protect sanitation professionals' health, working conditions, or social wellbeing (World Bank 2018; Mazars & Earwaker 2019). Attempts to recognize the rights of latrine emptiers and improve their working conditions are likely to be insufficient, so long as the sanitation system is predicated on manual excavation of excreta. CBS technologies can safely separate both the user and processor from direct contact with excreta, and the semi-centralization of CBS operations enables service providers to implement training, safety protocols, and healthcare support for workers. Adoption of CBS as part of urban sanitation systems may be a useful precedent to addressing the health and human rights risks associated with manual latrine emptying.

Users of Haiti's EkoLakay CBS service represent extremely low-income households, over a quarter of whom have no other sanitation option within or outside the home. These subscribers represent only a fraction of the fifty-thousand households in the Cap Haitien metro area who lack access to sanitation infrastructure. Seven years into its existence as a household service, EkoLakay demonstrated a high rate of user satisfaction, but it will remain inaccessible to many, without public support and incorporation into a CWIS framework. CBS is a promising tool to help LMICs aim for 100% urban sanitation coverage, CBS has been integrated successfully with urban sanitation service delivery in some major cities, but its global potential as a component of CWIS has only begun. With investment and regulatory support from the public sector, CBS could become an essential component of urban sanitation systems, creating safe and dignified livelihood alternatives to manual emptying, and extending the human right of safely managed sanitation to all for whom permanent infrastructure is not accessible, functional, or feasible. However, achieving this outcome will require approaches that effectively integrate the technical, regulatory, financial, and social components of urban sanitation systems; along with the public health goals of the public sanitation system, solutions will consider worker safety, social mores, affordability, climatological resilience, and technical reliability.

AUTHOR CONTRIBUTIONS

Froggi VanRiper: Conceptualization, Methodology, Validation, Formal analysis, Data Curation, Writing – Original Draft, Writing – Review & Editing, Visualization. Kory C. Russel: Conceptualization, Methodology, Supervision, Writing – Review & Editing. Daniel Tillias: Methodology, Supervision, Writing – Review & Editing. Jenna Tilt: Supervision, Writing – Review & Editing. Jessica Laporte: Methodology, Supervision, Project administration, Writing – Review & Editing.

FUNDING

This investigation was funded by research-targeted contributions to SOIL from the Schmidt Family Foundation's 11th Hour Project [grant G-1802-57466] and the Inter-American Development Bank's IDB Lab [grant ATN/ME-15767-HA].

CONFLICT OF INTEREST

The authors declare that the research was conducted in the absence of any commercial or financial relationships that could be construed as a potential conflict of interest.

DATA AVAILABILITY STATEMENT

All relevant data are included in the paper or its Supplementary Information.

REFERENCES

Andersson, K., Otoo, M. & Nolasco, M. 2017 Innovative sanitation approaches could address multiple development challenges. *Water Science and Technology* **77** (4), 855–858. https://doi.org/10.2166/wst.2017.600.

Bhagwan, J. N., Pillay, S. & Koné, D. 2019 Sanitation game changing: paradigm shift from end-of-pipe to off-grid solutions. *Water Practice and Technology* **14** (3), 497–506. https://doi.org/10.2166/wpt.2019.059.

Bischel, H. N., Caduff, L., Schindelholz, S., Kohn, T. & Julian, T. R. 2019 Health risks for sanitation service workers along a container-based urine collection system and resource recovery value chain. *Environmental Science & Technology* **53** (12), 7055–7067. https://doi.org/10.1021/acs.est.9b01092.

Boyer, K. K., Prud'homme, A. M. & Chung, W. 2009 The last mile challenge: evaluating the effects of customer density and delivery window patterns. *Journal of Business Logistics* **30** (1). Available from: https://onlinelibrary-wiley-com.ezproxy.proxy.library.oregonstate.edu/doi/epdf/10.1002/j.2158-1592.2009.tb00104.x

Carr, R. 2001 Excreta-related infections and the role of sanitation in the control of transmission. In: *Water Quality: Guidelines, Standards & Health* (Fewtrell, L. & Bartram, J. eds). IWA Publishing, London.

Clayton, T. & Radcliffe, N. 1996 *Sustainability: A Systems Approach*. Westview Press, Boulder, CO. https://doi.org/10.4324/9781315070711.

de Albuquerque, C. 2012 *Report of the Special Rapporteur on the Human Right to Safe Drinking Water and Sanitation: Stigma and the Realization of the Human Rights to Water and Sanitation*. UN General Assembly Human Rights Council.

Delaire, C., Peletz, R., Haji, S., Kones, J., Samuel, E., Easthope-Frazer, A., Charreyron, E., Wang, T., Feng, A., Mustafiz, R., Faria, I. J., Antwi-Agyei, P., Donkor, E., Adjei, K., Monney, I., Kisiangani, J., MacLeod, C., Mwangi, B. & Khush, R. 2021 How much will safe sanitation for all cost? Evidence from five cities. *Environmental Science & Technology* **55** (1), 767–777. https://doi.org/10.1021/acs.est.0c06348.

EY 2020 *How Cost Analysis Dispels Myths About Container-Based Sanitation*. London, UK. Available from: https://www.ey.com/en_gl/corporate-responsibility/how-cost-analysis-dispels-myths-about-container-based-sanitation

Fene, F., Ríos-Blancas, M. J., Lachaud, J., Razo, C., Lamadrid-Figueroa, H., Liu, M., Michel, J., Thermidor, R. & Lozano, R. 2020 Life expectancy, death, and disability in Haiti, 1990–2017: a systematic analysis from the Global Burden of Disease Study 2017. *Revista Panamericana de Salud Pública* 44. https://doi.org/10.26633/RPSP.2020.136.

Gambrill, M., Gilsdorf, R. J. & Kotwal, N. 2020 Citywide inclusive sanitation – business as unusual: shifting the paradigm by shifting minds. *Frontiers in Environmental Science* **7**. https://doi.org/10.3389/fenvs.2019.00201.

GHD 2018 *Portable Toilet Solutions Phase 2 Pilot Testing Report (No. 71/12564)*. Laguna AAA Water Corporation, Manila, Philippines. Available from: https://forum.susana.org/media/kunena/attachments/11616/2018-04-19FinalReportwithoutAppendix.pdf

Greene, J. C. 2007 *Mixed Methods in Social Inquiry*. Jossey-Bass, San Francisco.

Hyun, C., Burt, Z., Crider, Y., Nelson, K. L., Prasad, C. S. S., Rayasam, S. D. G., Tarpeh, W. & Ray, I. 2019 Sanitation for low-income regions: a cross-disciplinary review. *Annual Review of Environment and Resources* **44** (1), 287–318. https://doi.org/10.1146/annurev-environ-101718-033327.

IHSI 2015 *Population Totale, de 18 ans et Plus: Menages et Densites Estimes En 2015*. Institut Haïtien de Statistique et d'Informatique. Available from: https://web.archive.org/web/20151106110552/http://www.ihsi.ht/pdf/projection/Estimat_PopTotal_18ans_Menag2015.pdf

Institut Haïtien de l'Enfance – IHE/Haiti and ICF 2017 *Haiti Demographic and Health Survey 2016–17 [Dataset]*. HTPR71.SAV. Available from: https://dhsprogram.com/methodology/survey/survey-display-503.cfm

Jayathilake, N., Drechsel, P., Keraita, B., Fernando, S. & Hanjra, M. A. 2019 Guidelines and regulations for fecal sludge management from on-site sanitation facilities. *AgEcon Search* **14**, 57. https://doi.org/10.22004/ag.econ.296740.

Kabange, R. & Nkansah, A. 2019 A review of pit latrine emptying technologies for low-income densely-populated settlements of developing countries. *Current Trends in Civil & Structural Engineering* **1** (2), 1–5. https://doi.org/10.33552/CTCSE.2019.01.000510.

Kenya Ministry of Health 2016 *Kenya Environmental Sanitation and Hygiene Policy: 2016–2030*. Repulic of Keyna Ministry of Health, Division of Environmental Health. Available from: http://sanitationandwaterforall.org/wp-content/uploads/download-manager-files/KESH%20POLICY_1.pdf

Lozano-Gracia, N. & Garcia Lozano, M. 2018 *Haitian Cities: Actions for Today with an eye on Tomorrow (No. 122880)*. The World Bank, pp. 1–236. Available from: http://documents.worldbank.org/curated/en/709121516634280180/Haitian-cities-actions-for-today-with-an-eye-on-tomorrow

Mackinnon, E., Campos, L. C., Sawant, N., Ciric, L., Parikh, P. & Bohnert, K. 2018 Exploring exposure risk and safe management of container-based sanitation systems: a case study from Kenya. *Waterlines* **37** (4), 280–306. https://doi.org/10.3362/1756-3488.00016.

Mara, D. 2018 'Top-down' planning for scalable sustainable sanitation in high-density low-income urban areas: is it more appropriate than 'bottom-up' planning? *Journal of Water, Sanitation and Hygiene for Development* **8** (2), 165–175. https://doi.org/10.2166/washdev.2018.101.

Mazars, D. & Earwaker, P. 2019 Improving desludging in Haiti by building the capacity of local Bayakou (informal manual desludgers). *Presented at the 36th WEDC International Conference, WEDC*. Loughborough University, Nakuru, Kenya. Available from: /articles/conference_contribution/Improving_desludging_in_Haiti_by_building_the_capacity_of_local_Bayakou_informal_manual_desludgers_/9587831/1

McGranahan, G. 2015 Realizing the right to sanitation in deprived urban communities: meeting the challenges of collective action, coproduction, affordability, and housing tenure. *World Development* **68**, 242–253. https://doi.org/10.1016/j.worlddev.2014.12.008.

Neiburg, F. & Nicaise, N. 2011 *Garbage. Stigmatization, Commerce, Politics (Déchets. Stigmatisations, Commerces, Politiques)*. Interuniversity Institute for Research and Development, Rio de Janeiro.

R Core Team 2020 *R: A Language and Environment for Statistical Computing (Version 22.0)*. R Foundation for statistical computing, Vienna, Austria.

Reed, B. 2010 *Emergency Excreta Disposal Standards and Options for Haiti*. Direction Nationale de l'Eau Potable et de l'Assainissement (DINEPA) & Global WASH Cluster. Available from: http://www.susana.org/en/knowledge-hub/resources-and-publications/library/details/1549

Russel, K., Tilmans, S., Kramer, S., Sklar, R., Tillias, D. & Davis, J. 2015 User perceptions of and willingness to pay for household container-based sanitation services: experience from Cap Haitien, Haiti. *Environment & Urbanization* **27** (2), 525–540. https://doi.org/10.1177/0956247815596522.

Russel, K., Hughes, K., Roach, M., Auerbach, D., Foote, A., Kramer, S. & Briceño, R. 2019 Taking container-based sanitation to scale: opportunities and challenges. *Frontiers in Environmental Science* **7**. https://doi.org/10.3389/fenvs.2019.00190.

Sainati, T., Zakaria, F., Locatelli, G., Sleigh, P. A. & Evans, B. 2020 Understanding the costs of urban sanitation: towards a standard costing model. *Journal of Water, Sanitation and Hygiene for Development* **10** (4), 642–658. https://doi.org/10.2166/washdev.2020.093.

Santiago Ortiz-Correa, J., Resende Filho, M. & Dinar, A. 2016 Impact of access to water and sanitation services on educational attainment. *Water Resources and Economics* **14**, 31–43. https://doi.org/10.1016/j.wre.2015.11.002.

Schrecongost, A., Pedi, D., Rosenboom, J. W., Shrestha, R. & Ban, R. 2020 Citywide inclusive sanitation: a public service approach for reaching the urban sanitation SDGs. *Frontiers in Environmental Science* **8**, 19. https://doi.org/10.3389/fenvs.2020.00019.

Shaw, K., Kennedy, C. & Dorea, C. C. 2021 Non-sewered sanitation systems' global greenhouse gas emissions: balancing sustainable development goal tradeoffs to end open defecation. *Sustainability* **13** (21), 11884. https://doi.org/10.3390/su132111884.

Shepard, J., Stevens, C. & Mikhael, G. 2017 *The World Can't Wait for Sewers: Advancing Container-Based Sanitation Businesses as a Viable Answer to the Global Sanitation Crisis*. Water and Sanitation for the Urban Poor; EY, p. 11. Available from: https://www.susana.org/_resources/documents/default/3-2756-7-1490187791.pdf

Sklar, R. & Faustin, C. 2017 *Pit Latrines or Container Based Toilets? A Cost-Benefit Analysis Comparing Two Approaches to Improving Sanitation Access in Urban Areas of Haïti (Haiti Priorities)*. Copenhagen Consensus Center. Available from:

https://www.pseau.org/outils/ouvrages/copenhagen_consensus_center_pit_latrines_or_container_based_toilets_a_cost_benefit_analysis_comparing_two_approaches_to_improving_sanitation_access_in_urban_areas_of_haiti_2017.pdf

SOIL 2019 *Sanitation Safety Planning: Applying the WHO Methodology to SOIL's Operations in Northern Haiti*. Sustainable Organic Integrated Livelihoods. Available from: https://www.who.int/docs/default-source/wash-documents/sanitation-safety-planning-case-studies/haiti.pdf?sfvrsn=a055006e_4

Spuhler, D. & Lüthi, C. 2020 Review of frameworks and tools for urban strategic sanitation planning: considering technology innovations and sustainability. *Journal of Water, Sanitation and Hygiene for Development* **10** (4), 768–785. https://doi.org/10.2166/washdev.2020.062.

Tilmans, S., Russel, K., Sklar, R., Page, L., Kramer, S. & Davis, J. 2015 Container-based sanitation: assessing costs and effectiveness of excreta management in Cap Haitien, Haiti. *Environment and Urbanization* **27** (1), 89–104. https://doi.org/10.1177/0956247815572746.

Tolley, E. E., Ulin, P. R., Mack, N., Robinson, E. T. & Succop, S. M. 2016 *Qualitative Methods in Public Health: A Field Guide for Applied Research*. John Wiley & Sons, Incorporated, Hoboken, United States. Available from: http://ebookcentral.proquest.com/lib/principiacollege/detail.action?docID=4461564

UN Committee on Economic, Social and Cultural Rights Statement on the Right to Sanitation, Pub. L. No. E/C.12/2010/1 2010 Available from: http://www2.ohchr.org/english/bodies/cescr/docs/statements/E-C-12-2010-1.doc

UN General Assembly The Human Rights to Safe Drinking Water and Sanitation, Pub. L. No. A/RES/70/169 2016 Available from: http://docstore.ohchr.org/SelfServices/FilesHandler.ashx?enc=dtYoAzPhJ4NMy4Lu1TOebIM8c1X4GZjGEGHV9SBM9XSPnORaeC1ogH5%2BGl9s5XzMWx9cuWk6oqcJ%2F%2BZodiYw%2Fv6LDHaNS4Bkv79pl5XMZNjZxsPjPhUNTznL13N8C0Cz

UN General Assembly Transforming Our World: The 2030 Agenda for Sustainable Development., Pub. L. No. A/RES/70/1 2015 Available from: https://www.refworld.org/docid/57b6e3e44.html

UN Universal Declaration of Human Rights, Pub. L. No. Article 23 1948 Available from: https://www.un.org/en/universal-declaration-human-rights/

UNICEF 2019 *Progress on Household Drinking Water, Sanitation and Hygiene 2000–2017: Special Focus on Inequalities*. World Health Organization. Available from: http://www.who.int/water_sanitation_health/publications/jmp-report-2019/en/

USAID 2010 Haiti: Property Rights & Resource Governance. United States Agency for International Development. Retrieved from https://www.land-links.org/wp-content/uploads/2016/09/USAID_Land_Tenure_Haiti_Profile.pdf.

VanRiper, F. 2021 *The Viability of Container-Based Sanitation as A Tool for Sanitation Coverage in Urban Haiti. Doctoral Dissertation*, Oregon State University, Corvallis, OR. OSU Scholars Archive. Available from: https://ir.library.oregonstate.edu/concern/graduate_thesis_or_dissertations/7s75dk544.

WHO 2018 *Burden of Disease From Inadequate Sanitation in Low- and Middle-Income Countries [Dataset]*. Available from: https://www.who.int/data/gho/data/themes/topics/indicator-groups/indicator-group-details/GHO/burden-of-disease-from-inadequate-sanitation-in-low-and-middle-income-countries

WHO 2021 *Progress on Household Drinking Water, Sanitation and Hygiene 2000-2020: Five Years Into the SDGs*. World Health Organization, Geneva. Available from: https://apps.who.int/iris/bitstream/handle/10665/345081/9789240030848-eng.pdf?sequence=1

WHO Joint Monitoring Programme 2017 *Sanitation Service Levels [Dataset]*. Available from: https://washdata.org/data/household#!/dashboard/new

WHO/UNICEF 2017 *Progress on Drinking Water, Sanitation and Hygiene: 2017 Update and SDG Baselines*. World Health Organization (WHO) and the United Nations Children's Fund (UNICEF), Geneva, p. 114. Available from: https://www.unicef.org/publications/index_96611.html

World Bank 2018 *Looking Beyond Government-Led Delivery of Water Supply and Sanitation Services: The Market Choices and Practices of Haiti's Most Vulnerable People (WASH Poverty Diagnostic)*. International Bank for Reconstruction and Development, Washington, DC. Available from: http://hdl.handle.net/10986/28997

World Bank 2019 *Health, Safety and Dignity of Sanitation Workers: An Initial Assessment*. International Bank for Reconstruction and Development, Washington, DC. Available from: https://openknowledge.worldbank.org/handle/10986/32640

World Bank, World Development Indicators 2018a *People Practicing Open Defecation (% of Population) [Dataset]*. Available from: https://data.worldbank.org/indicator/SH.STA.ODFC.ZS

World Bank, World Development Indicators 2018b *Population Living in Slums (% of Urban Population) [Dataset]*. Available from: https://data.worldbank.org/topic/poverty?locations=HT

First received 31 January 2021; accepted in revised form 2 March 2022. Available online 15 March 2022

doi: 10.2166/h2oj.2022.032

Assessing mutual accountability to strengthen national WASH systems and achieve the SDG targets for water and sanitation

Sarah Dickin [a,*], Abu Syed [b], Nishrin Qowamuna[c], George Njoroge[d], Carla Liera[a], Mohamad Mova Al'Afghani[c], Sidratun Chowdhury [b], Zoraida Sanchez[e], Abdulwahab Moalin Salad[f], Keren Winterford[g], Erma Uijtewaal [h], Virginia Roaf[i], John Butterworth[h] and Juliet Willetts [g]

[a] Stockholm Environment Institute, Linnégatan 87D, Stockholm 104 51, Sweden
[b] Bangladesh Centre for Advanced Studies, Dhaka, Bangladesh
[c] Center for Regulation, Policy and Governance, Bogor, West Java, Indonesia
[d] Stockholm Environment Institute Africa Centre, Nairobi, Kenya
[e] Fundación Avina, Lima, Perú
[f] School of Public Health and Research, Somali National University, Mogadishu, Somalia
[g] Institute for Sustainable Futures, University of Technology Sydney, Sydney, Australia
[h] IRC WASH, The Hague, Netherlands
[i] Sanitation and Water for All, New York, NY, USA
*Corresponding author. E-mail: sarah.dickin@sei.org

SD, 0000-0003-0437-3755; AS, 0000-0003-4256-6793; SC, 0000-0003-4277-3290; EU, 0000-0002-7351-4542; JW, 0000-0002-3975-9642

ABSTRACT

Multi-stakeholder engagement is critical for making progress towards Sustainable Development Goal (SDG) 6 'Ensure access to water and sanitation for all', which is currently off track to be achieved by 2030. The aim of this paper was to investigate mutual accountability and multi-stakeholder platforms in the WASH sector in a diverse range of countries. Data were collected by Sanitation and Water for All (SWA) Research and Learning Constituency partners and collaborators in five SWA member countries: Bangladesh, Indonesia, Kenya, Peru and Somalia. Data collection involved document review, key informant interviews and workshops, and an online questionnaire. Across all the case study countries, there were no clear examples of mutual accountability mechanisms being widely used in the WASH sector. However, the findings indicate that some of the case study countries have active WASH multi-stakeholder platforms involving a range of actors from government, civil society and the private sector; however, these typically function as coordination and communication platforms rather than supporting mutual accountability. Other case study countries did not have multi-stakeholder platforms involving a diverse range of actors, and instead had platforms established for single stakeholder groups such as the private sector or civil society, leaving certain groups out of activities. Overall, the study highlights the importance of establishing strong multi-stakeholder processes and platforms that bring together a range of actors including government, civil society, private sector, research actors, and WASH external support organizations. Such platforms could provide a foundation to enable mutual accountability between these actors by providing a space to set commitments and monitor progress and have potential to strengthen WASH systems both nationally and globally.

Key words: accountability, governance, multi-stakeholder partnerships, sanitation, water

HIGHLIGHTS

- No examples of mutual accountability were found in the case study countries.
- Where active, multi-stakeholder platforms in the WASH sector typically function as coordination and communication platforms rather than for mutual accountability.
- Multi-stakeholder platforms could enable mutual accountability between actors by providing a space to set commitments and monitor progress.

INTRODUCTION

Multi-stakeholder engagement is critical for making progress towards Sustainable Development Goal (SDG) 6 to 'Ensure access to water and sanitation for all', which is currently off track to be achieved by 2030. Recent projections show that quadrupling the current rate of progress would be needed to achieve SDG 6 by 2030 (WHO and UNICEF 2021). A key part of the SDG Agenda is participatory multi-stakeholder processes, and

SDG 17 includes a number of targets that aim to support achievement of the other SDGs through a multi-stakeholder approach. Target 6.b for water and sanitation focuses on 'the participation of local communities in water and sanitation planning and management'. More broadly, target 17.16 aims to 'Enhance the global partnership for sustainable development, complemented by multi-stakeholder partnerships that mobilize and share knowledge, expertise, technology and financial resources, to support the achievement of the sustainable development goals in all countries'. This depends on increased international cooperation as well as strong partnership processes between different stakeholders and political will at a national level (Maltais *et al.* 2018). Effective mutual accountability, or accountability between stakeholders, is critical for driving multi-stakeholder action on SDG 6 with its targets on access to water, sanitation and hygiene (WASH). Mutual accountability refers to the ways that partners work together to build transparent and robust systems to hold each other to account (Ulimwengu *et al.* 2020; Willetts *et al.* 2020). Mutual accountability is one of several forms of accountability and draws on multi-stakeholder platforms and processes. A recent review of accountability in a WASH context found the need to further develop and invest in accountability and advocacy interventions in the sector (Hepworth *et al.* 2020). This is particularly needed in the case of mutual accountability in WASH systems, which has been under-researched compared with other forms of accountability.

To advance mutual accountability in the WASH sector, Sanitation and Water for All (SWA), a global multi-stakeholder partnership, developed a Mutual Accountability Mechanism (MAM) in 2018. The SWA MAM is intended to support partners to make commitments and hold each other accountable for progress towards the SDG targets, as well as reinforce country-level multi-stakeholder planning and review processes. However, to date, there is limited evidence on the extent of uptake of mutual accountability using the MAM or other tools, or the underlying multi-stakeholder processes that may enable or hinder mutual accountability. The aim of this paper was to investigate mutual accountability and multi-stakeholder processes in the WASH sector in a diverse range of selected countries.

Background: different forms of accountability in the WASH sector

Accountability is a complex idea which can be conceptualized in different ways but often involves elements of answerability and sanctions (Schedler 1999; Fox 2015). Hepworth *et al.* (2020) describe accountability in the water sector as a five-step cycle that includes rules, fulfilment of responsibilities, reporting, review of performance and effective reactions. We describe several forms of accountability that are applied in the WASH sector, including mutual accountability, which has received less attention.

One of the most common framings examines accountability between duty bearer and rights holder, sometimes referred to as vertical accountability (Mainwaring & Welna 2003). This is also the form of accountability referenced by the human right to water and sanitation, which includes aspects of responsibility, answerability and enforceability. For instance, in the context of WASH, governments are responsible for providing safe water and sanitation services for their citizens, so may monitor service provider compliance towards prevailing laws and regulations (Hofstetter *et al.* 2020). The Global Analysis and Assessment of Sanitation and Drinking-Water (GLAAS) is a UN Water report that provides insights on accountability in this regard using a survey distributed to national governments report on progress and challenges in WASH service delivery. Although GLAAS does not directly measure accountability, data include related information such as the existence of a formal mechanism exist to coordinate the work of different organizations with responsibilities for WASH, public provision of performance (e.g. quality of service) and customer satisfaction information, and procedures in laws or policies for participation by service users (e.g. households) and communities in planning services. Using GLAAS data, Jiménez *et al.* (2018) identified a number of ways to strengthen accountability such as improving access to information on the services provided, enacting participation policies and increasing the capacity of regulatory institutions. The UN Special Rapporteur has also collected country examples of accountability mechanisms in the WASH sector, such as processes of participation, monitoring, judicial and quasi-judicial, and political mechanisms (Heller 2018). Social accountability is a related process, whereby civil society organizations (CSOs) and citizens directly hold public officials and decision-makers accountable for their actions through direct engagement, dialogue and advocacy (UNDP-SIWI Water Governance Facility & UNICEF 2015). This can include the use of mechanisms such as citizen report cards and social audits (Jiménez *et al.* 2018).

A framing of accountability that is less common in the WASH sector is mutual accountability. Mutual accountability can be described as the process by which two (or multiple) partners hold each other accountable for the commitments they have voluntarily made to each other (Droop *et al.* 2008) and has emerged in response to power imbalance reflected in vertical models of accountability (Steer *et al.* 2009). It seeks to establish a more balanced partnership between stakeholders through shared commitments, agendas, values and ownership by establishing trust and understanding and working towards the achievement of shared objectives (Ulimwengu *et al.* 2020). For instance, relationships between donors and partner countries are often uneven, but mutual accountability suggests that all stakeholders must hold themselves and others accountable for development processes to succeed. Mutual accountability can also complement and support other types of accountability, including forms of vertical accountability (Willetts *et al.* 2020). For example, it can provide a tool to align limited resources in multi-stakeholder contexts to improve the efficiency of their use (Ulimwengu *et al.* 2020).

Accountability is an essential and fundamental concept in the WASH sector for several reasons. Firstly, it is a human rights principle and required for realization of the human right to water and sanitation. States, as duty bearers, are accountable for the WASH policy decisions and actions they make. Secondly, accountability is a key mechanism to strengthen WASH systems, though it is given scant or no mention in WASH systems strengthening academic literature to date (Hollander *et al.* 2020; Valcourt *et al.* 2020a, 2020b). Grey literature does make mention of it however, for instance. Gensch & Tillett (2019) mention accountability in relation to inclusive planning and also include a WASH system's building block titled 'regulation and accountability'. Huston & Moriarty (2018) refer to accountability in their Working Paper on WASH system building blocks, but with a limited scope focused on accountability of service providers. SWA includes a potentially broader scope in their WASH system building blocks, where 'clearly defined accountability mechanisms' is deemed part of the building block on 'planning, monitoring and review'.

Background: multi-stakeholder partnerships and platforms

Multi-stakeholder partnerships and platforms (MSPs) have the potential to contribute to good governance by allowing for both government leadership and meaningful stakeholder engagement and participation (Stewart & Gray 2006). This can increase effectiveness and efficiency of service delivery, increase governance legitimacy and allow ways to manage conflicting views and interests (Brinkerhoff 2007). More broadly, partnership can be described as a cooperative relationship characterized by a set of values (including trust, transparency, accountability, reciprocity and respect) that evolves through time by mutual learning, voluntary participation and commitment, towards common goals (Schaaf 2015).

In the context of WASH, MSPs exist at both national and global levels, for instance Sanitation and Water for All is a global MSP. However, there is limited research on MSPs relating to WASH and their effectiveness (e.g. Keatman 2012). In comparison, more extensive literature exists relating to MSPs in the broader water sector, such as in integrated water resource management where they have been used to manage competing claims to water and coordination challenges (Warner 2006; Hailu & Tolossa 2020; Sigalla *et al.* 2021). Further research is needed to explore the benefits of MSPs in the delivery of water and sanitation services because they are critical foundations for mutual accountability. Recent research proposed that an effective mutual accountability mechanism should be supported by functioning partnerships (Willetts *et al.* 2020). These authors described four main areas of partnership processes as important: (i) Shared vision, purpose, ownership of the partnership; (ii) Transparent, collaborative processes and rules of engagement; (iii) Shared decision-making, action and feedback amd (iv) Mutual learning to evolve partnership. They also identified key elements needed to provide the foundation of a partnership, including shared purpose, sufficient partner capability to partner, meaningful multi-stakeholder representation, and effective multi-stakeholder horizontal and vertical relationships in the partnership structure which can provide an equal playing field and bring together a wide range of affected stakeholders, including NGOs, the media, donors, multilateral organizations, private sector and governments.

METHODOLOGY

Case study description

This study was conducted between September and December 2020 by the SWA Research and Learning Constituency partners and collaborators from five member countries: Bangladesh, Indonesia, Kenya, Peru and Somalia. The choice of cases was motivated by geographical diversity and the presence of participating research partners. In addition, only countries that are members of SWA were eligible as cases, where there has been exposure to the SWA MAM. SWA is a global multi-stakeholder partnership that works towards achieving sanitation, hygiene and water for all (SDG 6) through coordinated collaborative action at global, national and sub-national levels. The

SWA MAM allows partners to make commitments and hold each other globally accountable for progress towards the SDG targets, as well as reinforcing country-level multi-stakeholder planning and review processes.

Data collection

Data for this study were collected through a mixed-methods approach using streamlined data collection tools that were implemented by researchers in each case study country who conducted data collection activities. In order to understand WASH mutual accountability processes and the multi-stakeholder partnerships in each of the selected case studies, the first step in data collection involved a document review. Key literature such as policy and planning documents relevant to the WASH sector and publications such as reviews and assessments on progress in WASH were identified and analysed for examples of multi-stakeholder platforms and mutual accountability mechanisms. Next, a stakeholder mapping of the WASH sector was conducted, and this enabled the country teams to identify a diverse range of key stakeholders divided into groups based on SWA constituencies: government, civil society, private sector, research and learning, and external support agencies such as donor organizations. Key stakeholders often included SWA focal points who are country stakeholders that act as SWA representatives. The stakeholder mapping was used for the next steps of data collection to ensure a range of perspectives were collected, with the aim to collect perspective from each constituency group, as well as to cover water, sanitation and hygiene, and differing levels of importance and influence. Next, a streamlined survey and key informant interview guide were developed for use in each case study country, which covered the following topics: multi-stakeholder platforms, accountability mechanism used in the WASH sector, including any mutual accountability mechanism, engagement with SWA, and the SWA MAM in particular and the impact of Covid-19 on existing accountability processes. An online questionnaire was sent to identify key informants in each country case. The online survey was centrally administered and common across all case study countries, ranging from 15 responses in Somalia to a maximum of 25 in Peru (Table 1). The information gathered from the survey served as the basis for the discussion during the key informant interviews. Qualitative data were collected using key informant interviews, where interviewees were selected to represent as many constituencies as possible. Interviews were conducted in person or online with a team of one or two interviewers and were recorded verbatim or recorded using detailed notes taken during the interviews. This information was then thematically analysed according to the main themes in the interview guide by each country team and through discussions with the overall multi-country research team. As part of the research findings validation process, a national workshop was held with key informants and other stakeholders that involved the presentation of initial findings and feedback from participants. The purpose of this workshop was to validate initial findings with participants and to obtain any missing information through a discussion with participants. The results of the study were then summarized by each research team in the form of a country brief and later synthesized into a report for SWA (Butterworth *et al.* 2021).

Table 1 | Summary of data collection activities and participants by case study country

Country	Key informant interviews	Survey respondents	Workshops participants
Bangladesh	10	16	56
Indonesia	7	16	25
Kenya	9	19	6
Peru	7	25	10
Somalia	6	15	9

RESULTS

Existing mutual accountability mechanisms

While there are a range of existing multi-stakeholder platforms in the case study countries that serve to facilitate coordination and communication between WASH stakeholders, they do not currently enable mutual accountability processes. No mutual accountability mechanisms were being employed in the case study countries as we define it in this study, where all actors set targets that other actors hold them accountable for achieving towards a common goal. Instead, mechanisms reported were those traditionally used for vertical accountability, such as national plans, sector progress or performance reports, committees, and meetings (see Butterworth *et al.* (2021) for more details on these).

The SWA MAM promotes the adoption of joint commitments to which WASH sector actors can hold each other accountable. While all case study countries participated in the SWA MAM, participation is still limited, and commitments have been made unilaterally by a limited group of stakeholders. In the case of Indonesia and Peru, commitments have been made only by governments (Table 2), while Bangladesh and Somalia have commitments from civil society and external support agencies, respectively, in addition to government. There has been a similar situation in Kenya, although recent efforts by SWA to support MAM commitment setting in Kenya have increased the engagement of different stakeholder groups.

Table 2 | SWA MAM commitments by country in 2020[a]

No. of existing commitments in the SWA MAM database	Bangladesh	Indonesia	Kenya	Peru	Somalia
Government	5	3		3	3
Civil Society	5				
Private sector			3		
External support agencies					3
Research and learning					

[a]This analysis was conducted in November 2020, and commitments will continue to be added. For instance, during the course of this study, Kenya was holding consultations to develop government and other stakeholder commitments, so SWA MAM commitments will soon increase.

In Bangladesh, Indonesia and Peru, accountability is seen as mainly an issue involving government as the duty bearer, limiting interest and uptake of mutual accountability mechanisms. For example, in Bangladesh, although CSOs play an important role in development activities, they do not have any legislative or constitutional provision or framework that supports their engagement in development planning and implementation processes with the government. This is seen to limit their power in a mutual accountability context, which means that although there are functioning multi-stakeholder platforms, they struggle to maintain critical government participation that is needed for mutual accountability. In Indonesia, vertical accountability is found within internal government institutions by which ministries/government agencies are required to submit performance reports (*Laporan Kinerja*/LAKIP), which are then made publicly available. The LAKIP reports, however, do not thoroughly discuss sectoral issues. In addition, development partners/donors also publish annual reports which cover their achievement and progress. Although development partners/donors were found to report their progress to government, feedback is not always received. Despite these existing forms of accountability, currently there is no clear mechanism that enables WASH stakeholders to hold each other accountable for their performance (mutual accountability).

The Peru case illustrated that functioning multi-stakeholder platforms are not enough to establish mutual accountability alone. There is an existing multi-stakeholder platform for poverty alleviation, which is a flagship government-led coordination platform, but accountability mechanisms are limited to government policy. Looking towards mutual accountability and the need for a common vision, respondents suggested that the National Sanitation Plan could be the focus of future efforts. Similarly in Indonesia, the Jejaring AMPL is a platform that facilitates coordination for a wide group of WASH stakeholders and is seen as inclusive but does not have mutual accountability mechanisms. However, in Peru, the SWA MAM was seen as a potential means to enable the government to open up accountability processes in the WASH sector to multiple actors.

In Somalia, the situation is characterized by a humanitarian focus due to civil war and an emergency context with limited government resources. Monthly meetings of the humanitarian coordination platform serve to provide a form of vertical accountability between donors and implementers, a 'soft accountability' rather than as mutual accountability between all actors including donors, government, civil society and the private sector.

There are potential opportunities to strengthen existing processes to incorporate mutual accountability. For instance, in Bangladesh, there are several national-level accountability mechanisms that allow input on government activities from other actors. For example, the Local Consultative Group (LCG) Water Supply and Sanitation, National Sanitation Taskforce, the National Policy Review Committee and the SDG tracker are forums where the government shares their progress on WASH goals and CSOs get to share their concerns regarding progress.

These offer ways for CSOs to participate in formulating action plans for each target. While these accountability mechanisms are more focused on vertical accountability and government responsibility alone, there is potential to expand on these processes for other partners to also be held accountable on their progress.

Likewise, in Indonesia, the existing WASH multi-stakeholder platforms such as National Pokja AMPL, Jejaring AMPL and SDG forum have the potential to be utilized to facilitate mutual accountability mechanism. National Pokja AMPL, for example, is led by the Bappenas/Ministry of Planning and has been facilitating coordination between government actors in charge of WASH – relevant stakeholders are also invited to their discussion. Jejaring AMPL has a broader membership which is open to, among others, government officials (in their personal capacity), NGOs, research and learning, and aid organizations. Additionally, SDG monitoring platform also emerges as a possible option where SDG targets, including WASH, are being tracked and reviewed.

In Kenya, there has been recent progress that took place after data collection, as a group of stakeholders comprising of government, civil society and the private sector came together in late 2020 to adopt a mutual accountability framework promoted by SWA. Through this framework, the government is expected to lead other stakeholders in making commitments on priority focus areas, while other stakeholders follow suit with their commitments that augment those made by government. Should these stakeholder binding commitments become successful, it presents a promising opportunity for enhancing mutual accountability in the sector.

WASH multi-stakeholder landscapes

As there were no findings on mutual accountability mechanisms, the remainder of the results focus on existing MSPs that could provide a foundation for establishing mutual accountability and their limitations. The findings in five case study countries show that MSPs formed at the national level exist in some form in all cases (Table 3). We defined multi-stakeholder platforms as those comprising more than one SWA 'constituency,' i.e., government, civil society, private sector, donors and external support or research and learning. A platform comprised only of civil society organizations such as NGOs was classified as a single stakeholder platform, and relevant examples are shown in Supplementary Material, Table S1. While the MSPs typically serve as communication and coordination platforms for their members, they also have diverse mandates in the WASH sector in their respective countries which includes sharing information and knowledge, influencing sector policy, playing a role in government accountability and transparency, enhancing the participation of various stakeholders, and developing capacity building and partnership. For example, in Indonesia, Jejaring AMPL is a platform that facilitates coordination for a wide group of WASH stakeholders, including governmental actors, development partners and donors, NGOs, private sector, as well as research and learning institutions. Its activities include advocacy, capacity building for its members and related actors, information sharing and partnership development. Key informants described Jejaring AMPL as inclusive and with a partnership spirit. In Peru, the working group on rural water and sanitation, part of the Multi-stakeholder Coordination Platform for Poverty Alleviation, aims to monitor WASH policies and make recommendations to government agencies to strengthen their WASH strategies. In countries with several MSPs, coordination between the different platforms was limited, indicating possible fragmentation between the activities occurring in different MSPs.

In several of the case study countries, active multi-stakeholder platforms and processes were more limited, and other types of single stakeholder platforms were more dominant. In Kenya, while platforms such as the Kenya Water and Sanitation Civil Society Network (KEWASNET) were reported to be active and hold regular meetings and activities, the broad-based multi-stakeholder processes like the Joint Sector Working Group were found to be either ad hoc or dormant. This meant that there were limited active processes coordinated by government actors to improve overall sector coordination with the potential to enable mutual accountability. However, it is important to note that KEWASNET is a strong partnership that could make valuable contributions to a MSP if it was successfully established. In Somalia and Indonesia, the Inter-ministerial WASH Steering Committee and National Pokja AMPL, respectively, focus more on inter-ministerial coordination rather than the involvement of diverse stakeholders. Somalia is emerging from a civil war, which explains the existence of limited MSPs, and instead an emergent inter-ministerial coordination platform and a WASH humanitarian response coordination forum. In the case of Indonesia, National Pokja AMPL receives feedback from relevant stakeholders, but strategic decisions are made by government agencies.

Gaps in multi-stakeholder platforms and processes

In the countries with active multi-stakeholder platforms, these have enabled coordination, development of networks among stakeholders and have influenced policy-making and implementation. For example, in Peru,

Table 3 | Examples of multi-stakeholder platforms in WASH

Country	Platform name	Main function	Composition	Status and regularity	Funding sources
Bangladesh	Local Consultative Group (LCG)	The LCG Forum serves at the national level as technical and policy support group. All government-related partners communicate through LCG. Decisions taken in this forum are then conveyed to the national forum for drinking water supply and sanitation where policy or strategies are based on this.	Is a multi-stakeholder forum for government, NGOs, CSOs, and external support agencies (e.g. ADB, World Bank, JICA, etc.) The Chair of this forum is Local Government Division (LGD) in the ministry of Local Government, Rural Development and Co-operatives and co-chair is UNICEF.	Ad hoc	Unknown
	Citizens Platform for SDGs	The platform provides an opportunity to track the progress of SDG delivery; sensitize policymakers towards challenges in implementation; bring transparency in the implementation process; and facilitate the exchange of information and coordination among all those working on the SDGs in Bangladesh.	Mainly consists of NGOs, CSOs, research organizations and the private sector. The Secretariat of the Platform is hosted by the Centre for Policy Dialogue (CPD).	Informal and active	Unknown
	Bangladesh Faecal Sludge Management Network	This is a common and collective platform for the sector actors to generate ideas, share views and create and focus more attention on faecal sludge management.	The platform is led by SNV and WaterAid in Bangladesh. This network engages government agencies such as the Department of Engineering, Dept. of Agriculture Extension, Dept. of Fisheries, Sustainable And Renewable Energy Development Authority (SREDA), relevant taskforces, networks and associations including National Sanitation Secretariat, National Forum for Water and Sanitation, Bangladesh Urban Forum (BUF), Municipal Association of Bangladesh (MAB), etc.	Ad hoc. Has regular conventions at least once a year	Unknown
Indonesia	Jejaring AMPL	A coordination and communication platform consisting of a wide group of WASH stakeholders. Jejaring is an independent organization. Jejaring activities include advocacy, capacity building for the members and relevant actors, as well as partnership development.	Jejaring consists of officials (in a personal capacity), NGOs, international organizations, research and learning organizations and donors. Jejaring is usually chaired by Bappenas officials in their personal capacity. Its steering committee also consists of representatives from relevant agencies in a personal capacity.	Informal. Meets every few months.	Funding from the members

	National Pokja AMPL (Drinking Water and Environmental Health Working Group)	A coordination and communication platform that ensures drinking water and sanitation development – from planning, implementation, monitoring, to the evaluation stage. It also enhances coordination between government actors that are in charge of drinking water and sanitation matters.	National Pokja AMPL officially consists of eight government ministries/institutions, namely Bappenas/Ministry of Planning (coordinator), Ministry of Public Works, Ministry of Health, Ministry of Home Affairs, Ministry of Environment, Ministry of Finance, Ministry of Education and Culture, and Statistics Indonesia. Although the official members consist of ministries/government agencies (which would define it as a single stakeholder platform), National Pokja AMPL also invites non-government stakeholders as appropriate.	Ad hoc. Used to be frequent, but lately infrequent. Usually hosts meetings based on needs.	State budget
	The Sanitation Partners Group (SPG) focused on urban sanitation	A platform to improve coordination between government and development partners/donors.	Development partners/donors, international organizations, government (Bappenas, Ministry of Public Works, Ministry of Health). World Bank and UNICEF are among the members.	Status: informal; Regularity: used to be frequent, lately infrequent.	
Kenya	Joint Sector Working Group	Sector coordination	Consists of all key stakeholders in WASH. Was coordinated by an Interagency Coordinating Committee (ICC) with several Technical Working Groups.	Dormant (annual review has not happened in 5 years)	NA
	National Steering Committee (NSC) on SDG 6	Coordinates and oversees reporting on SDG 6 progress to UN Water	Convened by the Ministry of Water and Sanitation and is composed of key government ministries, civil society organizations, academia, private sector and development partners.	Ad hoc	Unknown
	Water and Environmental Sanitation Coordination mechanism (WESCOORD)	To achieve a coordinated and integrated approach in the implementation of WASH emergency response in areas that are commonly affected by drought and floods	WESCOORD is chaired by MoWSI and co-chaired by UNICEF. The Ministry of Health also takes on a significant role in WESCOORD, chairing key TWGs in hygiene and sanitation sub-sectors. Membership comprises of WASH humanitarian agencies and national authorities.	Dysfunctional at a national level with ad hoc meetings during disasters. Only active in certain counties.	Unknown
Peru	Rural Water and Sanitation Working Group which is part of the multi-stakeholder	Their main function is to monitor the implementation of national public policies and produce policy	The rural water sanitation group is part of MCLCP, which brings together technical staff from various ministries,	Active with meetings once a month with the possibility of extraordinary sessions	Public financing, but largely dependent on

(*Continued.*)

Table 3 | Continued

Country	Platform name	Main function	Composition	Status and regularity	Funding sources
	platform for poverty alleviation (MCLCP)	recommendations on water and sanitation issues in rural context.	sub-national governments, civil society, NGOs and others. The MCLCP was created by the Supreme Decree 001-2001-PROMUDEH in 2001. It is currently part of the Ministry of Development and Social Inclusion (MIDIS).	in the case of urgent matters	stakeholders' willingness to engage
Somalia	Somali WASH cluster	A humanitarian platform that coordinates WASH activities	Mainly humanitarian organizations such as UN organizations, NGOs and the private sector, with very few R&L members. Co-chaired by UNICEF and another member organization and includes focal points for 13 regional WASH cluster organizations.	Monthly meetings	Somalia Common Humanitarian Fund

Up to three key MSPs are shown for each case study country. See Butterworth *et al.* (2021) for more detail on each country MSP landscape.

respondents described that the MCLCP through its Working Group on 'Rural Water and Sanitation' provided recommendations on the provision of drinking water targeting vulnerable population during the Covid-19 pandemic to the Ministry for Housing, Construction and Rural sanitation, and other government institutions. Such recommendations are the result of a series of evaluations and round table sessions between different stakeholders and government authorities. However, there are still gaps in these multi-stakeholder processes that hamper effective collaboration and implementation, and could thus limit opportunities for mutual accountability. One of the major issues is the ad hoc nature of some of the platforms, such as seen in the Bangladesh and Indonesia cases. Indonesia has the National Pokja AMPL, but due to its ad hoc nature it is dependent on the active participation of the members. The same issue is seen with the Local Consultative Group (LCG) of Bangladesh where meetings are irregular and usually held if a certain issue arises. Stakeholders sometimes do not actively participate through articulating their opinions and concerns in the multi-stakeholder process, which hampers constructive planning of policy-making for the sector.

Non-participation of key members is another issue weakening multi-stakeholder processes in some case studies. In Bangladesh, private sector actors have been traditionally less involved in planning and policy-making in multi-stakeholder processes in the WASH sector, so their views are under-represented. In Peru, there is limited integration of community organizations (COs) into existing MSPs as well as all other decision-making processes in the WASH sector. This stems from the fact that COs do not have a clear regulatory framework that ensures their sustainability and legal existence (Ñope Cueva 2019). However, these COs play a crucial role in the administration, operation and maintenance of WASH services in rural areas. According to DATASS (MVCS 2018), there are approximately 25,000 COs that provide sanitation services in rural areas.

In Indonesia, non-governmental actors are not officially members of the National Pokja AMPL. However, it is worth noting that non-governmental actors can be invited to participate in the Pokja's meetings and discussions. In Bangladesh, the results indicated that the sub-sector of hygiene is often left out of planning and policy-making, with the main focus on water and sanitation. This was observed in national platforms such as National forum for drinking water supply and sanitation. The under-representation or the absence of some stakeholders in multi-stakeholder platforms means that some critical issues may not be effectively presented or addressed in the sector planning and policy processes. This is also an issue among single stakeholder platforms, for example in Kenya, the Water Service Providers Association (WASPA) only draws membership from the water and sanitation utilities that are owned by county governments, leaving out other private players that are active in the sector.

A further important gap in many of the cases is lack of financial and human resources and capacity to provide strong government leadership in the promotion and implementation of multi-stakeholder engagement processes. In Peru, budget shortfalls due to the Covid-19 pandemic and a very high level of government turnover in the existing platform made continuity of the work and monitoring of activities almost non-existent. Related to this is a lack of resources and funds to support the functioning of the multi-stakeholder platforms, which was particularly highlighted in the Bangladesh, Kenya, Indonesia and Peru cases as a limiting factor. This hinders effective engagement and collaboration in the multi-stakeholder processes and gives more power to certain actors funding the activities. In Somalia, the Inter-ministerial WASH Steering Committee is supported by UNICEF with no sustainable government allocated budget, which means that it lacks legitimacy in terms of funding, human resources and enforcement power. In Kenya, a lack of financial resources for a government-led multi-stakeholder platform has resulted in ineffective coordination between the Ministry of Water and Sanitation and the Ministry of Health in their shared mandate to provide leadership and direction in the sector, particularly in the area of sanitation. As a result, in Kenya there was a perceived lack of knowledge and capacity among certain WASH stakeholders to get involved in multi-stakeholder processes.

Finally, in terms of supporting multi-stakeholder processes, a gap identified in the Bangladesh and Peru cases was the limited availability of a standardized and consolidated database for keeping track of progress towards national targets and SDG 6, including WASH services provided by implementers, with open data for all. In the Peru case study, such a system was identified as something that could help with monitoring and coordination of actions between different stakeholders, such as enabling civil society and even the private sector to share their action plans with the government, leading to greater synergy with the National Sanitation Plan. In Bangladesh this type of system was suggested as a way of reducing overlapping of services implemented by agencies.

Finally, in some countries such as Kenya and Peru, the Covid-19 pandemic has provided new opportunities to address these gaps through enhanced sector coordination and greater accountability through the use of electronic platforms and other online resources. In the case of Peru, the use of technologies increased due to the pandemic, this allowed greater participation of the actors that previously could not attend national meetings or forums because they were usually held in Lima, the capital. Nevertheless, limited access to internet and other communication technologies continues to be a barrier to overcome in several areas of the country.

DISCUSSION

As emphasized in SDG 17, effective partnerships are critical for achieving the SDGs. Promoting collaboration between stakeholders is widely considered to be critical for progress towards both national WASH targets and globally achieving SDG 6. It is important to ensure that multi-stakeholder platforms and partnership processes are as effective as possible to enable this progress. Multi-stakeholder platforms can serve as vehicles to enhance collaboration, innovation, and also have the potential to enable mutual accountability. However, across the cases in this study, we found no specific examples of mutual accountability mechanisms, despite examples of multi-stakeholder platforms and processes. Despite this, countries with effective multi-stakeholder platforms for coordination in the sector have a foundation to build on to establish such processes. As compared with other sectors, the concept of mutual accountability is yet to be widely taken up and the SWA MAM tool is still early in its development and wider application. As the focus in most case study countries was on vertical accountability, such as reviewing commitments set by government, greater awareness and capacity for mutual accountability that involves setting commitments across multiple actors may be needed. Furthermore, a full mutual accountability cycle requires some form of performative review and reaction when commitments are not met (Hepworth *et al.* 2020; Willetts *et al.* 2020).

Although there are few examples of research on MSPs or mutual accountability to draw on in the WASH sector, there are lessons from other sectors that have historically used multi-stakeholder platforms (Warner 2007; Hermans *et al.* 2017; Willetts *et al.* 2020). In watershed management, MSPs have been found to be helpful for communication, coordination, management of competing claims to water, and coalition-building and visioning but require clear goals and results to keep participants interested (Warner 2006; Hailu & Tolossa 2020). Other research on multi-stakeholder platforms has found that they are often characterized by ambiguity, uncertainty and complexity, where there are a range of different interests at stake (Kusters *et al.* 2018). This means that participants will often have different views on outcomes, and it is important to outline clear expectations and a common vision when establishing mutual accountability processes.

Active multi-stakeholder platforms were identified in the WASH sector in some of the country cases. In many of these cases, certain actors clearly held an unequal amount of power (e.g. donor or government-driven processes), such as in the case of the government-led MCLCP in Peru. As has been seen in other sectors, it is critical to address power relations among these different actors to ensure mutual interest, engagement and commitments are sustained (Brouwer *et al.* 2013). Establishing effective multi-stakeholder processes in these cases requires specifically reaching out to excluded groups, and establishing new inclusive ways of working where all actors' voices can be heard. This also has benefits as research on MSPs in other sectors has shown that a high level of knowledge exchange is needed to enable innovations, which results from a wide range of participants rather than limited participation of stakeholder groups (Hermans *et al.* 2017). Power analyses and social network analyses offer an opportunity to deepen understanding of many of the linkages and power relations between actors reported in this study (Brouwer *et al.* 2013).

In other cases, a focus on active single stakeholder platforms for external support partners, civil society or private actors contributed to a lack of strong multi-stakeholder engagement, such as in Kenya. This resulted in a fragmented approach to the sector, where certain stakeholder groups are frequently left out, thereby resulting in lack of coherence and also compromising opportunities for mutual accountability. Overall more work is needed to examine the effectiveness and different types of multi-stakeholder platforms for making progress in the sector and towards SDG 6. This includes identifying the most effective mechanisms in order to understand how they can be scaled up to strengthen WASH systems.

Opportunities for improving mutual accountability to strengthen WASH systems

This study shows the lack of use of mutual accountability in the WASH sector in the case study countries to date. Despite these gaps, mutual accountability offers opportunities to strengthen WASH systems and ensure gender and other inequalities are addressed in leaving no one behind (Mannan 2018). There are opportunities to draw on other applications of mutual accountability, and a review by Willetts *et al.* (2020) of four partnerships and one global accountability mechanism outside the WASH sector found five key elements in an effective cycle of mutual accountability: stakeholder responsibilities, obligations or commitments are clarified and communicated; stakeholders share information with one another on their performance; performance by each stakeholder is explained, discussed and evaluated; a reputational or relational consequence (of performance or non-performance); and corrective action is undertaken by stakeholders to better achieve partnership goals. Consequences are an important part of mutual accountability that differentiate it from partnerships alone. In the case of the Universal Periodic Review of the UN Human Rights Council, this works on the basis of learning rather than on actual naming and shaming. This review also found that in some cases, there were examples of partial cycles, where regular meetings and reports on progress contributed to performance review, but were missing next steps of behaviour change if commitments were not met.

Since this study data were collected, Kenya has taken steps to apply the SWA MAM tool more widely, so this is a promising case to observe going forward. To better enable mutual accountability, it is important to focus on strengthening foundational elements of multi-stakeholder platforms and processes in the WASH sector as a first step. The findings indicate the importance of establishing strong multi-stakeholder processes and platforms that are cross-sectoral and bring together government, external support organizations, civil society (including larger organizations as well as small community-based organizations), the private sector and research actors on the same level. Global-level processes such as the SWA partnership and the MAM can play a key role in adding legitimacy and collective scrutiny to track performance against commitments.

A challenge to consider may be the resource intensive structures required to ensure equitable, effective and sustainable collaboration between actors (MacDonald *et al.* 2018). For instance, MSPs require certain actors to coordinate and drive the process forward on a regular basis with dedicated resources, rather than on an ad hoc basis. Donor-driven multi-sectoral platforms may not be sustainable in the long run to enable mutual accountability. In the WASH sector, this role would ideally be taken up by government actors, which is in line with what Sanitation and Water for All terms 'Collaborative Behaviours' for the sector, which includes 'Enhance government leadership of sector planning processes'.

CONCLUSIONS

This study examined multi-stakeholder platforms and mutual accountability in the WASH sector across a series of cases in Bangladesh, Kenya, Indonesia, Peru and Somalia. While many of the case study countries have existing multi-stakeholder platforms in the WASH sector, these typically function as coordination and communication platforms rather than as platforms for mutual accountability. There were no mutual accountability mechanisms being widely used in the WASH sector that involve a full accountability cycle, and this is an emerging area for further research and policy engagement.

Improving multi-stakeholder platforms and processes is a clear next step to lay the foundations for mutual accountability. A national-level focus is needed to integrate context-specific mutual accountability mechanisms into existing or strengthened multi-stakeholder platforms. This includes ensuring a wide range of stakeholders are represented, recognizing and addressing power imbalances where possible and putting in place well-resourced government coordinated processes. Ensuring more actors have an equitable seat at the table can contribute to the good governance needed to strengthen WASH systems and achieve SDG 6 targets for water and sanitation. To support this, further research to examine causal pathways and connections between effective multi-stakeholder platforms and improved service delivery is needed to operationalize mutual accountability more widely.

DATA AVAILABILITY STATEMENT

Data cannot be made publicly available; readers should contact the corresponding author for details.

REFERENCES

Brinkerhoff, J. M. 2007 Partnership as a means to good governance: towards an evaluation framework. In: P. Glasbergen, F. Biermann & A. P. J. (eds). *Partnerships, Governance and Sustainable Development: Reflections on Theory and Practice.* Edward Elgar, Cheltenham, p. 68.

Brouwer, H., Hiemstra, W., van Vugt, S. & Walters, H. 2013 Analysing stakeholder power dynamics in multi-stakeholder processes: insights of practice from Africa and Asia. *Knowledge Management for Development Journal* **9** (3), 11–31.

Butterworth, J., Uytewaal, E., Rozenburg, E., Adjagodo, A., Al Afghani, M. M., Chowdhury, S. N., Dickin, S., Gaddam, S. R., Iman, A. M. S., Jimale, L. H., Liera, C., Njoroge, G., Qowamuna, N., Sanchez, Z., Syed, A. M., Willetts, J., Winterford, K. &

Zongo, B. 2021 *Strengthening Mutual Accountability in Partnerships for WASH: Part 2 – Summary of Six Country Case Studies*. Available from: https://www.sanitationandwaterforall.org/sites/default/files/2021-06/FINALSWAActivity2reportfinal.pdf.

Droop, J., Isenman, P. & Mlalazi, B. 2008 *Paris Declaration on aid Effectiveness: Study of Existing Mechanisms to Promote Mutual Accountability (MA) Between Donors and Partner Countries at the International Level: A Study Report*. Oxford Policy Management, London, UK, p. 35.

Fox, J. A. 2015 Social accountability: what does the evidence really say? *World Development* **72**, 346–361. https://doi.org/10.1016/j.worlddev.2015.03.011.

Gensch, R. & Tillett, W. 2019 *Strengthening Sanitation and Hygiene in the WASH Systems Conceptual Framework*. Available from: https://app.germantoilet.org/media/pages/politik/politik-ssi/diskussionspapier-zur-starkung-der-sanitarversorgung-und-hygiene-im-wash-systemansatz/2421490622-1579528646/ssi-discussion-paper-sanitation-hygiene-fv.pdf.

Hailu, R. & Tolossa, D. 2020 Multi-stakeholder platforms: institutional options to achieve water security in the awash basin of Ethiopia. *World Development Perspectives* **19**, 100213. https://doi.org/10.1016/j.wdp.2020.100213.

Heller, L. 2018 *The Principle of Accountability in the Context of the Human Rights to Safe Drinking Water and Sanitation: Report*. Available from: https://undocs.org/A/73/162.

Hepworth, N. D., Brown, B. D. & Brewer, T. 2020 *Accountability and Advocacy Interventions Witness, in the Water Sector: A Review of Global Evidence. Part 1. Overview and Summary Results*.

Hermans, F., Sartas, M., van Schagen, B., van Asten, P. & Schut, M. 2017 Social network analysis of multi-stakeholder platforms in agricultural research for development: opportunities and constraints for innovation and scaling. *PLoS ONE* **12** (2), e0169634. https://doi.org/10.1371/journal.pone.0169634.

Hofstetter, M., Bolding, A. & van Koppen, B. 2020 Addressing failed water infrastructure delivery through increased accountability and end-user agency: the case of the sekhukhune district, South Africa. *Water Alternatives* **13** (3), 843–863.

Hollander, D., Ajroud, B., Thomas, E., Peabody, S., Jordan, E., Javernick-Will, A. & Linden, K. 2020 Monitoring methods for systems-strengthening activities toward sustainable water and sanitation services in low-income settings. *Sustainability* **12** (17), 7044.

Huston, A. & Moriarty, P. 2018 *Building Strong WASH Systems for the SDGs: Understanding the WASH System and its Building Blocks*. Available from: https://www.ircwash.org/resources/understanding-wash-system-and-its-building-blocks.

Jiménez, A., Livsey, J., Åhlén, I., Scharp, C. & Takane, M. 2018 Global assessment of accountability in water and sanitation services using GLAAS data. *Water Alternatives* **11** (2), 238.

Keatman, T. 2012 *Silent Partner: the Role of the Environment in Multi-Stakeholder Partnerships for Urban and Peri-Urban Water and Sanitation Supply*. BPD Water & Sanitation, London, GB.

Kusters, K., Buck, L., de Graaf, M., Minang, P., van Oosten, C. & Zagt, R. 2018 Participatory planning, monitoring and evaluation of multi-stakeholder platforms in integrated landscape initiatives. *Environmental Management* **62** (1 , 170–181. https://doi.org/10.1007/s00267-017-0847-y.

MacDonald, A., Clarke, A., Huang, L., Roseland, M. & Seitanidi, M. M. 2018 Multi-stakeholder partnerships (SDG #17) as a means of achieving sustainable communities and cities (SDG #11) BT. In: *Handbook of Sustainability Science and Research* (Leal Filho, W. eds). Springer International Publishing, Cham, pp. 193–209. https://doi.org/10.1007/978-3-319-63007-6_12.

Mainwaring, S. & Welna, C. 2003 *Democratic Accountability in Latin America*. OUP, Oxford, UK.

Maltais, A., Weitz, N. & Persson, Å. 2018 *SDG 17: Partnerships for the Goals. A Review of Research Needs*.

Mannan, F. 2018 Are organizations accountable? Disconnect between gender and sanitation in Bangladesh. *Environment and Urbanization ASIA* **9** (1), 101–106. https://doi.org/10.1177/0975425317748534.

Ministerio de Vivienda, Construcción y Saneamiento (MVCS). 2018. DATASS - Sistema de Diagnóstico sobre Abastecimiento de Agua y Saneamiento en el Ámbito Rural. Retrieved June 2021, from https://datass.vivienda.gob.pe/.

Ñope Cueva, M. 2019 *Planeamiento Estratégico para Organizaciones Comunitarias de Agua Potable en Nuevas Zonas Periurbanas sin Servicio Público Caso Pachacútec*.

Schaaf, R. 2015 The rhetoric and reality of partnerships for international development. *Geography Compass* **9** (2), 68–80. https://doi.org/https://doi.org/10.1111/gec3.12198.

Schedler, A. 1999 Conceptualizing accountability. In: A. Schedler, L. Diamond & M. F. Plattner, eds. *The Self-Restraining State: Power and Accountability in New Democracies*. Lynne Rienner Publishers, Boulder, CO, pp. 13–29.

Sigalla, O. Z., Tumbo, M. & Joseph, J. 2021 Multi-stakeholder platform in water resources management: a critical analysis of stakeholders' participation for sustainable water resources. *Sustainability*. https://doi.org/10.3390/su13169260.

Steer, L., Wathne, C. & Driscoll, R. 2009 *Mutual Accountability at the Country Level: A Concept and Emerging Good Practice Paper*. Centre for Aid and Public Expenditure. ODI, London.

Stewart, A. & Gray, T. 2006 The authenticity of 'type two'multistakeholder partnerships for water and sanitation in Africa: when is a stakeholder a partner? *Environmental Politics* **15** (03), 362–378.

Ulimwengu, J. M., Matchaya, G., Makombe, T. & Oehmke, J. F. 2020 Chapter 15, Mutual accountability in African agricultural transformation. In: D. Resnick, X. Diao, & G. Tadesse, eds. 2020 Annual Trends and Outlook Report: Sustaining Africa's Agrifood System Transformation: The Role of Public Policies. pp. 182–194. International Food Policy Research Institute (IFPRI) and AKADEMIYA2063, Washington, DC, and Kigali.

UNDP-SIWI Water Governance Facility & UNICEF 2015 *WASH and Accountability: Explaining the Concept*. Available from: https://www.watergovernance.org/resources/accountability-in-wash-explaining-the-concept/.

Valcourt, N., Javernick-Will, A., Walters, J. & Linden, K. 2020a System approaches to water, sanitation, and hygiene: a systematic literature review. *International Journal of Environmental Research and Public Health*. https://doi.org/10.3390/ijerph17030702.

Valcourt, N., Walters, J., Javernick-Will, A., Linden, K. & Hailegiorgis, B. 2020b Understanding rural water services as a complex system: an assessment of key factors as potential leverage points for improved service sustainability. *Sustainability* **12** (3), 1243.

Warner, J. 2006 More sustainable participation? Multi-stakeholder platforms for integrated catchment management. *Water Resources Development* **22** (1), 15–35.

Warner, J. 2007 *Multi-stakeholder Platforms for Integrated Water Management*. Ashgate Publishing, Farnham, UK.

WHO and UNICEF 2021 *Progress on Household Drinking Water, Sanitation and Hygiene 2000–2020: Five Years Into the SDGs*. Available from: https://washdata.org/sites/default/files/2021-06/jmp-2021-wash-households-LAUNCH-VERSION.pdf.

Willetts, J., Winterford, K., Liera, C. & Dickin, S. 2020 *Strengthening Mutual Accountability in Partnerships for WASH: Part 1 – Literature Review and Learning From Other Sectors*. Available from: https://www.sanitationandwaterforall.org/sites/default/files/2021-06/SWAStrengtheningMutualAccountabilityinPartnershipsforWASH_Part1_April2021pdf.

First received 30 January 2021; accepted in revised form 23 March 2022. Available online 5 April 2022

© 2022 The Authors
doi: 10.2166/h2oj.2022.076

Jakarta water supply provision strategy based on supply and demand analysis

N. Ardhianie [a,*], D. Daniel[b], P. Purwanto[a] and K. Kismartini[c]

[a] Doctorate Program of Environmental Science, School of Postgraduate Studies, Universitas Diponegoro, Semarang 50275, Indonesia
[b] Department of Health Behaviour, Environment, and Social Medicine, Gadjah Mada University, Yogyakarta, Indonesia
[c] Department of Public Administration, Universitas Diponegoro, Semarang 50275, Indonesia
*Corresponding author. E-mail: nilaardhianie@gmail.com; nilaardhianie@students.undip.ac.id

NA, 0000-0002-9573-794X

ABSTRACT

Robust water service delivery systems require an understanding of resource dynamics as a basis for planning sustainable solutions for Jakarta, the capital with challenging water service provision performance. A significant gap needed to be closed with a service coverage of 64.06% and non-revenue for water at 45.06%. However, there is no detailed information on Jakarta's water supply and demand, which is essential to understand the resources situation. This study estimates Jakarta's total water supply and demand for 5 years. Furthermore, we analyze the natural and socio-economic factors shaping the current supply and demand using thematic analysis based on the DPSIR (Driver, Pressure, Status, Impact, and Response) framework. This paper contributes to two building blocks in the system strengthening approach to water and sanitation, i.e., (1) the sector policy strategy and (2) planning, monitoring, and review. Key drivers, pressures, status, impacts, and responses were critical to informing the system-wide understanding of the problem and suggesting evidence-informed explanations for a better water supply provision strategy in Jakarta.

Key words: Jakarta, supply and demand, urban water planning, water provision

HIGHLIGHTS

- Jakarta does not have a thorough supply and demand analysis available for the public.
- DPSIR (Driver, Pressure, Status, Impact, and Response) analysis to show the ways in which water and society are interlinked.

INTRODUCTION

The United Nations General Assembly explicitly recognized the human right to water and sanitation. It acknowledged that clean drinking water and sanitation are essential to realizing all human rights through Resolution 64/292 in 2010 (The Human Rights to Water & Sanitation 2010). In November 2002, the Committee on Economic, Social, and Cultural Rights of Indonesia adopted General Comment No. 15 that defined the right to water as the right of everyone to sufficient, safe, acceptable, and physically accessible and affordable water for personal and domestic uses (United Nations 2002). However, many households in Jakarta, the capital of Indonesia, do not have access to piped water or sewer networks (Figure 1).

Jakarta, a delta on the North Coast of Java Island, faces an extraordinarily complex water supply challenge. In 2020, the megapolitan with a population of 11 million required roughly 1.2 billion cubic meters of water, and the piped water system can support less than half of it. Most of the population relies on groundwater and uses it without any treatment for their daily needs, except for drinking, where people usually boil it before drinking. The water quality test by the Environment Agency in five areas in 2018 showed that 64.6% of the samples were contaminated with detergent and *Escherichia coli* above the quality standard threshold (Dinas Lingkungan Hidup Jakarta 2018), all samples from rivers in Jakarta are polluted, 1% light polluted, 20% moderate, and 79% highly polluted (Dinas Lingkungan Hidup Jakarta 2021) which is also supported by other studies (Irawan *et al.* 2015; Padawangi & Douglass 2015; Kooy & Walter 2019).

One of the reasons for high-level groundwater contamination occurs because the current sewer network only covers 4% of the city (KPPIP 2019). All Jakarta's surface water is heavily polluted by greywater from households, commercial buildings and discharges from industries, chemicals from agricultural soil, solid waste, and fecal

Figure 1 | Location of Jakarta in Indonesia's map.

matter from overflowing or leaking septic tanks (Apip *et al.* 2015). Another study revealed a complex mixture of organic contaminants comprising more than 100 structurally heterogenic compounds was detected in a river affected by discharges from a large industrial manufacturing center and municipal sources. It is estimated that 5–17 tons of pollution from municipal sources are carried by just one urban river, the Ciliwung, into the Jakarta Bay each year. These various contaminants might seep and contaminate the groundwater (Dsikowitzky *et al.* 2017).

The heavy reliance on groundwater also poses serious environmental risks such as land subsidence, damage to buildings, increased risk of flooding, and seawater intrusion (Neise 2019). Jakarta sinks about 1–15 cm a year, while some northern parts sink by as much as 28 cm a year (Abidin *et al.* 2011). Thus, Jakarta is the fastest sinking city in the world (BBC 2019).

The provision of clean water via piped water in Jakarta started long ago. The practice of providing water through a centralized system dated to the colonial period when Jakarta was still called Batavia, and the government established the Water Leidinggen Bedriff van Batavia (local water supply enterprise) in 1918 (Kooy 2014). The era of modern management began in 1957, marked by the operation of the Pejompongan I Water Treatment Plant, which treated water from the Jatiluhur Dam and flowed through an open channel of 70 km (Nasution 2016). Since 1997, the piped water system has been managed under a concession agreement between PAM Jaya, the water utility owned by Jakarta Provincial Government and two private operators (PAMJAYA 2020). Since 1997, the piped water system has been managed under a concession agreement between PAM Jaya, the water utility owned by Jakarta Provincial Government and two private operators (Nasution 2016).

To provide piped water access to all residents, the authorities need to develop a comprehensive master plan or strategic planning based on a detailed and accurate analysis of demand and supply for the megapolitan. Equal access to water and sanitation services requires a political choice on the part of the State, coupled with well-considered planning, budgeting, and financing, not just for capital cost but also for operation and long-term capital maintenance (De Albuquerque & Roaf 2012). An instruction by the Regional Secretary in 2021 stated that the process should start in 2021 (INSEKDA NO 1/2021 2021). The planning should be based on analysis of the availability and the needs so that the imbalance between water demand and supply can be identified, and solutions to close that gap can be conducted. Water balance across various space and time scales is needed to facilitate water management planning and ultimately ensure human and environmental health (Vörösmarty *et al.* 2000). This need is critical in a region with rapid population growth and where the sustainability of water supply is uncertain (Loucks & Van Beek 2017).

Soentoro (2011) analyzed the raw water problem in Jakarta using DPSIR (Driver, Pressure, Status, Impact, and Response) and concluded that the rapid growth of urban and industrial demand could not be fulfilled due to limited local raw water and inadequate funding to build new regional raw water infrastructure. This leads to finding the most accessible solution: to use groundwater, including issuing permits from the government to use groundwater. However, after several decades, the policy has depleted the aquifer, and to overcome the problem, an integrated approach should be applied. Still, its success depends on the political will of the government.

The complexity of the problem Jakarta currently faced, i.e., underinvestment, limited involvement of government, especially in terms of technical performance, resource management and despite Indonesia having a regulation on how to develop strategic planning or a master plan, showed a lack of system approach in the city water delivery system. Research by WaterAid in four countries showed a similar result to Jakarta. When monitoring was a weak area, it accentuated deficiencies in planning and financing, weak coordination, unclear institutional arrangements, insufficient resource allocation, and lack of prioritization of resources (WaterAid 2020). Given the characteristics of the water and sanitation system that are multiple connections, dependencies and relations among different actors and factors and the interconnected nature of financial, institutional, environmental, technological, and social aspects that influence sustained service delivery (Valcourt *et al.* 2020), a system approach is proposed to facilitate better service delivery for Jakarta.

System strengthening can be used to understand the complex and interrelated social, political, environmental, institutional, and technical factors like Jakarta. System strengthening is increasingly being adopted by water and sanitation practitioners (Neely 2019). This study aims to focus on some elements as part of a robust water service delivery system by answering two research questions: (1) what the current status/pattern of water supply and demand in Jakarta is, and (2) what and how different aspects/elements influence Jakarta's current supply and demand status/pattern. We use an equation developed based on previous studies to estimate the demand and supply. The estimation covers local and imported water. To analyze interconnected aspects influencing the water supply and demand status/pattern, we utilized the DPSIR framework. The results were applied to develop suggestions for water planning in Jakarta.

METHODS

The data were collected from the desktop study (statistics and archives), interviews with 10 key resource persons, and two focus group discussions in 2019 and early 2020; all respondents and participants of the discussion are working with Jakarta water-related agencies. Prior to the interview, a set of general questions relating to sources of water for Jakarta and its volume, how water comes to the city, institutional arrangement, master plan for the city water provision, and the problem with the city water system provision were prepared. The interviews were recorded, and a step-by-step thematic analysis was then performed based on DPSIR classifications. Thematic analysis is a method for identifying, analyzing, organizing, describing, and reporting themes found within a data set (Braun & Clarke 2008).

To estimate the demand and supply in Jakarta, approaches by Ahmadi (2019), Sharvelle *et al.* (2017), and Li *et al.* (2019) were used to develop an equation that fits with the availability of data in Jakarta. Ahmadi (2019) calculated cities' utility net water demand by quantifying the urban population and average water per capita demand and gross water demand by dividing the net water by 100 and subtracting with the physical water loss percentage. The supply is quantified by identifying the nature of water supply sources (i.e., groundwater and surface water), and the percentage of water from these resources for each city was calculated. Sharvelle *et al.* (2017) estimated detailed household water use both indoor and outdoor. The indoor residential water use is estimated using demand profiles that relate the household size to average daily indoor water use and household units. Li *et al.* (2019) define domestic water demand as the total population multiplied with domestic water consumption per capita and surface water supply as the sum of groundwater supply, wastewater reuse, and rainwater utilization. Due to the limited data available, we explained the equation for Jakarta as follows:

$$\begin{aligned}&\text{Domestic demand} + \text{Commercial demand} + \text{Agriculture demand}\\&\quad = \text{Local groundwater} + \text{Local surface water} + \text{Import surface water}\end{aligned} \tag{1}$$

In this equation, demand consists of domestic demand, equivalent to the total amount of water consumption based on per capita average water use: commercial demand or the total amount of non-domestic water use; and agricultural use. Supply is obtained from local groundwater (the total amount of underground water, the shallow and deep groundwater use), local surface water indicating the total amount of water used from rivers in Jakarta, Import surface water shows the total amount of water diverted from outside Jakarta and the desalination water through a Sea Water Reverse Osmosis (SWRO) process.

We estimated demand and supply for five consecutive years from 2016 to 2020; each represents the primary input and output of water volume relative to one-year data. Total input from piped water is billed meter or water consumed by the customer and does not count the non-revenue water (NRW). The calculation is equivalent to the estimate for groundwater use. Domestic water demand is calculated by multiplying the number of populations with 175 liters/person/day for residents and 120 liters/person/day for commuters. The estimation for commercial demand is based on PAM Jaya's 21 years of historical data on average domestic and commercial consumption. Water for agriculture is calculated based on wetland use data from the Ministry of Agriculture (Kementrian Pertanian 2020) and multiple with average water need for cultivating land (Juliardi & Ruskandar 2006). The bulk water supply for piped water is taken from PAM Jaya's data. The groundwater supply is the total water needs deducted from the total supply for piped water consumption.

The DPSIR framework is used to understand the cause and effect relations contributing to the current supply and demand pattern. DPSIR is a simple qualitative causal framework for understanding and describing people's interactions with the environment, the impact of human activities on the environment and vice versa (Pahl-Wostl 2015; Kapetas *et al.* 2019). The DPSIR framework helps describe the relationship between environmental problems' origins and consequences and understand their dynamics (Kristensen 2004). The framework has been used frequently in water issues and demonstrated to better understand the interaction between water and people. DPSIR provides a practical analytical framework to support problem scoping and structuring. DPSIR also provides the starting point for hypothesizing the relationship among the system components, bridges the gaps between ecological, social science, and economic disciplines, and represents relationships between humans and the environment (Zare *et al.* 2019; Zhao *et al.* 2021). However, DPSIR also has some criticisms, including how the indicators are static and cannot consider the dynamics of the system under discussion and cannot offer a clear cause–effect association of a complex system (Song & Frostell 2012).

Drivers are societal, economic, and population forces that cause pressure on the system; *Pressures* are direct stresses that affect the system states; *States* denote the system's condition; *Impact* reflects observable changes in system conditions; and *Responses* are the institutional responses to change the system (Sun *et al.* 2016; Zare *et al.* 2019). It must be applied to a system with boundaries, inputs, and outputs (Akmalah 2010).

Studies have employed DPSIR to analyze Jakarta's water, but none use it to analyze the supply and demand. Lubis (2018) examines the issues and problems of urban hydrogeology in Jakarta and makes suggestions on improving the groundwater management strategy. Jensen & Khalis (2020) used PSIR (Pressures, State, Impact, and Responses) to develop indicators for urban water security, Akmalah (2010) evaluated the flood issues in Jakarta and described the causes and relationships related to the flood, and Soentoro (2011) evaluated the cause–effect for the raw water problem.

We use the DPSIR framework to understand the contextual understanding of cause and effect association, identifying the key pressure that 'if address' would offer the highest benefit and recognize the gaps in control management or responses. Elements of DPSIR were acquired from interviews and focus group discussions and put under categorization based on category definition (Sun *et al.* 2016; Zare *et al.* 2019) and the ways in which they are interlinked. We identified an initial set of 42 DPSIR, and later focused on 18 statements. We put aspects related to the social, demographic, and economic developments in societies under the *Drivers* categorization; activities affecting the socio-ecological system are classified as *Pressures*; state of water service provision and state of water resources as *States*; cascading social, environmental, or economic changes as *Impact*; and institutional policies and programs to reduce and control the problem as *Responses*, all are validated by literature reviews and our demand and supply analysis.

RESULTS AND DISCUSSION

Supply and demand analysis

Jakarta relied on multiple sources (supply): groundwater, surface water (local and imported), and seawater (Pradafitri *et al.* 2018; Pratiwi & Herdiansyah 2018; Jensen & Khalis 2020). We used the data from Jakarta water utility as a baseline to calculate the supply for five consecutive years (BPS-Statistic of DKI Jakarta Province 2017, 2018, 2019, 2020, 2021). This study categorizes supply into three groups: surface water consisting of both local and imported water, groundwater, and seawater. Water use (demand) consists of domestic, commercial, and agricultural water. To estimate the domestic water use, we multiplied the population by the standard of daily water use of 175 liters per day and 120 liters per day for commuters. We used water utility historical data to calculate the commercial demand.

Yearly total water demand from 2016 to 2020 did not change significantly; the demand is around 1.2 billion cubic meters per year (Table 1). The increase in demand from year to year occurred with a relatively small volume. Average domestic demand leads the number with 61.5%, followed by commercial demand of 35.5%. Domestic demand is between 61 and 62% of total demand, with a volume of 748 million cubic meters in 2016 and 769 million cubic meters in 2020. Commercial water need is around 440 billion cubic meters or 35–36%

Table 1 | Jakarta water supply and demand 2016–2020 (m^3)

	2016	2017	2018	2019	2020
Demand (m^3)					
Residents	656.226.548	668.356.260	668.358.112	674.116.169	674.389.319
Commuters	92.498.652	94.208.400	94.208.661	95.020.290	95.058.792
Domestic – D	748.725.200	762.564.660	762.566.773	769.136.459	769.448.111
Commercial – C	434.260.616	442.287.503	442.288.728	446.099.146	446.279.904
Agriculture – A	37.918.686	29.782.441	26.111.670	23.576.189	35.345.362
Total Demand D+C+A	**1.220.904.502**	**1.234.634.604**	**1.230.967.171**	**1.238.811.793**	**1.251.073.377**
Supply (m^3)					
Piped/surface water – P	341.601.198	352.007.000	362.371.000	372.189.000	342.401.394
Groundwater – G	879.303.304	882.548.764	868.517.331	866.086.680	908.135.870
SWRS – S	–	78.840	78.840	536.113	536.113
Total Supply P+G+S	**1.220.904.502**	**1.234.634.604**	**1.230.967.171**	**1.238.811.793**	**1.251.073.377**

of total yearly demand. Most of the supply is provided by groundwater, as much as 70–72.6%, while piped water using surface water sources contributed 28% in 2016, 28.51% in 2017, 29.43% in 2018, 30.04% in 2019, and 27.37% in 2020. SWRO, although small (less than 1%), started to contribute to supply in 2017 (Figure 2). The percentage of piped water supply is comparable with data from The Water Supply Development Supporting Agency which said that the piped water service coverage in Jakarta was 28.05% in 2018 (BPPSPAM 2019).

We also examined further the origin of the water to explain the composition between local water and water obtained from other areas. Data from PAM Jaya, the public water company, showed that they bought bulk water from Jatiluhur Dam in Purwakarta, West Java, and treated water from PDAM Tangerang in West Java daily. The bought surface water, i.e., imported water, accounted for around 96% of the whole piped water system. Dependence on other regions for most of the piped water supply showed that Jakarta's water security is weak and gradually reduced. Every day around 17.300 litres/second of bulk water is directly transferred daily from Jatiluhur Dam, transported using a 72 km open channel, and treated water from PDAM Kerta

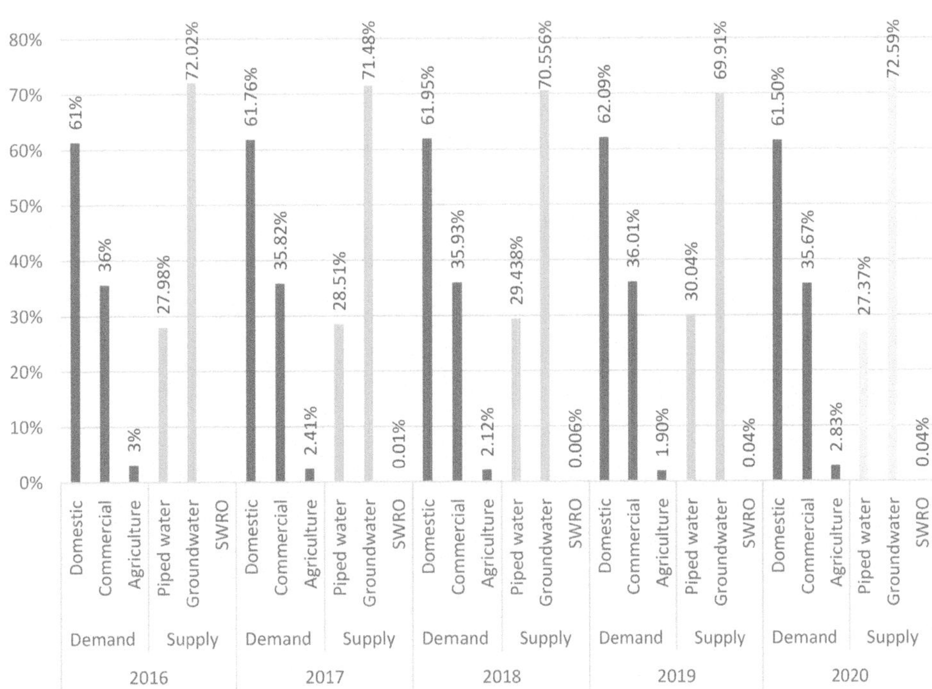

Figure 2 | Jakarta water supply and demand 2016–2020 (%).

Rajasa in Tangerang Regency using a 30-km pipeline. Only 4% of the water from Jakarta's rivers is utilized for piped water (PAMJAYA 2020). Compared with the total water supply for the whole city, the imported water is approximately 28%. However, this imported water is responsible for 96% supply for all piped water systems (Figure 3).

All imported water is treated through a complex infrastructure system consisting of seven water installation plants with varied capacities from 150 litres/second in Taman Kota to 4.000 litres/second in Pulo Gadung, some Distribution Central Reservoir (DCR), and many other smaller processing facilities. The entire system is divided into two subsystems: the east and west system; each is managed by different private operators under public water company supervision (PAMJAYA 2020). The imported water is essential to the piped water system in Jakarta, while groundwater is essential for self-supply practice by households and commercials. The high percentage dependency on groundwater is aligned with other countries, whereas nearly 85% of the Indian population uses groundwater, as well as sub-Saharian countries where groundwater is the primary drinking water source for between 70 and 90% of domestic water users (Grönwall & Danert 2020).

Groundwater comes from the Jakarta groundwater basin, an area of 1,439 km^2 that covers the entire DKI Jakarta Province, as well as some parts of the Banten and West Java province (Pemprov DKI 2021). People can use groundwater for daily needs without restriction, while its use for commercial purposes or large amounts is requires a permit and is subjected to a groundwater tax. While regulation has set gradually stricter requirements for permits and higher tariffs, a study reported that there are figures of 4,000 illegal deep wells in Jakarta, in addition to the registered users who are also under-reporting their actual consumption (Agustinus 2016).

Following (Samsuhadi 2009) calculation of groundwater potential capacity of 1,170–4,600 m^3/year for shallow groundwater, 62–230 m^3/year for deep groundwater, and the suggestion that when the utilization rate of water resources exceed 60%, the ecological environment will deteriorate (An *et al.* 2021), we understand that groundwater utilization in Jakarta is already at an alarming rate which is 76% of its potential capacity.

DPSIR analysis

This study identifies that increased demand for domestic and commercial activities, and population growth are the main *Drivers* that give escalation to *Pressure* (water pollution, inadequate/low investment in infrastructure, and privatization). The *Pressures* then resulted in a *State*, i.e., limited access to piped water and groundwater over abstraction. *The Impact* is water insecurity, falling groundwater table, land subsidence, water-related disease, and low productivity. *Responses* are demand management measures, desalination, supply from other places (trans-basin water transfer), wastewater treatment, law enforcement, and capacity building of institutions working on the water-related issue (Figure 4).

Regarding the *Drivers*, we argue that the increased demand for commercial needs is more significant than the population growth. The population in 2000 was 8.3 and 11 million in 2019, which increased approximately less than twice the population size in 2000. While the regional income tax received by the province rose 19 times. In 2000, the regional income of DKI Jakarta Province was IDR 2.1 trillion and rose to IDR 40.2 trillion in 2019 (BPS-Statistic of DKI Jakarta Province 2001, 2020). Regional income is a tax obtained from various economic activities, including motor vehicles, hotels, restaurants, billboards, parking, entertainment, and buildings.

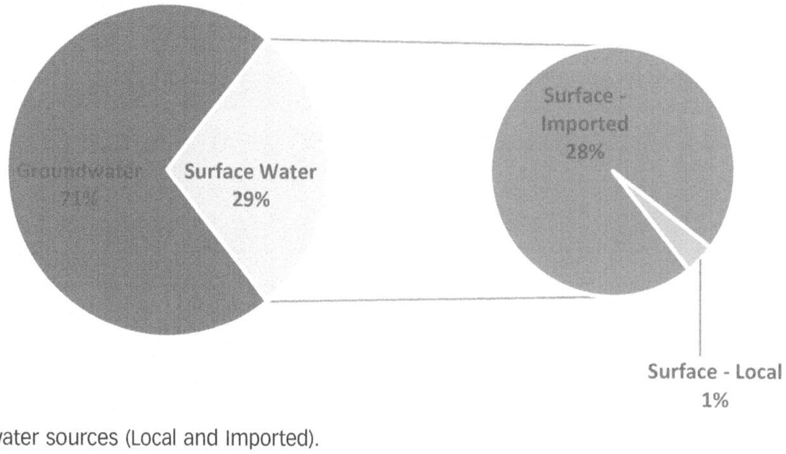

Figure 3 | Jakarta water sources (Local and Imported).

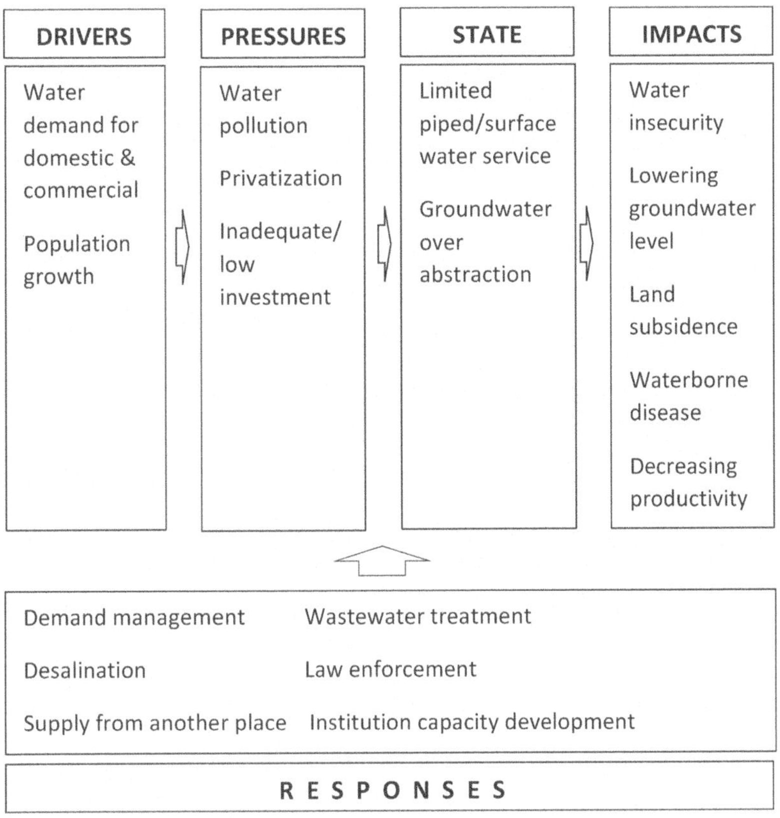

Figure 4 | DPSIR for Jakarta water supply and demand.

Data from the consecutive 10 years showed the same phenomenon. The three years with the most considerable rise were 2010, 2012, and 2019. The total regional tax increase in 2010 was 41.6%, 2012 was 31.9%, and 2019 with an increase of 24.5%, while the increases of population in those years were 1.16%, 1.09%, and 0.04% (Figure 5). The average rise of regional tax for 10 years is 17%, and the average population growth is 1%. The fact that the growth of the commercial demand is big showed two opportunities: (1) cross-subsidy tariff from commercial consumer to low-income families is available and (2) if the commercial consumers use water-saving technologies, the amount of water that can be saved is significant.

Unfortunately, the increasing water demand that goes hand in hand with the increasing economic activities is not supported with the piped water supply capacity. The Jakarta water treatment production capacity, volume

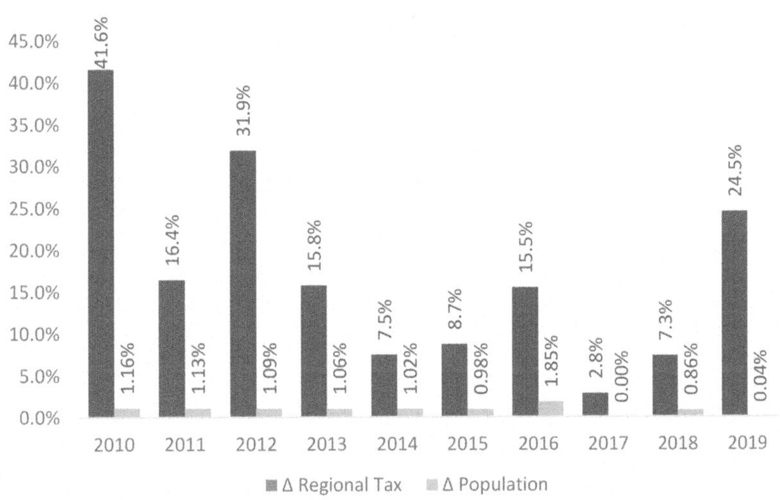

Figure 5 | Comparison between population growth and regional tax 2010–2019.

sold, and the number of piped water customers in Jakarta only increased by less than double its quantity from 2000 to 2019, whereas the business activities rose 19 times. The production capacity in 2000 was 465 million m^3 and increased 34% by 2019 into 622 million m^3. The volume sold in 2000 is 228 million m^3 and increased 59% to 362 million m^3 in 2019 (BPS-Statistic of DKI Jakarta Province 2020). Nasution (2016) showed that investment in the system decreased from USD 160 million in 1997 to around USD 30 million in 1998 when privatization started. Since then, the investment has continued to be in the same range and has even fallen in recent years.

This situation produced the *Status* of limited pipe water services and gave no alternative except to resort to groundwater as the business sector typically requires a significant amount of water (Neise 2019). This aligns with the supply and demand calculation above that showed groundwater supplied around 70% of Jakarta's total water needs. Since access to groundwater is quite easy with relatively low installation cost, especially for illegal wells, it has caused severe problems for Jakarta (Irawan *et al.* 2015); the excessive groundwater abstraction and imbalance in replenishment have led to a decrease in the groundwater level and indirectly led to land subsidence at an alarming rate, i.e., up to 20 cm per year in some locations in Jakarta (Delinom *et al.* 2015). Massive land conversion also adds more pressure as recharging water into the ground becomes more difficult.

All these *Pressures* have turned Jakarta into experiencing *Impacts* such as water insecurity (Danielaini *et al.* 2019), lower groundwater levels (Soentoro 2011), land subsidence (Deltares 2016), waterborne disease (Mungkasa 2017), and low productivity (Argo 1999). Different *Responses* to the corresponding *Drivers*, *Pressures*, and *Impact* are in place, but some of the *Responses* given by the government have not been adequate to solve the urgent water problem in Jakarta. Some *Responses*, such as fragmented management and frequent change of responsibilities within water-related agencies to some extent obfuscate Jakarta's efforts to solve its water problems.

A previous study indicated that the institutional aspects is one of the most critical aspects to sustain water services in Indonesia (Octavianti & Charles 2018; Daniel *et al.* 2021). Jensen & Khalis (2020) stated that, currently, governance of the water sector in Jakarta is complex and highly fragmented: water supply, water resources management, groundwater, wastewater, and flood management are all under the responsibility of different government agencies. Moreover, the institutional responsibility changes frequently as articulated by a focus group discussion's participant: 'Last year the management of groundwater was under the agency where I worked, recently it transferred to another agency. I was starting to understand how it works; now I have to learn new things again'.

Indonesia has decentralized the water supply provision to the local government. The fragmentation can also be seen from the weak law enforcement which leads to high rates of groundwater and piped water theft, the frequent changes of name, structure, and scope of work of these institutions. Another thing that reflects the suboptimal performance of water management institutions is the low number of piped water and wastewater coverage. An improvement has been identified though, as starting from 2020, the management of groundwater, surface, desalination, and flooding are under one agency i.e., Dinas Sumber Daya Air (Water Resource Agency).

A significant measure for the future of Jakarta is strategic planning or a master plan. The government aims to have the master plan as soon as possible, 'Other cities have developed the master plan and the government has made guidelines on the thing that must be included in the master plan. We will include the budget to develop a master plan in the next year's budget', as said by a senior government official in an interview.

Strategic planning is important to make fundamental changes. Planning includes analyzing the future as a different reality from the *status quo* by acknowledging that human beings and changes are inevitable; thus, fundamental changes are likely to happen. Planning aims to make a difference by preparing for, contributing to, and facilitating change (Hagen & Higdem 2020). Strategic planning can also be understood as a valuable tool to direct the city's overall development to address challenges and benefit from opportunities (Thi Bich Ngoc 2011). Urban water strategic planning should comprise a wastewater strategy, clean water planning, recycling strategies, and prioritizing local water supply (Najar & Persson 2019).

Key points for water planning in Jakarta

Traditionally, strategic planning has gained insignificant attention in the infrastructure sectors as competition in infrastructure is low or non-existent and captive customers or the public can accept the investment risks. On the other hand, there has been insufficient concern in the literature on how to improve the performance of public organizations. However, the urgency and interest in the need for adequate strategic planning have been growing in the field of infrastructure-based services, and several planning methods to inform strategy making have recently earned attention (Dominguez *et al.* 2009).

Sanitation and Water for All identified five building blocks needed for the well-functioning water and sanitation sector, i.e., (1) sector policy strategy, (2) institutional arrangement, (3) financing, (4) planning, monitoring, and review, and (5) capacity development (Sanitation & Water for All 2021). Strategic planning is associated with two building blocks: the sector policy strategy and planning, monitoring, and review. For planning, it is important to understand the resource situation, and demand and supply analysis is one solid foundation for it as the evaluation can provide information about who, when, from where and for what purposes the water in a specific area is utilized.

The above demand and supply calculations and the DPSIR analysis showed that due to limited piped water access, Jakarta's groundwater basin supplies most of the total water needs at the level that puts Jakarta at risk of experiencing critical land subsidence. Therefore, a program to reduce reliance on untreated groundwater use and control groundwater over-abstraction by developing new infrastructure and increasing raw water supply to increase household access to piped water is important.

For groundwater management in Jakarta, this study has three recommendations. First, Jakarta must prioritize the management of the groundwater basin by improving its quality, especially because many residents use the contaminated groundwater and are potentially exposed to contaminants, such as harmful chemicals and microorganisms daily (Luo *et al.* 2019) and programs to improve access to wastewater services need to be seriously developed and executed (PALJAYA 2017). Communities need to understand the benefits of processing wastewater and not dumping it directly into the water bodies. Jakarta already has Governor Decree No. 45/1992, which stipulates that every building in an area with a sewage pipe installed must dispose of its wastewater into the pipe through a parcel connection pipe. The provision of a piping system/wastewater management facility can reduce contamination of groundwater and surface water. Wastewater from grey and blackwater will flow through a piping system and be treated to meet the quality standards for disposal to the receiving water body, i.e., river. Some water can go further through a process to improve its quality in order to meet the standard for bulk piped water supply.

Second, increasing supply from surface water by decentralizing the supply system and utilizing water from the river that meets the health and environmental standard. Mini-plants can be built near the selected river to process and supply water to a specific area (Mungkasa 2017). Developing a smaller structure under the management of the current system does not need huge investment, so it is an urgent priority to help the residents that indeed need clean water. The benefits are more significant than the costs (Tsegaye *et al.* 2020).

Also, rainwater reservoirs made from light material complete with small water treatment can be built to provide services in certain areas. Light materials are needed to reduce soil loads and do not increase the pressure on land subsidence. Increasing the capacity of water treatment facilities should be developed through both decentralized and centralized systems while reducing the high NRW of the piped water system. It should be applied as it will help increase the supply significantly. The Jakarta non-revenue for water in 2020 is 45.7% or around 280 million m^3 (PAMJAYA 2021). This is above the average of NRW in Indonesia, i.e., 37% (El-Ahmady & Sembiring 2014). If the leakage can be reduced to half of the current figure, there will be approximately 140 million m^3 of water for daily consumption. However, it is important to calculate the economic level of water losses to make sure that the strategies to reduce the NRW are cost-efficient (Heryanto *et al.* 2021).

Third, measures to control the groundwater use by big industries especially those located in industrial parks are important as a large amount of deep groundwater withdrawal can lead to shallow groundwater decrease due to the lowering water table. This situation has implications for the equity of access as shallow groundwater is used for people's daily consumption. Incentives in the form of tax reduction can be given to the companies willing to report their illegal wells and change to water-saving equipment or technology (Amundsen *et al.* 2014; Montginoul *et al.* 2016). At the same time, disincentive, i.e., additional charges can be applied to the companies using groundwater in areas with piped water connections.

Other management measures targeted to surface water are needed through encouraging consumer demand for water-saving technologies at commercial, government buildings, and at household levels as well as system maintenance and renovation. A study in four European cities found that investments in renovation and maintenance of networks are the most important measures to conserve water, followed by regulations, individual meters, and public campaigns promoting the use of water-saving technologies and water conservation practices (Biswas & Tortajada 2019). All measures need to be combined with an affordable tariff scheme for every stratum of society which also should be strengthened by implementing a policy that provides basic free water for daily needs (minimum standard of water supply) targeted to the unprivileged. This subsidy policy means that even if a customer is not capable of paying the tariff due to lack of means, they should not be disconnected from a minimum amount of basic water service, which is to ensure the right to water of all people is protected. Disconnections due to nonpayment are only acceptable if it can be shown that the beneficiaries do not pay the water fee when they are able to pay.

In early 2023, the clean water supply and service improvement for the western and eastern part of Jakarta between PAM Jaya and PT Palyja and PT Aetra covering all Jakarta areas except Seribu Island will end; strategic planning following the new setting should be developed soon. The new institutional setting, capacity building, and new governance framework are required to manage a system that can provide safe and affordable water for all residents. The water resource understanding provided by this study can be used as a foundation for planning the strengthening of the Jakarta water system to achieve equitable access for all.

Study limitation

We used a simple approach to estimate the water balance. Further research is needed including research that covers climatological data and water needs for the environment to better estimate the water balance. Furthermore, the results concluded from this study are in the early stages yet are important to assist the planning process for clean water supply in the capital city.

CONCLUSION

Jakarta is experiencing rapid economic and social growth with limited water resources and infrastructure. Moreover, there is no comprehensive data on the water demand and supply in Jakarta which is challenging for proper water planning. This study estimated that the yearly total water demand for Jakarta in 2016–2020 was around 1.2 billion m^3. Furthermore, Jakarta's local water resources account for about 73% of the final water demand, and neighboring provinces support the remaining 27%. However, this imported water is responsible for more than 90% of raw water for the piped water system, indicating that the sustainability of water services in Jakarta depends on the areas tens of kilometers away from the city. The dependence on groundwater is already at 76% of its potential capacity, causing land subsidence at a hazardous level, decreasing the groundwater table and risking Jakarta's water supply. The DPSIR framework reveals that the water scarcity problem results in other problems, such as uneven water access and waterborne disease. There are some responses from the government of Jakarta, but the implementations are still not sufficient to eliminate water problems. This study also underlines some factors related to institutional aspects that should be improved in Jakarta. Finally, we also provided some recommendations to improve water management in Jakarta.

DATA AVAILABILITY STATEMENT

Data cannot be made publicly available; readers should contact the corresponding author for details.

REFERENCES

Abidin, H. Z., Andreas, H., Gumilar, I., Fukuda, Y., Pohan, Y. E. & Deguchi, T. 2011 Land subsidence of Jakarta (Indonesia) and its relation with urban development. *Natural Hazards* **59** (3), 1753–1771. https://doi.org/10.1007/s11069-011-9866-9.

Agustinus, R. B. 2016 *Understanding the Implementation of Groundwater Regulations and Policies for the Commercial Deep Well Users in Jakarta, Indonesia.* UNESCO-IHE Institute for Water Education, Delft.

Ahmadi, M. S. 2019 *Assessment of Water Demand Changes in Several Rapidly Growing Global Cities.* IHE Delft Institute for Water Education, Delft, the Netherlands.

Akmalah, E. 2010 Integrated flood management model: a socio-technical systems approach to overcome institutional problems in Jakarta. In *Itinerario*, Vol. 5. Colorado State University. https://doi.org/10.1017/S0165115300023299.

Amundsen, E. S., Jensen, F. & Hansen, L. G. 2014 Regulating groundwater use in developing countries: a feasible instrument for public intervention. *Journal of Institutional and Theoretical Economics* **170** (2), 317. https://doi.org/10.1628/093245614X13867465880983.

An, M., Fan, L., Huang, J., Yang, W., Wu, H., Wang, X. & Khanal, R. 2021 The gap of water supply-demand and its driving factors: from water footprint view in Huaihe River Basin. *PLoS ONE* **16**, 1–16. https://doi.org/10.1371/journal.pone.0247604.

Apip, Sagala, S. A. & Pingping, L. 2015 Overview of Jakarta water-related environmental challenges. Water and Urban Initiative Working Paper Series 4, pp. 1–5.

Argo, T. A. 1999 Thirsty Downstream: *The* Provision of Clean Water in Jakarta, Indonesia. University of British Columbia.

BBC 2019 Jakarta, the fastest-sinking city in the world. *BBC News.* Available from: https://www.bbc.com/news/world-asia-44636934

Biswas, A. K. & Tortajada, C. 2019 Water crisis and water wars: myths and realities. *International Journal of Water Resources Development* **35** (5), 727–731. https://doi.org/10.1080/07900627.2019.1636502.

BPPSPAM 2019 *Buku Kinerja PDAM 2019 Wilayah II*, Vol. 1. BPPSPAM, Jakarta.

BPS-Statistic of DKI Jakarta Province 2001 *Jakarta Dalam Angka 2001.*

BPS-Statistic of DKI Jakarta Province 2017 *Jakarta Dalam Angka 2017.*

BPS-Statistic of DKI Jakarta Province 2018 *Jakarta Dalam Angka 2018.*

BPS-Statistic of DKI Jakarta Province 2019 *Provinsi DKI Jakarta Dalam Angka 2019.*

BPS-Statistic of DKI Jakarta Province 2020 *Provinsi DKI Jakarta Dalam Angka 2020.*

BPS-Statistic of DKI Jakarta Province 2021 *Provinsi DKI Jakarta Dalam Angka 2021.*

Braun, V. & Clarke, V. 2008 Using thematic analysis in psychology, qualitative research in psychology. *Journal of Chemical Information and Modeling* **3** (2), 77–101. http://dx.doi.org/10.1191/1478088706qp063oa.

Daniel, D., Djohan, D. & Nastiti, A. 2021 Interaction of factors influencing the sustainability of water, sanitation, and hygiene (WASH) services in rural Indonesia: evidence from small surveys of wash-related stakeholders in Indonesia. *Water (Switzerland)* **13** (3), 1–11. https://doi.org/10.3390/w13030314.

Danielaini, T. T., Maheshwari, B. & Hagare, D. 2019 An assessment of household water insecurity in a rapidly developing coastal metropolitan region of Indonesia. *Sustainable Cities and Society* **46**, 101382. https://doi.org/10.1016/j.scs.2018.12.010.

De Albuquerque, C. & Roaf, V. 2012 *On the Right Track. Good Practices in Realising the Rights to Water and Sanitation.* Available from: http://www.ohchr.org/Documents/Issues/Water/BookonGoodPractices_en.pdf

Delinom, R. M., Lubis, R. F., Bakti, H. & Imron, M. 2015 In: *Ancaman Bawah Permukaan Jakarta* (Delinom, R. M. ed.). LIPI Press, Jakarta.

Deltares. 2016 *Sinking Jakarta Causes & Remedies (2016).*

Dinas Lingkungan Hidup Jakarta. 2018 *Air Tanah di Jakarta Sudah Tercemar Detergen Dan E. coli.* Available from: http://www.pdamtirtabenteng.co.id/berita/air-tanah-di-jakarta-sudah-tercemar-detergen-dan-ecoli

Dinas Lingkungan Hidup Jakarta. 2021 Laporan Akhir Pemantauan Kualitas Lingkungan Air Tanah Dki Jakarta Tahun Anggaran 2020. In *Dinas Lingkungan Hidup* (Issue 6).

Dominguez, D., Worch, H., Markard, J. & Guier, W. 2009 Closing the capability gap: strategic planning for the infrastructure sector. *California Management Review University of California Berkeley* **51** (2).

Dsikowitzky, L., Hagemann, L., Dwiyitno, Ariyani, F., Irianto, H. E. & Schwarzbauer, J. 2017 Complex organic pollutant mixtures originating from industrial and municipal emissions in surface waters of the megacity Jakarta – an example of a water pollution problem in emerging economies. *Environmental Science and Pollution Research* **24** (35), 27539–27552. https://doi.org/10.1007/s11356-017-0164-2.

El-Ahmady, I. I. & Sembiring, E. 2014 Pemilihan program pengendalian kehilangan air peningkatan pendapatan PDAM. *Jurnal Teknik Lingkungan* **20**, 142–151.

Grönwall, J. & Danert, K. 2020 Regarding groundwater and drinking water access through a human rights lens: self-supply as a norm. *Water* **12** (2), 419. https://doi.org/10.3390/w12020419.

Hagen, A. & Higdem, U. 2020 *Innovation in Public Planning Calculate Communicate and Innovate.* Palgrave Macmillan, London.

Heryanto, T., Sharma, S. K., Daniel, D. & Kennedy, M. 2021 Estimating the economic level of water losses (ELWL) in the water distribution system of the city of Malang, Indonesia. *Sustainability (Switzerland)* **13** (12). https://doi.org/10.3390/su13126604.

INSEKDA NO 1/2021 2021

Irawan, D. E., Silaen, H., Sumintadireja, P., Lubis, R. F., Brahmantyo, B. & Puradimaja, D. J. 2015 Groundwater–surface water interactions of Ciliwung River streams, segment Bogor–Jakarta, Indonesia. *Environmental Earth Sciences* **73** (3), 1295–1302. https://doi.org/10.1007/s12665-014-3482-4.

Jensen, O. & Khalis, A. 2020 Urban water systems: development of microlevel indicators to support integrated policy. *PLoS ONE* **15** (2). https://doi.org/10.1371/journal.pone.0228295.

Juliardi, I. & Ruskandar, A. 2006 *Teknik Mengairi Padi: Kalau Macak-Macak Cukup, Mengapa Harus Digenang*, p. 3213024. Available from: http://www.Pustaka-Deptan.Go.Id/Publikasi/.Pdf [18 Juni 2015].

Kapetas, L., Kazakis, N., Voudouris, K. & McNicholl, D. 2019 Water allocation and governance in multi-stakeholder environments: insight from Axios Delta, Greece. *Science of The Total Environment* **695**, 133831. https://doi.org/10.1016/j.scitotenv.2019.133831.

Kementrian Pertanian 2020 Pusat Data dan Sistem Informasi Pertanian. Sekretariat Jenderal-Kementerian Pertanian. Center for Agriculture Data and Information System. Secretariat General-Ministry of Agriculture, pp. 1–203.

Kooy, M. 2014 Developing informality: the production of Jakarta's urban waterscape. *Water Alternatives* **7** (1), 35–53.

Kooy, M. & Walter, C. T. 2019 Towards a situated urban political ecology analysis of packaged drinking water supply. *Water (Switzerland)* **11** (2). https://doi.org/10.3390/w11020225.

KPPIP. 2019 *Pengolahan Air Limbah Jakarta.* KPPIP. Available from: https://kppip.go.id/proyek-prioritas/air-dan-sanitasi/pengolahan-air-limbah-jakarta/

Kristensen, P. 2004 *The DPSIR Framework. A Comprehensive/Detailed Assessment of the Vulnerability of Water Resources to Environmental Change in Africa Using River Basin Approach*, pp. 1–10. Available from: http://enviro.lclark.edu:8002/rid=1145949501662_742777852_522/DPSIROverview.pdf

Li, T., Yang, S. & Tan, M. 2019 Simulation and optimization of water supply and demand balance in Shenzhen: a system dynamics approach. *Journal of Cleaner Production*. https://doi.org/10.1016/j.jclepro.2018.10.052.

Loucks, D. P. & Van Beek, E. 2017 *Water Resource Systems Planning and Management: An Introduction to Methods, Models, and Applications*. Springer International Publishing. https://doi.org/10.1007/978-3-319-44234-1.

Lubis, R. F. 2018 Urban hydrogeology in Indonesia: a highlight from Jakarta. *IOP Conference Series: Earth and Environmental Science* **118** (1). https://doi.org/10.1088/1755-1315/118/1/012022.

Luo, P., Kang, S., Apip, Zhou, M., Lyu, J., Aisyah, S., Binaya, M., Regmi, R. K. & Nover, D. 2019 Water quality trend assessment in Jakarta: a rapidly growing Asian megacity. *PLoS ONE* **14** (7). https://doi.org/10.1371/journal.pone.0219009.

Montginoul, M., Rinaudo, J. D., Brozović, N. & Donoso, G. 2016 Controlling groundwater exploitation through economic instruments: current practices, challenges and innovative approaches. In: *Integrated Groundwater Management: Concepts, Approaches and Challenges*. Springer International Publishing, pp. 551–581. https://doi.org/10.1007/978-3-319-23576-9_22.

Mungkasa, O. M. 2017 *Desain Besar Air Minum dan Limbah DKI Jakarta*. Available from: http://tarulh.com/wp-content/uploads/2018/01/3.-Desain-Besar-Air-Minum-dan-Limbah-Domestik.pdf

Najar, N. & Persson, K. M. 2019 Strategies, processes, and results for the future water supply of the Växjö municipality: evaluation of an evidence-based case study of long-term strategies within the water and wastewater sector in Sweden. *Water (Switzerland)* **11** (10). https://doi.org/10.3390/w11102150.

Nasution, N. 2016 *The Dynamics of Piped Water and Sewer Development in Jakarta, Indonesia (1945–2015): A Case Study Using Multilevel Perspective*. The University of North Carolina, Chapel Hill, NC.

Neely, K. 2019 *Systems Thinking and WASH*. https://doi.org/10.3362/9781780447483.

Neise, T. 2019 *Firms' Adaptation Strategies to Floods and Their Potential Implication on Regional Economic Development-Insights From Jakarta and Semarang, Indonesia. The Effect of Natural Disasters on FDI Attraction – A Sector-Based Analysis Over Time and Space View Project*. University of Cologne. Available from: https://www.researchgate.net/publication/333146677

Octavianti, T. & Charles, K. 2018 The evolution of Jakarta's flood policy over the past 400 years. *Environment and Planning C: Politics and Space* **37** (6), 1–30.

Padawangi, R. & Douglass, M. 2015 Water, water everywhere: toward participatory solutions to chronic urban flooding in Jakarta. *Pacific Affairs* **88** (3), 517–550. https://doi.org/10.5509/2015883517.

Pahl-Wostl, C. 2015 *Water Governance – Concepts, Methods, and Practice*. Available from: http://www.springer.com/series/13400

PALJAYA 2017 *Menuju Layanan Pengolahan Air Limbah Handal dan Terintegrasi di DKI Jakarta [Towards Reliable and Integrated Waste Water Treatment Service in DKI Jakarta]*.

PAMJAYA 2020 *Laporan Tahunan PAM JAYA 2019*.

PAMJAYA 2021 *Surat Penyampaian Kajian Subsidi Pelayanan Air Minum DKI Jakarta ke DSDA*. PAM Jaya.

Pemprov, D. K. I. 2021 *Rencana Kerja Pemerintah Daerah 2021*.

Pradafitri, W. S., Moersidik, S. S. & Abdini, C. 2018 East Canal Flood as PDAM water resource DKI Jakarta. In: *E3S Web of Conferences*, p. 74. https://doi.org/10.1051/e3sconf/20187409001.

Pratiwi, D. M. & Herdiansyah, H. 2018 Development of desalination technology using reverse osmosis membrane for the provision of clean water in DKI Jakarta. *AIP Conference Proceedings* **2026**. https://doi.org/10.1063/1.5064987.

Samsuhadi 2009 Pemanfaatan air tanah Jakarta. *JAI* **5** (1), 9–22.

Sanitation and Water for All 2021 *Five Building Blocks*. Available from: https://ebookcentral-proquest-com.aib.idm.oclc.org

Sharvelle, S., Dozier, A., Arabi, M. & Reichel, B. 2017 A geospatially-enabled web tool for urban water demand forecasting and assessment of alternative urban water management strategies. *Environmental Modelling and Software* **97**, 213–228. https://doi.org/10.1016/j.envsoft.2017.08.009.

Soentoro, E. A. 2011 *Integrated Decision-Making for Urban Raw Water Supply in Developing Countries*. Colorado State University, Fort Collins, CO.

Song, X. & Frostell, B. 2012 The DPSIR framework and a pressure-oriented water quality monitoring approach to ecological river restoration. *Water (Switzerland)* **4** (3), 670–682. https://doi.org/10.3390/w4030670.

Sun, S., Wang, Y., Liu, J., Cai, H., Wu, P., Geng, Q. & Xu, L. 2016 Sustainability assessment of regional water resources under the DPSIR framework. *Journal of Hydrology* **532**, 140–148. https://doi.org/10.1016/j.jhydrol.2015.11.028.

The Human Rights to Water and Sanitation 2010 *UN Resolution 64/292*. https://doi.org/10.4324/9781315471532-2.

Thi Bich Ngoc, D. 2011 *Application of City Water Balance Model to Develop Strategies for Urban Water Management in Hanoi, Water Balance, Contaminant Fluxes*. UNESCO-IHE Institute for Water Education.

Tsegaye, S., Missimer, T. M., Kim, J. Y. & Hock, J. 2020 A clustered, decentralized approach to urban water management. *Water (Switzerland)* **12** (1). https://doi.org/10.3390/w12010185.

United Nations 2002 General comment No. 15: The right to water (arts. 11 and 12 of the Covenant) - Twenty-ninth session (2002). *Human Rights Instruments, Vol. I. Compilation of General Comments and General Recommendations Adopted by Human Rights Treaty Bodies – HRI/GEN/1/Rev.9*, 2003(15), 97–113.

Valcourt, N., Javernick-will, A., Walters, J. & Linden, K. 2020 System approaches to water, sanitation, and hygiene: a systematic literature review. *International Journal of Environmental Research and Public Health* **17** (3). https://doi.org/10.3390/ijerph17030702.

Vörösmarty, C. J., Green, P., Salisbury, J. & Lammers, R. B. 2000 Global water resources: vulnerability from climate change and population growth. *Science* **289** (5477), 284–288. https://doi.org/10.1126/science.289.5477.284.

WaterAid 2020 *System Strengthening for Inclusive, Lasting WASH That Transforms People's Lives*, pp. 1–92. Available from: https://washmatters.wateraid.org/sites/g/files/jkxoof256/files/suswash-global-learning-report.pdf

Zare, F., Elsawah, S., Bagheri, A., Nabavi, E. & Jakeman, A. J. 2019 Improved integrated water resource modelling by combining DPSIR and system dynamics conceptual modelling techniques. *Journal of Environmental Management* **246**, 27–41. https://doi.org/10.1016/j.jenvman.2019.05.033.

Zhao, R., Fang, C., Liu, H. & Liu, X. 2021 Evaluating urban ecosystem resilience using the DPSIR framework and the ENA model: a case study of 35 cities in China. *Sustainable Cities and Society* **72**, 102997. https://doi.org/10.1016/j.scs.2021.102997.

First received 4 February 2021; accepted in revised form 31 March 2022. Available online 12 April 2022

doi: 10.2166/h2oj.2022.073

Analysis of experience using human rights to accelerate WASH access in four countries

Louisa Gosling [a], Tripti Rai [b], Pedi Obani [c,*], Moussa Alou Traore [d], Landry Ouangre [e], Fauzia Aliu [f] and Sonu Kumar Shah [b]

[a] WaterAid, London, UK
[b] WaterAid, Lalitpur, Nepal
[c] School of Law, University of Bradford, Bradford, UK
[d] WaterAid, Bamako, Mali
[e] WaterAid, Ouagadouga, Burkina Faso
[f] WaterAid, Accra, Ghana
*Corresponding author. E-mail: pedi.obani@gmail.com; p.obani@bradford.ac.uk

LG, 0000-0002-4024-4590; PO, 0000-0002-6863-980X

ABSTRACT

Human rights to water and sanitation have been widely recognised in legal instruments at the international, regional, and national levels of governance. More awareness of states' obligations has provided additional impetus to promote human rights in policy advocacy. The international non-governmental organisation WaterAid, as a non-state actor specialising in the water, sanitation, and hygiene services (WASH) sector, adopts a human rights-based approach (HRBA) to WASH programming. This paper draws on the experiences from WaterAid WASH projects in four countries – Nepal, Ghana, Mali, and Burkina Faso, to evaluate the practical impacts of the HRBA to ensure that governments fulfil their responsibility to realise universal access to water and sanitation services in different contexts. The outcomes highlight that three important contributions of the HRBA: (1) generates greater awareness among rightsholders and duty bearers about responsibilities and entitlements over safe drinking water and sanitation; (2) promotes constructive engagement between the government and rightsholders; and (3) equips people with the motivation, skills, and agency which are critical dimensions to work on sustainable WASH. The WASH sector should, therefore, embrace the power of human rights and invest in the specific activities and frameworks to integrate human rights into systems strengthening the WASH sector, while continuing to analyse and learn how to adapt and improve the approach in different contexts.

Key words: human rights-based approach, human rights to water and sanitation, WASH practice

HIGHLIGHTS

- The human rights to water and sanitation can add value to other interventions when used as a driver for change, policy content, and people-centred approach.
- This paper analyses the experience of using human rights to advance WASH coverage in Nepal, Ghana, Burkina Faso, and Mali, to assess the impact and examine the potential of using human rights to water and sanitation in different low-income countries.

INTRODUCTION

The international human rights to water and sanitation were affirmed as legally binding by the Human Rights Council in September 2010. The rights have been increasingly domesticated through national constitutions and laws and recognised in case of law across countries. The nature of economic, social, and cultural rights requires states to commit their maximum available resources to the progressive realisation of the rights to water and sanitation for all without discrimination and to avoid retrogression. Relying on the recognition of the rights to water and sanitation in international and domestic legal instruments, unserved and underserved populations around the world have engaged in activism, advocacy, and adjudication, among other strategies for holding states, and where relevant non-state actors, to account and promote improved access to water and sanitation services.

Many states that have either endorsed the rights to water and sanitation at the international level or integrated the same in their domestic legal frameworks nonetheless lag behind in the water, sanitation, and hygiene (WASH) services coverage within their jurisdictions. This demonstrates that the formal recognition of the rights to water and sanitation does not automatically translate into improvements in access, particularly for populations in vulnerable situations, unless active measures are taken to embed the rights in policy and practice. Unsurprisingly, the world is off-track to meet the Sustainable Development Goal (SDG) to ensure universal access to water and sanitation, including hygiene – SDG6 which complements the human rights to water and sanitation to some extent (Obani 2020; WHO & UNICEF 2021). The lack of adequate action to back the political commitments to ensure universal access to WASH is due to weak political will at all levels of governance and inadequate financing combined with threats to water from climate change, increasing demands for water, and ecosystem degradation (Obani & Gupta 2016). At local levels, the challenges are exacerbated by a lack of autonomy over funding to match the decentralised responsibility for WASH services, weak prioritisation in development plans, and weak accountability mechanisms (Obani & Gupta 2016). In most situations, water and sanitation are under-valued public services that are rarely understood or articulated as fundamental human rights. More so, access to water and sanitation is often considered to be a privilege rather than a human right obligation of the state.

Increasing awareness of states' obligations as the primary duty bearer has provided additional impetus to promote human rights in policy advocacy, just as the relevance and usefulness of applying the human rights to water and sanitation in different contexts has been increasingly examined in academic literature (Angel 2019; Fantini 2019; Faulkner *et al.* 2021). As a non-state actor specialising in WASH, the international non-governmental organisation (INGO) WaterAid has developed a human rights-based approach (HRBA) to WASH programming that evolved over several years of experience in applying human rights principles to its work in different countries (Gosling 2014). The HRBA has a focus on equality and non-discrimination, accountability of duty bearers, empowerment and participation of rights holders, and strengthening platforms and systems for engagement. This paper reflects on the use of the HRBA in WASH programming by WaterAid as a tool to ensure that governments fulfil their responsibility to realise universal access to water and sanitation services in different contexts. It examines how the tool is being applied and to what effect, through cross-country analysis, drawing on the project experiences of WaterAid in WASH services delivery in four country programmes: Nepal, Ghana, Mali, and Burkina Faso. The method section of the paper covers WaterAid's approach and the data collection and analysis. The results and discussion section of the paper examines the impacts of the HRBA as applied by WaterAid and its partners and demonstrates factors that affect progress based on the highlighted projects. The paper finds that while there are legal, political, institutional, and other limitations to the human rights to water and sanitation in different contexts, the HRBA remains a significant tool in the struggle for access to WASH for people in vulnerable situations.

METHODS

Overview of WaterAid's HRBA

WaterAid designs its programmes in each country based on the understanding that WASH exists in complex systems with many component parts and within different social, economic, political, and environmental contexts. This systems approach involves a process of analysis, implementation, adaptation, and learning in order to address the barriers to the achievement of inclusive, sustainable, universal access to WASH. Applying human rights principles and standards brings a particular focus on equality and non-discrimination and accountability, achieved through sensitising rights holders and duty bearers on the human rights to water and sanitation and strengthening the engagement between rights holders and duty bearers for accountability and inclusion at various levels. These efforts are carried out in collaboration with other stakeholders including civil society organisations and networks, communities, National Human Rights Institutions, parliamentarians, and the media. WaterAid always works closely with local government and communities to strengthen community ownership, integrate WASH into local development plans, mobilise resources, and hold duty bearers accountable for delivering on commitments. While there are many common elements of the HRBA in the selected countries, each project is designed to address context-specific barriers, opportunities, and priorities.

Data collection and analysis

This paper is based on the experience of the authors, five of whom are directly involved in managing the highlighted WASH projects in this paper, supported by the review of project documentation including learning briefs, evaluations, media reports, and changes in government policy and plans. The contents of the project documentation have been analysed to identify any commonalities or differences across the countries. It is difficult to analyse all the legal and policy instruments across WASH and related sectors such as housing, education, and health, which could be relevant to promote the realisation of the human rights to water and sanitation, mainly due to the lack of codification of the laws applicable to the sector. However, there have been some recent attempts at the legal mapping of the extent of domestication of the international commitments that provide an overall picture applied in the paper (WaterLex 2018; UNU-INRI 2020). Following this, the outcomes are presented through a three-dimensional framework that was developed by the former United Nations Special Rapporteur on the human rights to safe drinking water and sanitation, which identifies the benefits of human rights as a driver, as a policy tool, and as a people-centric approach (Heller 2020).

Human rights as a driver indicates the importance of a HRBA to water and sanitation in order to identify groups in vulnerable situations who lack access to water and sanitation services and understand the underlying 'causes of the attributes and the structural determinants, or drivers, of the exclusion' (Heller 2020). Based on this analysis, WASH stakeholders can address the causes and drivers of exclusion, working with both rightsholders and duty bearers to eliminate the obstacles to progress. The idea of human rights is also closely associated with concepts of justice and fairness which motivate, mobilise, and unite people to strive for change. Hence, the concept of human rights as a driver is explored in the paper to explore the extent to which the WaterAid programmes have interrogated the structural causes of inequalities and the mobilising power of rights to address them.

Human rights in policies focus on the incorporation of the rights to water and sanitation in the legal, political, institutional, and regulatory 'dimensions of rights' as well as evaluation of the human rights gaps in each dimension. The human rights in policies lens are therefore used in the paper to explore how WaterAid has sought to identify gaps and opportunities to strengthen the policy and legal frameworks needed to respect, protect, and fulfil the human rights to water and sanitation.

Human rights and a people-centric approach emphasise the need to mainstream rightsholders and rights protection. The paper adopts this lens to explore how the highlighted programmes have worked directly with different groups and increased the focus of governments on identified marginalised groups.

RIGHTS-BASED APPROACHES IN PRACTICE: THE FOUR COUNTRY PROGRAMMES

This section briefly describes how WaterAid has applied a HRBA in selected projects in four of the countries in which it works.

In Nepal, the sustainable WASH project (HSBC Water Programme II) from June 2017 to December 2019 was explicitly based on human rights, from the context analysis through to design, implementation, and monitoring. The overall aim was to contribute to the realisation of human rights to water and sanitation as enshrined in the Constitution of Nepal (2015). WaterAid Nepal worked with different partners: Karnali Integrated Rural Development and Research Center (KIRDARC) in Jumla and Kalikot districts, the Centre for Integrated Urban Development (CIUD) in Kavre District, and for advocacy at the national level collaborated with the NGO Federation of Nepal and the Federation of Drinking Water and Sanitation Users Nepal (FEDWASUN).

The project was based on a systematic process to identify marginalised communities with poor access to WASH. A study by Overseas Development Institute (ODI), commissioned by WaterAid, analysed possible pathways to reduce inequalities in access to WASH in Nepal (Sarwar 2017). This was followed by Participatory Action Research (PAR) in Kavre, Kalikot, and Bardiya districts (Sanoj Tulachan 2019) to identify the marginalised and most deprived groups, analyse their situation in relation to rights to water and sanitation, and identify ways to work with them to overcome the barriers. WaterAid Nepal selected the project areas in consultation with district and local authorities, also considering the needs of the communities, remoteness, marginalisation, and the presence of socially discriminated or disadvantaged populations.

WaterAid then did a rapid assessment of the contexts, based on human rights norms, and in collaboration with the implementing partners, assessed the capacities of rights holders and duty bearers to claim and fulfil the rights to water and sanitation, respectively. Based on these analyses of the challenges, stakeholders, and available resources, interventions were designed to address the identified problems.

WaterAid and partners provided training and orientation for rights holders (including water user committees and community members) on the human rights to water and sanitation and on social accountability mechanisms. They also provided training for duty bearers (municipal and provincial leaders) on their obligations with respect to human rights. WaterAid constructed some WASH facilities as models that meet human rights standards that can be replicated by duty bearers. Social accountability mechanisms were established to monitor the status of WASH and hold duty bearers accountable for their obligation towards rights holders. These tools include Jal Kachhari (public audit), a local participatory platform organised by municipal authorities where people, including those who do not have access to services such as water and sanitation, can share their views and raise concerns to the local government. This tool has been instrumental in informing the community people about the WASH budget plan and expenditure of their municipality, as well as increasing their understanding of programme quality and monitoring.

As WaterAid and partners engage with duty bearers and rights holders, they actively promote the fact that the human rights to water and sanitation are enshrined in the Constitution of Nepal of 2015. They have produced several manuals and training materials in local languages to explain this and show how the rights are relevant to the situation of communities.

In Ghana, WaterAid's WASH for Public Health (WASH4PH) project funded by Global Affairs Canada (GAC) (2015–2019) was designed to improve the health outcomes of boys, girls, women, and men in Ghana's Upper East Region, with a specific focus on Bongo and Kassena Nankana West Districts. This was done using an integrated approach to public health that incorporated improved WASH into district-wide development and health service delivery planning, implementation, and management. The district-wide approach involved strengthening systems, plans, finances, skills, knowledge, coordination, and accountability mechanisms to achieve sustainable and universal access to WASH services. The project used an endogenous development and a HRBA in its implementation process. Partners were Widows and Orphans Movement (WOM), Centre for Indigenous Knowledge and Organizational Development (CIKOD), and Global Platform Ghana.

The project partners work through rooted advocacy in the districts, strengthening the capacities of communities, via individuals and institutions to advocate for their rights. This involves a general sensitisation on the rights to WASH for partner communities, capacity strengthening on advocacy for individuals identified by the community, and the subsequent development of community advocacy teams. Advocacy groups in communities are supported to mobilise around services (quality, quantity, and safety) – holding themselves, the government, and other duty bearers accountable for sustainable services. The project activities include capacity strengthening on rights, leadership, team building, hygiene, entrepreneurship, and on evidence-based advocacy for the advocacy groups and for women's groups in the communities. The specific focus on women's groups supports more active participation of women in the community advocacy groups and provides them with more confidence and capacity to take advantage of other development opportunities. WaterAid also carried out training with the groups on the Make Rights Real approach (Carrard 2020) on how best to work collaboratively with duty bearers, aligning demands with their incentives and responsibilities.

WaterAid also works directly with duty bearers, supporting them to recognise their obligations to ensure access to sustainable and equitable WASH services for all and to be accountable to rightsholders. This includes building the capacity of local government authorities – Metropolitan, Municipal, and District Assemblies (MMDAs), Ghana Health Service (GHS), and Ghana Education Service (GES) who are at the frontline of WASH service delivery. WaterAid Ghana (WAG) recognises that the supply side must also be strengthened to facilitate equitable and efficient service delivery systems, which are accountable to rightsholders. Support includes providing demonstration WASH facilities in health centres, schools, and communities. It also includes training and coaching on the obligations of MMDAs towards fulfilling the rights of its people (with a focus on right to WASH). A total of 28 advocacy groups in three Municipal and Districts assemblies (Wa Municipal, Bongo, and Kassena Nankana West) with a membership of 431 members and 90 MMDA staff have been strengthened in the three MMDAs.

At the national level, WaterAid uses human rights content in advocacy with a national advocacy platform: the Coalition of NGOs in the Water and Sanitation sector (CONIWAS), and forming alliances and networks with other stakeholders in the sector to ensure coordinated efforts to influence politicians and bureaucrats to effect change, based on an action plan to implement the human rights to water and sanitation prepared by the Centre for Housing Rights and Eviction, WaterAid, and CONIWAS (CONIWAS 2015).

In Mali, the 'Ji Ni Beseya' project, funded by WaterAid and One Drop, aims to increase access to safe water and sanitation services in nine rural communes in the districts of Kati and Bla, collaborating on behaviour change with the social arts organisation Centre Culturel Kôrè. A HRBA was used with local authorities, citizens, and technical services. The project was carried out with partners Association Libre pour la Promotion de l'Habitat et du Logement (ALPHALOG), Association for the Promotion of Women and Children in Mali (APROFEM), NGO CARP, and La Société de Développement International (SDI-Mali).

Another project in Bamako, 'Improving Access to Sanitation', involved the establishment of the citizen jury, capacity building of authorities and citizens on human rights, budget monitoring, advocacy and contributed to the institutionalisation of a framework for dialogue between citizens and authorities of the municipality, and improvement of the budget of the municipality allocated to sanitation and water. The project was implemented in collaboration with the partner Association Malienne pour la Sauvegarde du Bien-être Familial (AMASBIF).

In these projects, WaterAid Mali and its partners have built the capacity of citizens and authorities on the HRBA, advocacy, and leadership. The Ji Ni Beseya project organised spaces for questioning, and citizen watch committees, women's and youth groups were set up and strengthened on the HRBA and leadership to improve their participation in the planning and budgeting process of development actions. The project has also facilitated the renewal of local development plans and the integration of WASH into planning tools in all the project intervention communes.

They have also helped to establish accountability and dialogue frameworks at the local level to address inadequate WASH in Bamako. These include spaces for democratic interpellation, Citizen Juries, joint dialogue, and public hearings. WaterAid has supported their local partner Coalition pour l'Accès à l'Eau Potable, l'Hygiène et l'Assainissement (CAEPHA) to monitor and provide evidence on cases of violations of rights to water and sanitation. They were able to mobilise community members in Gouana, near Bamako airport where the pit emptying association was dumping faecal sludge. The community members gathered evidence that was used in collaborative advocacy with National Human Rights Institutions (CNDH) and included in Mali's report to the UN on human rights violations. Additional advocacy with parliamentarians and the media helped increase demand for government response which has now committed to allocate land for a faecal sludge treatment station.

The advocacy aims to influence governments, parliamentarians, and local elected officials to change laws, policies, and strategies in favour of WASH rights using the following concepts: respect for the principles of equity, equality, and non-discrimination; protection of the rights of the most vulnerable; allocation of sufficient resources; responsibility and leadership of authorities; accountability of officials and advocacy for the reduction of inequalities, including gender inequalities.

In Burkina Faso, all WaterAid's projects integrate the promotion of human rights to water and sanitation and include components related to the HRBA. In particular, the 'Mission Fas'Eau' project (2017–2020) was designed to achieve two outputs: For the Ministry in charge of water to ensure the respect and implementation of the right to water and sanitation in the investment decision-making processes, and access to WASH services and water resources management. And for public institutions and civil society organisations to ensure the protection and defence of the human rights through developing information and capacities and actions to monitor and influence public action in Integrated Water Resources Management (IWRM) and WASH.

The Fas'Eau[1] mission, funded by the Ministry of Foreign Affairs of Denmark (DANIDA) and Oxfam, is implemented by a consortium of 11 civil society organisations led by WaterAid and IRC WASH[2], and five media organisations.[3] This project uses several means to enable citizens to claim their rights to water and sanitation. The first is by raising awareness among rights holders and duty bearers using training modules developed at the national level to build the capacity of different actors (national and local elected officials, state services, civil society organisations, community agents, etc.) on the HRBA. The training is based on the commitments made in the Burkinabe constitution of November 2015 and the President's commitments to achieve the zero-hours water fetching vision.

The capacity building enables civil society organisations and other actors, such as media professionals, local water committees, water user associations, and other community-based organisations, to better understand

---

[1] https://faseau.com/.

[2] Association of Women Lawyers of Burkina Faso (AFJ/BF), Centre pour la Gouvernance de l'Eau et de l'Environnement (CG2E), Comité National du Livre Bleu (CNLB), Eau Vive, Ligue des Consommateurs du Burkina (LCB), Lay Volunteers International Association (LVIA), Mouvement Burkinabè des Droits de l'Homme et des Peuples (MBDHP), Musée de l'Eau, Secrétariat Permanent des ONG, Associations de Développement (SPONG).

[3] Fasozine, Lefaso.net, Radio Oméga, Radio Télévision du Burkina (RTB), Radio Savane FM.

and advocate for the rights to water and sanitation. The training is reinforced by public campaigns to promote the human rights to water and sanitation through popular events such as the Tour du Faso or the Pan-African Film Festival of Ouagadougou (FESPACO), which attract large numbers of people; through events such as the celebration of World Water Day; by influencing national and local elections to ensure that candidates include water and sanitation in their programmes; and finally, through working with the media at all levels to disseminate messages and raise awareness. These actions reinforce other efforts to inform citizens about their rights.

The project has included the creation, animation, and promotion of accountability frameworks on the effectiveness of the rights to water and sanitation. These frameworks are created at both national and local levels and allow citizens to dialogue, question, and raise issues related to the implementation of the human rights to water and sanitation.

WaterAid has supported five studies that contribute to understanding the rights in Burkina Faso: the definition of indicators for the analysis of the WASH sector performance with regard to HR principles (Mission de Gestion du Fonds DANIDA 2018); the inadequacies and limitations of the WASH and IWRM consultation and decision-making platforms: the consideration of HRBA in water and sanitation legislation (WaterAid 2018); the socio-cultural levers of HRBA in water and sanitation (FAS'EAU 2018): and the justiciability of human rights violations to water and sanitation (WaterAid Burkina Faso 2018).

The study on justiciability was presented to the presidents of the High Courts (TGI) and executives of the Ministry of Justice, Human Rights, and Civic Promotion. This action allowed for reflection on the one hand on the reasons why citizens are not in the habit of referring to the courts in case of violations of the rights to water and sanitation and on the other hand on the need to strengthen the awareness of citizens on the possibility of referring to the courts in case of violation of their human rights to water and sanitation.

RESULTS AND DISCUSSION

In this section, the three-dimensional framework described above is applied to look in more detail at the effect of using rights in the selected countries and projects as previously presented, demonstrating what factors enable more or less progress to be made.

Human rights as a driver

All four countries show tangible evidence of productive engagement between rightsholders and duty bearers resulting in commitments and resources from authorities and communities to overcome barriers to WASH access for marginalised communities.

In Nepal, the project interventions were based on analysis of causes of exclusion to ensure that marginalised and vulnerable people are able to participate actively in the planning process using tools like community score cards that enable users and user committees to assess the WASH situation as a basis for planning improvements to meet human rights standards. The discussions with municipal authorities at the 'Jal Kachahari' platforms have enabled them to agree how to address the demands related to water and sanitation and resulted in commitments from duty bearers to prioritise WASH in the municipal budget allocation. As reported in the Jumla district-based newspaper (Mountain Today Daily 2019) on 26th November 2019, *'the Municipal authority committed to organize JALKACHAHARI twice a year, Allocate at least 20% of their budget for WASH, and ensure that at least 33% of community participants at the hearings should be marginalized and discriminated people'*. This commitment, to continue holding Jal Kachahari beyond the project period, was considered a strong sign of sustainability of the project impact, although fewer people are currently able to attend the meetings due to Covid restrictions.

Through the public hearings, the local authorities have realised their responsibility for implementing the one house one tap policy of the government and have incorporated this into their plans. Meanwhile, WaterAid has provided demonstration schemes that fulfil this policy in communities and institutions that comply with human rights standards of accessibility and acceptability and that can be replicated by the authorities.

In Ghana, the first step to realising rights is being aware of rights and people are now moving beyond this to engage and demand. We are seeing this in the communities we are working in. The community advocates were trained on human rights concepts using the 'make rights real' approach (Carrard 2020) designed to make human rights relevant and actionable for local authorities with a process of constructive engagement between civil society and local government professionals. Community members who started using this approach for WASH advocacy are now using the knowledge and skills gained to engage with duty bearers for electricity, healthcare centres, and other needs: representatives of the community have liaised with the community to identify priority issues, written to the assembly to ask for things they need, and kept a record of requests for follow-up and to hold the District Assembly (DA) to account. This has already produced some tangible results in a response from the DA, complemented by community contributions, including the provision of boreholes in two communities; the building of a kindergarten with three rooms that are conducive for learning, the provision of benches for pregnant women waiting at health centres, and the repair of a fridge for storing medicines in the Chief's compound. Furthermore, the community groups are now an important resource for on-going work in the region.

As a result of the sustained engagement between communities and duty bearers at the district level, the budget allocation is being prioritised. Although the national allocation for WASH has been reduced overall, the Bongo District Metropolitan, Municipal and District Assembly (MMDA), where community advocates have been engaging persistently with the technocrats and Assembly Members, has increased the allocation and reported at the National Level Exchange on District-Wide Initiatives Platform on WASH that this was due to increased demand from communities (Ayinbisa 2020). There have also been changes at the national level where WaterAid has contributed to the influencing work of CONIWAS and the Vice president has made a commitment to establish a sanitation authority.

The outcome is that the communities have become capable of making change independently of WAG. For instance, Kalvio-Gugoro advocacy group gathered WASH evidence on their Community-Based Health Planning and Services (CHPS) compound and actively supported the community to organise a health forum. This enabled them to engage with the District Health Services in providing services, such as electricity, drugs, a refrigerator, and water, as well as sanitation facilities, to make the facility fully functional upon completion.

Results in Mali include an increase in participation of women and young people in the decision-making process in rural areas in the Cercle of Bla where the HRBA work has included support for women's groups to grow and be more involved in WASH decision-making. In urban areas, citizens are involved in the budget process at the local level, resulting in budget increases for WASH through the local authorities' planning document Programme de Développement Économique Social et Culturel (PDSEC) – which is revised every 5 years. Citizens and authorities discuss the problems in WASH and where the resources can come from. Often the municipality does not have enough money to meet the communities' needs in which case the municipality and communities together advocate the state for support, following up to make sure that money allocated to budgets is, in fact, transferred.

In a separate activity, as described above, WaterAid supported the mobilisation of the citizens of Gouana, a marginalised community where the Pit Emptiers' Association were emptying sludge, to claim their rights to sanitation by organising sit-ins, interactive radio programmes, advocacy meetings, and submitting evidence to the report of the National Human Rights Institution. As a result, the state accepted their responsibility as duty bearer and committed to supply land for a faecal sludge management station.

WaterAid has also seen progress in embedding principles of accountability at the national level through the establishment of an annual consultation framework for the sector; increased citizen participation in the planning process of the local budget and the restitution of the administrative account, and improved governance in favour of sanitation where there has been a more than 46% increase in the budget allocated to sanitation in the 5th commune of the District of Bamako (CVI) (Traore 2019).

In Burkina Faso, one of the major results is that there is now a critical mass of citizens who understand the scope and stakes of the recognition of the human rights to water and sanitation. A lot of attention is being paid to the issue by actors who are raising awareness and mobilising other citizens. More and more citizens are demanding that these human rights be respected, that the authorities invest more in the sector and that water and sanitation be prioritised, for example in this news report about a protest by women in Gaoua about broken water sources (lefaso.net 2019).

From the perspective of the authorities, it is clear that the discourse of responsibilities has changed. There is evidence in policy and strategy of greater targeting of services towards the most marginalised, and the Minister for Water and Sanitation is held to account through sessions hosted by Presimetre, the platform for dialogue and citizen accountability (Ouedraogo 2020).

From the citizens' perspective, the demand is stronger and authorities are more responsive. For example, officials who are not performing are replaced. Officials who previously avoided direct contact with communities are now sitting with them each year to agree how to mobilise resources for WASH. This includes the annual contributions made by community members to community organisations to maintain the WASH facilities and increases the accountability of those community organisations.

These experiences show how the focus on raising awareness about rights and responsibilities among rights holders and duty bearers is helping to address one of the key causes of exclusion: the lack of political will to reach the marginalised communities. The consistent engagement between people and authorities is strengthening the mobilisation, planning, and accountability for resources to back up the political commitments for universal WASH.

Human rights content and principles in policy and applied to services

The four countries have, respectively, signed up to international treaties that commit them to the realisation of human rights to water and sanitation. While each country recognises the human rights to water and sanitation in domestic laws and policies to different extents, the enabling institutional framework required to enforce the rights is not complete in any of the four countries.

In Nepal, the conversation about human rights to water and sanitation has grown over the last 15–20 years, backed by civil society organisations, water user federations, and others. In 2015, the Constitution of Nepal was passed recognising the right to water and sanitation as fundamental rights of its citizens, a nationally owned commitment from which to advocate for further domestication and application of the rights. Organisations like WaterAid are now helping the government to put this policy commitment into practice and push for universal access. The Constitution provides a high-level commitment that needs to be translated into laws guaranteeing the rights to education and health and rights to water and sanitation for all citizens of Nepal. It provides the space for duty bearers and civil society to work on it, and WaterAid Nepal commissioned a policy analysis of human rights to water and sanitation to inform and direct its advocacy.

However, the specific laws required to enforce these constitutional rights are yet to be formulated. A water-related bill was drafted by the government a few years back, but it did not cover all aspects of the right to water and sanitation. WaterAid Nepal is part of the core technical committee for drafting and finalising the bill 2018–2019 ensuring the inclusion of all the elements of rights related to water and sanitation. WaterAid has also supported the Federation of Drinking Water and Sanitation Users to lobby with parliamentarians to incorporate their agenda in the water and sanitation bill and has oriented the Chair and members of Development and Technical Committee of Parliament on the principles and contents of human rights to water and sanitation. As a result, the current water and sanitation bill reflects the content of human rights to water and sanitation and when (if) passed will provide a strong legal framework for holding the state to account. This means that if a community or person gets dirty water from service providers, they can charge them. The bill informs the discussion about tariffs for equitable access, and the draft bill is already being used as leverage to lobby the local governments (which are autonomous) to draft local acts that align to constitutional provisions. However, it has not been given due priority to finalise it and present it to the parliament. The bill was presented in the parliament in 2020, but the theoretical discussion was postponed due to the pandemic and political instability in Nepal.

Meanwhile, WaterAid and others have been able to base advocacy for WASH on the constitutional commitment and have seen the government's budgetary allocation to WASH increasing incrementally in spite of the restricted budget overall; and the government commitment to provide one tap per house in 5 years. The implementation of these commitments is compromised by the restructuring of government and the lack of governance mechanisms. Nevertheless, the constitutional commitment has enabled government prioritisation of WASH funding at the local level. The federal government releases resources to rural municipalities for allocation to different activities in the sector. The Jal Kachahari report (Mountain Today Daily 2019) shows that more than 20% of its resources have been allocated to reaching water to vulnerable people. The government is seeking to fill the resource gap by coordinating with the contributions of development organisations such as WaterAid that has demonstrated 23 schemes in three rural municipalities. The challenge now is not the intention to deliver, but the institutional capacity. The main barriers include weaknesses in policies, plans, human capacity, and supply chains.

On the demand side, the constitutional commitment has shifted the focus of communities to quality and availability. Furthermore, the discussion on rights and responsibilities has generated greater incentive for them to take better care of the resource, whereas traditionally water has previously been considered as an abundant natural resource to be accessed for free. A typical attitude can be expressed as 'god has given us so much water – we think it is always there and free'.

Similarly, in Ghana, the human rights to water and sanitation are not explicit in the 1992 constitution. The closest reference is 'Article 35 [...] 3. The State shall promote just and reasonable access by all citizens to public facilities and services in accordance with law'. Ghana adopted several international resolutions and co-sponsored the General Assembly resolution 68/157 in 2013, the first resolution where all UN Member States affirmed that the human right to safe drinking water and sanitation is legally binding in international law. The Government of Ghana and its Development Partners (DPs) have acknowledged that access to water and sanitation should be treated as rights, and some efforts have been made to integrate rights to water and sanitation in key strategic documents to serve as a basis for a progressive realisation of the rights. However, not having the rights to water and sanitation explicitly in the constitution is seen as a limitation. While these rights could be derived from international treaties and from the African Charter on Human and Peoples Rights, it does not accord the same legitimacy as recognition in the Constitution; it is more subject to interpretation, the legal framework for the protection of citizens is lacking, with a risk of politicisation of issues along party lines and potential victimisation of people demanding rights.

WAG supported and pushed for WASH integration into the Nationally Determined Contributions (NDCs) to address Climate Change challenges and WASH/IPS in Healthcare Facilities national guidelines and development of the National NTDs Master Plan and investment case through our influencing work and serving on various national technical committees. Furthermore, as part of integrating HRBA into WaterAid Ghana's intervention, conscious efforts have been made to build the capacity of communities to demand for rights enforcement.

The government of Burkina Faso, like Nepal, has included the human rights to water and sanitation in its constitution. WaterAid was one of the key organisations lobbying for this in 2015 when the constitution was drafted. To aid implementation, WaterAid Burkina Faso commissioned an additional analysis of relevant frameworks and mechanisms at the start of its Mission Fas'Eau project in 2018 to identify gaps to work on. The human rights to water and sanitation are now explicitly recognised in various policy documents, originating from the constitution, and further recognised in the national Water Policy and national water and sanitation sector programmes. Further legislation and regulation are underway, and in March 2019 Burkina Faso adopted a decree defining standards, criteria and indicators for access to water and sanitation derived from these rights.

WaterAid and other WASH actors have used human rights principles to inform their participation in the evaluation of the performance of the water and sanitation sector and in the development and review of WASH sector policy and strategy documents (lefaso.net 2021), the development and validation of the HRBA gender mainstreaming manual and its water and sanitation sub-sector action plan; the selection of disadvantaged areas and localities that do not have access to water and sanitation services for investment; the adoption of the Integrated Monitoring and Evaluation System (DISE) of the Ministry of Water and Sanitation programmes; and the definition of HRBA indicators for the five national programmes of the water and sanitation sector.

The obligations in the constitution provide leverage to push authorities towards more action. WaterAid represents civil society in a periodic report to the UN, ensuring that it incorporates a certain number of points on water and sanitation. Citizens can also claim because these are constitutional rights. Moreso, there is increasingly greater awareness of accountability for providing access to WASH as a legal obligation on the part of the government and more budgetary allocation to the sector.

In Mali, the human rights to water and sanitation have been ratified in international treaties and conventions and are considered through various national policies, action plans, programmes, and institutional arrangements for the development of water services and sanitation. However, they are not embedded in the constitution and are not integrated explicitly into the legal framework for the sector.

The WaterAid teams use the international commitments to advocate for human rights standards and principles to be applied to WASH services through legislation, policy review processes, engaging with parliamentarians, and through the monitoring and planning processes at local and national levels. For example, working with local authorities to review and prioritise WASH in the 2017–2021 Programme de Développement Économique Social et Culturel (PDSEC) in our intervention areas (PDSEC Commune VI 2017–2021). The prioritisation of ODD6 in the PDSECs has allowed WaterAid to advocate for an increase in the communal budget allocated to WASH at the local level. For example, in the CVI of the District of Bamako, the rate has increased by more than 40%. Also, local authorities have committed to allocating 15% of the authorised budget to WASH in all WaterAid intervention communes. Monitoring of this commitment is underway to ensure compliance. In 6/16 communes of intervention of the Beseya Blon project, we have seen an increase in the budget rate to over 10%. However, the challenge remains the low capacity of local authorities to mobilise resources. Other results include the availability of WASH-sensitive PDSECs in all Beseya Blon project intervention communes (16 communes). Currently, WaterAid is working with the authorities to revise the expired Sector Wide Development Plans by prioritising the SDG6 targets and integrating human rights principles and standards.

WaterAid and partners also support duty bearers by carrying out vulnerability studies and developing WASH services that demonstrate how the standards and principles can be applied in practice. They have also supported specific challenges on rights violations that are addressed through policy change. For example, an investigation into the alleged violation of human rights in relation to the dumping of faecal sludge in the community of Gouana, described above, elicited a government response, committing to allocate land for the safe treatment of faecal sludge. Recognising the human rights to water and sanitation in the constitution would further enable advocacy.

Human rights as people-centred – focus on equality and non-discrimination

WaterAid's focus has always been on hard-to-reach communities, applying human rights principles of equality and non-discrimination, and of participation. The projects all involve participatory work with marginalised people and communities, women's rights, and disability organisations to strengthen their capacity and ensure their voices are heard and influence service provision. The HRBA considers people as agents of change and that increased capacity, agency, and leadership strengthens their position in all spheres of their life. WaterAid's revised equality, inclusion, and rights framework (WaterAid 2020) emphasises the importance of working through inclusive towards empowering and transformative approaches that foster meaningful partnerships and ultimately transform the power imbalances that stand in the way of equality.

In Nepal, WaterAid followed a systematic process to understand the dynamics of marginalisation in Nepal and to identify project areas that would require them to work with very marginalised communities to overcome barriers to WASH. The community sessions and training approaches on human rights to water and sanitation were designed and executed to empower people so they can claim their rights. The Jal Kachahari was organised in a way to raise the voice and concerns of local people related to water and sanitation and strengthen their relationships with the authorities, to work together towards ensuring access.

In Ghana, WaterAid and its partners invested time and resources to develop the capacity of advocacy groups and women's groups in the communities using participatory methods. The tools used for this include community institution and resource mapping (CIRM); community visioning and planning (CVAP) to visualise what the group want to become or see themselves in 10–20 years, community organization self-assessment (COSA) to assess capacity, weaknesses and aspirations; and capacity institutional strengthening (CIS) to identify and strengthen gaps in capacity. The project also used the Make Rights Real approach (Carrard 2020) to strengthen the awareness of rights to water and sanitation among these groups and fuel their engagement with local authorities.

In Mali, there were specific activities that focused on the empowerment of marginalised community members, women, and youth, for example helping to establish income-generating businesses such as soap making and market gardening, ensuring access to capital and training in entrepreneurship, helping to carry out governance assessment in grassroots community organisations, and the establishment of citizen watch committees in health centres and at the urban level. The women's groups are now more actively involved in supporting services and wider development through the training in rights and leadership. One woman is now managing a cooperative of more than 300 men and women, with increased power and influence over decisions.

In Burkina Faso, there were specific capacity-building activities for people living with disabilities at the local level, intentional involvement of citizens and civil society organisations in the development and implementation of projects, and diversification of partnerships with local organisations, including women's organisations. There was also a focus on innovations to reach everyone with our WASH services, for example subsidised latrines for the most vulnerable, developing and promoting latrines for people with disabilities, and managing menstrual hygiene in schools.

WaterAid in Burkina Faso also actively involves citizens in service delivery projects, recognising that empowerment is about being at the centre of actions and becoming a true actor of change. Some notable examples are the involvement of women's and youth associations at the local level to carry out our sanitation projects through training and community outreach, which strengthens their knowledge about WASH and they become advocates in their communities. In addition, it supports the organisational development of their associations, which sometimes leads them to participate in other activities, such as forming cooperatives. The fact that they are members of the community allows for sustainability.

Project managers have seen government authorities prioritising vulnerable people in target selection, giving greater consideration of their needs and constraints in investments and increased awareness of the need for their involvement in WASH service delivery, for example through concerted efforts to integrate gender into the 2021–2025 action plan of the Ministry of Water and Sanitation (Ministere de l'Eau et de l'Assainissment 2021).

Enabling and constraining factors

The use of human rights as a tool is enabled where there is a wider discourse on human rights between people and governments. Collaboration with other organisations with experience on human rights and HRBA is a critical enabling factor. While the specific application of rights is still quite limited among stakeholders in the WASH sector, there is much wider adoption of a rights-based approach in areas such as health, education, gender, and disability. In the selected projects, WaterAid was collaborating with a range of local (and some international) partners and coalitions who used their experience of HRBAs and legal aspects of human rights. Some partners had a specific experience of WASH but not all. Mobilising this range of expertise is having a demonstrable effect on rights holders and duty bearers, enabling more constructive and participatory decision-making to find solutions and resources to address WASH needs.

The examples shared here also demonstrate the advantage of having human rights to water and sanitation in the national constitution. Advocates in Nepal and Burkina Faso can use these commitments to engage with all different stakeholders on roles and responsibilities to realise rights. It provides a strong basis to add the relevant content into policies, legislation, plans, and monitoring tools. Where human rights are derived from international commitments (as in Ghana and Mali) it is still possible, but more difficult to bring all stakeholders on board.

Social and customary norms are both enabling and constraining factors. They are enablers where the human rights principles converge with traditional practice traditions for water, as identified in the study into the socio-cultural levers favourable to the promotion of human rights to water and sanitation, carried out to inform the design of the project in Burkina Faso (FAS'EAU 2018). This makes it easier to raise awareness and interest in the principles and content of human rights by building on existing practices. As an example, in relation to non-discrimination and equality, the study found that the elderly, children, people living with disabilities and other vulnerable people (sick people, pregnant women, etc.) are exempt from community work. However, they are entitled to the benefits of the work, including to use the water and sanitation facilities in the same way as others. Another finding is in relation to participation: every member of the community (man/woman) can participate in community activities. This is irrespective of social strata. Freedom is left to the different strata to participate according to their means. For example, shopkeepers and farmers participate with cash, foodstuffs, or an animal (sheep, goat, etc.) for cooking. Older people make blessings and prayers in order to find water.

On the other hand, social norms are often a constraint, especially in issues of equality and non-discrimination. Women are generally disadvantaged in all questions of governance, direct participation and management, and suffer wider discrimination, for example, due to taboos around menstruation. However, through the increasing awareness, advocacy, organisation, and capacity strengthening on HRBA, women's participation and the inclusion of other marginalised groups are increasing. Persons with disabilities are also widely stigmatised in traditional practices and excluded from decision-making. Where there has been a focus on their participation supported by partnerships with organisations of persons with disabilities, this has helped to address attitudinal barriers to inclusion.

Another major constraint to the effectiveness of human rights as a tool is the generally low commitment among political leaders to increase funding allocation for WASH, beyond paying lip service to rights. As a result, local authorities, with limited resources and competing priorities, are not able to respond to community demands. Notwithstanding, as illustrated above, all the projects have helped to mobilise resources and increase allocation towards marginalised groups, while also strengthening the accountability for budget expenditure.

The limited motivation of community members to participate in projects can be a constraint, especially where communities have previously tried and failed to get a response from the authorities. The results above show the potential benefits, but community members, especially women, have many competing priorities for their time. It requires time and skilled facilitation by committed and experienced organisations to build trust and capacity in a way that is truly empowering. Indeed, this is the main investment required.

The limited capacity of the state and low trust between state and people are common across all countries. This relationship is further stressed by increasing challenges through impacts of climate change, Covid, and insecurity, which governments globally are failing to address in a way that satisfies citizens. While this is a constraint to the impact of the rights-based approach, it could be argued that the approach is even more important in such situations. Where governments are using these crises to renege on commitments to support marginalised people, the human rights provide some leverage for people to continue to demand support. For example, the low security in the Sahel has resulted in governments abandoning their responsibilities for communities in areas dominated by terrorist groups, where displaced people are moving to different villages, exacerbating WASH needs. Communities with more awareness of human rights are using their understanding to call on the state to fulfil their responsibilities for basic rights to WASH and for providing security. Terrorists are not taking responsibilities for the Communities they occupy; in fact, they are often destroying water supplies to increase their control.

The awareness of human rights in the water and sanitation sector is relatively low, resulting in limited commitment and support for addressing human rights issues among technical and financial partners for WASH who still prioritise investment in infrastructure. Although some donors, including Danida, Global affairs Canada, and Sida consistently champion a HRBA (Sida 2020), it is still difficult to access the funding needed to integrate human rights principles in WASH projects and programmes through the activities described here. There is limited knowledge and capacity among WASH actors on how to design services and facilities that meet human rights standards of quality, accessibility, acceptability, affordability, and availability. For example, this guide on female-friendly public and community toilets (WaterAid 2018) was developed to address the dearth of public toilets that are accessible and acceptable for women and people with disabilities.

Indicators and tools needed to provide a rigorous and objective assessment of the extent to which interventions apply a HRBA are not yet widely available for WASH interventions and have not been used on the projects described here. The conceptual and methodological framework of the Office of the High Commissioner of Human Rights (OHCHR n.d.) recommends the development of indicators to assess commitments and acceptance of international human rights standards (structural indicators); to the efforts made to meet the obligations that flow from the standards (process indicators); and in the results of these efforts (outcome indicators). These aspects have been considered throughout this paper but without applying a clear evaluation frame it is not possible to give a comprehensive analysis of progress. For example, without disaggregated data, it is not possible to tell the extent to which WASH facilities are accessible and acceptable to women, children, and people with disabilities, who are enabled to participate in decision-making and who now has the capacity to maintain services or to hold duty bearers to account. WASH evaluations are heavily weighted towards measuring outcomes, for example the evaluation of the WASH4PH project in Ghana (Frempah-Yeboah 2020). This tends to skew effort and funding away from the process-focused interventions discussed in this paper.

CONCLUSION AND WIDER IMPLICATIONS FOR WASH SYSTEMS AND HUMAN RIGHTS

The experiences documented here demonstrate that a HRBA to WASH interventions helps to generate greater awareness among rights holders and duty bearers about responsibilities and entitlements to safe drinking water and sanitation. Constructive engagement between the state and people increases the resource mobilisation and accountability needed ensure sustainable and inclusive services, especially to marginalised populations with limited political and economic power. This is supported by strengthening the human rights content in legal and policy frameworks and plans. All these aspects are deepened through collaboration with other actors focused on human rights and governance. There are many examples of these qualitative changes, although the lack of an evaluation framework makes it impossible to demonstrate conclusively.

Rights alone are not enough, but the HRBA adds a sustained value to any attempts to strengthen systems for WASH by providing leverage, policy content, and by providing people with the motivation, skills, and agency required the human rights add critical dimensions to work on sustainable WASH. WASH sector should, therefore, embrace the power of human rights and invest in the specific activities and frameworks to integrate human rights into systems strengthening of the WASH sector, while continuing to analyse and learn how to adapt and improve the approach in different contexts.

DATA AVAILABILITY STATEMENT

All relevant data are included in the paper or its Supplementary Information.

REFERENCES

Angel, J. A. 2019 With-against-and-beyond the human right to water. *Geoforum* 206–213. https://doi.org/10.1016/j.geoforum.2017.05.002.

Ayinbisa, P. 2020 *National Level Exchange on District Wide Initiatives Platform on WASH*. District Chief Executive. Bongo.

Carrard, N. 2020 Designing human rights for duty bearers: making the human rights to water and sanitation part of everyday practice at the local government level. *Water* **12**, 378. doi:10.3390/w12020378.

CONIWAS 2015 *Concept Note for Development of a National Action Plan*. The Coalition of NGOs in Water and Sanitation (CONIWAS).

Fantini, E. 2019 An introduction to the human right to water: law, politics and beyond. *WIREs Water*. https://doi.org/10.1002/wat2.1405.

FAS'EAU 2018 *Etude D'identification Des Leviers Socioculturels Favorables a La Promotion Des Droits Humains a L'eau Et a L'assainissement Au Burkina Faso*.

Faulkner, C. M., Lambert, J. E., Wilson, B. M. & Faulkner, M. S. 2021 The human right to water and sanitation: using natural language processing to uncover patterns in academic publishing. *Water* **13**, 3501. https://doi.org/10.3390/w13243501.

Frempah-Yeboah, K. 2020 *Summative Evaluation of the WASH 4PH Project Submitted to Dept of Foreign Affairs, Trade and Development*, Canada.

Gosling, L. 2014 Moving along the right track? The experience of developing a rights-based approach at WaterAid. *Waterlines* **33** (4), 357–374.

Heller, L. 2020 *Progress Towards the Realisaton of the Human Rights to Water and Sanitation (2010–2020)*. United Nations, OHCHR, Human Rights Council, New York. Available from: https://undocs.org/A/HRC/45/11 (accessed 15 June 2021).

lefaso.net 2019 *Gaoua: Des femmes dans la rue pour réclamer de l'eau*. lefaso.net. Available from: https://lefaso.net/spip.php?article88059. Accessed 16 June 2021.

lefaso.net 2021 *Eau et assainissement: Vers la validation d'une version actualisée du Dispositif intégré de suivi-évaluation des programmes*. lefaso.net. Available from: https://lefaso.net/spip.php?article105951. Accessed 16 June 2021.

Ministere de l'Eau et de l'Assainissment 2021 *Droit d'accès à l'eau et à l'assainissement: la Cellule genre s'approprie les documents de base*. Available from: https://www.mea.gov.bf/informations/details?tx_news_pi1%5Baction%5D=detail&tx_news_pi1%5Bcontroller%5D=News&tx_news_pi1%5Bnews%5D=280&cHash=d690b7e6449ce53fbfe1b41664085b13. Accessed 16 June 2021.

Mission de Gestion du Fonds DANIDA 2018 *Mission de formulation des indicateurs AFDH dans le cadre du suivi des performances du secteur Eau et Assainissement au Burkina Faso*.

Mountain Today Daily 2019 *JALKACHAHARI (Public Hearing) Event Completed with 10 Points Declaration*.

Obani, P. 2020 SDG 6.2 and the right to sanitation: exploring the complementarities and incoherence. *UNIPORT Journal of International and Comparative Law 1*, 1–17.

Obani, P. & Gupta, J. 2016 Human security and access to water, sanitation and hygiene: exploring the nexus and the drivers. Chapter 12. In: *Handbook on Water Security* (Pahl-Wostl, C., Bhaduri, A. & Gupta, J., eds.). Edward Elgar, Cheltenham and Northampton.

OHCHR n.d. *Human Rights Indicators – Main Features of OHCHR Conceptual and Methodological Framework*. United Nations Human Rights Office of the High Commissioner. Available from: https://www.ohchr.org/EN/Issues/Indicators/Pages/framework.aspx.

Ouedraogo, N. A. 2020 *Dialogue Citoyen No. 33 avec le Ministre de l'Eau et de l'Assainissement, Monsieur Niouga Ambroise OUEDRAOGO*. Presimetre. Available from: https://www.presimetre.bf/fr/dialogue/dialogue-citoyen-ndeg33-avec-le-ministre-de-leau-et-de-lassainissement-monsieur-niouga, Accessed 11 March 2021.

Sanoj Tulachan, S. R. 2019 *Participatory Action Research to Strengthen Approaches to Local Planning for WASH: Approach, Method and Process*.

Sarwar, M. 2017 *How to Reduce Inequalities in Access to WASH*. Overseas Development Institute. Available from: https://cdn.odi.org/media/documents/11607.pdf, Accessed 11 March 2021.

Sida 2020 *Evaluation of the Application and Effects of a Human Rights Based Approach to Development: Lessons Learnt From Swedish Development Cooperation. What Works Well, Less Well and Why?* Sida. Available from: https://www.sida.se/en/publications/evaluation-of-the-application-and-effects-of-a-human-rights-based-approach-to-development-lessons-learnt-from-swedish-development-cooperation-what-works-well-less-well-and-why, Accessed 11 March 2021.

Traore, M. A. 2019 *Gestion Des Connaissances: Participation Citoyenne Et La Fourniture Des Services D'assainissement*.

UNU-INRI 2020 *Country Report – Mali. Review of the Status of the Domestication of the Human Rights to Water and Sanitation, and Measures to Leave No-One Behind in WASH in Eight Countries in Africa, Asia and the Pen-European Region*. United Nations University – Institute for Natural Resources in Africa.

WaterAid 2018 *Etude Diagnostic De La Prise En Compte De L'afdh Dans La Legislation En Matiere D'eau Et D'assainissement Au Burkina Faso*.

WaterAid 2020 *Equality, Inclusion and Rights*. Available from: https://washmatters.wateraid.org/sites/g/files/jkxoof256/files/equality-inclusion-and-rights-framework.pdf, Accessed 15 March 2021.

WaterAid Burkina Faso 2018 *Etude Sur La Justiciabilite Des Droits Humains a L'eau Et a L'assainissement Au Burkina Faso*.

WaterLex 2018 *Country Legal Mapping*. HumanRight2Water. Available from: http://humanright2water.org/country-legal-mapping/, Accessed 11 March 2021.

World Health Organization (WHO) and United Nations Children's Fund (UNICEF) 2021 *Progress on Household Drinking Water, Sanitation and Hygiene 2000–2020: Five Years into the SDGs*. WHO and UNICEF, Geneva. Available from: https://www.unwater.org/publications/who-unicef-joint-monitoring-program-for-water-supply-sanitation-and-hygiene-jmp-progress-on-household-drinking-water-sanitation-and-hygiene-2000-2020/, Accessed 5 July 2021.

First received 4 February 2021; accepted in revised form 4 March 2022. Available online 4 April 2022

doi: 10.2166/h2oj.2022.063

Strengthening water, sanitation and hygiene governance: a critical review of Zimbabwe's WASH sector institutional arrangements

L. Dhoba

College of Business, Peace Leadership & Governance, Africa University, P.O. BOX 1320, Mutare, Zimbabwe
E-mail: dhobal@africau.edu

LD, 0000-0003-2964-6633

ABSTRACT

This paper sought to review the structural and operational modalities for Zimbabwe's water, sanitation and hygiene (WASH) sector coordination architecture with an objective to identify gaps and recommend areas for further strengthening. Data were collected using a mixed-method approach involving a questionnaire survey, focus group discussions, key-informant interviews and literature review. The study established that the institution of the National Action Committee (NAC) remains incapacitated to effectively deliver on its mandate. Its various committees are failing to convene meetings on regular basis, hence issues take longer to resolve. Decision-making and role clarity among members still remain indistinct, triggering undesirable conflicts. The legitimacy of the institution of the NAC also remains highly questionable, lacking legal bases to back its operations. In terms of information management, structures and systems are in place, only requiring support to make sure they thrive. Finally, financing including intersectoral coordination remains very weak and would need to be built-on, considering the potential of the coordination mechanism to enhance service delivery. A key recommendation emerging from this study is the urgent need for government to commission a review of the current multi-stakeholder platform in the best interest of aligning the coordination structure to the 2013 national constitution and national water policy.

Key words: coordination, governance, National Action Committee, service delivery, WASH sector, Zimbabwe

HIGHLIGHTS

- On paper, Zimbabwe has an impressive WASH sector governance arrangement.
- The institution of the National Action Committee on WASH remains weak in its operations.
- WASH service delivery in Zimbabwe continues to regress.
- Systems strengthening for sustainable WASH service delivery urgently required to reverse current negative trend.

GRAPHICAL ABSTRACT

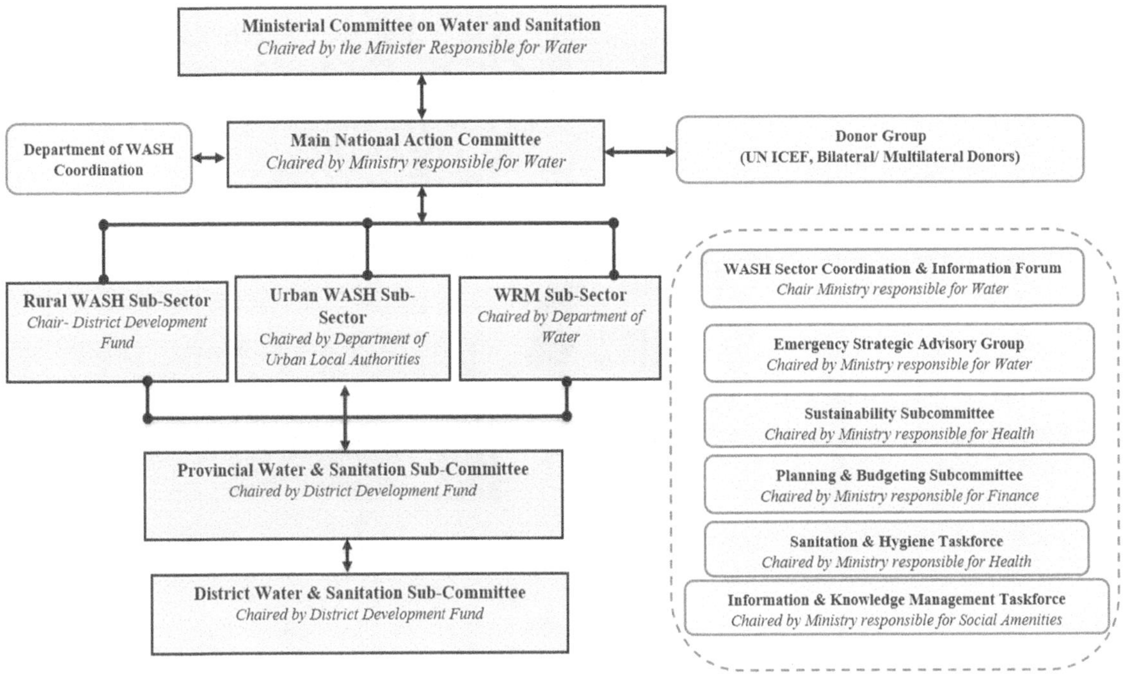

INTRODUCTION

The past 50 years have seen development partners injecting billions of dollars into the water, sanitation and hygiene (WASH) sector across the globe. These investments largely prioritized the development of new WASH infrastructure with the hope that governments will take over and sustain services (Moriarty *et al.* 2013). Post-implementation evaluations, however, reveal the opposite. Failure rates of rural water supply systems remain high across Africa, with various systems tumbling into a state of disrepair far sooner than their design life (Hollander *et al.* 2020). Local funding and accountability for service delivery continue to fall short of what is required to sustain infrastructure, resulting in a high degree of service disruptions.

In September 1990 and as part of the International Drinking Water and Sanitation Decade, African countries adopted the National Action Committee (NAC) for the WASH sector (GoZ 1991). This move according to Butterworth *et al.* (2010) was meant to plug the glaring gap between infrastructure investment and sustainable services through looking beyond infrastructure provision to systems (Figure 1).

In Zimbabwe, the NAC for the Integrated Rural Water Supply and Sanitation Programme was established in the late 1980s with an objective to address the complex challenge of sustaining basic WASH services. This marked the design and implementation of interventions explicitly aimed at addressing system weaknesses (Huston & Moriarty 2018). These interventions sought to understand, engage with and positively influence the network of actors and the interacting factors that deliver services ultimately leading to improved service delivery and sustainability (SWS 2020).

In 1998, a new Water Act was enacted following the rationalization of water resource management approaches resulting in the establishment of the Zimbabwe National Water Authority (ZINWA) and the seven catchment councils. In 2004, a Rural WASH Policy document was drafted for cabinet consideration. A subsequent review of the document in 2009 picked major gaps in the policy, a key part of which was the need for clarity on sector leadership and climate change issues (NAC 2010). These institutional issues were raised immediately after the harsh lessons learnt in 2008/2009 when Zimbabwe experienced one of its worst cholera outbreaks in the living history of mankind (Mason 2009). A ministerial retreat was held in Nyanga in June 2010 for the key sector ministries to review and rejuvenate the sector coordination mechanism (Jiménez *et al.* 2020). The outcome of this ministerial retreat was a coordination framework bringing together the three sub-sectors of rural WASH, urban WASH and the water resource management.

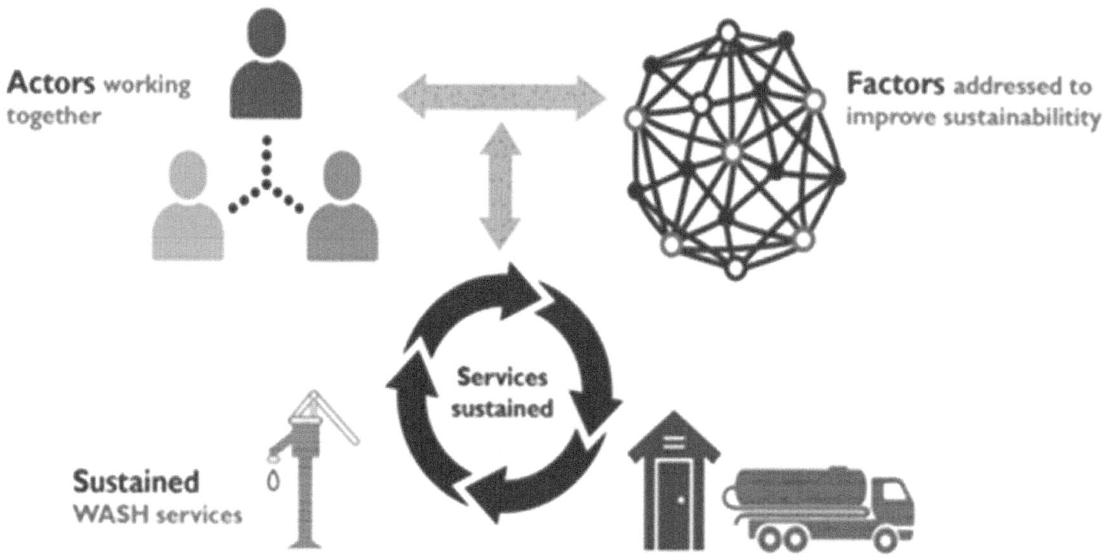

Figure 1 | System for sustaining WASH services (Hollander & Cunningham 2020).

STATEMENT OF THE PROBLEM

The success of the 2030 Agenda requires a clear institutional apparatus to lead the coordination, implementation and monitoring of individual sustainable development goals (SDGs) across sectors (UNDP 2015). Ensuring the existence of vibrant institutional frameworks is critical for countries to ascertain whether these goals can successfully be mainstreamed into national policies. Obscured institutional clarity, roles and responsibilities among stakeholders have always been a challenge hindering the successful implementation of SDGs. While traditional measurements of an activity's successes and impacts focus directly on recipients served (Drinking Water 2020), new metrics are required to monitor and evaluate the impacts of the system's approach. Existing indicator frameworks and metrics, such as the SDG 6 for WASH, fail to capture the changes in a system and its strength although SDGs 16 and 17 make an attempt to capture the same.

Zimbabwe lately adopted a new national blueprint, the National Development Strategy (NDS 1) meant to propel the country into an upper middle-income economy by 2030. However, with 21% of the national population still practicing open defaecation (ZIMSTAT & UNICEF 2019), this remains a pipe dream. While political commitment and leadership are key to redressing this, responsibility for implementation lies with the sector institutional coordination mechanism in place. The recurring WASH challenges including the slow pace at which the sector is reducing the population without access to safely managed services point to a weakened institutional arrangement. It is against this background that this paper commissioned a critical review of the structural set-up and operational modalities for Zimbabwe's WASH sector coordination mechanism with an objective to identify and recommend areas requiring further strengthening.

LITERATURE REVIEW

WASH sector governance arrangement in Zimbabwe

Governance remains a contested concept despite years of academic debate (Glass & Newig 2019). According to the contemporary literature, governance involves a plurality of public and private stakeholders, hybrid practices and is considered to be multi-jurisdictional (Bevir 2011). Consequently, governance is theorized as a multi-dimensional concept covering different actors, processes and institutions involved in political decision-making (Driessen *et al.* 2012). Essentially good governance relies more on a diversity of participating actors.

Zimbabwe's WASH sector has a well-defined mutual accountability mechanism, the NAC. This inter-ministerial structure incorporating the three sub-sectors of urban WASH, rural WASH and water resource management was put in place by the government in 2010 (GoZ 2019). This was mooted as a direct response to contemporary challenges exacerbated by recurring WASH emergencies. Since then, the NAC has been undergoing regular reviews, adapting it to effectively support various transitions (UNGA 2015). Figure 2 gives a pictorial summary of the NAC's structure.

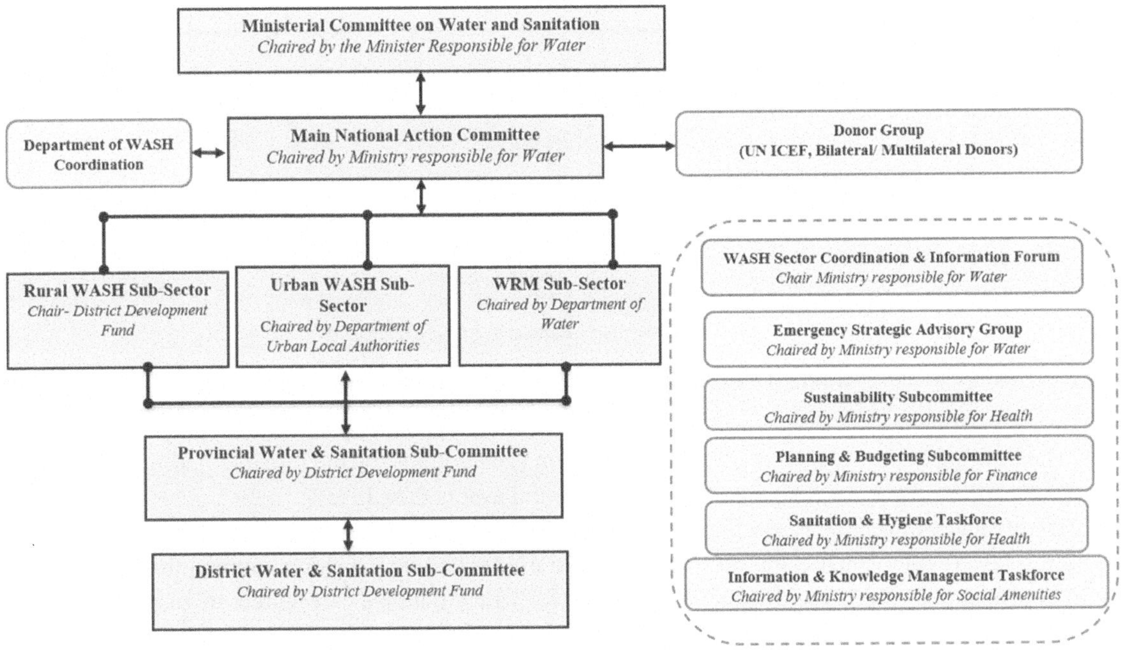

Figure 2 | Sector's mutual accountability mechanisms (adapted from NAC 2010).

Perret *et al.* (2006) argued that providing people with access to essential services such as WASH demands a combination of solutions in the face of a wide range of problems. Non-governmental organizations (NGOs) today view sector governance as a fundamental condition for establishing lasting access to WASH services (Kemp *et al.* 2005). Unfortunately, past studies have tended to concentrate on assessing the suitability of WASH infrastructure with the governance component receiving less attention from many scholars (Dube 2012). There is, therefore, a need to explore this field further to get an appreciation of the decision-making process in the sector (Harper *et al.* 2017). Improvement in institutional arrangements for WASH is bound to result in improved effectiveness of WASH assistance provided to people.

Zimbabwe's WASH sector has key policies and strategies providing sector direction and clarification of roles, namely National Water Policy (2013), Water Act (Chapter 20:24), ZINWA Act (Chapter 20:25), Rural District Act (Chapter 29:13), Urban Councils Act (Chapter 29:15), Public Health Act (Chapter 15:17) and the National Sanitation and Hygiene Strategy (2018–2022). The National Sanitation and Hygiene Policy and the National Water Resource Master Plan are still under development. The Water Act established seven catchment councils and set out ZINWA's responsibility for water resource management. The Water Act also established the water fund as a common pool of funds that the minister responsible for water can use in areas of greatest need. However, WASH services in Zimbabwe have been deteriorating despite this enabling environment. The level of service is not being expanded to boost reach. Existing investments are not being effectively maintained over time, resulting in performance slippage. Fewer people have access to basic water services than 20 years ago. According to JMP (2021), the population accessing at least basic services has declined from 72% in 2000 to 63% in 2021. Essentially, more people are being pushed into reliance on unsafe sources. On sanitation, access to safely-managed sanitation declined from 30% in 2000 to 26% in 2020.

The literature also reveals that recurrent expenditure continues to dominate the central government budget. Public outlay growth has been skewed towards recurrent expenditure, making up more than 90% of total public spending between 2011 and 2018 thus leaving less than 10% of the public budget for capital expenditure (Figure 3). The role of wage expenditure in the public budget limits the government's fiscal policy options.

METHODOLOGY

Research design

Research design is the overall plan for answering the main research question which often includes how, where and when data for a research study are to be collected and analyzed. Teddlie & Yu (2007) cited cross-sectional, longitudinal, experimental and case study as the commonly used types of research designs in governance-related

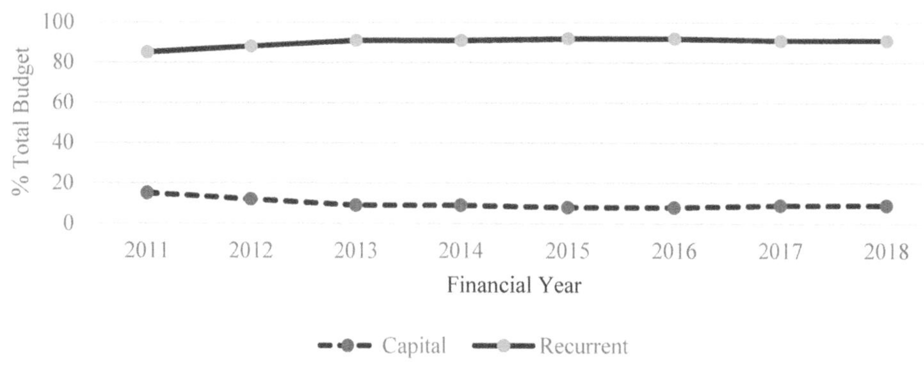

Figure 3 | Government total budget (2011–2018) (adapted from Government Blue Books).

studies. The design that any researcher ultimately settles on is determined by the research questions, the focus of the study and the ability of the design to control behavioural events (Yin 1993).

The current study utilized the case study research design. Although a case is often at a fairly small scale, this study used a larger scale which is the case of Zimbabwe's NAC and its structures running from national right down to the community level. The literature has shown that where multiple case studies are used, they are combined with cross-sectional design elements (Whittington *et al.* 2009). The power of a cross-sectional design lies in its ability to give a snapshot of a situation at a given point in time, a critical element in governance studies (Sedgwick 2014).

Conceptual framework for analysing Zimbabwe's WASH sector institutional arrangements

The conceptual approach to this study (Figure 4) is largely informed by Hollander's system approach. This theory provides guidance on effective perspectives through which local systems can be understood and strengthened to sustain outcomes. Differing rubrics exist around how to segment and understand systems and how to engage with and seek to change them (Hollander 2020).

The strength, weakness, opportunities and threats (SWOT) analytical framework is also going to be evoked to help understand underlying issues and how to further strengthen the system. In assessing multi-stakeholder partnerships' performance, building blocks and governance functions are also going to be used to enable an evidence-based assessment of performance and effectiveness (Pattberg & Widerberg 2014). Building blocks outline key aspects of successful governance through a multi-stakeholder partnership. Governance functions help to identify in more detail how partnerships ought to be designed and later implemented. For WASH, besides structural and institutional factors, Jiménez *et al.* (2016) and SWA (2019) identified several building blocks and governance functions that are key in defining the performance of any system. These include sector policy/strategy, institutional arrangements, sector financing, monitoring, review, and capacity development (Figure 5).

Study population

The study population was pooled from the sector's six key constituencies, namely government, development partners, civil society organizations, private sector players, utilities, and research and learning institutions. These six

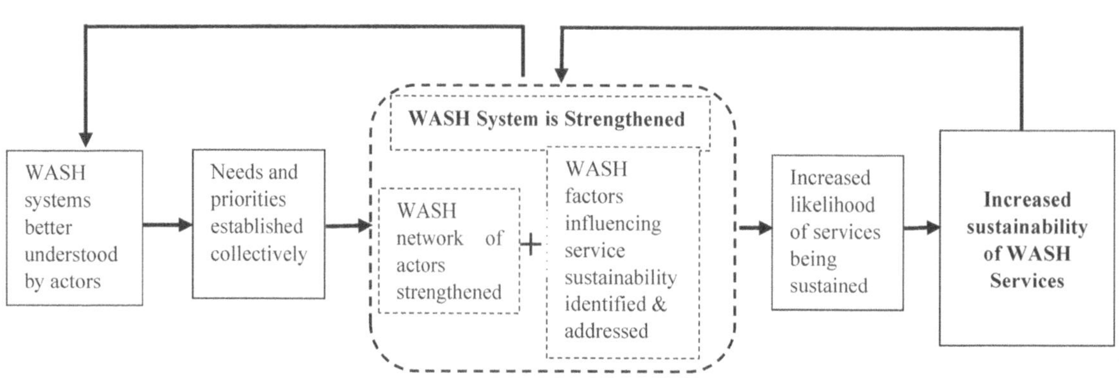

Figure 4 | System approach (adapted from Hollander 2020).

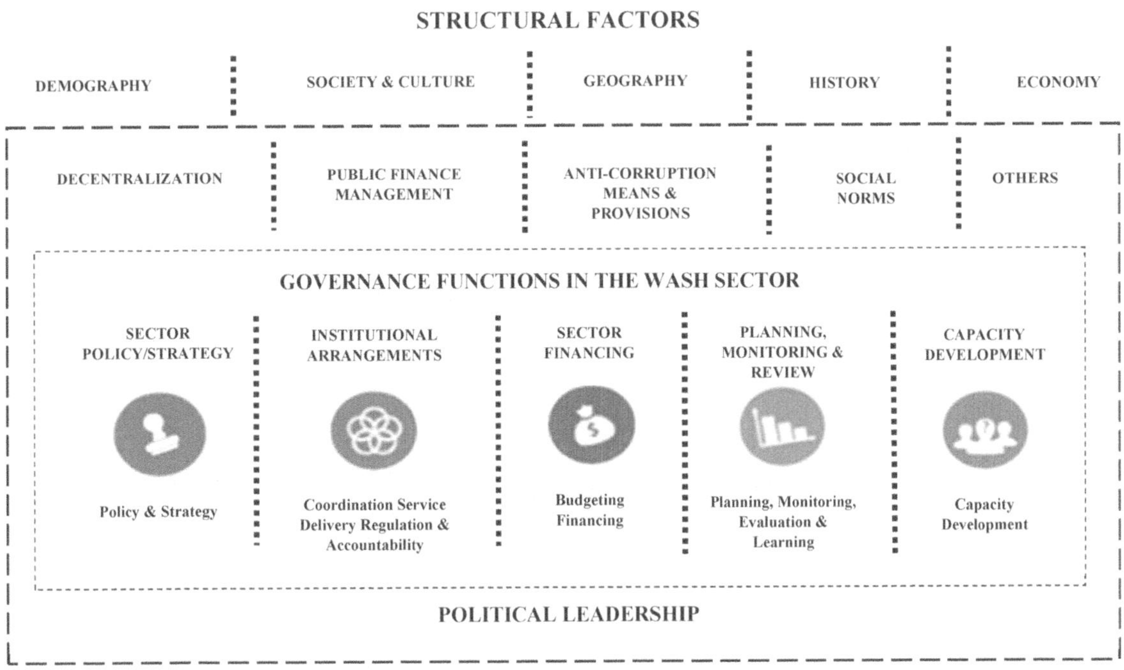

Figure 5 | WASH enabling environment framework (adapted from UNICEF 2016).

constituencies together constitute the WASH sector's mutual accountability mechanism which according to SWA (2020) provides a platform for all partners to tender commitments and hold each other accountable.

Study sample

The study sample was pooled from the sector's WASH Sector Coordination and Information Forum, a multi-stakeholder platform that brings together sector players from all the six sector constituencies constituting the WASH sector's mutual accountability mechanism. In total, 43 key sector organizations were reached out to through the forum's mailing list. Due to limited time and resources, convenience sampling was employed in selecting a study sample from the population.

Data collection tools

In this study, a structured questionnaire was designed as the main data collection tool circulated through an online survey to members of the WASH Sector Coordination and Information Forum (WSCIF). This helped gather data on existing capacities of the sector's coordination mechanism to deliver. Responses were made available for analysis as soon as the respondents clicked the submit button at the end of the survey questionnaire.

A follow-up key informant interview was conducted with the NAC secretariat followed by two focus group discussions, one with chairs of the various committees of the NAC and the other one with Provincial Focal Agencies responsible for the 10 administrative provinces. A mini desk-study analysis was also conducted with an objective to get primary data on the sector coordination mechanism, its policies, strategies and financing models.

STUDY FINDINGS

There was an overwhelming response to the questionnaire circulated virtually to sector partners through an online survey as the main study tool. The study reached out to 43 active organizations under the WSCIF out of which a total of 40 agencies responded and 37 ultimately managed to fully complete the questionnaire giving an overall questionnaire response rate of 86% as summarized in Figure 6.

Respondents were pooled from across all the six constituencies constituting the sector's mutual accountability mechanism, namely government, private sector, utilities, research and learning institutions, civil society and development partners. The mailing list for the study sample was made available by the NAC Secretariat, a designated institution with the responsibility to coordinate all players in the sector.

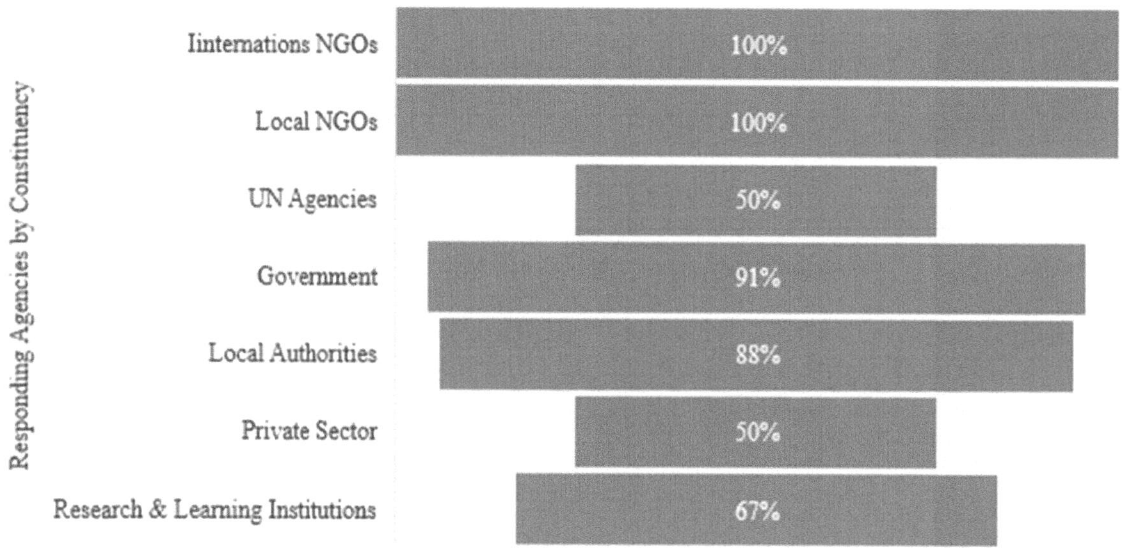

Figure 6 | Questionnaire response rate by constituency.

Institutional issues and their bearing on sector coordination

WASH sector governance is guided by the 2013 National Water Policy. However, the bulk of the provisions of this sector guiding document is still to be fully implemented 9 years after its promulgation. The policy recommended the setting up of a National Water Supply Services Utility expected to take over the provision of water supply from the ZINWA.

> *'However, this next phase of ZINWA's evolution has not happened, and there appears to be limited appetite within central government to effect such a change which means ZINWA has a long way to go to help recover costs and adopt a modern customer focus.'* (Focus Group Discussion, female respondent)

In the same manner, the National Water Policy proposed the creation of a Water and Wastewater Services Regulatory Unit to monitor all water supply and sanitation services, receive and assess tariff applications in collaboration with relevant ministries and oversee the licensing of water service providers by water services authorities.

> *'While progress has been made towards the creation of this strategic institution, the more general reluctance by Government to establish additional new institutions has overall slowed the pace leading to the setting up of the Water and Waste Water Regulatory Unit.'* (Key Informant Interview, male respondent)

Respondents indicated that delay in the operationalization of the National Water Policy has seen the ZINWA continuing to play regulator and service provider roles, local authorities not able to freely choose service providers, rampant and uncontrolled groundwater mining, and total disregard for WASH service delivery standards. It was also cited that continued delay in the approval of the 2016 draft National Sanitation and Hygiene Policy has resulted in a lack of direction in the implementation of life-saving sanitation and hygiene-related interventions.

Outline of roles for committees constituting the coordination architecture

The NAC for WASH operates at four levels, namely national, provincial, district and community level. From the questionnaire survey, 65% of responding organizations confirmed that the roles and responsibilities of the NAC and its standing committees are clear to comprehend. The coordination mechanism document was cited by 80% of the respondents as a key reference document for clarity on roles and responsibilities including membership of the various committees. Regarding effectiveness of the structure in discharging its mandate, 75% of the responding organizations indicated that the various committees of the NAC still remain very weak as more than 80% of the committees struggle to convene coordination meetings as per their set terms of reference. In the past 8 months (July 2020–February 2021), the urban WASH sub-committee only meet once with the other two committees of rural WASH and water resource management all reporting nil meetings. The research established that many of the officers constituting various committees of the NAC lack the motivation to work.

'We are dealing with civil servants already demoralized by the current remuneration package. Most members are either incapacitated to come to work or when they do come, they are seized with personal businesses meant to subsidize their meager salaries.' (Focus Group Discussion, female respondent)

Participation of committee members in the NAC business is motivated by incentives and envisaged personal benefits.

'Members avail themselves to WASH business only if there is a monetary incentive attached to the activity. More often than not, monthly coordination meetings at sub-national level are delayed or totally called off due to poor attendance by members.' (Focus Group Discussion, female respondent)

All these highlighted developments cripple NAC's sector leadership and coordination capacity.

Clarity of roles and responsibilities for individual NAC member institutions

When asked about the clarity of roles and responsibilities of individual agencies constituting NAC, 88% of stakeholders from line ministries responded with a 'Yes', meaning that the mandates were clear. However, the rest of the other six constituencies reported that the roles and responsibilities were indistinct (Figure 7). The District Development Fund, for example, was mentioned as misplaced to chair the rural WASH sub-committee and its sub-national level structures.

'...the District Development Fund is a technical arm in terms of water supply and chairmanship of the coordination structure at various levels ought to be left to Ministry responsible for Local Government which is better placed to play the coordination role through its Provincial and District Development Coordinators' offices.' (Focus Group Discussion, female respondent)

The study also confirmed that:

'In the context of devolution, Local Government has convening powers crucial when calling for coordination meetings' (Key Informant Interview-male respondent)

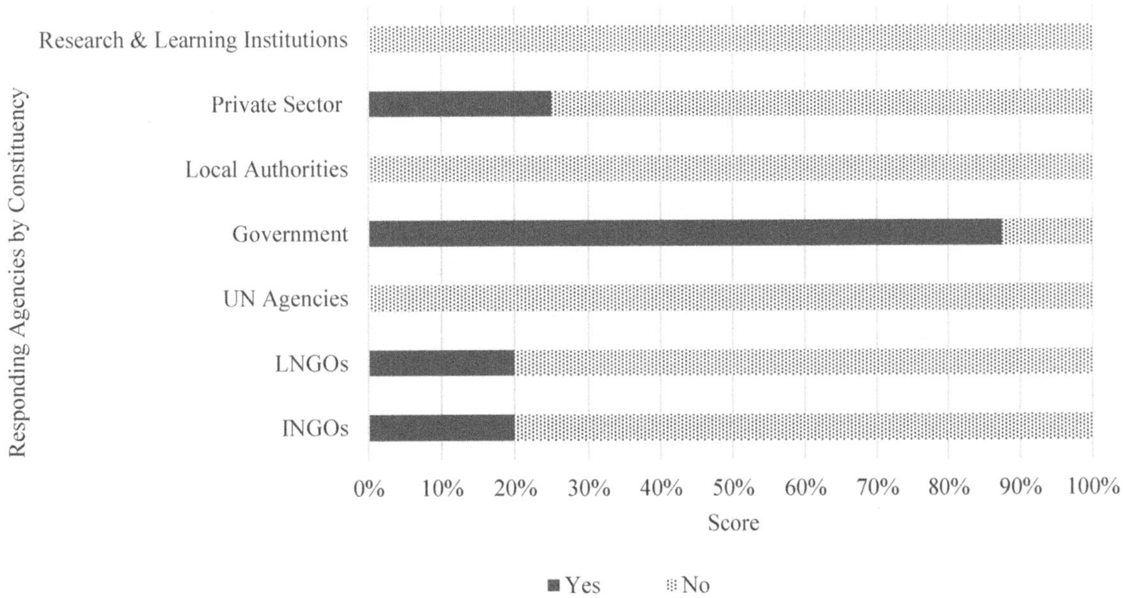

Figure 7 | Are roles and responsibilities of individual NAC members clear?

Conflict of constitutional mandates was also mentioned in one of the Focus Group Discussions.

'Conflict exists between government agencies like the Department of Water, Local Authorities, ZINWA and District Development Fund when it comes to water supply; Department of WASH Coordination and the Ministry of Health when it comes to implementation of the sanitation and hygiene component; government agencies and civil society partners when it comes to programme implementation at community level.' (Focus Group Discussion-male respondent)

Part of the reasons why institutions continue to perform overlapping functions arose from the reality of non-implementation of some of the National Water Policy (2013) proposals. Full implementation would have helped realignment of institutional roles and responsibilities. All these conflicting roles derail programme implementation often resulting in targets missing timelines and certain planned works getting suspended. The study also revealed weakened WASH sector coordination arrangement for members. Participation in WASH business by line ministries remains voluntary and with limited consequences for non-cooperation and performance.

Legitimacy of the institutional arrangement

When asked about the legitimacy of the NAC and its sub-national level structures, questionnaire respondents expressed mixed feelings (Figure 8).

One key informant reported the need for government to legalize the institution of the NAC, thus giving it legal-authority and obligation to lead and direct the sector.

'No matter how good the arrangement is, the fact that NAC was not established through an act of parliament leaves the whole coordination structure as a club whose membership remains voluntary.' (Key Informant Interview, male respondent)

'This accounts for the reason why committees of NAC have not been able to meet over time, and why sub-sectors operate without annual work plans and budgets.' (Focus Group Discussion, male respondent)

The fact that both national and sub-national level structures of the NAC are not recognized at law is a drawback on their level of sector authority, thus affecting their effectiveness when discharging set mandates.

Planning capacity

The questionnaire survey revealed that the NAC has a dedicated taskforce responsible for sector planning and budgeting and chaired by the Ministry of Finance and Economic Development. Its key mandate lies in guiding the development of sector strategic and operational plans including resource mobilization for operationalization

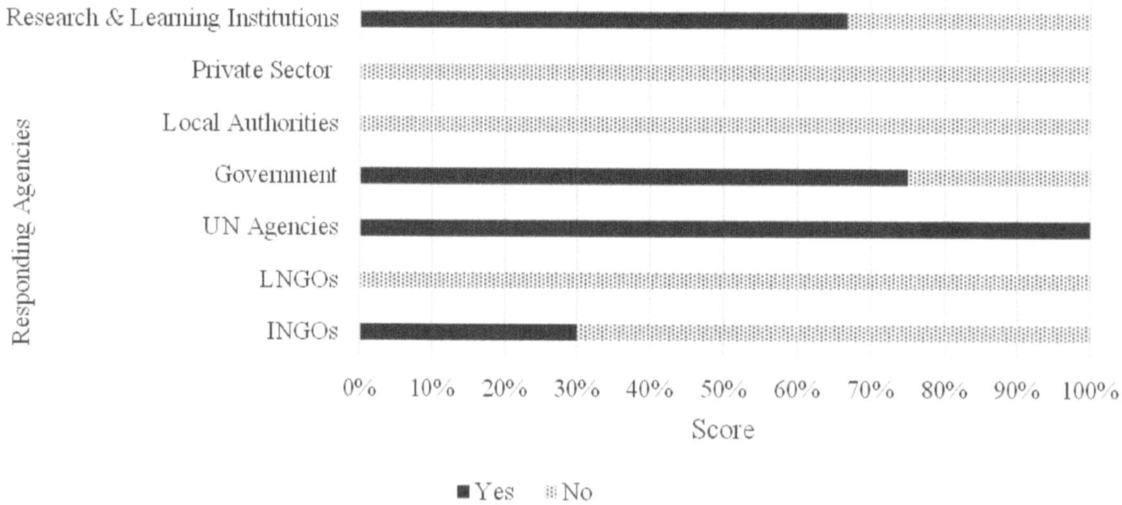

Figure 8 | In your own view is the institution of the national action committee legitimate?

of the same. One Focus Group Discussion held, however, revealed that this taskforce has not been meeting since 2010.

> *'The Ministry of Finance which is supposed to chair this Taskforce is not bothered whether members meet or not. The desk officer responsible for WASH always moans of being overwhelmed.'* (Focus Group Discussion, female respondent)

The planning function has therefore been wholly relegated to the Department of WASH Coordination which develops annual sector plans and budgets that it exclusively presents to various donors including treasury.

> *'This status quo has killed the multi-sectoral approach that the WASH sector is proud of, thus stifling issues of mutual accountability.'* (Focus Group Discussion, male female respondent)

The study also established that strategic planning is only happening under the WASH emergency arm with the Emergency Strategic Advisory Group spearheading the initiative. From the questionnaire administered, 64% of the respondents felt that there is a good level of multi-stakeholder strategic planning under the emergency preparedness and response wing with facilitation from UNICEF. Two WASH contingency plans, the Zimbabwe WASH Cholera Contingency plan (November 2019–July 2021) and the Zimbabwe WASH Sector COVID-19 Contingency Plan (March 2020–December 2021), were mentioned.

Decision-making and inter-sectoral coordination

Study findings revealed that the institution of the NAC has a very weak decision-making system. This is partly because of the passiveness of its committees many of which have not been meeting for some time now. In that regard, many issues of a policy nature requiring government position and leadership remain unresolved.

> *'The sanitation and hygiene policy remains a draft document since its development in 2016; the water policy is still to be fully operationalized since 2013 and research in the sector remains under funded regardless of the existence of a committee dedicated to research and development.'* (Key Informant Interview, male respondent)

Government constituency acknowledged that the appointment of the institution of the Provincial Focal Agency has helped re-activate WASH sector coordination at a sub-national level.

> *'...weekly provincial update reports and meeting minutes with very clear action points are now being compiled and shared both horizontally and vertically to influence policy.'* (Focus Group Discussion, male respondent)

This alone has helped strengthen emergency response, and coordination at the sub-national level although this is limited to the WASH emergency wing.

When it comes to inter-sectoral coordination, 80% of responding organisations felt that intersectoral coordination is rather weak requiring urgent attention at all levels. From the responses to questions on the WASH sector tools and guideline documents that sector agencies indicated were familiar with, not even a single tool or guideline with a thrust on intersectoral coordination, assessment, planning or programming was mentioned.

Sector information and knowledge management

90% of the questionnaire respondents answered 'yes' to the question on whether an enabling environment exists for data generation and knowledge management to inform evidence-based sector programming. Some of the enablers reported with high frequencies of above 60% include the existence of the Information and Knowledge Management Taskforce and the existence of sub-sector information management systems including a platform for knowledge management. On assessing the robustness of this enabling environment through Focus Group Discussions, the respondents however indicated that the Information and Knowledge Management Taskforce is poorly resourced and comprised of individuals with very little if any background and qualifications in information management (human resource capacity gap). These two key issues have great implications for the success of any initiative developed under this Taskforce. In one of the Focus Group Discussions, it emerged that:

'Regardless of how incredible the Rural WASH Information Management System (RWIMS) is and the demand that it commands sector-wide, its sustainability across districts still remains uncertain. The WASH Connector regardless of how strategic it was in documenting and sharing sector knowledge prematurely succumbed to poor resourcing.' (Focus Group Discussion, male respondent)

Overall, it emerged that Rural Local Authorities are benefitting a lot from planning data from the Rural WASH Information Management System (RWIMS), although many have been struggling to keep the system afloat. For Service Level Benchmarking, Urban Local Authorities also continue to finance annual peer review meetings to update their reports although the best arrangement would have been to convert the process into a system for ease of access by the sector, including management and updating of performance data moving forward.

Financing of the institutional arrangement

Focus Group Discussions held with stakeholders at both national and sub-national levels on sector funding revealed that government spending in WASH has been on an increase over the past 5 years, although with some notable yearly fluctuations. For the financial year 2021, 'real WASH' expenditure totaled US$186 million; a rise by 4% from the 3.6% (US$91 million) reported in 2020 (Figure 9).

From the Focus Group discussions held, it emerged that,

'As a percentage of the national budget, WASH expenditure has averaged 3.3% over the last four years; well below the SWA commitment of 7% per financial year.' (Key Information Interview, male respondent)

Using data gathered through the national fiscus, WASH spending as a percentage of Gross Domestic Product (GDP) averages 0.6% between 2017 and 2020, and 0.5 percentage points from 0.6% in 2020 to 1.1% in 2021 (see Figure 10).

One of the key informants revealed that this is mainly due to increased planned outlays for 'big WASH' (dam construction) and improved allocation for environmental health sub-programmes that are critical for promoting

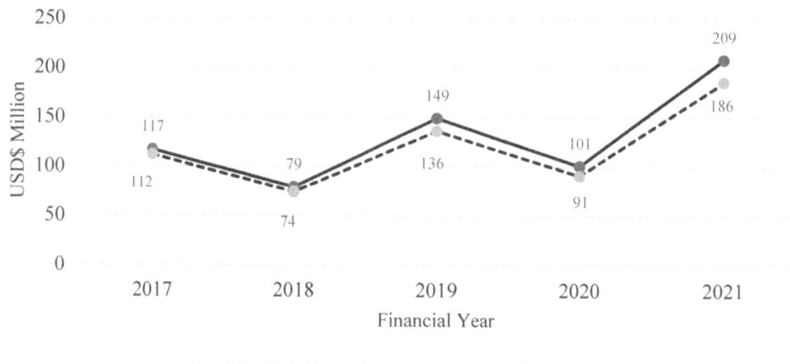

Figure 9 | Nominal and real WASH sector (Spending Trends, 2017–2021).

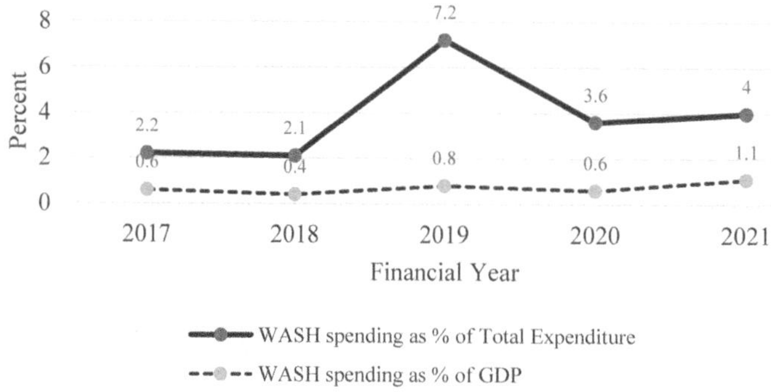

Figure 10 | WASH spending (as percentage of total spending and percentage of gross domestic product).

hygiene and monitoring of water quality standards in light of the current COVID-19 pandemic. It was also further claimed that the national fiscus budget for the sector remains principally a 'big WASH-focused budget' which tends to swallow 'real WASH' issues.

Follow-up Focus Group Discussions revealed that, institutionally, the ZINWA administers the largest proportion of the WASH sector budget, with an average of 70% for the period 2017–2021. This is mainly due to the so-called 'big WASH' implemented by the water authority. The Departments of Environmental Health and Local Authorities account for the second and third largest shares at 8.8 and 8.6% of total WASH spending in 2021 (Figure 11).

DISCUSSION

Findings from the current study established that WASH services constitute basic human rights that the Zimbabwean government is obligated to deliver and the NAC provides a whole-of-government framework for fulfilling this obligation. However, even though this coordination structure has evolved over time, absence of a clear guiding framework on how this structure is expected to link with other sectors to leverage from the comparative advantage poses a very big limiting factor. In his study of the Ugandan WASH sector, Handley (2009) argued that limited sectoral coordination and dialogue across sectors, national sector actors and development partners have resulted in duplication, high transaction costs and fragmented monitoring and reporting. In support of Handley's argument, UNICEF (2021) contended that with emerging developments propelled by COVID-19 and climate change, multi-sectoral coordination is one of the ways in which some progressive countries are able to better develop, finance, and sustain sector interventions for results and impact. These inter- and across-linkages will help explain how WASH can best align with and feeds into broader national health and climate rationale.

The core committees of the NAC are manned by line ministries with an interest in WASH. Members constituting these various committees have their mandates in the NAC business derived from respective acts and statutes establishing them. However, the reported conflict of roles and responsibilities and in some cases misplaced roles accorded to some agencies all affect the efficiency and effectiveness of the coordination machinery. This also affects issues of social capital among members, accountability, and transparency, all of which are critical governance success factors. Pswarayi-Riddihough (2018) echoed that good governance requires credible and trustworthy institutions built on principles of transparency and accountability.

Beyond the issue of membership, roles and responsibilities, the NAC ought to be accorded legal rights through an act of parliament for it to attract necessary authority and command to coordinate the sector. This according to Kativhu *et al.* (2021) is in line with the recommendations of the International Drinking Water and Sanitation Decade which dictated the need to make sure that beyond the formation of National Action Committees for WASH, parties go ahead and regularize this institution through linking it to an act of parliament. Zimbabwe had such an opportunity in 2013 when it developed its National Water Policy. A policy statement on the formation of the NAC was going to help regularize, legalize and give power to the structure. Unfortunately, the policy went ahead to propose new institutions, one such institution being the Water and Wastewater Regulatory Unit (GoZ 2013).

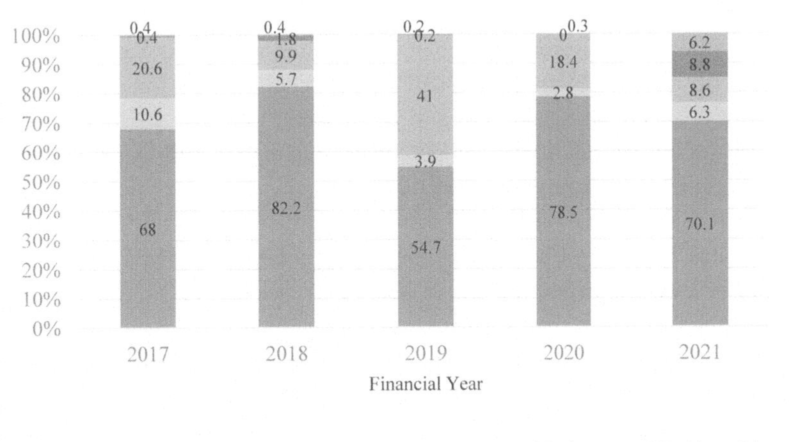

Figure 11 | WASH sector spending by institution, financial year 2017–2021.

Regarding sector planning and financing, the significant capacity challenges noted through this study affect the performance of the NAC in its WASH service delivery role. This same argument was put forward by Dahlstrom *et al.* (2010) that lack of resources which is an indicator of poor planning hinders smooth coordination of WASH structures and interventions. The fact that the resource mobilization machinery of the NAC, the planning and budgeting sub-committee is struggling with elementary things like convening coordination meetings makes the challenge insurmountable. Muyambo & Klaassen (2015) noted that due to weakened planning capacity, many towns and utilities in Zimbabwe are operating without current and validated master plans. According to AMCOW (2011), resource constraints have also grossly undermined the NAC's short- to medium-term efforts to rejuvenate the sector and make it responsive to the pressing rural and urban WASH service demands. Performance of the whole institutional structure thus always remains sub-optimal. Current sector financing falls far short of what national governments committed at the 2014 SWA investment conference. At this conference, governments committed to allocate 7% of their individual annual budgets to WASH. The current allocation is also below the ideal funding for WASH proposed under the Foster & Briceno-Garmendia (2010) requiring Sub-Saharan African countries to allocate 0.9% of GDP to achieve improved WASH outcomes.

Strengthened sector information and knowledge management remain key in ensuring the existence of an effective coordination mechanism. Steyn & Van Belle (2015) argued that this element is a core component of development strategies employed by governments, multilateral agencies, NGOs and academia yelling for success. Schaub-Jones (2013) argued that harnessing information and communication technologies brings about radical improvements in the way that health, education and other sectors function. The RWIMS developed for the rural WASH sub-sector in Zimbabwe always stands out as a reference point. This system has been useful in directing evidence-based sector planning, monitoring reporting, and resource mobilization. A lot however still needs to be done to perfect the system and ensure that it operates efficiently. Currently, the institution of the Rural District Council is seized with a huge budget required to operate, maintain and sustain the system.

RECOMMENDATIONS

Based on the above study findings, closer integration of the sector's WASH components is needed, further reinforced by streamlining of agency responsibilities particularly for state institutions based on constitutionally aligned competences. This will also help frame role clarity for WASH functions between central and local government institutions, state agencies and external development partners, and between state and non-state actors. This will help provide an appropriate basis for the development of a national WASH sector investment strategy, and Local Authority level WASH investment plans thus making WASH financing easier to frame in terms of public–private sector and civil society resources.

Government through the ministry responsible for water also ought to urgently commission a review of the current WASH sector coordination mechanism which was last rejuvenated in 2010. This ought to be done with a view to further strengthen the coordination mechanism in line with the tenets of the 2013 national water policy and the 2013 national constitution. The policy in one of its clauses recommending the setting up of additional WASH management institutions, key among which is the water and wastewater regulatory authority. To strengthen and make effective operations of the NAC Secretariat, the Department of National WASH Coordination (DWASHC), there is an urgent need to expand and decentralize DWASHC. This expansion will be in terms of functions where sector regulation and standards are anticipated additions to the current focus on just coordination as per the proposed structure in Figure 12.

The proposed DWASHC structure is subject to further sector discussions to better refine it. In refining the structure some of the focus may be on addressing the tensions between 'big-WASH/dams' and 'small/real-WASH'. The lead ministry remains principally a 'big-WASH focused institution' and on budgets 'real-WASH' issues tend to be swallowed.

There is also urgent need to review terms of reference and membership for all existing committees and taskforces of the NAC with an objective to strengthen the structure. This will see new terms of reference developed where they don't exist. The many political and development structures coming in with devolution are all key enablers for a strengthened sector coordination mechanism to back this review. This move will thus be one attempt to re-align the current coordination mechanism to changes in the political landscape. The 'new-look structure' ought to be multi-stakeholder and multi-sectoral allowing all standing sector constituencies

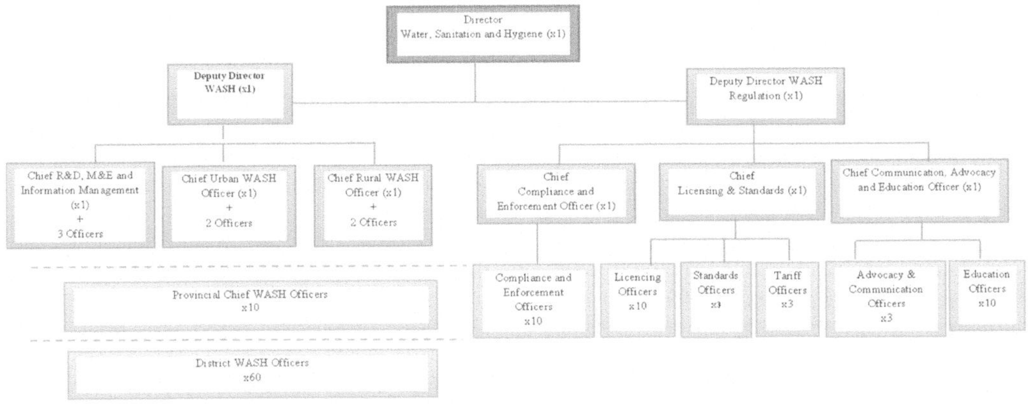

Figure 12 | Proposed structure for a strengthened department of national WASH coordination.

to fully subscribe to the coordination mechanism and mandated with clear responsibilities. Additionally, a strategic operational framework also ought to be mooted to anchor the coordination structure.

Along with the above, the lead ministry also ought to capitalize on the review of the sector coordination mechanism to make sure that the 'new-look structure' is regularized before the law. This will allow the NAC structure to emerge as a legal establishment thus giving it more power, recognition and authority required for effective sector coordination, decision-making and leadership. The enabling statutes are already there through the 2013 National Water Policy and the Water Act (Chapter 20:24) to cause that to happen, especially now when the Water Act is undergoing review to incorporate the water and wastewater regulatory functions previously held under the ZINWA Act (Chapter 20:25).

CONCLUSION

The slow pace at which the country continues to reduce the number of people without access to safely managed WASH services is evidence of a weakened coordination mechanism. Sustainable WASH services require a delivery structure and process where central government, local authorities, civil society organizations, research and learning institutions, private sector and rights holders fully engage each other, have mutual trust and effectively deliver on their roles. This local compact ought to fully own its service delivery challenges and be committed to work together to solve them with public sector support through a robust financing mechanism. A more enabling public sector system (governance and leadership) such as this one creates an appropriate environment within which the capacities of different stakeholders can be deployed effectively, at higher levels of value for money and in ways that are not overly 'projectized'. The institution of the National Action Committee fittingly reviewed and capacitated as guided by the above recommendations is a potential and strategic game changer with great potential to revolutionize WASH service delivery and enhance the attainment of SDG6 on clean water and sanitation by 2030.

DATA AVAILABILITY STATEMENT

All relevant data are included in the paper or its Supplementary Information.

REFERENCES

AMCOW 2011 *An AMCOW country status overview: water supply and sanitation in Zimbabwe-turning finance into services for 2015 and beyond.* Water and Sanitation Programme, Harare. Available from: https://www.wsp.org/sites/wsp/files/publications/CSO-Zimbabwe.pdf. (accessed 17 November 2020).

Bevir, M. 2011 Governance as theory, practice and dilemma. In: *The SAGE Handbook of Governance.* http://dx.doi.org/10.4135/9781446200964.n1.

Butterworth, J., Warner, J., Moriarty, P., Smits, S. & Batchelor, C. 2010 Finding practical approaches to integrated water resources management. *Water Altern* **22** (3), 68–81.

Dahlstrom, C., Peters, B. G. & Pierre, J. 2010 *Steering From the Centre: Strengthening Political Control in Western Democracies.* University of Toronto Press Driessen, Toronto.

Driessen, P. P. J., Dieperink, C., van Laerhoven, F., Runhaar, H. A. C. & Vermeulen, W. J. V. 2012 Towards a conceptual framework for the study of shifts in modes of environmental governance: experiences from The Netherlands. *Env. Pol. Gov.* **22**, 143–160.

Drinking Water 2020 Public water supplies for England and Wales. Drinking Water Inspectorate, London. Available from: http://www.dwi.defra.gov.uk

Dube, T. 2012 Emerging issues on the sustainability of the community based rural water resources management approach in Zimbabwe: a case study of Gwanda district. *International Journal of Development and Sustainability* **1** (3), 644–655. Available from: http://isdsnet.com/ijds-v1n3-3.pdf. (accessed 16 December 2020)

Foster, V. & Briceno-Garmendia, C. 2010 *Africa's Infrastructure: A Time for Transformation.* Africa Development Forum. World Bank, Washington, DC. Available from: https://openknowledge.worldbank.org/handle/10986/2692

Glass, L. M. & Newig, J. 2019 Governance for achieving the sustainable development goals: how important are participation, policy coherence, reflexivity, adaptation and democratic institutions? *Earth System Governance* **2**. https://doi.org/10.1016/j.esg.2019.100031.

Government of Zimbabwe (GoZ) 1991 *Evaluation of the Integrated Rural Water Supply and Sanitation Project: Mt Darwin District.* Government of Zimbabwe, Harare.

Government of Zimbabwe (GoZ) 2013 *National Water Policy.* National Action Committee, Harare.

Government of Zimbabwe (GoZ) 2019 *WASH Joint Sector Review.* National Action Committee, Harare.

Handley, G. 2009 *Sector Budget Support in Practice – Literature Review.* ODI and Mokoro, London. (November). Available from: http://www.odi.org/sites/odi.org.uk/files/odiassets/publications-opinion-files/5575.pdf (accessed 10 December 2020).

Harper, S. L., Wright, C., Masina, S. & Coggins, S. 2017 Climate change, water, and human health research in the Arctic. *Water Security* **10**, 100062. https://doi.org/10.1016/j.wasec.2020.100062.

Hollander, J. A. & Cunningham, J. 2020 Empowerment self-defense training in a community population. *Psychology of Women Quarterly* **44** (2), 187–202.

Hollander, D., Ajroud, B., Thomas, E., Peabody, S., Jordan, E., Javernick-Will, A. & Linden, K. 2020 Monitoring methods for systems strengthening activities towards sustainable water and sanitation services in low-income settings. *Sustainability* **12** (17), 7044. http://dx.doi.org/10.3390/su12177044.

Huston, A. & Moriarty, P. 2018 *Building Strong WASH Systems for the SDGs: Understanding the WASH System and its Building Blocks.* IRC, The Hague.

Jiménez, A., LeDeunff, H., Avello, P. & Scharp, C. 2016 Enabling environment and water governance: a conceptual framework. In *Accountability for Sustainability Partnership.*

Jiménez, A., Saikia, P., Gine, R., Avello, P., Leten, J., Liss Lymer, B., Schneider, K. & Ward, R. 2020 Unpacking water governance: a framework for practitioners. *Water* **12** (3), 827.

JMP 2021 Progress on household drinking water, sanitation and hygiene 2000-2020: Five years into the SDGs. World Health Organization (WHO) and the United Nations Children's Fund (UNICEF), Geneva, Switzerland.

Kativhu, T., Madzivanyika, T., Nunu, W., Macherera, M. & Chinyama, A. 2021 Sustainability of water facilities under community based management in Zimbabwe. *Journal of Water Supply: Research and Technology-Aqua* **71**. doi:10.2166/aqua.2021.089.

Kemp, R., Parto, S. & Gibson, R. B. 2005 Governance for sustainable development: moving from theory to practice. *International Journal of Sustainable Development* **8** (2), 12–30.

Mason, P. R. 2009 Zimbabwe experiences the worst epidemic of cholera in Africa. *Journal of Infection in Developing Countries* **3** (2), 148–151. https://doi.org/10.3855/jidc.62.

Moriarty, P., Smits, S., Butterworth, J. & Franceys, R. 2013 Trends in rural water supply: towards a service delivery approach. *Water Altern* **30** (6), 329–349.

Muyambo, A. & Klaassen, W. 2015 Capacity building for Zimbabwean Urban Local Authorities in water supply needs assessment and business opportunities. In: *Study Mission Commissioned by the Netherlands Enterprise Agency.*

National Action Committee 2010 Water, Sanitation and Hygiene (WASH) Sector Coordination Mechanisms. Government of Zimbabwe, Harare.

Pattberg, P. & Widerberg, O. 2014 *Transnational Multi-Stakeholder Partnerships for Sustainable Development: Building Blocks for Success. Report R-14/31.* IVM Institute for Environmental Studies.

Perret, S., Farolfi, S. & Hassan, R. 2006 *Water governance for sustainable development.* Earthscan, London. Available from: https://hdl.handle.net/10568/37398

Pswarayi-Riddihough, I. 2018 Keynote address: Transparency and public accountability is at the core of good corporate governance. In: *Launch of the Handbook on Good Governance for Chairman and Boards of Directors of Public Enterprises* (V. Kanagasabapathy, ed.). The Institute of Chartered Accountants of Sri Lanka: CA Sri Lanka, Colombo.

Schaub-Jones, D. 2013 *Using a simple ICT solution to support better water supply in urban Angola.* SeeSaw, Luanda.

Sedgwick, P. 2014 Cross sectional studies: advantages and disadvantages. *BMJ* **348**, g2276. https://doi.org/10.1136/bmj.g2276.

Steyn, J. & Van Belle, J. P. 2015 Beyond development. Time for a new ICT4D paradigm? In: *Proceedings of the 9th IDIA Conference, IDIA2015*, Nungwi, Zanzibar. pp. 266–285. ISBN: 978-0-620-68395-1

SWA 2019 Sector Ministers' Meeting. Available from: https://www.sanitationandwaterforall.org/meeting/sector-ministers-meeting-2019

SWA 2020 Finance Ministers' Meeting. Available from: https://www.sanitationandwaterforall.org/meeting/2020-finance-ministers-meetings

SWS 2020 Sustainable WASH Systems Learning Partnership: End of Project Report. University of Colorado Boulder, Boulder, CO. Available from: https://www.globalwaters.org/sites/default/files/sws_eop_annexes_for_508.pdf (accessed 15 March 2020).

Teddlie, C. & Yu, F. 2007 Mixed methods sampling: a typology with examples. *Journal of Mixed Methods Research* **1**, 77–100. https://doi.org/10.1177/2345678906292430.

UNDP 2015 WASH and accountability: Explaining the concept Accountability for Sustainability Partnership: UNDP Water Governance Facility at SIWI and UNICEF. Stockholm and New York. Available from: http://www.watergovernance.org/ (accessed 12 November 2020).

UNICEF 2016 *Strengthening Enabling Environment for Water, Sanitation and Hygiene (WASH): Guidance Note*. United Nations Children's Fund (UNICEF), New York. Available from: https://www.unicef.org/wash/files/WASH_guidance_note_draft_10_3_hr.pdf (accessed 5 January 2021).

UNICEF 2021 The State of the World's Children. United Nations Children's Fund (UNICEF), New York. Available from: https://www.unicef.org/reports/state-worlds-children-2021

United Nations General Assembly 2015 Resolution adopted by the General Assembly on 25 September 2015. In: *Proceedings of the Seventieth Session of the UN General Assembly*. Available from: https://www.un.org/en/development/desa/population/migration/generalassembly/docs/globalcompact/A_RES_70_1_E.pdf (accessed 5 January 2021).

Whittington, J. L., Coker, R. H., Goodwin, V. L., Ickes, W. & Murray, B. 2009 Transactional leadership revisited: self other agreement and its consequences. *Journal of Applied Social Psychology* **39** (8), 1860–1886.

Yin, R. K. 1993 *Applications of case study research*. Sage Publications, Thousand Oaks, CA.

ZIMSTAT & UNICEF 2019 Zimbabwe Multiple Indicator Cluster Survey 2019, Snapshots of Key Findings. ZIMSTAT and UNICEF, Harare.

First received 31 January 2021; accepted in revised form 24 April 2022. Available online 6 May 2022

© 2022 The Authors
doi: 10.2166/h2oj.2022.060

Evidenced transformations: using data to demonstrate improvement in school-based access to water, sanitation and hygiene in Uttar Pradesh, India

Kumar Bikram and Ram Manohar Mishra*
United Nations Children's Fund (UNICEF), Lucknow, India
*Corresponding author. E-mail: rmmishra@unicef.org

KB, 0000-0003-2412-9696

ABSTRACT

The availability and access to water, sanitation and hygiene (WASH) facilities in schools have been associated with enhanced learning and nutritional outcomes among children in India and elsewhere. However, there is limited documentation of successful scaled-up models to improve WASH facilities in schools. This paper describes an innovative 'transformation campaign' to strengthen WASH facilities in government-run elementary schools in Uttar Pradesh, the most populous state in India. The campaign was launched in 2018 followed by an online monitoring portal using data-informed feedback and role modeling in 2020. A comprehensive baseline assessment was conducted in 114,790 government elementary schools in Uttar Pradesh, during which school focal persons ranked their schools' access to WASH services against the set benchmarks. Concurrently, high-ranking schools were established as 'learning labs' for neighboring schools and were provided with on-site capacity building opportunities. More than 150,000 teachers were systematically engaged through regular virtual training and automated calls that tracked progress, all parts of a centralized online portal. This champion-led transformation of school (CLTS) approach was adopted by the government to recognize best practices supported by students and teachers. Initiatives of teachers and village-elected representatives (*Gram Pradhan*), who were identified as 'champions', were recognized and disseminated to build an enabling environment that will sustain access and practices. As of July 2021, 65.8% of the assessed schools across 75 districts have improved their basic infrastructure through repairing/retrofitting defunct facilities and construction of new ones as per the gaps identified. They become 'WASH compliant' and leveraged around INR 45.71 billion of public financing from multiple sources in the last 3–4 years. Now, the state is rolling out a similar transformation campaign for 188,997 pre-school centers that will cover 5.6 million children between the ages of 3 and 6 years. The implementation process, challenges and learnings of this campaign can be used to formulate strategies and design scaled-up interventions to improve WASH facilities in schools.

Key words: evidence, hygiene, sanitation, schools, transformations, water

HIGHLIGHT

- WASH facilities in schools are critical for learning achievements for children and creating child-friendly environments in schools. Governments can achieve this at scale even in the challenging context through partnerships, convergence, leveraging resources, capacity building and a comprehensive online monitoring system.
- The innovative approach of using ICT-based tools to generate evidence, facilitate planning and monitor quality of implementation, along with a strong enabling environment is critical for transformation of water, sanitation and hygiene services in the schools.

INTRODUCTION

Children are entitled to the human rights of sanitation, safe water, good health and quality education (United Nations 1989). In the Indian context, the right to water is an integral part of Article 21 (right to life) of the Indian constitution wherein safe and clean drinking water as well as sanitation have been recognized as basic human rights necessary for the enjoyment of life (Wahi 2022). The Right of Children to Free and Compulsory Education Act has set norms and standards for adequate and quality water, sanitation and hygiene (WASH) facilities in schools.

In order to comply with the constitutional mandate, the Government of India has expressed its commitment to improve WASH facilities in government-run elementary schools through various programs. However, access to adequate WASH facilities in government-run elementary schools continues to be a challenge in Uttar Pradesh, which has more than 130,000 government elementary schools with an enrollment of more than 15 million children (U-DISE, 2019–2020, Ministry of Education). A number of research studies suggest that availability and access to WASH facilities, together with child-friendly classrooms and campus ambience, are essential for creating an enabling environment – a learning ecosystem for both children and teachers. In India, the availability of gender-specific toilets in schools can substantially catalyze enrollment, particularly among girls (Adukia 2017). Moreover, improved water and sanitation facilities in school have led to a marked reduction in diarrheal and other gastro-intestinal diseases (Jasper Le & Bartram 2012; McMichael 2019).

This paper dwells on a school transformation campaign known as '*Operation Vidyalaya Kayakalp*' (OVK) launched in 2018 in the state of Uttar Pradesh. The campaign focuses on ensuring availability of adequate and safe WASH facilities in government-run elementary schools. This paper examines the most outstanding results achieved by the OVK campaign and reflects upon the gaps, challenges and lessons learned during the process.

WASH *VIS-À-VIS* EDUCATION AND NUTRITION

The commonplace experience remains that WASH services in schools are necessary for ensuring unhindered participation of everyone in the conduct of classrooms and teaching–learning processes particularly, for girls and female teachers. The necessity of WASH facilities such as drinking water, sanitation, hand washing station, incinerator, for instance, is easy to identify. Not so very long ago, when these basic facilities were not common, children would leave the class when they needed water to drink or wash, many times not returning. The situation was even more challenging for girls due to lack of facilities for menstrual hygiene management in the schools. Several studies have shown that poor WASH facilities in schools is a major obstacle in the achievement of good-quality education (Alexander *et al.* 2014; Fuller *et al.* 2015; Vishnupriya *et al.* 2015). The need of hand washing facilities was not felt with equal intensity earlier. But, of late, children are beginning to understand the importance of hand washing with soap after the construction of child-friendly group hand washing stations in their schools. For adequate WASH facilities in schools, there should be:

- A separate toilet complex for girls and boys in each school with one toilet seat and three urinals per 80 students.
- One toilet for children with special needs.
- At least 10–12 taps in the ratio of 1 tap per 10–12 children for group hand washing stations.
- Running water supply in toilets and hand washing stations.
- More than 1.5 L of safe/treated drinking water per child per day.
- Water quality tested twice a year (before and after monsoon).

Nutrition is largely seen as an outcome of WASH, in addition to other factors. Safe environment ensured by WASH practices can contribute explicitly to generate awareness and, subsequently, develop informed behavior change among school-going children, thus having a direct bearing on their nutritional status (Toetzeke *et al.* 2017). Therefore, a nutrient-rich diet along with WASH can drastically improve the overall health of children in India. These have become more critical during the COVID pandemic as well.

Dean Spears (2012) found that the height of children in India is directly linked to availability of sanitation and hygiene at home and in their neighborhood, comprising the school as well. He further argues that open defecation is one of the key reasons for stunting. However, on the contrary, popular perception holds that only poor quality and quantity of food are important determinants of undernutrition among children. Robert Chambers with Medeazza (2013) takes the argument further by referring to the term 'Asian Enigma' which is the persistence of high levels of undernutrition among children despite an adequate amount of food. Ramalingaswami *et al.* (1996) note that poor living environment and meagre access to safe drinking water and sanitation together constitute the 'poverty syndrome' that leads to undernutrition among children as well as women.

UNICEF'S RENDEZVOUS WITH WASH IN SCHOOLS (WINS)

UNICEF's engagement with WASH in schools builds on the commitment to ensure a clean and safe environment for all children. It considers access to clean water, basic sanitation and good hygiene practices as one of the fundamental rights of every child, which keeps them thriving and give them a healthier start in life right from the beginning of their schooling years. WASH in schools is also reflected in the Sustainable Development Goal (SDG-6), which aims to achieve access to adequate and equitable sanitation and hygiene as well as universal and equitable access to safe and affordable drinking water for all by 2030. As children spend a significant portion of their day at school, the inclusion of WinS in the SDG-6 turns out to be a testimony to their standing as key components of a 'safe, non-violent, inclusive and effective learning environment' (United Nations 1989).

WASH in school programs in Uttar Pradesh dates back to 2010–2012, when more than 7,000 schools in 10 districts were made WASH compliant by the *Panchayati Raj* Department in partnership with UNICEF. Two years later in 2014, '*Swachh Bharat, Swachh Vidyalaya*' (Clean India, Clean School), a nation-wide campaign, was launched by the Government of India to support the ambitious national flagship program on sanitation '*Swachh Bharat Mission*' (Clean India Mission). This campaign was launched to set up well-functioning and well-maintained WASH facilities in schools in the country (Ministry of Human Resource Development 2014).

Clean school awards (*Swachh Vidyalaya Puraskar*)

In 2016, the central government started '*Swachh Vidyalaya Puraskar* (SVP)' (clean school award) that gave impetus to the Clean India, Clean School campaign for developing critical WASH facilities in the schools as per the standardized child-friendly norms. However, the interventions carried out under Clean India, Clean School campaign were primarily targeted at physical achievements and failed to put much emphasis on behavioral practices of children, quality and sustainability of available WASH services which led to significantly limited outcomes (Borthakur & Baruah 2019). The SVP was rolled out again in 2017 by the Ministry of Human Resource Development (MHRD), Government of India. In 2017, *Samagra Shiksha Abhiyan* (SmSA) (comprehensive education campaign) of Department of Basic Education of Uttar Pradesh in partnership with UNICEF, Field Office of Uttar Pradesh and district administrations extended its support to government-run elementary schools to participate in the award competition. The explicit purpose of this award is to honor those schools that have taken significant steps towards fulfilling the mandate of the Clean India, Clean School campaign (Swachh Vidyalaya Puraskar 2017-18).

Thus, there was an opportunity for schools to strengthen their WASH facilities that comprised the following five key components, namely (i) water, (ii) toilet, (iii) hand washing with soap, (iv) operation and maintenance of WASH facilities and (v) behavior change and capacity building of users, including the teachers and staff. Indicators related to each of these components were assigned a certain weightage ranging from 1 to 4 depending upon basic to improved services on a scale of 100, that is, 22, 28, 20, 15 and 15, respectively.

Weightage is assigned on the basis of SVP guidelines that standardize the benchmarking for WASH facilities and practices in government-run schools. Following this, a system of '5-Star' ranking was also introduced at national level by the MHRD for clean school award. In India, the 5-Star approach has been designed to improve safe water supplies, sanitation and hygiene in schools. It complements UNICEF's broader child-friendly schools initiative and GIZ's 'Fit for School' global approach, which aim to institutionalize health-promoting behaviors among school-going children. GIZ is a federal enterprise working for improvement of the education system and is owned by the Federal Republic of Germany (Stavem 2019).

Schools scoring 90–100% were to be ranked '5-Star' (green), 75–89% as '4-Star' (blue), 51–74% as '3-Star' (yellow'), 35–50% as '2-Star' (orange) and below 35% as '1-Star' (red). Akin to the ranking scheme, an assessment checklist (encompassing the five components enumerated above) was developed to conduct the online survey of WASH facilities and related components in participating schools. An online dashboard and validation protocols for '5-Star' and '4-Star' schools at district, state and national level through third parties were also institutionalized for tracking progress, ensuring accountability and transparency and validation of reported data of schools eligible for the award under different categories (SVP, MHRD 2017–2018).

Operation Vidyalaya Kayakalp

Drawing learnings from these nation-wide initiatives, a district-wide mega school transformation campaign took off in 2016–2017 in two districts of Uttar Pradesh, namely Mirzapur and Shravasti, led by the respective district administration in collaboration with UNICEF, which finally evolved into a state-wide mission named OVK (operation school transformation or OVK) led by the honorable Chief Minister of the state in the year 2018.

A standard operating procedure (SOP) was also developed to strengthen WASH facilities, services and behavior practices in schools, which highlighted the standardization of technical norms and designs, capacity building of stakeholders, training of masons and plumbers, monitoring of ongoing school activities and progress reviews to accelerate implementation. For illustration, when the SVP was launched in 2016, only 2,741 schools managed to participate and complete their award assessments. But the next year in 2017, the participation of schools for the award astonishingly shot up to 72,155. However, only 52,917 schools managed to qualify for the final registration by providing complete information on 39 indicators before the deadline. Of which, 6,320 schools ranked '1-Star', 18,067 schools got '2-Star' and 26,366 schools ranked '3-Star', whereas 1,722 schools ranked '4-Star' and 142 schools achieved '5-Star' status, and two of them won the national award.

This was in essence the outcome of collective efforts of Department of Basic Education's SmSA, *Panchayati Raj* Department, district administration and UNICEF that helped in building enabling environments and program perspective of a wide range of stakeholders to prioritize WinS that led to increased participation of schools in SVP. It was an important step towards the institutional strengthening of WASH for OVK.

METHODS

The foundations were laid by the *Panchayati Raj* Department in the year 2017–2018 and this was continued in 2019 by SmSA and district administrations integrating a number of programs focusing on WASH in schools, namely *Swachh Bharat* Mission, *Swachh Baharat Swachh Vidyalaya*, SVP and Operation *Kayakalp*. The Director General, School Education of SmSA, with technical support from UNICEF worked together to take forward OVK to a state-wide campaign for transformation of WASH and basic infrastructure in schools. This campaign was further aligned with another innovative campaign of SmSA named Mission *Prerna* that works to improve the quality of elementary education and learning outcomes in all government elementary schools. The program took off with an online rapid assessment to ascertain the status of WASH facilities and basic infrastructure in elementary schools across 75 districts, in November–December 2019. It began with the field testing of online assessment tools and the IT system by gauging the capacity of identified assessors and their means of verification. Based on the analysis of preliminary data, technical errors occurring in online tools were rectified, capacities of field staff were enhanced and an interactive online OVK dashboard was developed. Thus, the mammoth exercise of baseline assessment of WASH facilities and basic infrastructure in schools was successfully accomplished by using an android-based mobile app in February–March 2020.

The baseline assessment included a comprehensive checklist of 28 indicators (contextualized for the state based on SVP indicators) administered in 114,790 schools across 75 districts of the state using mobile application in February–March 2020. This online checklist had a list of questions regarding assessment of availability and functionality of 19 key WASH and basic infrastructure in schools. The assessment was conducted by a team of 2,007 special educators officially designated as Itinerant Resource Teachers (IRTs). The IRTs were trained in the use of the mobile app, along with an orientation on WASH components and their significance for educational and nutritional outcomes of children as mandated under the Right to Education Act, 2009.

The midline assessment was conducted during March–July 2021, covering 131,388 schools across 75 districts. The duration of the midline extended due to the local body election in the state- and nation-wide lockdown of the second wave of the COVID-19 pandemic. The total number of schools assessed were increased due to improved programming in terms of planning, training, IT services, monitoring and follow-up support. The number of assessors also increased from 2,007 to 3,987. An interactive OVK dashboard, hosted by SmSA in the Mission *Prerna* portal at www.prernaup.in, was put in place to reflect consolidated assessment findings of block, district and state. The geo location of the schools and photographs of available WASH and basic infrastructure were also captured in the mobile app. The number of indicators was increased to 39 (by including an additional 11 indicators on costing, financing and implementation for informed decision-making and improved programming). Midline assessment was conducted by a trained team of 3,987 State Resource Group members (SRGs) and Academic Resource Persons (ARPs) of SmSA (checklist available at https://www.basicshikshakparivar.com/2021/03/assessment-checklist-for-mediline.html). These 39 indicators capture the augmentation and compliance status of 19 key WASH and basic infrastructural items comprising water, toilets, urinals, hand washing facilities, classrooms, mid-day-meal kitchen, accessibility, electrification, painting and school premises. These 19 key items

Table 1 | Indicator-wise description

Indicator	Description[a]
Safe drinking water	At least one drinking water source; either India Mark-II handpump or deep borewell inside the school premises or connected with village piped water supply scheme (PWSS).
Running water supply	Village or standalone PWSS sourced from India Mark-II handpump or deep borewell. Daily provision of adequate water for toilets, hand washing, school cleaning, food preparation and cooking.
Boys' and girls' toilets (including urinals)	Separate toilets for boys and girls, with one unit generally having one toilet (WC) plus three urinals. A hand washing unit with soap must be installed inside/nearby the toilets. There should be a minimum distance of 10 m between toilets and drinking water source to avoid contamination.
Toilets with tiles	Stone or ceramic tiled floor with adequate slope towards the pan and maintainable durable finish.
CWSN toilets	One toilet with special fixtures such as ramp, wide door, matte-finish tiles, handrails and handlebars for use by the children with special needs (CWSN)/differently abled persons. A hand washing unit with soap must be installed inside the toilet.
Multiple hand washing units	Sufficient group hand washing facilities allowing 10–12 children to wash hands at the same time. The hand washing units should be child-friendly, simple, scalable as well as sustainable, relying on minimum water. These hand washing facilities can be developed using local materials.
Classrooms with tiles	Stone or ceramic tiled floor with adequate slope and maintainable durable finish in each classroom.
Black/green/white boards	One black/green/white board of adequate size preferably 2,100 mm × 1,300 mm (sill 800 mm) and durable quality inside each classroom.
Renovated kitchens	Cement, stone or ceramic tile floor with adequate slope and maintainable durable finish, safe drinking water point, hand washing point, elevated platform and an outlet for grey water towards a nearby kitchen garden or soak pit or drainage.
Whitewash/painting of school building	School building, classrooms and WASH facilities painted with attractive multiple colors, displaying various messages related to knowledge and personality development, following the concept of Building as Learning Aid (BaLA).
Ramps and railings	Ramps with a gradient of 1:12, handrails on both sides and maintainable durable matte-finish or dotted tiles.
Electrical appliances	All classrooms and toilets with functional electrical wiring and electrical points for lights and fans.
Electric connections	Regular power supply either government electricity connection or solar power or both.
School furniture	Sufficient quantity of desks and benches of adequate height in each classroom for all enrolled children.
Boundary wall	School premises surrounded by plastered brick wall of adequate height with a main gate and iron grilled panels.

[a]*Source*: National *Swachh Bharat Swachh Vidyalaya* guidelines and OVK, Technical Design Manual U.P, India (available at: https://www.prernaup.in/).

(mentioned in Table 1) were prioritized for universal coverage of WASH facilities and basic infrastructure in schools. Looking at the identified gaps and funding availability, 14 infrastructural items were prioritized for universal coverage by the end of March 2021, whereas it was decided that the remaining five indicators, which include installation of submersible pumps for running water supply, augmentation of boys' and girls' urinals, furniture (desk-bench) in classrooms and boundary wall, were to be covered by March 2022 (Operation Kayakalp Archives n.d).

Summary of school assessment coverage under OVK

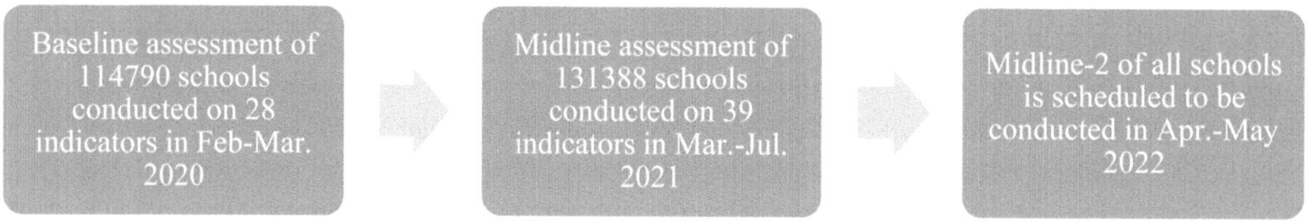

An online planning template was added in the Management Information System (MIS) of *Prerna Portal* to facilitate a uniform OVK planning on 14 priority infrastructural items (as mentioned above) of 833 blocks of the state. The progress of the OVK plan was also monitored through an online monthly progress report (MPR). The mechanism of Interactive Voice Response System (IVRS) call was employed from June 2020 onwards, for real-time tracking of work progress status and inter-departmental convergence. The headteachers of schools were contacted on a weekly/fortnightly basis according to the critical operational issues identified through multiple sources. The average response rates of IVRS calls were 75–80%. District performance rating was prepared based on the data received through call responses and shared by the Director General, School Education to all 75 District Magistrates (DMs) and District Education Officers for remedial actions.

An online Project Monitoring and Information System (PMIS) was also instituted to track physical and financial progress against the approved civil work of SmSA. There is a provision in the PMIS to upload the overall fund utilization certificate of the district on each activity to supplement the progress reported. An online Data Capture Form was also integrated in the MPR window for rapid collection of specific data on electrification, additional classrooms and fund leveraging, etc. to support decision making at state level. Funds have been leveraged from composite grants and civil works of *SmSA*, finance commission grants of *Panchayati Raj* Department, urban local bodies, district mineral fund and corporate social responsibility. Moreover, the data progressively made available through PMIS, MPR and IVRS calls were primarily employed in block and district performance ranking. This also served the critical purpose of course-correction at the district level and promoting healthy competition among the stakeholders engaged in transformation efforts under the leadership of respective DM. The 'star-ranking' approach outlined above was followed and an equal weightage of '1' assigned to each indicator for uniformity. Lowest performing schools were assigned '1-Star', whereas the best performing schools received a '5-Star' rating. Schools achieving 5-Star and 4-Star were adjudged 'transformed' and evolved into 'learning labs' for neighboring schools.

Learning labs

The idea of a '*Learning Lab*' as a methodology has been adopted keeping in mind the enormity of scale and physical explication of locale-specific and child-friendly structural design of WASH facilities. The schools designated as learning labs served the purpose of a local working model and this was capitalized on successfully for on-site training of key stakeholders such as teachers, elected head of the village council (*Gram Pradhan*), secretary and masons and recognizing the efforts of champions for scaling up.

Capacity building

In March 2020, lockdown began due to the pandemic. This disrupted the progress of the campaign as all schools were closed and administrative machinery came to a halt. The Chief Secretary of State (administrative head of state government) ordered on 29 April 2020 that essential WASH facilities should be constructed in all the schools while the schools were closed and migrant laborers could be employed in this process. This was a win-win scenario where the migrant laborers got employment and schools became WASH compliant and construction work could be carried out unhindered.

Making judicious use of emerging online platforms, *SmSA* and UNICEF took on the primary task of connecting key stakeholders across levels for providing technical and operational assistance, in terms of training, motivating through champions, monitoring, documenting best practices and advocacy, etc. through virtual platforms to keep the ball rolling:

- Between 21 May 2020 and 3 July 2020, a number of online refresher trainings on child-friendly structural design of WASH facilities were conducted, wherein 63,184 key stakeholders including block officials, teachers, elected village representatives and village council secretaries across 16 districts were oriented. This led to replication of child-friendly designs and best practices at the school level. Composite progress of each school on 14 priority items was regularly uploaded on the portal.
- The second round of divisional meets was conducted online throughout the state, chaired by the Director General, School Education in the months of June and July 2020 (post lockdown) to reaffirm the mandate of WASH in schools as a priority for district administration. In order to sustain the focus at scale, newly appointed DMs, Chief Development Officers (CDOs) and Block Education Officers (BEOs) were also oriented through online platforms on the broader idea and technical nuances of the school transformation campaign. They were also guided by the state authorities and development partners through regular meetings and contacts.

- Based on the analysis of MIS data and district performance ranking, the Director General, School Education conducted online face-to-face meetings to nudge DMs and CDOs of 10 poorly performing districts on a weekly basis.
- A positive deviance was brought about by providing platforms to real change-makers and champions. Online monthly progress review meetings were capitalized on to facilitate cross-learning and dialogue among District Education Officers of 10 best progressing and 10 slowest progressing districts, as well as BEOs of the 50 best performing and 50 slowest performing blocks.
- The Department of Basic Education and *SmSA*, on the other hand, proactively ensured the highest level of political and bureaucratic support for this program based on the successful evidence generated from the demonstration of a district-wide approach in selected districts, led by the DMs. From this standpoint, the DMs of focus districts emerged as the real-time champions!

Statistical analysis

In this paper, we have presented the percentage coverage of 19 key WASH and basic infrastructural items in the baseline and midline assessment. *Z*-statistic has been used to test the significance of differences in their values from baseline to midline. Table 1 provides a description of each of the 19 items included in the analysis. Microsoft Excel software was used to compute value of the test-statistic and the corresponding *p*-values.

RESULTS

Table 2 enumerates baseline and midline assessment on 19 key priority infrastructure items and their percentage coverage. In the midline survey, significant changes were observed in coverage of 15 of 19 items. The significant and substantial improvements were observed in the coverage of running water supply (from 19.1 to 68.3%, $p < 0.001$), multiple hand washing units (from 8.3 to 62.9%, $p < 0.001$) and toilets with running water (from 14 to 58.6%, $p < 0.001$). Due to lockdown and consequent closure of schools, it was not possible to assess the impact of improvement in school infrastructure, particularly WASH facilities, on children's attendance, their learning outcomes and their behavioral change with respect to the use of WASH services.

Table 2 | Comparison of baseline and midline coverage findings

Indicators	Baseline data, February–March 2020 (%) (*n* = 114,790)	Midline data, March–July 2021 (%) (*n* = 131,388)	*p* value	Status of statistical significance
Safe drinking water	77,076 (67.2)	112,429 (85.6)	0.0005	Significant
Running water supply	21,961 (19.1)	89,781 (68.3)	< 0.0001	Significant
Boys' toilets	68,578 (59.6)	105,938 (80.6)	0.0001	Significant
Girls' toilets	70,463 (61.4)	107,773 (82.0)	< 0.0001	Significant
Boys' urinals	30,471 (26.5)	66,771 (50.8)	< 0.0001	Significant
Girls' urinals	36,210 (31.5)	71,975 (54.8)	< 0.0001	Significant
Toilets with running water	16,035 (14.0)	76,928 (58.6)	< 0.0001	Significant
Toilets with tile	41,150 (35.9)	73,669 (56.1)	0.0002	Significant
CWSN toilets	6,635 (5.0)	21,233 (15.9)	< 0.0001	Significant
Multiple hand washing units	9,508 (8.3)	82,591 (62.9)	< 0.0001	Significant
Classrooms with tiles	27,869 (21.3)	40,368 (30.7)	0.0278	Not Significant
Black/green/white boards	94,077 (82.0)	119,346 (90.8)	0.0357	Not Significant
Renovated kitchens	62,374 (47.1)	104,489 (79.8)	< 0.0001	Significant
Whitewash/painting of school building	64,061 (55.8)	117,778 (89.6)	< 0.0001	Significant
Ramps and railings	44,201 (38.5)	88,044 (67.0)	0.0001	Significant
Electrical appliances	33,161 (28.9)	98,432 (74.9)	< 0.0001	Significant
Electric connections	56,565 (49.3)	92,255 (70.2)	0.0001	Significant
School furniture	21,233 (18.5)	23,093 (17.6)	0.5870	Not Significant
Boundary wall	63,384 (55.2)	86,446 (65.8)	0.0277	Not Significant

Source: https://www.prernaup.in/.

The key reasons behind successful implementation of the school transformative initiative are summarized as follows:

- An enabling environment created by the proactive political and bureaucratic leadership at the state and districts.
- Use of comprehensive MIS with digital tools for assessment, planning, budgeting and monitoring.
- Capacity building of stakeholders at all levels; use of technical manual, tools and visual aids to ensure child-friendly facilities and quality work. Use of virtual platforms to reach out to large numbers of stakeholders for training and follow-up support.
- Inter-departmental coordination for resource leveraging. Healthy competition among elected representatives of the villages and teachers for development of better schools for their children.
- Partnerships with development agencies; promotion and appreciation of innovative actions to address emerging bottlenecks.
- Media coverage of best practices/success stories along with felicitation of champions by the higher authorities and elected representative.

The comprehensive OVK-MIS and the use of online platforms for assessment, planning, capacity building and monitoring played a crucial role in informed decision-making and results-based programming. The collaborative OVK campaign of Uttar Pradesh has successfully leveraged around INR 45.71 billion (634.95 million USD) from multiple sources of public financing to improve school-based access to WASH facilities in the state.

Learnings

The school transformation campaign (OVK) implementation process offers four discernible areas of learning. First, the thrust on *gram panchayats* (village council) and support of teachers as the crux of the program was a rational choice marked by the commitment to the idea of community involvement. Second, the MIS was made justifiable through a process of transparency and accountability as the basic tenets of governance. Slowly and steadily, the approach generated confidence among the teachers, community members and their representatives, in addition to other key stakeholders. Third, the overtly hierarchical system of bureaucracy was made more open by the personalist functioning of the majority of team leaders and the requirements of structural wisdom, brought in by the Director General, School Education. The experience, however, highlighted the fact that any such transformation initiative could be made effective with reliance on people's participation and their sense of ownership. Fourth, the program teaches how a sense of hope can be invoked among the people in general and children in particular in the government schooling system. One comes across scores of motivated education officers at all levels, Gram Pradhans (*elected village heads*) and head teachers, enthusiastic about the prospects of an 'enrollment explosion'.

Challenges

The coronavirus pandemic was the biggest stumbling block in leading the program to the grand accomplishments as envisaged; however, hand washing with soap at household level increased due to COVID response interventions of the state (Jatav *et al.* 2021). The instances of school closure for 16 months (from 22 March 2020 to 4 February, 2021 and again from 23 March 2021 to 31 August 2021) withered away the prospects of children's participation and prevented the community and teachers from playing their entrenched role in the physical process of transformation and bringing about behavioral change in observance of WASH practices among the school incumbents. The opportunities to demonstrate and establish the mechanism of operation & maintenance and sustainability through community of practices, including alumni engagement, were put on hold due to the pandemic. In addition, leveraging finances from *gram panchayats* was also constrained in the last two quarters of the financial year 2020–2021 due to shifting priorities in the event of their elections. Finally, poor internet connectivity in remote rural areas and limited IT expertise of some frontline functionaries created unwanted delays in monitoring and reporting and thus affected informed decision-making.

Mitigation measures

Several mitigation measures were adopted in view of above-mentioned challenges. First, the programs of '*Swachh Bharat* Mission' and 'SVP' were restrained under the prevailing conditions of uncertainties to provide a sense of direction to the program and setting up priorities from afresh. A government order of 30 October, 2019, however, clarified the obligations, with categorical emphasis on universal coverage of child-friendly WASH infrastructure in all the government schools (*letter no. 2706/33–03-2019 dated 30.10.2019* issued by the Chief Secretary, *Uttar Pradesh*). It was essentially the use of Information and Communication Technology (ICT) which kept the wheels moving during the lockdown. Work-from-home enabled by ICT and the opening of digital platforms ushered in a new era of virtual human interaction – webinars, online meetings and capacity building sessions along with online review and assessment processes as well as the use of WhatsApp groups at all levels were very effective in accomplishing the set targets. Second, the online database of PMIS and PFMS (*SmSA* and PRD) were capitalized on for mapping and accessing unused funds of composite funds of SmSA and finance commission grants of *gram panchayat*. This further helped in employing incoming migrants in school transformation work. Third, the intense campaign during the period following the month of June 2020 registered phenomenal achievements. Stepped up monitoring comprising regular online monthly meetings, consistent review of progress recorded by each *gram panchayat* and schools, steady follow-up, block-level meetings of *gram pradhans* and teachers, establishment of district- and block-level 'war rooms' for regular follow-up of frontline functionaries – these together brought the progress close to the target. Fourth, the WASH team of UNICEF took on the primary task of providing 'Technical Assistance' – comprising training, planning, handholding support, connecting champions and facilitating cross sharing and evidence-based advocacy, etc. at state, district and block levels through virtual platforms and on-site support.

DISCUSSION

We found increasingly high participation of schools in OVK and moderate levels of adherence to child-friendly WASH standards by the participating school. While we acknowledge that OVK is still being implemented and a true measure of its impact on sustained availability of WASH facilities and education outcomes may only be available in future, our initial findings corroborate with the results obtained from interventions for improving WASH infrastructure in schools in Laos (Chard & Freeman 2018), Mali (Garn *et al.* 2017) and Kenya (Garn *et al.* 2016). These studies found that interventions could reach a large proportion of targeted schools; however, improvements in the availability of WASH facilities were either moderate or low mainly due to the lack of availability of running water for hand washing or toilet flushing. Insights from our concurrent monitoring are also consistent with the results obtained from a similar 'Three-Star Approach' implemented in a few countries (DepEd 2021; UNICEF 2021). The 'Three-Star Approach' also resulted in significant improvement in the school's compliance with the critical WASH indicators while struggling with the indicators which require higher investments for sanitation infrastructure and their operation and maintenance for sustainability.

OVK has built upon the learnings and challenges observed from other similar interventions implemented at scale. Like the innovative Three-Star Approach implemented in the pacific (UNICEF 2021) and school-based interventions in Rural Tanzaina (Hetherington *et al.* 2017), OVK also engaged and empowered schools and communities so that they can support improvements to WASH facilities in schools using local resources. The implementation of 'learning labs' was an additional and critical strategy not only for identifying local solutions to the local problems in a given context, but also to create promoting healthy competition among the stakeholders. The culture of cross-learning was further strengthened by using online monthly review meetings as a platform to connect district- and block-level officers from better performing areas and from areas that were lagging. Additionally, OVK addressed two critical limitations related to availability of resources for sustainability and lack of robust M&E systems in the implementation of the Three-Star Approach (UNICEF 2021). First, OVK had a strong component of inter-departmental convergence for leveraging public financing from multiple sources for greater community ownerships, accountability and sustainability of interventions. Lack of adequate budgetary allocations for the recurrent costs of consumable supplies has been identified as one of the key bottlenecks in other school-based interventions as well (Garn *et al.* 2017). Since OVK leveraged funds from multiple sources of public financing, we not only believe we will see more improvements even in the indicators with higher resource needs, but also sustained achievements. Second, the monitoring and reporting mechanisms were kept robust and transparent through the use of advanced ICT tools such as new-age mobile-based data collection tools to collect images and geolocations, use of IVRS, linking data from several sources (baseline, improvement plans, midline assessment, monthly reports, funds utilization) to transform them into information in the form of an interactive dashboard. The robust monitoring systems were key to ensure intervention fidelity which has been reported as one of the key drivers of successful school-based interventions in the literature (Chard & Freeman 2018).

CONCLUSIONS

The innovative approach of using ICT-based tools to generate evidence, facilitate planning and monitor quality of implementation, along with a strong enabling environment resulted in achieving the desired results in a short span of time. Prompted by the success of the OVK program, the state is now all set to roll out a new transformation campaign for pre-schooling centers (*Aganwadi Kendras)* with UNICEF's technical support, that provide basic essential nutritional services to 5.6 million children belonging to the age-group of 3–6 years. This will be an important policy level initiative by the state government aimed at initiating young children to WASH facilities during their formative years.

DATA AVAILABILITY STATEMENT

Data cannot be made publicly available; readers should contact the corresponding author for details.

REFERENCES

Adukia, A. 2017 Sanitation and education. *American Economic Journal: Applied Economics* **9** (2), 23–59. Available from: https://www.aeaweb.org/articles?id=10.1257/app.20150083

Alexander, K. T., Oduor, C., Nyothach, E., Laserson, K. F., Amek, N., Eleveld, A., Mason, L., Rheingans, R., Beynon, C., Mohammed, A., Ombok, M., Obor, D., Odhiambo, F., Quick, R. & Phillips-Howard, P. A. 2014 Water, sanitation and hygiene conditions in Kenyan rural schools: are schools meeting the needs of menstruating girls? *Water* **6** (5), 1453–1466.

Borthakur, M. & Baruah, J. 2019 Swachh Vidyalaya Abhiyan: findings from an empirical analysis. *Social Change and Development* **130**. Available from: https://www.researchgate.net/publication/334653820_Swachh_Vidyalaya_Abhiyan_Findings_from_an_Empirical_Analysis

Chambers, R. & Von Medeazza, G. 2013 Sanitation and stunting in India: undernutrition's blind spot. *Economic and Political Weekly* 15–18. Available from: https://www.epw.in/journal/2013/25/commentary/sanitation-and-stunting-india.html

Chard, N. C. & Freeman, M. C. 2018 Design, intervention fidelity, and behavioral outcomes of a school-based water, sanitation, and hygiene cluster-randomized trial in laos. *International Journal of Environmental Research and Public Health* **15** (4), 570. doi: 10.3390/ijerph15040570.

Fuller, J. A., Westphal, J. A., Kenney, B. & Eisenberg, J. N. 2015 The joint effects of water and sanitation on diarrhoeal disease: a multicountry analysis of the Demographic and Health Surveys. *Tropical Medicine & International Health* **20** (3), 284–292.

Garn, J. V., Brumback, B. A., Drews-Botsch, C. D., Lash, T. L., Kramer, M. R. & Freeman, M. C. 2016 Estimating the effect of school water, sanitation, and hygiene improvements on pupil health outcomes. *Epidemiology (Cambridge, Mass.)* **27** (5), 752–760.

Garn, J. V., Trinies, V., Toubkiss, J. & Freeman, M. C. 2017 The role of adherence on the impact of a school-based water, sanitation, and hygiene intervention in Mali. *The American Journal of Tropical Medicine and Hygiene* **96** (4), 984–993. https://doi.org/10.4269/ajtmh.16-0558.

Hetherington, E., Eggers, M., Wamoyi, J., Hatfield, J., Manyama, M., Kutz, S. & Bastien, S. 2017 Participatory science and innovation for improved sanitation and hygiene: process and outcome evaluation of project SHINE, a school-based intervention in Rural Tanzania. *BMC Public Health* **17** (1), 172. https://doi.org/10.1186/s12889-017-4100-7.

Jasper, C., Le, T. T. & Bartram, J. 2012 Water and sanitation in schools: a systematic review of the health and educational outcomes. *International Journal of Environmental Research and Public Health* **9** (8), 2772–2787. Available from: https://www.ncbi.nlm.nih.gov/pmc/articles/PMC3447586/

Jatav, S. S., Nayak, S., Meher, S. & Narang, S. 2021 Coping to COVID-19 in.Uttar Pradesh, India: evidence from NSSO 76th round data. *Current Urban Studies* **9** (2), 206–217. Available from: https://www.scirp.org/journal/paperinformation.aspx?paperid=109460

McMichael, C. 2019 Water, sanitation and hygiene (WASH) in schools in low-income countries: a review of evidence of impact. *International Journal of Environmental Research and Public Health* **16** (3), 359. Available from: https://www.ncbi.nlm.nih.gov/pmc/articles/PMC6388361/

Ministry of Human Resource Development 2014 Swachh Bharat Swachh Vidyalay: A National Mission, Government of India. Available from: https://smartnet.niua.org/sites/default/files/resources/Swachch%20Bharat%20Swachch%20Vidhalaya%20Mission%20Guidelines_0.pdf

Operation Kayakalp Archives. n.d. Available from: http://basicshikshak.com/tag/operation-kayakalp/

Philippine Department of Education (DepEd) 2021 *WinS Monitoring Results: School Year 2017–18 to School Year 2019–20*. Available from: https://wins.deped.gov.ph/wp-content/uploads/2020/12/TSA-WinS-Monitoring-Booklet-Web-2020-11-27.pdf

Ramalingaswami, V., Jonsson, U. & Rodhe, J. 1996 *The Asian Enigma: The Progress of Nations*. UNICEF, NY, pp. 10–17.

Spears, D. 2012 *Effects of Rural Sanitation on Infant Mortality and Human Capital: Evidence From India's Total Sanitation Campaign*. Princeton University, Princeton, NJ. Available from: https://www.dartmouth.edu/neudc2012/docs/paper_86.pdf

Stavem, S. M. 2019 When the walls in the school building can speak. In: *Educational Architecture – Education, Heritage, Challenges* (A. Alegre, T. Heitor, M. Bacharel & A. Fernandes, eds.). Conference Proceedings, Instituto Superior Técnico, Departamento de Engenharia Civil, Arquitectura e Georrecursos, Lisboa, p. 143.

Swachh Vidyalaya Puraskar 2017–18 *Ministry of Human Resource Development*. Available from: http://www.swachhvidyalaya.com/.

Technical Design Manual. Operation Vidyalay Kayakalp, Government of Uttar Pradesh.

Toetzke, J., Rück, J. & Panzerbieter, T. 2017 *2+6=17 Linking WASH and Nutrition – A Blueprint for Living SDG 17*. German WASH Network. Available from: http://www.washnet.de/wp-content/uploads/washnet17_linking-wash-nutrition_web_170721_jr.pdf

United Nations 1989 *Convention on the Rights of the Child*. Office of High Commissioner for Human Rights (OHCHR), Geneva. Available from: https://www.ohchr.org/EN/ProfessionalInterest/Pages/CRC.aspx

United Nations Children's Fund (UNICEF) 2021 *Formative Evaluation of UNICEF Three Star Approach for WASH in Schools in the Pacific*. Available from: https://www.unicef.org/pacificislands/media/3051/file

Vishnupriya, S., Prasad, S., Kasav, J. B., Trout, K., Murthy, S., Surapaneni, K. M. & Joshi, A. 2015 Water and sanitation hygiene knowledge, attitudes and practices among school settings in rural Chennai. *Journal of Water, Sanitation and Hygiene for Development* **5** (2), 192–200.

Wahi, N. 2022 The evolution of the right to water in India. *Water* **14** (3), 398. https://doi.org/10.3390/w14030398.

First received 31 January 2021; accepted in revised form 8 May 2022. Available online 18 May 2022

doi: 10.2166/h2oj.2022.062

Accountability and advocacy interventions in the water sector: a global evidence review

N. D. Hepworth, T. Brewer *, B. D. Brown, M. Atela, J. Katomero, J. Kones and M. Gashaw
Accountability for Water Programme, c/o Water Witness International, 3/2 Boroughloch Square, Edinburgh EH8 9NJ, UK
*Corresponding author. E-mail: timbrewer@waterwitness.org

TB, 0000-0002-9189-6361

ABSTRACT

This paper reports on a global review of evidence on the outcomes of accountability and advocacy interventions for improved water and sanitation service delivery, water resource management (WRM) sector governance, and the factors which influence their performance. Mapping current knowledge will help guide the community of practice, policy, and research working to strengthen water, sanitation, and hygiene (WASH) service delivery systems. We summarise the state of the evidence, key insights, and implications. Available evidence is relatively limited (151 papers) and is skewed towards India and East Africa (c. 50%), water service delivery (62%), and social accountability monitoring (54%). Most included articles (80%) associate accountability and advocacy with positive outcomes for sector performance. Positive outcomes are achieved through, for example: improved operational maintenance; access to data; representation and inclusion; political will; changes in policy, law, and process; new investment; and organisational performance. Impact is less frequently reported (32%) and attribution of impacts to interventions is a recurrent challenge. We identify 28 factors that determine the interventions' performance. These are used to populate a theory of change for accountability and advocacy interventions. The results of the review are important for those working on water sector system strengthening. They suggest the potential of accountability interventions to accelerate delivery of the Sustainable Development Goals (SDGs) on WASH, and provide an organising framework to strengthen intervention design and delivery, and future research.

Key words: accountability, advocacy, evidence, system strengthening, WASH, water governance

HIGHLIGHTS

- A global review of the evidence about accountability and advocacy interventions on water, including their outcomes and factors shaping performance.
- Provides a theory of change and knowledge base to guide theory, policy, and practice.
- Validates new research to explore community, governance, and contextual dynamics that constrain or stimulate accountability in the water sector.

GRAPHICAL ABSTRACT

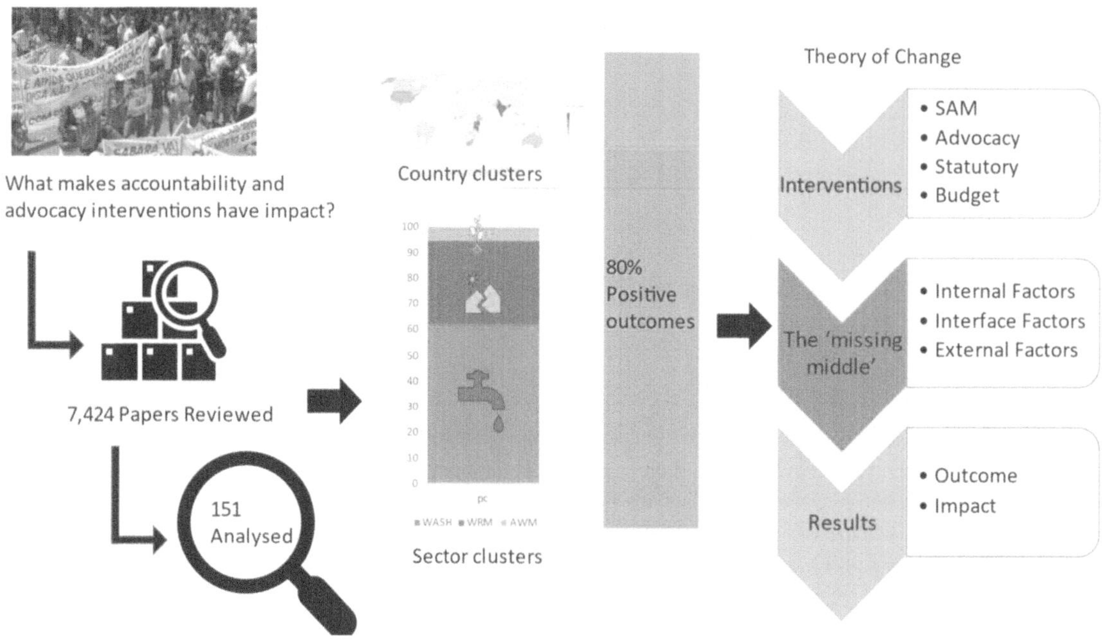

INTRODUCTION

This paper summarises the results of a global review of evidence on the outcomes of accountability and advocacy interventions for improved water and sanitation service delivery, water resource management (WRM) and water governance, and the factors which influence their performance. It was conducted by Water Witness International as an output of the Accountability for Water Programme, a collaboration which develops knowledge and practice to strengthen the accountability of all water stakeholders for the delivery of the water Sustainable Development Goal (SDG 6).[1]

The global water crisis means one in four people are without access to a safely managed water service, and one in three are without access to safely managed sanitation services (WHO/UNICEF 2020, p. 7). Four billion people face severe water scarcity each year (Mekonnen & Hoekstra 2016, p. 3), and the climate emergency may double the numbers facing severe water stress by 2050 (Munia *et al.* 2020, p. 9). Millions of vulnerable people face chronic water insecurity (UNESCO 2019, p. 142). This incorporates lack of protection from flooding and drought, and 'economic water scarcity' (p. 142) which can reflect the lack of security of water tenure, and lack of protection against pollution, water grabs, and resource degradation. Without mitigation, this in turn threatens deepening gender and social inequality and human suffering.

Our review is based on two foundational propositions, the first of which is that to ensure water security for all, the most important change is to improve governance.[2] The root cause of the water crisis is poor governance rather than absolute scarcity, or an absence of technological or other solutions (UNDP 2006; UNESCO 2019). By governance we mean the policies, institutions, rules, processes, and practices which articulate interests, take decisions, regulate human behaviour, and exercise power. In the water sector, these make up the systems that determine who received water and sanitation services, when, how, and why. Our second proposition is that stronger accountability leads to better governance and can prevent or resolve service delivery failures on water.

Formal accountability is defined here as the ability to review, explain, and report performance against rules, responsibilities, and obligations, and to react constructively to improve performance through sanctions, incentives, or corrective measures. We also incorporate informal accountability, 'socially shared rules that are

[1] The Accountability for Water consortium includes End Water Poverty, the Institute of Development Studies (IDS), Kenya Water and Sanitation Network, Oxfam, Partnership for African Social and Governance Research (PASGR), Shahidi wa Maji, WaterAid, Water Integrity Network (WIN), Water Witness and is funded by the William and Flora Hewlett Foundation.

[2] See Ernstorfer and Stockmayer (2009) for a discussion.

created, communicated and enforced outside official channels' (Helmke & Levitsky 2004). Advocacy was included in this review because it is often joined with accountability interventions, particularly where formal opportunities for accountability are weak, or where a need for systemic change is identified. We use a simple definition of advocacy as 'the organised attempt to change policy, practice and attitudes by presenting evidence and arguments for how and why change is needed' (Open Societies Foundation 2010, p. 3). Advocacy more broadly understood encompases a far wider range of definitions and contexts, and this review was limited to where advocacy is for, or combined with, accountability. A typology of accountability interventions was used to organise the evidence into four categories: social accountability, whereby community members are mobilised to call duty-bearers to account; budget analysis, whereby non-governmental organisations (NGOs) or other bodies review budget and/or spending as a means of claiming accountability; evidence-based advocacy, whereby individuals or groups gather evidence of service failures or unmet needs and use this to lobby, either directly or using media or other means; and finally statutory accountability, using formal processes such as ombudsmen, complaint mechanisms, or public-interest litigation to claim accountability.

Accountability and advocacy interventions have secured positive results, in terms of government responsiveness, policy change, and inclusion of marginalised communities in decision making, but important knowledge gaps and unanswered questions persist (Hepworth 2016; Farrow 2018). These centre on the contextual and design factors which determine why, when, and how accountability and advocacy interventions deliver positive impacts, and who benefits. Practitioners need to understand the factors that determine the performance of accountability interventions, particularly their benefits for marginalised people. Better characterising and understanding these factors will support the practice and policy community to design, deliver, support, and resource future interventions that maximise positive impact in multiple settings. This review responds to the knowledge needs of local and global stakeholders in government, utilities, regulators, communities, civil society organisations, international non-governmental organisations (INGOs), and bilateral and multilateral donors. Since 2016, through global[3] and regional meetings,[4] over 500 practitioners and decision-makers have shared insights and reflected on the bottlenecks and priority knowledge gaps which hold back accountability for water.

The need to better understand the determinants of success for accountability interventions is reflected in academic literature and meta-evaluations from beyond the water sector. For example, McGee & Gaventa (2011, p. 12) conclude that 'under some conditions, some TAIs (transparency and accountability initiatives) create opportunities for citizens and states to interact constructively, contributing to … better budget utilisation, improved service delivery, greater state responsiveness, the creation of spaces for citizen engagement and the empowerment of local voices.' This review investigates what those conditions are. Similarly, ITAD's (2016, p. 1) review for the UK Department for International Development (DFID) found that 'accountability processes almost always lead to better services, accessibility and staff attendance' but need to be adaptive and responsive to different contexts to get results. Fox (2015, p356) found potential in strategic approaches to accountability where a conducive, enabling environment[5] requires institutions and authorities to listen, removes the perceived risks and costs of speaking up, and emboldens people to exercise voice and urgency. This in turn triggers reform and encourages further engagement. However, the dynamics that drive this cycle are poorly understood, and research and theory lag well behind practice. It is these challenges, and this 'lag' that this evidence review addresses. This review presents and analyses the current evidence on accountability. The analysis shows the factors that determine whether accountability interventions contribute to positive outcomes and impacts. It also highlights where additional research is needed.

Purpose, scope, and objectives

The review aims to help researchers, practitioners, and policymakers who together comprise the community of practice across the water and sanitation sectors to access, interpret, and use the current evidence relating to accountability for water and sanitation services. Our scope is inclusive of rural and urban water, sanitation,

---

[3] Including Stockholm Water Week (2016, 2017, 2018, 2019), the Global Partnership for Social Accountability (GPSA) Forum (2017, 2019), the International Water Association Congress (2019), the Accountability for Water Thinkshop (2018).

[4] Ethiopia (2019), Kenya (2019), Tajikistan (2017), Tanzania (2018, 2019) and Zambia (2017, 2020)

[5] Fox (ibid.) recognises that term is rarely defined with precision and proposes that in the context of accountability interventions, it refers to actions by external allies which (a) reduce the actual and perceived risks and costs inherent in taking action and (b) bolster the actual and perceived efficacy of taking action by increasing the likelihood and/or degree of positive institutional response.

and hygiene (WASH), WRM, including flood and drought risk and agricultural water management (AWM). This reflects the highly interdependent nature of different water uses. To ensure that we capture the full range of potential evidence, studies, and literature on WASH from different parts of the world and at any scale of intervention have been included. This review provides an objective analysis of the state of the literature, a guide to its key insights and the knowledge gaps which persist. We map the extent and nature of the evidence by country and region, by intervention type, by sub-sector focus, by outcomes and impacts discussed, and the 'factors' which influence these. Simple descriptive statistics are used to organise the evidence by the outcomes and factors observed. These are then organised into An Accountability for Water – Theory of Change (ToC) (see Figure 5).[6]

The review identifies and draws on available knowledge to populate, test, critique, and validate the ToC, which can then be used as a strategic resource for research, policy, and practice on the topic. In particular, we aim to understand the 'missing middle' of the ToC, the factors that explain why accountability and advocacy interventions, do, or do not deliver desirable outcomes and impact. The ToC is central to our approach and frames the literature of interest. This analysis uses inductive 'open' coding of the available literature to identify key factors and processes that either support or undermine successful achievement of outcomes or impact through accountability interventions.

In summary, the objectives for undertaking this review are fourfold:

- To map and review current knowledge relating to the question:
  What evidence is available on the outcomes of accountability and advocacy in the water sector and the factors which determine their impact?
- To draw on the available evidence to populate and develop a ToC which helps to explain the outcomes and impacts of accountability and advocacy interventions.
- To support a comprehensive knowledge platform that can be used by the community of practice, policy, and research to strengthen accountability in the water sector.
- To validate and guide future research by setting out the extent of contemporary knowledge, contested areas, case studies, concepts and theories for testing and further development, available methodologies and centres of research expertise.

METHODOLOGY

Our approach draws on rapid evidence review, systematic review, and theory-based review methodologies (Shadish & Myers 2004; Hammerstrøm *et al.* 2010; Higgins & Green 2011; Snilstveit *et al.* 2012; EPPI 2019). The latter are particularly useful for mapping causal links and factors which influence outcomes. The rigour of systematic review was adopted within the search, screening, and data-extraction stages, with a more flexible approach taken within the synthesis stage, which allowed us to present formative evidence in accessible and useful ways to the end user. The results describe the current literature on the topic, and the discussions, associations, and insights within it.

The steps involved to ensure a comprehensive, replicable, and valid review that avoids bias are set out here, with supplementary material available in the appendices.

Question formulation

The following review question was developed with support from the review advisory group (Appendix 2):

> *What evidence is available on the outcomes of accountability and advocacy in the water sector, and the factors which determine their impact?*

The review provides a comprehensive search, collation, organisation, and synthesis of the available evidence, documented in peer-reviewed academic journal papers, book chapters, organisational and grey literature, in response to this question. The definitions of terms used in the question are presented in Appendix 3.

---

[6] Theory of Change (ToC) is a specific type of methodology for planning, participation, and evaluation that is used in companies, philanthropy, not-for-profit, and government sectors to promote social change. ToC defines long-term goals and then maps backward to identify necessary preconditions. This ToC was developed through an iterative process comprising review of the literature, focussed through detailed and in depth engagement with key sector actors and validated by an iterative consultation process that captured the voices of a diverse group.

Search strategy

The pool of potentially relevant articles was found by searching:

- Academic databases, specifically: Web of Knowledge; SCOPUS; Econ Lit; and, Science Direct, using the following Boolean search string developed and tested through steps set out in Appendix 4.
    - *Water AND (Accountability* OR Advocacy OR Accountability Monitoring OR Social Accountability Monitoring) AND (Governance OR Service)*
- Specific journals, namely the *Journal of Development Studies* and *Water Alternatives*.
- Google and Google Scholar for unpublished material and grey literature with the first 50 hits checked with further examination if highly relevant studies are found.
- Websites of 72 relevant organisations, networks, and initiatives identified by the review advisors.
- Direct requests to sector experts: additional materials were obtained following consultations with review advisors, a process that generated a set of papers used to check whether the automated literature searches are returning the most relevant papers.

The definitions of terms used in the question are presented in Appendix 3.

Inclusion criteria and screening

Identified titles and abstracts were downloaded for review and were assessed for relevance and inclusion using a two-stage process, first by reviewing the title and abstract, and, for those selected, via full text screening. Multiple reviewers worked on this process (and subsequent coding) and to support a consistent approach, training, coaching, and internal checks were provided against the inclusion or exclusion criteria set out in Table 1. Lead reviewers audited a 10% random sub-sample of title and abstracts, then 10% of full texts to ensure consistency. Where inconsistencies were found, these were discussed to support unambiguous interpretation and application of the criteria. A full list of articles excluded at full text screening can be provided along with reasons for exclusion. All included papers were used in the coding exercise.

Data coding, extraction, and analysis

For each of the included studies, key characteristics were coded to support mapping and review. Standardised coding was supported by EPPI Reviewer software and data extracted to an Excel spreadsheet for analysis. The following information in relation to each paper was coded for.

- Full bibliographical reference/citation
- Year of publication
- URL link
- Publication type (peer review journal article, conference/working paper/book chapter/organisational report, or evaluation)
- Sub-sectoral focus (WRM, urban or rural WASH, AWM)
- Methodology (qualitative/quantitative)
- Geography – country of study
- Type of advocacy/accountability intervention (social accountability monitoring, budgets, etc. with sub-category)
- Outcomes and impacts associated with intervention and nature of effect
- Factors discussed and nature of effect (see Annex 5 for the Factors Codebook – developed iteratively based on the reviewed papers).

Included papers were mapped geographically and simple statistical analysis undertaken to describe the available evidence. Data on the nature of outcomes, impacts, and determinants of performance have been used to populate the ToC. The frequency with which each is discussed alongside the nature of effect was also analysed with the findings represented graphically.

LIMITATIONS OF THE REVIEW

The key limitations of the review are set out below:

- *Only English language literature since 1999.* The language skills and time available meant that we limited to the literatures that were written in English since 1999. The search therefore potentially misses valuable work published in non-English languages or in the last century.

Table 1 | Inclusion and exclusion criteria

Question element	Description	Criteria
Population	Scale	We are interested in accountability and advocacy mechanisms that are applied to improve water services and governance for any human population. We are interested in accountability and advocacy mechanisms that are applied to improve WRM or WASH services and governance at any scale (household/community/municipality/basin/region/country). To prevent exclusion of potentially relevant data, scale will not be used as an exclusion criterion.
Interventions		Interventions will be grouped into four categories as set out in the ToC: 1. Social accountability monitoring and initiatives (a) Citizen report cards (b) Social audit (c) Rights activation (d) Community scorecards (e) Participatory situation analysis (f) Strengthening citizen voice 2. Budget analysis, tracking, and reporting (a) Public expenditure tracking surveys (b) Participatory budget processes (c) Participatory budget analysis and expenditure review 3. Evidenced-based advocacy (a) Research and analysis (b) Freedom of information campaigns (c) Media campaigns (d) Lobbying (e) Public hearings, debate, and dialogue processes (f) Integrity audits 4. Statutory accountability mechanisms (a) Public complaint and grievance mechanisms (b) Citizen oversight panels (c) Formal audit and disclosure (d) Ombudsmen services (e) Customer service charters (f) Public interest litigation An additional category of 'other' was included to avoid exclusion of articles considered relevant but not captured by the above terms.
Outcome	Qualitative and quantitative measures of outcomes traced to advocacy and accountability mechanisms	To be included, studies and literature must link interventions to outcomes of relevance for improved water services and WRM. These are organised and defined as follows: 1. Changes in WRM/WASH services (aligned with JMP/WHO indicators of interest – See Annex 3 for definitions): (a) Availability (b) Accessibility (c) Quality (d) Affordability (e) Acceptability (f) Sustainability (g) Hygiene (h) Equity 2. Improved WRM/WASH governance (see OECD indicators in Annex 3) via existence/implementation/functioning of: (a) Roles and responsibilities (b) Appropriate scale (c) Policy coherence (d) Capacity (e) Data and information (f) Financing (g) Regulatory frameworks (h) Innovative governance (i) Integrity and transparency (j) Stakeholder engagement

(*Continued.*)

Table 1 | Continued

Question element	Description	Criteria
		(k) Trade-offs across users and generations (l) Monitoring and evaluation 3. Water security defined as the reliable availability of an acceptable quantity and quality of water for production, livelihoods, health and ecosystems, coupled with an acceptable level of risk from hazards including droughts, floods, pollution and conflicts.

- *Absence of full quality and validity assessment.* Our scoping work revealed a low number of highly rigorous studies. Had we applied strict quality criteria then the final sample would have been very limited in size and scope. To capture the maximum breadth of useful material, a proxy measure of the quality and replicability of studies was used, which was to exclude studies which provided no methodological description.
- *Preferential reporting of successful interventions.* A potential bias is the tendency for preferential reporting of successful interventions, particularly in the grey, institutional literature. To counter this we included the requirement for studies to link interventions to measurable outcomes and impacts as an inclusion criterion.
- *Restricted access to some literature.* We identified some literature which merited inclusion based on the abstract, or information available, but which was inaccessible as full text (e.g. prohibitively expensive, or not accessible through the research portals available to us.)

RESULTS AND DISCUSSION

The results of the review are presented in three sub-sections. First, we set out the state and nature of the literature, its distribution, and focus. Second, we provide the insights and evidence the literature provides about outcomes and determinants of performance. Finally, we conclude by populating the ToC.

Initial search identified 7,424 potentially relevant papers and reports. Out of these, 372 were duplicates and were thus dropped. 6,430 further papers were excluded because they were not in English, were written before 1999, or did not concern interventions in the water sector which linked interventions of interest to outcomes. 622 papers remained. Of these, 471 did not provide a description of methods used. In total, 2% of the original sample met our inclusion criteria. These 151 papers were kept for data extraction, mapping, and analysis. The majority (62%) of selected publications are peer-reviewed journals followed by organisational reports and evaluations (20%), and smaller numbers of working papers, book chapters, and conference papers. A full list of included references is provided in Appendix 6, and is available as a searchable database on www.accountability-forwater.org.

The geographical distribution of the evidence found is uneven (see Figure 1). Evidence identified is clustered in particular regions while others were under-represented, particularly West Asia, Eastern Europe, Middle East and North Africa (MENA), and Central Africa. Sub-Saharan Africa accounted for the largest share of the available literature (41%), followed by East Asia and Pacific (17%), South Asia (15%), Latin America and the Caribbean (10%), Europe and Central Asia (10%), and North America (5%).

Within these regions, evidence is not spread evenly, with a small set of countries accounting for most papers. For example, India was the geographical focus for over 40% of literature from the South Asian region, and just three countries, Kenya, Tanzania, and Uganda, together accounted for over 40% of studies in Sub-Saharan Africa. While this can be explained in part by the scale and severity of water issues in these countries, it also suggests a pattern in the way accountability interventions and research are targeted and resourced. Some countries or regions may be overlooked or excluded, and the resultant global knowledge base skewed towards a narrow set of contexts. Examining the reasons for this was beyond the scope of this review.

Analysing the evidence by sector shows it is heavily skewed towards WASH and water service delivery (62%), and within this primarily drinking water. Only six of the papers reviewed dealt exclusively with sanitation and hygiene, with a further 20 including substantive discussion of sanitation and hygiene alongside drinking water. Fewer papers examine water resources (32%) or AWM (5%). This might be explained by the larger number of donor commissioned reports and assessments in the grey literature that evaluate the outcomes of WASH accountability or advocacy interventions. The lack of research on accountability for AWM and water resources is

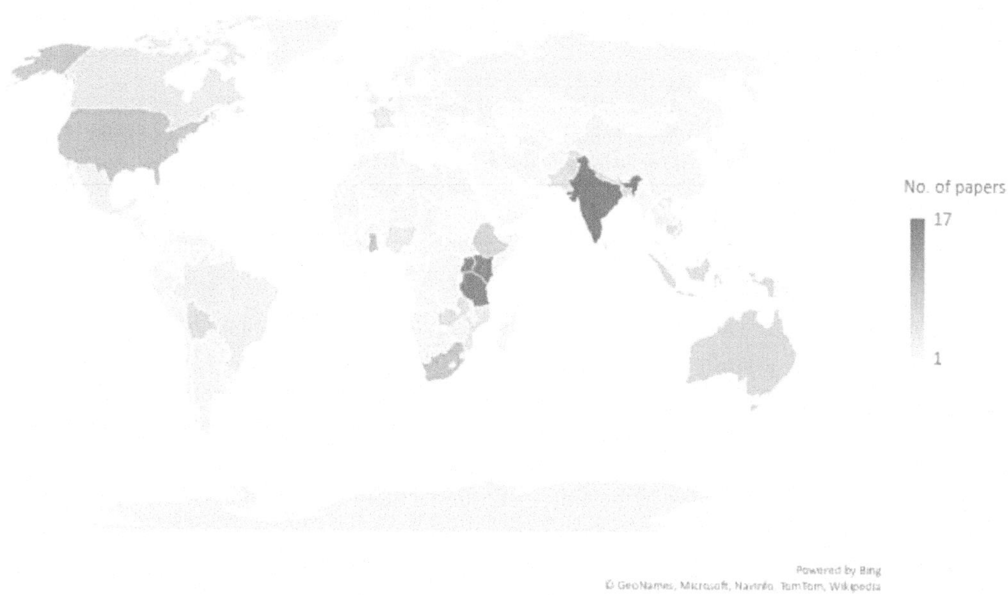

Figure 1 | Global distribution of the literature identified by the review indicating number of papers per country.

important, given the essential roles of agriculture and domestic supply on the sustainability of the underlying water resource.

Figure 2 shows the distribution of intervention types studied within all articles, grouped into the following four main intervention types: (1) social accountability monitoring; (2) budget analysis, tracking and reporting; (3) evidence-based advocacy; and (4) statutory accountability mechanisms. Social accountability monitoring initiatives are reported on most often, and feature in over half of the papers reviewed. Within these, strengthening of citizen voice was the most common intervention sub-type (43% of cases), followed by participatory situation

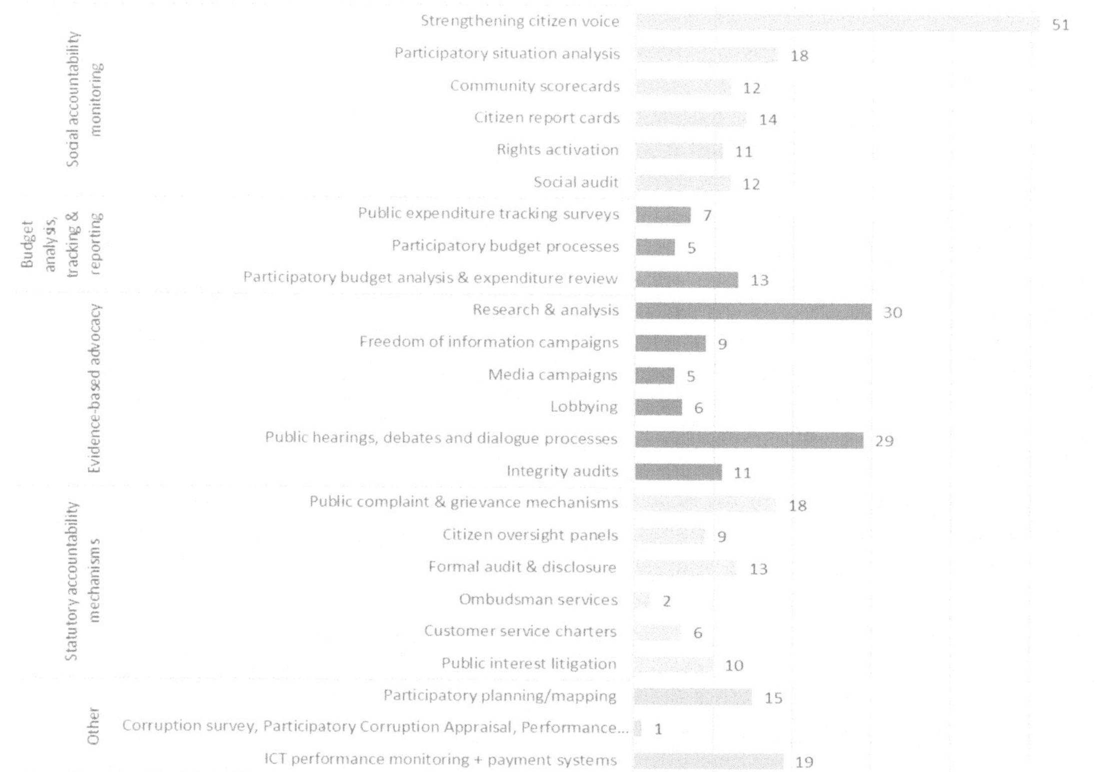

Figure 2 | Intervention sub-types examined in the literature, showing number of articles featuring each.

analysis (15%), citizen report cards (12%), community scorecards and social audits (10% each), and rights activation (9%). These findings reflect a predominant interest in participatory, community-based accountability mechanisms, and forms of collective action to amplify common grievances. Some patterns emerge in terms of the way interventions link to sub-sectors, for example, report cards, scorecards and social audits are more commonly applied to WASH, while others, such as community mapping, are more commonly associated with WRM.

Evidence-based advocacy features in 44% of the papers. Within this subset, the most frequent approaches examined are research and analysis (33%), followed by public hearings, debate and dialogue processes (32%), integrity audits (12%), freedom of information campaigns (10%), lobbying (7%), and media campaigns (6%).

Statutory accountability mechanisms feature in 28% of the papers. Within this, public complaint and grievance mechanisms featured in 31% of papers, followed by formal audit and disclosure (22%), public-interest litigation (17%), citizen oversight panels, where citizens have a formal role overseeing service providers (16%), customer service charters (10%), and ombudsman services (3%).

Budget analysis, tracking and reporting are the focus in 11% of papers. Just over half of these concern participatory budget analysis and expenditure review (52%), a little over one quarter deal with public expenditure tracking surveys (PETS) (28%), and a fifth with participatory budget processes (20%).

Other interventions featured in 16% of papers. Within this subset, a majority looked at Information, and Communications Technology (ICT)-based monitoring and payment systems (54%), with a smaller number examining participatory planning and mapping (43%), and a single paper each addressing corruption surveys, participatory corruption appraisal, and performance benchmarking.

What does the evidence tell us?

This section presents the associations made in the literature between interventions, outcomes and impacts. It does not weight or critique the evidence, so 'more evidence' does not necessarily mean 'better evidence' or stronger causation. Comparisons between interventions represent differences in the evidence, not necessarily differences in efficacy.

80% of the articles reviewed associate positive change for water governance at outcome level with accountability and advocacy interventions. Positive change at impact level is less frequently reported, in three out of ten articles. Illustrative case studies from the included literature which demonstrate these are set out in Appendix 8.

The evidence associates accountability and advocacy interventions with positive change across a range of important outcomes for improved water governance and services (see Figure 3). The outcomes most frequently associated with interventions are improved operational maintenance, monitoring and reporting (featuring in 40 papers), followed by access to data and information, representation and inclusion, and political recognition and will. The frequency and nature of change reported in the literature is presented in Figure 3 which indicates where the literature claims a positive effect, where no effect or a negative effect was observed, where a mixed effect was reported, or where insufficient data were presented to determine the nature of the effect.

A wide range of impacts are associated with accountability and advocacy interventions although the linkages and effects are not as well articulated in the literature as they are for outcomes, perhaps because they are harder to measure. The greater focus on WASH interventions in the literature corresponds to higher numbers of articles with associated WASH impacts. There were more studies on WASH-focused interventions that reported on impact than there were in other sub-sectors. Changes in accessibility were discussed most frequently, followed by impacts on quality, availability, equity, sustainability and affordability. Across all impacts, precise data to validate and demonstrate claims was often lacking. For example, our coding identified impact claims across many of these variables, but the code 'insufficient data' shows cases where an impact is mentioned but not measured in the paper. Interventions on water resources are most frequently associated with impacts on sustainable use, followed by conflicts, pollution, floods and droughts, these too were limited by inadequacy of data. There are few studies on flood and drought impacts.

Factors associated with the performance of accountability and advocacy interventions

Our coding of the literature reveals 28 factors associated with the performance of accountability and advocacy interventions, with multiple factors mentioned in most accounts reviewed. For readability, the factors are listed as enabling, but in each case the reverse is a disabling factor and both positive and negative examples were coded together. Factors including a constructive approach (37), training, human resource availability and professionalism (28), leadership and champions (28), taking a strategic approach (27), good public

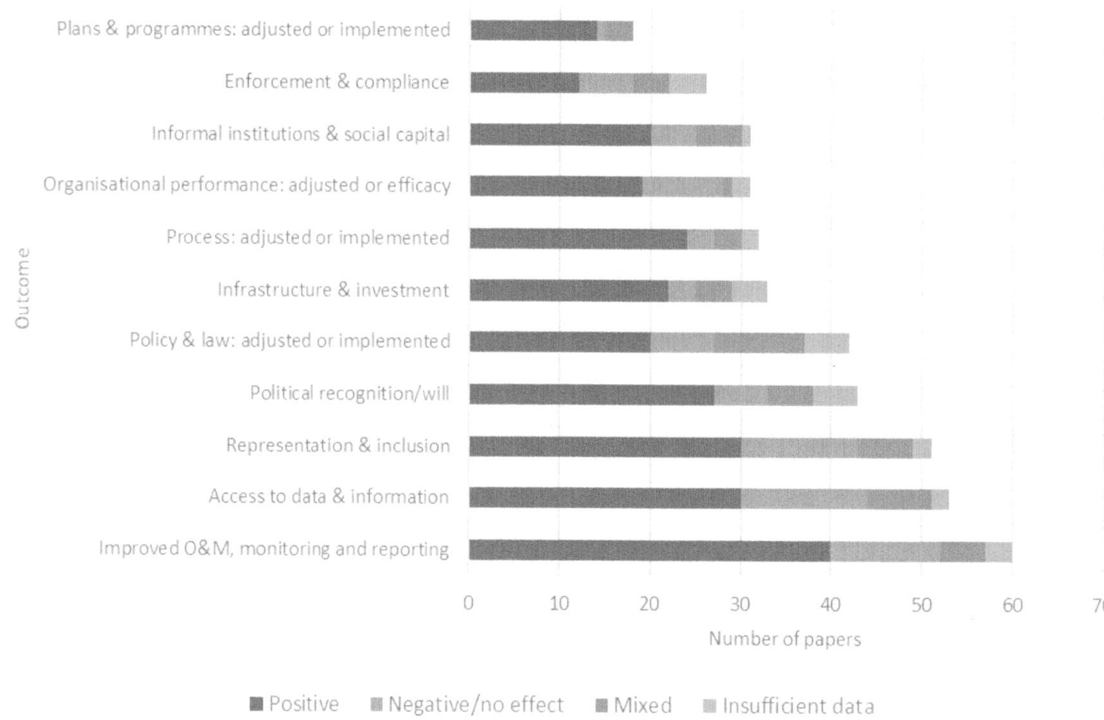

Figure 3 | Outcomes associated with accountability and advocacy interventions in the water sector: frequency in the literature and nature of effect.

communications (23), dialogue and convening (23) and trust and legitimacy (21) are most frequently associated with positive outcomes. A poor understanding of community and power dynamics (28), ambiguous institutional responsibilities (17), weak inter-agency accountability (17), a challenging water use context (17), limited financial resources (16), and trust and legitimacy (15) are factors most frequently associated with negative or no outcomes. Full explanations of these factors are provided in the codebook in Appendix 5.

To address the complexity of these multiple factors and to make the data accessible, we have classified and grouped these factors.[7] Our typology draws on foundational analysis of water institutions by Saleth & Dinar (2005) which characterises factors that influence performance as being either endogenous (internal) or exogenous (external) in origin. The factors are therefore grouped into three groups:

- 'internal' factors – those that are to do with intervention design and implementation, or internal organisational dynamics of implementing agencies such as its longevity, training and skills, or gender sensitivity. They can be directly controlled or are controllable by the practitioners and decision-makers.
- 'external' factors – those that are to do with the context in which the intervention takes place. For example, the level of government capacity, or literacy rates in the population. These are beyond the direct reach or influence of the intervention designer, decision maker or practitioner.
- 'interface' factors – those that are to do with how internal and external factors interact or are managed. In these cases there is some capacity to manage or control the factor through careful attention to the external context and appropriate intervention design.

This typology can be used to organise the literature and its insights for further development. The results of applying this typology to the literature are presented in Figure 4. Internal (Figure 4a), external (Figure 4b), and interface factors (4c), have been ranked based on the frequency cited in the literature, together with an indication of the nature of the effect reported.

[7] The approach draws on the concept of 'locus of causality', borrowed from the field of psychology where it is a dimension used in attribution theory that relates to the perception of the cause of success or failure. The locus of causality may be internal (i.e. based on the one's own characteristics, such as ability or effort) or external (i.e. due to factors such as luck, outside the control of the individual).

The factors are described in more detail in the factor code book in Appendix 5, together with examples from the literature. Internal factors are more often associated with positive influence on outcomes than are external and interface factors, which are more often associated with a negative influence on outcomes. The online dataset (www.accountabilityforwater.org) allows practitioners to cross reference intervention types with factors, to improve and focus detailed planning.

Developing the ToC

The coding of the literature presented above was used to redevelop a ToC for accountability and advocacy interventions on WRM and WASH. An outline ToC was used to develop the evidence review strategy, but this outline was found to have a 'missing middle' – the factors that determine the success or failure of interventions. Analysis of the evidence entailed coding papers according to their main findings. This enabled the 'missing middle' to be populated with the internal, interface and external factors driving success or failure. Figure 5 presents the ToC, with interventions, and their associated outcomes and factors set out from left to right based on the frequency with which they appear in the literature. The figure presents a multi-pathway process from the core issue – lack of water security and human rights to water and sanitation, through various accountability interventions that are mediated by internal, interface and external factors. The influence of these factors affects whether an outcome occurs, and, in turn, an impact on WRM or WASH.

This ToC presents a conceptual framework that can support analysis, prior to intervention design, of which factors are most likely to be influential in success or failure. Tools such as bottleneck analysis or political economy analysis can be used to investigate which factors to prioritise and adjustments to make to the intervention design. Ultimately, this means adjusting internal factors (i.e. those within the control of the intervention design and delivery) to mitigate for interface and external factors. This may explain why internal factors are more often found to be associated with positive outcomes, while external factors are more often associated with negative outcomes (Figure 4(a) and 4(b)).

Implications for policy and practice

Given the scale of the global water crisis and the central role of water security for social and economic progress, the available evidence of 151 papers is modest in size and limited in coverage. India, Tanzania, Kenya, and Uganda account for almost half (45%) of all literature on the topic, but even here, there are fewer than 20 papers focused on each country.[8] The topical and thematic range of the evidence is also limited. Thematic areas that are infrequently mentioned, but critical for global and local water security, climate resilience and sustainable development include: AWM, pollution control, climate change related impacts such as droughts and flooding, and the relevance of gender. The depth of the literature is also limited, with few papers that provide detail on the material, long-term impacts of interventions as compared to shorter-term outcomes.

The majority of evidence comes from academic research, with three times as many accounts (92) drawn from journal articles than organisational reports and evaluations (30). These figures suggest that NGOs and the practitioners of accountability and advocacy have little capacity for, or are placing little priority on evaluating, learning from and sharing the results of their work.

This review identifies a challenge for policymakers and practitioners interested in stimulating stronger accountability for improved water governance. While there are numerous case studies of interventions driving positive change, our current knowledge about what works, why, and for whom is limited. Although important and useful insights are available, as presented above, the evidence is insufficient for detailed conclusions about causation that can safely be transferred between contexts. The results of the review lend support to those seeking to improve governance and service delivery in the water sector. The evidence suggests that accountability and advocacy interventions have an overwhelmingly positive influence. In eight out of ten included accounts, interventions are associated with positive change in outcomes including: operation, maintenance and monitoring; data availability; representation; political will; investment; policy; law; processes; organisational performance; social capital; enforcement and compliance; programmes; and plans. The potential for harnessing these interventions to accelerate delivery of the SDGs is therefore apparent. By strengthening accountability

[8] Detailed findings from East Africa are discussed in country-level evidence summaries in Brown *et al.* 2021 – available on the Accountability for Water website: www.accountabilityforwater.org.

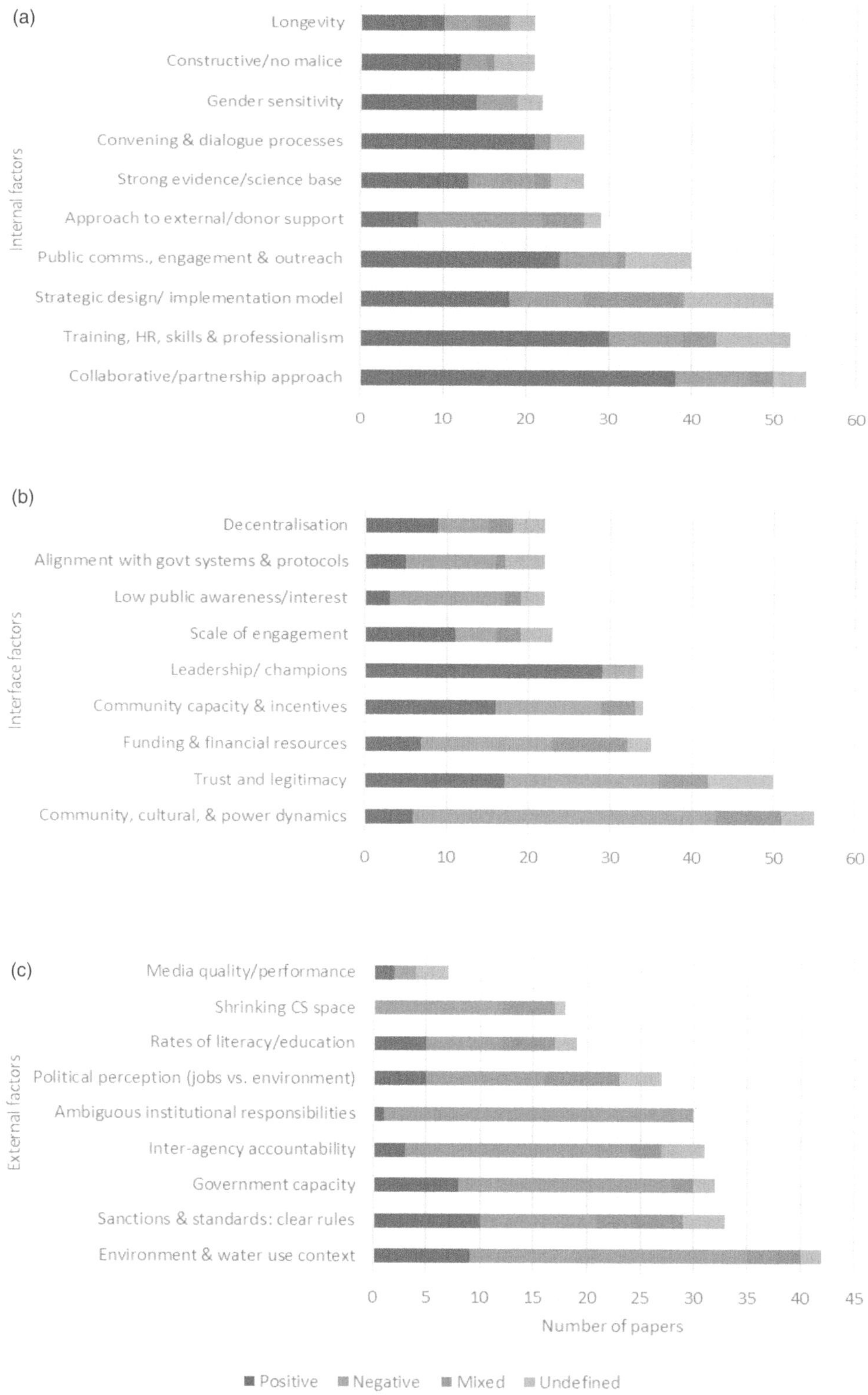

Figure 4 | Factors influencing accountability and advocacy interventions. Frequency of discussion and nature of effect: (a) internal factors; (b) external factors; and (c) interface factors.

processes, a positive feedback loop is created in the governance system, increasing the incentive and the ability of duty-bearers to deliver progressive realisation of human rights.

Positive impacts are anticipated in all the included papers, but detailed accounts of positive impact through material changes in parameters such as water access, quality, availability, affordability, sustainable use, levels of pollution, conflict, droughts, and flooding are limited to only one in three papers. This means that better

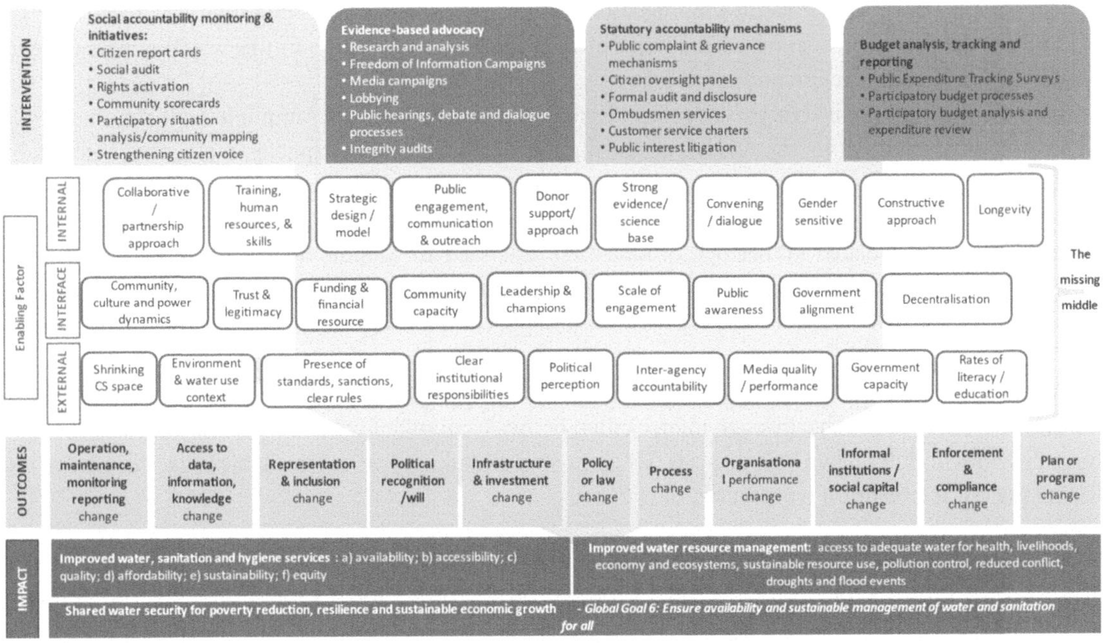

Figure 5 | Accountability for Water – ToC.

reporting and research on impacts remains a challenge. Further development and investment in accountability and advocacy interventions is a worthwhile endeavour for those seeking better water governance and water services and should be pursued as a priority.

How practitioners can use the evidence

The evidence presented highlights the range of factors which can influence the success of accountability interventions. The diversity of socio-economic, political, cultural, and environmental contexts that accountability interventions must respond to means that optimal design and delivery of interventions should be based on local situation and political economy analysis. A key value of this review is to highlight factors to be examined as part of such analysis. Important determinants of performance, either positive or negative, can then be understood at the outset and harnessed for positive influence, or mitigated against where they act as a constraint.

The internal factors mapped by the review are within the direct control of the practitioner or decision maker and can be shaped towards optimal design and delivery. Key internal factors identified were: investing in human resources, skills and training; strategic designs; a collaborative versus disruptive approach; the influence of donor support; convening and dialogue; foregrounding gender issues; generating strong evidence; public communications; and longevity and tenacity of engagement.

Interface factors such as poor understanding of community and power dynamics, and the recruitment of strong leadership and champions appear to be particularly important. Interface factors are partially within the purview and influence of the practitioner, providing that adequate groundwork is done to understand and respond to them. For example, the literature suggests that understanding community dynamics and champions, levels of trust and legitimacy, the financing context, levels of community awareness and interest, and alignment with government systems are important, and that each require consideration.

External factors provide a menu of the contextual challenges which enable or disable positive outcomes, and again they warrant careful consideration and response. They include issues such as absence or ambiguity of rules and statutory responsibilities; limited government capacity; prevailing political perceptions; restrictions on civil society and freedoms, levels of literacy and media performance. Their identification can improve the risk management and mitigation strategies for accountability programmes and planning. Improved understanding of how these structural or systemic issues influence accountability and advocacy interventions will aid their successful navigation.

In some instances, factors labelled as external may be malleable to positive action by sector actors. For example, government or other influential stakeholders can shape the wider landscape for accountability by initiating or funding reform processes which establish new standards or systems, which shape capacity and interagency accountability, or which resolve ambiguous or overlapping statutory responsibilities. This latter factor emerges across the literature as an important barrier to stronger accountability. In this way the review identifies issues for additional focus, action and advocacy.

This review offers utility and support to the policymaker and practitioner in multiple ways. It offers collated evidence on topics ranging from the roles of donor support and government capacity in effective accountability, to the relative performance of constructive versus disruptive engagement. It can be used to review the performance of specific interventions and factors associated with them, to inform planning. The factor 'checklist', the populated ToC and associated knowledge products are intended to support and guide more effective policy and practice in future. Rich and relevant insights are therefore available through deeper analysis of subsets of the literature. For the senior civil servant it points to the opportunities for reform and modified governance processes, for the civil society protagonist it supports design and practice guidance, and for the external support agent and donor it flags issues for reflection, development and investment. All the originally included references and further analysis are available through interrogation of the Accountability for Water Knowledge Platform. The platform provides an interactive hub through which to search and browse the available data, disaggregating literature by region, intervention type, factor, or outcome.

Implications for research

In characterising the existing spread and focus of knowledge on the topic we reveal its relatively limited geographical and thematic coverage. Coding of the available literature has enabled the identification of common themes, outcomes, factors and associations between them. This invites further conceptual and theoretical development and new research to explore causality and priorities within different contexts. By organising the current evidence base and presenting a ToC the review provides a resource to aid and stimulate future inquiry. We also bring together knowledge about suitable methods and approaches for research into these issues.

The review highlights priority knowledge gaps facing water and development stakeholders. A range of gaps and issues for further investigation emerge, reflecting the methodological and conceptual heterogeneity that was a feature of the evidence reviewed:

- Extending the geographical coverage beyond the small number of core countries highlighted, and developing a better understanding of community and government dynamics and the enabling environment.
- Examining the performance and role of accountability for water in agriculture, ecosystem protection and in the management and mitigation of climate impacts including drought and flood events.
- Unpicking the gender dynamics and social equity dimensions of accountability and advocacy.
- Evaluating the performance and outcomes of budget analysis and tracking.
- Clarifying the potential interplay between the media and public awareness.

Exploring these themes in greater depth will support more effective design and strategic support and resourcing for interventions. However, this is not a straightforward exercise and will require focused research effort over extended timescales, and comparative contexts. Long-term engagement is one of the factors associated with efficacy in the literature, and this should be reflected in research design. The ToC provides an initial framework for testing the relationships, assumptions and causality, and invites critical reflection to strengthen this initial framework.

CONCLUSIONS

The identification and organisation of global literature on the outcomes of accountability and advocacy interventions on water, and the factors which influence their performance, is a significant undertaking. By screening the initial dataset from 7,424 to the final 151 included papers, and subsequent coding and synthesis, this exercise sets out the current, English-language knowledge base on the topic.

Describing the evidence geographically, topically and thematically will help researchers and practitioners make sense of this field of work and take it forward. It confirms that the pool of knowledge is formative and emphasises the value of future collaboration between researchers, practitioners and policymakers, particularly given the importance of improved water governance and water service delivery.

The analysis finds that in eight out of 10 available studies, accountability and advocacy interventions are associated with positive outcomes for water governance, but that positive impacts on the ground are reported less frequently. We identify 28 factors which help explain the performance of accountability interventions, and organise them in terms of the frequency with which they appear, and within a typology based on their origin: internal to the intervention, external, within the wider context of its delivery, or at the interface between the two.

We develop a ToC for further elaboration and testing. The review is supplemented by an interactive Knowledge Platform: a searchable database of the collated evidence has been produced to support and supplement future work.[9] In reviewing current knowledge and organising it for further interrogation and use, this work contributes towards strengthening of accountability for water – to unlock a fairer water future for all.

While there is some literature on accountability for WASH it is under-represented. Perhaps, one of the reasons for this is that sanitation and hygiene have less of a political profile and priority compared to water. Given that public health gains depend on all elements of WASH, this is an area that demands greater efforts for research and development in the sector. Donors could also direct their attention to this area, which is often viewed as an individual or community concern, failing to recognise the need for systemic accountability and fundamental obligations of duty bearers. In the same vein, agricultural use of water is not adequately covered by the global water sector accountability and advocacy literature. Climate change impacts and the need to grow more food in the future, and manage water resources across the demands of drinking water, agriculture and production make strengthening accountability for this a high priority.

Recommendations

- The available evidence base should be deepened and widened through a learning-focused approach within future interventions. Decision-makers and practitioners should proceed with caution when designing and delivering interventions. Collaboration with researchers to ensure rigorous knowledge is generated across appropriate timescales should be embedded. While eight in ten interventions with accountability and advocacy components reveal positive water sector outcomes, further research is in order, to establish if all the positive relations are causative or non-causative. This could be context specific so that such and related considerations need to be factored in while designing and executing accountability and advocacy research in this regard. Better understanding of this causation is also likely to require a comparative and quasi-experimental research design.
- Further development and investment in accountability and evidence-based advocacy interventions is a worthwhile endeavor for those seeking better water governance and water services and should be pursued as a priority.
- The evidence review and knowledge platform can be drawn on as a 'one stop shop' for evidence relating to accountability and advocacy interventions and their ToC. Insights provided can be drawn on by all stakeholders to improve the design and delivery of accountability and advocacy interventions. Factors identified as influencing the way interventions perform can be used as a checklist for improvement, as the basis for risk management or to target collective action to understand and address strategic and systemic barriers to change. The reference material and deeper analysis provided on an array of topics and issues, should be drawn and added to through further exploration via the Accountability for Water Knowledge Platform.
- Additional research, and monitoring and evaluation effort is needed to extend the geographical and thematic coverage, and to deepen the global knowledge base on accountability for water. Priority needs to be given to testing the relationships and addressing the knowledge gaps identified by this review to enable the policy and practice community to better harness, deploy, and realise the benefits of stronger accountability for water. Opportunities to resource and undertake collaborative research between practitioners and researchers over timescales of an appropriate length to trace interventions to impact are a priority.
- The 28 factors we identified are likely to interact to a different degree in response to differing contexts. The degree of their interaction is also likely to affect their outcomes, be it in the positive or negative. Future research could focus on establishing more understanding about which factors are more important in what context and how and what could be done to augment their positive accountability and advocacy impacts. Taking note of these factors, we develop a ToC for further elaboration and testing.
- The review is supplemented by an interactive Knowledge Platform: a searchable database of the collated evidence has been produced to support and supplement future work.[10] In reviewing current knowledge and organising it for further interrogation and use, this work contributes towards strengthening of accountability for water – to unlock a fairer water future for all.

---

[9] Available at https://www.accountabilityforwater.org/knowledge-hub-overview.

[10] Available at https://www.accountabilityforwater.org/knowledge-hub-overview.

- Two specific factors, 'longevity' (i.e. sustained engagement over a long period) and 'trust and legitimacy' (i.e. being perceived as an authentic and legitimate agent to demand accountability) both suggest that the broader goals of 'decolonising' research and aid, ensuring funding reaches local organisations who are empowered to lead programme development, will also improve accountability outcomes.

DATA AVAILABILITY STATEMENT

All relevant data are included in the paper or its Supplementary Information.

REFERENCES

Brown, B. D., Brewer, T. & Hepworth, N. D. 2021 *Accountability and Advocacy Interventions in the Water Sector: A Review of Global Evidence. Part 2. Country-Level Evidence Summaries.* Water Witness International, Edinburgh, UK.

EPPI 2019 *What is A Systematic Review?* Social Science Research Unit, UCL Institute of Education, London. Available from: https://eppi.ioe.ac.uk/cms/Default.aspx?tabid=67 (accessed 15 April 2020).

Ernstorfer, A. & Stockmayer, A. 2009 *Capacity Building Development for Good Governance.* Nomos Verlagsgesellschaft Mbh & Co, Baden-Baden.

Farrow, T. 2018 *Water and Accountability Thinkshop: Event Summary and Evaluation Report.* Landmark Hotel, Dar es Salaam, Tanzania. March 2018. Water Witness International, Edinburgh.

Fox, J. 2015 Social accountability: what does the evidence really say? *World Development* **72**, 346–361.

Hammerstrøm, K., Wade, A., Hanz, K. & Jørgensen, A. K. 2010 Searching for studies: information retrieval methods group policy brief. *A Campbell Collaboration Policy Brief.* Available from: https://citeseerx.ist.psu.edu/viewdoc/download?doi=10.1.1.366.294&rep=rep1&type=pdf (accessed 28 July 2020).

Helmke, G. & Levitsky, S. 2004 Informal institutions and comparative politics: a research agenda. *Perspectives on Politics* **2** (4), 725–740. http://www.jstor.org/stable/3688540.

Hepworth, N. 2016 *Social Accountability for A Water-Secure Future: Knowledge, Practice and Priorities.* Discussion paper for Stockholm International Water Week.

Higgins, J. P. T. & Green, S. 2011 *Cochrane Handbook for Systematic Reviews of Interventions. Version 5.1.0. The Cochrane Collaboration.* Available from: http://handbook-5-1.cochrane.org (accessed 06 August 2020).

ITAD/ePact 2016 *Empowerment and Accountability Annual Technical Report 2016: What Works for Social Accountability.* ITAD.

McGee, R. & Gaventa, J. 2011 Shifting power? impact of transparency and accountability initiatives. *IDS Working Paper* **2011**, 383.

Mekonnen, M. M. & Hoekstra, A. Y. 2016 Sustainability: four billion people facing severe water scarcity. *Science Advances* **2** (2), 1–7. https://doi.org/10.1126/sciadv.1500323.

Munia, H. A., Guillaume, J. H. A., Wada, Y., Veldkamp, T., Virkki, V. & Kummu, M. 2020 Future transboundary water stress and its drivers under climate change: a global study. *Earth's Future* **8**, e2019EF001321. https://doi.org/10.1029/2019EF001321.

Open Society Foundation 2010 *Evidence, Messages, Change! An Introductory Guide to Successful Advocacy.* OSF, London.

Saleth, R. M. & Dinar, A. 2005 Water institutional reforms: theory and practice. *Water Policy* **7** (1), 1–16.

Shadish, W. & Myers, D. 2004 *Research Design Policy Brief.* Available from: https://citeseerx.ist.psu.edu/viewdoc/download?doi=10.1.1.174.1744&rep=rep1&type=pdf (accessed 25 July 2020)

Snilstveit, B., Oliver, S. & Vojtkova, M. 2012 Narrative approaches to systematic review and synthesis of evidence for international development policy and practice. *Journal of Development Effectiveness* **4** (3), 409–429. doi:10.1080/19439342.2012.710641.

UNDP 2006 *Human Development Report 2006. Beyond Scarcity: Power, Poverty and the Global Water Crisis.* Available from: http://hdr.undp.org/sites/default/files/reports/267/hdr06-complete.pdf (accessed 12 July 2020)

UNESCO 2019 *Leaving No-one Behind: The World Water Development Report.* Available from: https://unesdoc.unesco.org/ark:/48223/pf0000367306 (accessed 12 July 2020).

WHO/UNICEF 2020 *Progress On Household Drinking Water, Sanitation And Hygiene 2000-2020: Five Years Into The SDGs.* WHO/UNICEF 2020 Joint Monitoring Programme For Water Supply, Sanitation And Hygiene, Geneva, Switzerland.

First received 31 January 2021; accepted in revised form 9 May 2022. Available online 19 May 2022

doi: 10.2166/h2oj.2022.029

Achieving handwashing with Social Art for Behaviour Change: the experience of the Lazos de Agua programme in Latin America

Anna Zisa*, Kristina Nilsson, Raisa Mirza and Tania Vachon

One Drop Foundation, Montreal, Canada

*Corresponding author. E-mail: anna.zisa@ondrop.org

AZ, 0000-0001-6399-7387; TV, 0000-0002-0440-3849

ABSTRACT

The effectiveness of different hygiene behaviour change approaches is inconsistent. Proven effective elements of behaviour promotion include the following: involving the community, adding psychosocial theory-derived elements and using interpersonal communication with active teaching methods and innovative and culturally sensitive messaging. The One Drop Foundation Social Art for Behaviour Change (SABC) approach encompasses those elements and is embedded in a system-strengthening approach involving users, service providers and policymakers within the Lazos de Agua Programme. Halfway into the programme, the SABC approach has been implemented in more than 280 rural and urban communities in five Latin American countries. According to its midline outcome measurement, the programme's efforts have contributed to a 15% point increase in the population practising proper handwashing within intervention areas. Story-based interviews revealed that SABC interventions are believed to have caused lasting changes in behaviour, perception and skills which transcend beyond the individual and are felt at the household and community levels. While the SABC approach is mostly limited to addressing psychosocial factors, the experience of this programme proves that artists can serve as behaviour change facilitators to accompany water, sanitation and hygiene (WASH) system projects. The SABC approach builds capacity, both within artist groups as permanent local institutions which can act as behaviour change facilitation service providers, and within service users, who become empowered and can continue influencing behaviour change among their peers.

Key words: behaviour change, community involvement, handwashing, hygiene, monitoring and evaluation, WASH system-strengthening

HIGHLIGHTS

- SABC is a system approach for behaviour change, involving users, service providers and policymakers.
- The SABC approach in Latin America has shown it has the potential to be an effective approach towards increasing handwashing with soap.
- Through the SABC approach, capacity is built within artist groups as permanent local institutions that can act as behaviour change service providers.

GRAPHICAL ABSTRACT

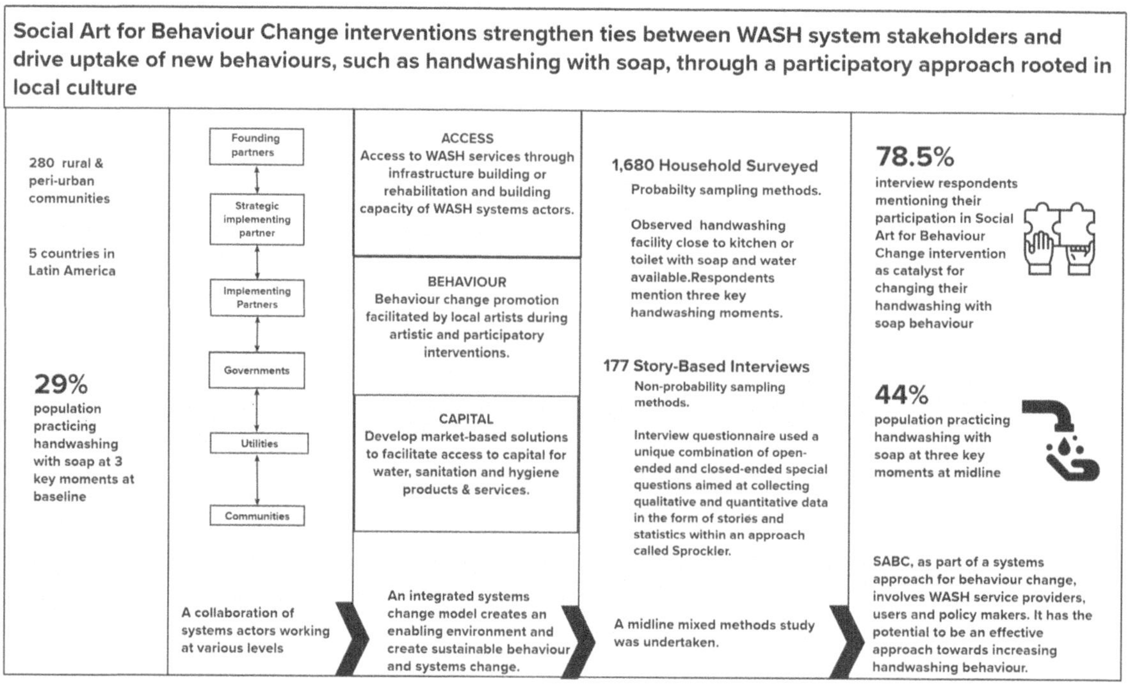

INTRODUCTION

Handwashing with soap remains the most simple and powerful hygiene practice to curb the spread of viruses and other infectious diseases that weaken or kill millions of people around the world every year (Prüss-Üstün *et al.* 2016; World Health Organization 2020). Although the solution seems basic, the alarming reality is that 30% of the world's population lack handwashing facilities with soap and water at home, and that critical settings such as health care facilities (at least 16%) and schools (44%) around the world lack basic handwashing services (WHO and UNICEF 2018, 2019, 2021). As expressed by the Global Handwashing Partnership (2017), 'behaviour change efforts are critical to ensure handwashing with soap becomes habitual'. Handwashing-promotion efforts that address key identified enablers and barriers have been found to be more effective (Kalam *et al.* 2021). However, determinants associated with the practice of handwashing with soap are varied and differ depending on the context. Empirical evidence has identified these determinants to include the following among others: the availability of handwashing facilities, access to soap and water, social and cultural norms and broader social and structural factors (White *et al.* 2020). This suggests that tackling handwashing determinants should involve several actors and factors to be working together, including the actors and factors required for ensuring the continuous delivery of water services which are the subject of many water, sanitation and hygiene (WASH) programmes across low-income settings.

Even when behaviour change efforts are part of WASH programmes, evidence of sustained behavioural uptake is mixed (Hulland *et al.* 2015; Taylor *et al.* 2015; De Buck *et al.* 2017). While some observational studies and numerous impact evaluations using experimental or quasi-experimental designs show significant increases in behavioural outcomes (Biran *et al.* 2014; Briceño Coville & Martinez 2015; Pickering *et al.* 2015), others have shown no significant effect of behaviour change efforts on behavioural outcomes compared to control groups (Huda *et al.* 2012) or have shown an attenuation of desired effects over time (Parker *et al.* 2006). Inconsistencies in the effectiveness of behaviour change interventions may be rooted in the type of approach used for promoting behaviour change and the understanding and tackling of determinants that influence WASH behaviour change (De Buck *et al.* 2017). De Buck *et al.* (2017), identified that effective elements of behaviour promotion include the following: involving the community, psychosocial theory-derived elements and using interpersonal communication with active teaching methods and messaging that is innovative and culturally sensitive. The One Drop Foundation Social Art for Behaviour Change™ (SABC) approach includes these effective elements and has been integrated into a WASH system-strengthening programme, the Lazos de Agua Programme (LAP).

Implementation of the LAP goes from 2016 to the end of 2022 and is carried out in Colombia, Guatemala, Mexico, Nicaragua and Paraguay by Plan Colombia, Water For People, Living Water International, WaterAid and Fundación Moisés Bertoni, respectively.

This article presents how handwashing is encouraged through the SABC approach, embedded within a system-strengthening programme. It argues why SABC has the potential to be an effective approach towards sustaining handwashing behaviour based on the progress achieved halfway into the LAP and shares the limitations and learnings of its implementation in five countries in Latin America.

METHODOLOGY

System-strengthening model

Designing and implementing the LAP involves coordinating and involving multiple actors, including funding agencies (foundations and multilateral development banks); national-level authorities (ministries of water, health, education), service authorities (local governments at various levels), service providers (often voluntary water user committees), civil society (implementing NGOs and artist groups) and users (community members). The LAP follows the One Drop Foundation's A•B•C for Sustainability™ model, a systemic approach to support holistic programme implementation where A stands for Access, B for Behaviour and C for Capital:

- The A (Access) seeks to increase the proportion of the population using at least basic levels of water and sanitation services as defined by the Joint Monitoring Programme (WHO and UNICEF 2021) through the construction and rehabilitation of infrastructure and strengthening the governance, operations and maintenance capacities of water user committees and other entities responsible for providing WASH services.
- The B (Behaviour) seeks to increase the proportion of the population practising desired WASH behaviours through the implementation of the One Drop Foundation Social Art for Behaviour Change™ (SABC). The SABC™ approach integrates psychosocial theory-derived elements from the Transtheoretical Model of Behaviour Change developed by Prochaska & DiClemente (1983) and the Integrated Behavioural Model for WASH (IBM-WASH) developed by Dreibelbis *et al.* (2013). The latter framework identifies three dimensions that influence WASH behaviours, notably contextual, psychosocial and technological. SABC is a participatory and creative approach undertaken in collaboration with intervention communities to bring about behaviour change. By engaging participants through social art, SABC interventions connect and empower community members, improving their leadership and other interpersonal skills and giving them a sense of ownership and agency over their WASH services.
- The C (Capital) component is intended to develop market-based solutions – including financial products, income-generating activities and other services – to increase the pace of making safe water and sanitation accessible to all.

Each country-based project part of the LAP has its own set of actors and factors that need to be addressed to create a transformative WASH programme. Each project uses a system approach aiming to strengthen the enabling environment that provides support and creates coordination mechanisms between the various actors within the system. The projects collaborate and support various parts of this system to promote long-term and sustainable access to WASH services through an integration of the multiple layers of government, community and value chain actors. Given that key determinants of handwashing with soap are access to water, soap and a handwashing station at home, handwashing promotion in the LAP is married to the promotion of another behaviour: the payment of water tariff by the users, which allows the maintenance of the water supply infrastructure, and thereby sustained access to water for handwashing. Institutional involvement has been instrumental in helping the uptake of both the handwashing and the water tariff payment behaviours.

Behavioural change intervention design and implementation

SABC interventions are designed and implemented by local artists trained on the SABC approach by the One Drop Foundation's country-based implementing partners. Before implementation, formative research reveals the profile of the priority groups who ought to practise the desired WASH behaviour, and the behavioural determinants that influence whether these groups engage or not in the behaviour. This step uses Dreibelbis *et al.* (2013) IBM-WASH model to acknowledge and investigate behavioural determinants belonging to contextual, psychosocial and technological dimensions that operate on five levels: structural, community, household, individual and habitual.

Once these behavioural determinants have been identified and prioritized, artists start ideating and prototyping SABC interventions, including folklore and art techniques, alongside priority group members from the communities. The resulting SABC interventions are diverse but all have common elements: they are artistic, participatory, based on the cultural and behavioural realities of the intervention communities and aim at encouraging the desired behaviour. Each community hosts multiple interventions including one-time and multi-session interventions. One-time interventions invite the community at large thereby involving numerous participants: for example, theatre shows followed by a facilitated discussion where attendees express a commitment to a concrete solution. One-time interventions are either led by artists, or by community members, who were involved in multi-session interventions. Multi-session interventions are led by artists and involve fewer participants actively involved in an artistic process to co-create: for example, short films, songs, murals or puppet shows. At the end of a multi-session intervention, the participants, referred to as leaders of change, lead a one-time intervention, referred to as a *replica*, accompanied by artists thereby presenting their creation to the community at large.

Outcome measurement

While the LAP aims at changing various WASH behaviours, this article will focus on handwashing with soap. To measure the prevalence of handwashing behaviour (as well as other WASH outcomes which are not the subject of this paper), a probabilistically representative sample of households was surveyed in each project, totalling over 1,680 households among four projects, at their respective baseline (2017–2019) and again at midline, 22 months later on average and before the start of the COVID-19 pandemic except for the project in Colombia (see Table 1). The project in Guatemala started before the other four projects and had a different monitoring and evaluation system and is therefore excluded from these results. The endline household survey in late 2022 has yet to occur. The number of households surveyed at baseline or midline ranged from 306 up to 615 depending on the project (see Table 2). In Colombia, Paraguay and Mexico, the sampling design was multi-stage, stratified first and then random. In Nicaragua, households from all communities of intervention were sampled randomly and proportionately to the communities' population. For all projects, sample sizes were calculated at a 95% confidence level and a 5% margin of error. The household questionnaire used can be found in the Supplementary Material.

Households were considered to practise handwashing if they met all of the following criteria: the enumerator observed a handwashing station (fixed or mobile) within 10 paces of the cooking area or the toilet, that had water and soap available, and the respondent mentioned at least the three handwashing key moments encouraged by the Programme – after defaecating, before eating and before preparing food – when asked: 'On what occasions do

Table 1 | Household survey data collection dates

	Baseline	Midline	Time elapsed between measurements
Colombia	2018-12	2020-12	24 months
Mexico	2018-03	2019-09	18 months
Nicaragua	2018-01	2019-08	18 months
Paraguay	2017-05	2019-08	27 months

Table 2 | Number of households surveyed and sample universe

	Number of households surveyed		
	Baseline	Midline	Population of intervention communities (sample universe)
Colombia	418	358	20,000
Mexico	306	414	62,500
Nicaragua	584	615	24,400
Paraguay	377	403	87,200
Total	1,685	1,790	194,000

members of your household wash their hands with soap and water?' without answer options given by the enumerator.

In addition to household surveys, story-based interviews were conducted during the second semester of 2020. Story-based interviews complemented household surveys for measuring handwashing by providing depth and isolating the effect of SABC interventions among other programme efforts which include building infrastructure and capacity building of WASH service providers. A story-based data collection approach called Sprockler, developed by the organization Perspectivity, was used. Within the Sprockler approach, the interview questionnaire uses a unique combination of open-ended and closed-ended special questions aimed at collecting qualitative and quantitative data in the form of stories and statistics. At the core of the questionnaire in the Sprockler approach, the respondent is invited to share a story after which closed-ended questions follow through which the respondent self-interprets the meaning of their story. Chances of misinterpretation are reduced by asking respondents questions about their story which also allows them to own and reflect upon their stories. Sprockler uses special questions such as bipoles or tripoles, which trigger more intuitive and instinctive responses and therefore help the self-interpretation set-up to provide more accurate and nuanced findings. Within the Sprockler approach, the importance of stories and the natural tendency and skill humans have for telling stories is not only acknowledged but embraced. Respondents share their experiences through stories, and by collecting many of such experiences and visualizing responses from close-ended questions, recognition of emerging patterns that tell something about the world surrounding the respondents is enabled (Perspectivity n.d.). Voicing experiences through stories can contribute to decreasing participant bias and is particularly suitable among highly indigenous and other population groups where storytelling and oral forms of knowledge transmission are a common means of communication.

In total, 177 community members across four projects who had participated in SABC interventions were interviewed with the Sprockler approach, ranging between 30 and 63 respondents per project (Table 3). Respondents were sampled using non-probability sampling methods to represent the diversity of SABC participants in terms of gender, age, occupation, type of SABC intervention they participated in and time elapsed since the last SABC intervention they participated in. The interview questionnaire can be found in the Supplementary Material. Two story-based questions were at the core of the questionnaire. The first invited the respondent to share a memory about a moment that happened when they participated in SABC interventions, and the second asked them to share and describe any changes, big or small, that they have noticed in their life or community since participating in SABC interventions.

Table 3 | Socio-demographic information about SABC participants interviewed

		Colombia	Mexico	Nicaragua	Paraguay	Total
Total		34	50	63	30	177
Gender	Woman	32	40	41	17	130
	Man	2	10	22	13	47
Age	<13	2	0	0	0	2
	13–17	8	2	1	6	17
	18–30	11	7	28	7	53
	31–40	3	15	11	4	33
	41–50	3	21	16	10	50
	51–60	4	3	5	2	14
	>60	3	2	2	1	8
Occupation	Domestic worker	15	27	26	3	71
	Student	15	3	4	10	32
	Teacher	1	3	4	7	15
	Farmer/rancher/fisherman	1	3	16	1	21
	Member of water committee	0	9	2	6	17
	Other	2	5	11	3	21

RESULTS

According to household surveys, the population practising handwashing with soap at three key moments increased from 29 to 44% between baseline and midline for the four projects combined (weighted average), an increase that was observable among them all except for project Nicaragua (Figure 1).

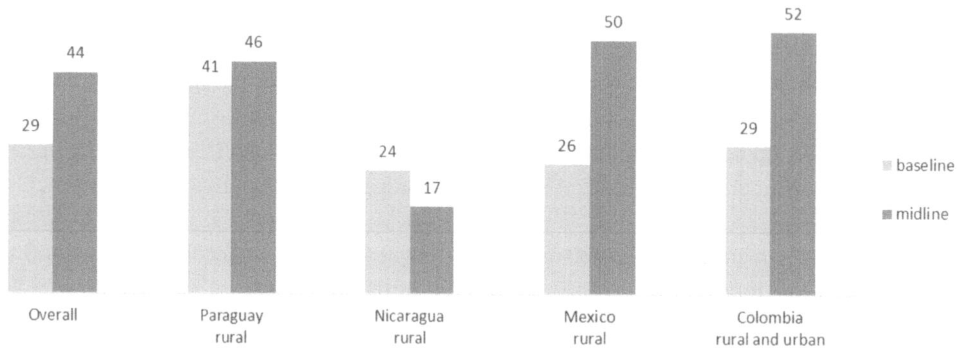

Figure 1 | Proportion of the population handwashing with soap within the intervention areas per project and overall across projects at baseline and midline (%).

Various programme efforts could have contributed to this increase, from the improvement of drinking water and sanitation services for households to the SABC interventions encouraging this behaviour. As seen in Figure 2, there was an increase between baseline and midline in the proportion of households meeting each of the criteria for handwashing, being contextual (a handwashing station close to the kitchen or toilet, and water availability), psychosocial (mentioning three key moments) and technological determinants (soap availability). According to story-based interviews, SABC interventions contributed to this change with 78.5% out of the 177 respondents identifying handwashing with soap at key moments among the changes they have noticed since participating in SABC interventions. In particular, some respondents from story-based interviews mentioned adding soap to their handwashing practice since participating in SABC interventions:

The change that we have had in the community is in handwashing. People did not use to care if they washed their hands only with water. Now we have had the experience that one has to wash with soap and water at all times. (Nicaraguan domestic worker, participant in multiple SABC interventions, the most recent occurring 6–12 months before the interview)

...sometimes we used to wash our hands only with water and we would say our hands were clean and I learned that in order to eliminate all the bacteria we have to wash our hands with plenty of water and soap and I think I didn't do that before, and after the workshop yes. (Nicaraguan domestic worker, participant in multiple SABC interventions, the most recent occurring 6–12 months before the interview)

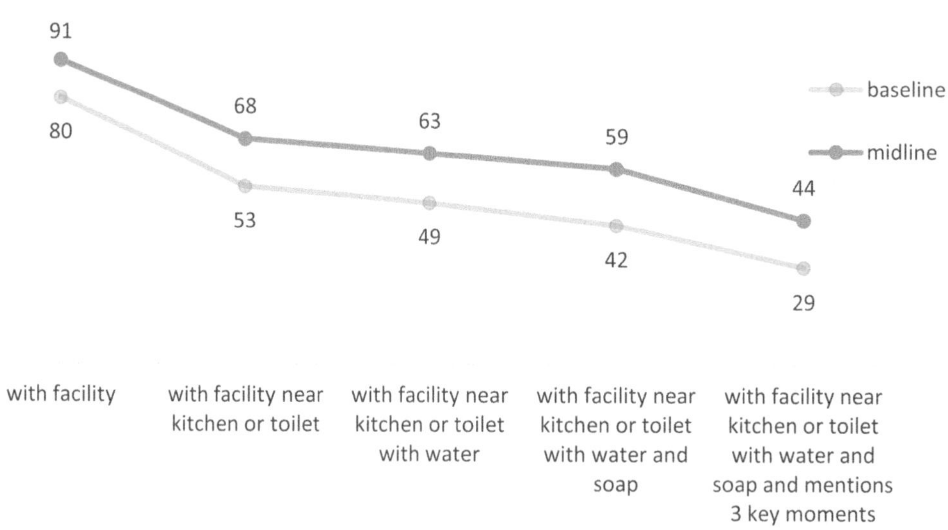

Figure 2 | Proportion of the population within the intervention areas meeting the different criteria for handwashing at baseline and midline (%).

Other responses from interviews referred to building handwashing facilities and using running water:

...before handwashing was done in a little tray, now we know that it is with running water, fluid water that falls, so those who cannot afford a container with a tap build it themselves. Young people and children have built handmade artisanal containers that allow water to fall while they wash their hands, that is very interesting. (Colombian teacher, participant in multiple SABC interventions, the most recent one occurring less than 6 months before the interview)

We wash our hands radically different than we used to because before you would take the pot and wash your hands, even if it was just one person, they would wash their hands inside the pot, but from then on we don't do that anymore, we don't put our hands inside the pot, we instead let the water fall, with water running and soap. (Colombian domestic worker, participant in a SABC workshop occurring less than 6 months before the interview)

According to story-based interviews, changes perceived by respondents since participating in SABC interventions transcend the individual and are felt at the household and community level (Figure 3).

The first change I noticed happened in my institution. The kids wash their hands and now with the pandemic it worked out that they have practised how to handwash properly. (Paraguayan teacher, participant in a SABC workshop that took place one or two years before the interview)

At the community level, beyond increasing handwashing, the LAP has strengthened the role of artists as mediators of change by bringing together water user committee members and water users to improve their collaboration: this mediation has improved discussions between water users and water user committee members, which gives users more confidence in the committee's actions and contributes to them becoming more assiduous in paying the water service tariff.

Those of us on the (water user) committee are from three different communities, and what we have felt is that the communities have become more united since then, since this activity brought us together, the three of us are

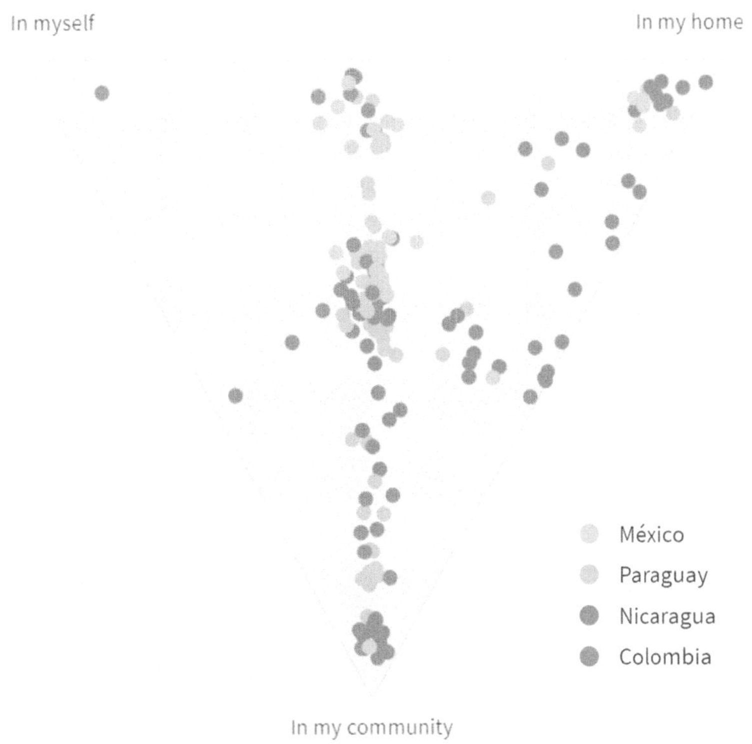

Figure 3 | Story-based interview responses to the question 'Where do you recognize this change?' referring to the change respondents mentioned since participating in SABC interventions.

> *more united, before, each one of us was on our own. We dialogue more. Before, nobody gave their opinion out of shyness, people have become more open and ask questions and we like that.* (Mexican farmer, participant in a SABC activity that took place less than six months before the interview)

> *As a user, I see a big change in the understanding, there is more dialogue and people know their rights. We as a family have learned to value hygiene, to wash our hands at key times, and I personally have improved a lot, I have grown as a person. Before I couldn't speak in public and now I can do it thanks to the constant training I receive.* (Paraguayan member of the water user committee, participant in a SABC activity that took place less than six months before the interview)

A group of 10 women of the village San José de Manantiales, in the state of Guanajuato, Mexico, who initially met through the SABC intervention of a puppet theatre production gained the confidence to have their voices heard and replaced the members of their water user committee by themselves, a role that was previously dominated by men.

At the state level, the community engagement from the initial SABC interventions were enough to convince the Guanajuato State Water Commission (in Spanish Comisión Estatal del Agua de Guanajuato or CEAG) to incorporate elements of the SABC approach as part of their social assistance programmes and engagement with water user committees. In particular, the CEAG has added participatory and artistic methods to the role traditionally played by community engagement officers referred to as 'water promoters'. Following training and support by the State Water Commission, with technical support from the LAP, these promoters now have new SABC tools and ways of involving communities in the management of water systems.

Finally, at the national level, officials from CONAGUA, the Mexican National Water Commission, supported the State Water Commission's efforts by funding their WASH programmes, including social assistance interventions which included SABC training workshops at the municipal level. Tapping into the collective intelligence, connecting various stakeholders through creative and participatory processes has been appreciated by the CEAG and CONGAGUA organizations and is inspiring them to modify their approaches in other locations. This led to the creation of a 'Coordination Table' (*mesa operativa*) which brings together several actors of the water services provision system to define clear ways on how some elements of the SABC approach can be integrated into the social assistance programmes carried out by CONAGUA at the federal level.

DISCUSSION

While the SABC approach includes effective elements of behaviour promotion identified by De Buck *et al.* (2017) and is contributing to an increase in handwashing, it has limitations. The SABC approach cannot address contextual or technological determinants as framed in the IBM-WASH by Dreibelbis *et al.* (2013). SABC interventions are most apt for addressing psychosocial determinants at the community, household and individual levels such as self-efficacy, social integration, cultural identity and social norms, but are limited in their ability to change infrastructure gaps and broader sector systems weaknesses. These gaps are an argument for embedding this approach within larger WASH system-strengthening approaches that could tackle other determinants. This was done with the LAP, which takes a systemic approach to behaviour change by bringing together actors for ensuring Access, Behaviour and Capital to communities. SABC interventions work within this framework to strengthen ties between stakeholders and to drive uptake of new behaviours through a participatory approach rooted in the local culture. Besides handwashing with soap and water tariff payment, the SABC approach can be tailored to address different WASH behaviours, as relevant in different contexts, such as latrine use in areas where open defaecation is common. For example, other projects co-financed by the One Drop Foundation have addressed handwashing with soap and latrine use with SABC in Mali, Burkina Faso, Malawi and India that could provide learnings to behaviour change efforts in those areas. Since the design of SABC interventions relies on context-specific evidence and local arts, their implementation can be adapted to any region. Through SABC, the acceptability by users of newly developed sanitation systems could be addressed, which could potentially benefit the sustainability of sanitation systems by lifting the score of the social sub-index of the Sanitation Sustainability Index developed by Hashemi (2020).

Burton *et al.* (2021) argue that in order to better design and evaluate 'transformative WASH' programmes, a mixed-methods toolkit utilizing human-centred design practices, and proxy methods such as 'participatory design' or 'behaviour-centred design theory' should be prioritized in order to overcome a siloed approach in delivering long-term impact. As part of the monitoring and evaluation of the LAP, the use of mixed-methods combining household surveys with story-based inquiry through Sprockler has emerged as a dynamic way to capture important data around participation in SABC interventions, and the uptake of the behaviours. These interactive and relational elements allow opportunities for iteration of programming to better fit the needs of users.

Another challenge in implementing the SABC is that it can be complex for artists to design artistic interventions that also aim to change perceptions around taboo topics, increase confidence in services or products or improve self-efficacy in practising a behaviour. The development of the intervention's prototypes may require substantial external support and time early in the projects affecting their cost-effectiveness. Nonetheless, artists have shared their appreciation when involved in the design process from the beginning instead of responding to isolated tasks. The experience of the LAP proves that through the SABC approach, artists can serve as behaviour change facilitators to accompany WASH projects and trigger behaviour change. Through their deep connection to the local context and their imagination and creativity, artists implementing SABC interventions generate a fun and enabling environment by translating complex and often taboo issues into accessible formats that inspire, activate and sustain change. This further empowers people in communities to take charge of their water and sanitation services, participate in water user committees and add soap to their handwashing routine. Examples such as Guanajuato, Mexico, show that SABC interventions can be used as a platform to drive broader behaviour changes among various actors in the WASH value chain and sector, to train, explore, elaborate, innovate and share their experiences, learnings and insights.

Results obtained until now in the uptake of handwashing with the SABC approach are encouraging, as is the institutionalization of the approach by the Mexican Government at the municipal, state and national levels. In the other LAP programme countries, implementing partners continue to collaborate closely with municipal, state and national governments in order to coordinate and support the continuous delivery of WASH services through strong WASH systems after the end of the programme. An endline household survey is planned for the LAP in the last quarter of 2022 which will provide additional opportunities to examine the factors that support a resilient WASH system and will bring further insights about the sustainability of this systemic programme. This new data will be made available at the end of the project, which could inform the potential of scaling the implementation of SABC interventions to the wider Latin American region.

CONCLUSIONS

SABC, as part of a system approach for behaviour change, involves WASH service providers, users and policymakers. It has the potential to be an effective approach for increasing handwashing behaviour. The programme works with relevant sector actors at multiple levels and seeks to modify the factors that influence sustained WASH service delivery and use, through multi-sectoral collaboration and system-strengthening, while keeping community users at the centre of the programme.

Ultimately, this aims to create an enabling environment that addresses the Access, Behaviour and Capital needs of the various users in the system. In Guanajuato, Mexico, the government has adopted elements of the SABC approach to strengthen their pre-existing social assistance programmes with a focus on ensuring that all households pay their monthly water tariffs, an essential behaviour to ensure smooth water service provision which is needed for households to continue handwashing with soap. Mixed-methods evaluation of the programme through household surveys and story-based inquiries support the conclusion that SABC interventions were memorable to participants, even after 6–12 months of their last participation in an intervention, and that the vast majority believed the changes in the communities would be long-lasting. Ongoing monitoring and an endline evaluation of the programme will provide further insights on the sustainability of the outcomes.

FUNDING

The Lazos de Agua programme (LAP) is funded through a multi-stakeholder partnership including the founding partners – the Coca Cola Foundation, the Inter-American Development Bank, FEMSA Foundation and the One Drop Foundation – national and local governments, implementing organizations and communities. The One Drop Foundation is also the LAP's Strategic Implementing Partner, coordinating the programme and providing technical assistance to the implementing organizations in various approaches including Social Art For Behaviour Change, value chain, system-strengthening, integration of the components of the ABC for sustainability model, monitoring, evaluation and learning, and communications. An evaluation was undertaken as part of the LAP, funded by the founding partners, and conducted in countries by monitoring and evaluation staff of the implementing organizations. Funding for writing this article has been provided as part of the ongoing staff salary of the One Drop Foundation, as part of its Knowledge and Innovation Unit.

DATA AVAILABILITY STATEMENT

All relevant data are included in the paper or its Supplementary Information.

REFERENCES

Biran, A., Schmidt, W. P., Varadharajan, K. S., Rajaraman, D., Kumar, R., Greenland, K., Gopalan, B., Aunger, R. & Curtis, V. 2014 Effect of a behaviour-change intervention on handwashing with soap in India (SuperAmma): a cluster-randomised trial. *The Lancet Global Health* **2** (3), e145–e154.

Briceño, B., Coville, A. & Martinez, S. 2015 Promoting handwashing and sanitation: evidence from a large-scale randomized trial in rural Tanzania. *World Bank Policy Research Working Paper*, (7164).

Burton, J., Patel, D., Landry, G., Anderson, S. M. & Rary, E. 2021 Failure of the 'gold standard': the role of a mixed methods research toolkit and human-centered design in transformative WASH. *Environmental Health Insights* **15**, 11786302211018391.

De Buck, E., Van Remoortel, H., Hannes, K., Govender, T., Naidoo, S., Avau, B., Veegaete, A. V., Musekiwa, A., Lutje, V., Cargo, M. & Mosler, H. J. 2017 Approaches to promote handwashing and sanitation behaviour change in low-and middle-income countries: a mixed method systematic review. *Campbell Systematic Reviews* **13** (1), 1–447.

Dreibelbis, R., Winch, P. J., Leontsini, E., Hulland, K. R., Ram, P. K., Unicomb, L. & Luby, S. P. 2013 The integrated behavioural model for water, sanitation, and hygiene: a systematic review of behavioural models and a framework for designing and evaluating behaviour change interventions in infrastructure-restricted settings. *BMC Public Health* **13** (1), 1–13.

Global Handwashing Partnership 2017 Handwashing – Vital for Sustainable Development. https://Globalhandwashing.Org/. Retrieved April 3, 2021, from https://globalhandwashing.org/wp-content/uploads/2017/09/AdvocacyBrief-Vital-for-Sustainable-Development.pdf.

Hashemi, S. 2020 Sanitation sustainability index: a pilot approach to develop a community-based indicator for evaluating sustainability of sanitation systems. *Sustainability* **12** (17), 6937.

Huda, T. M. N., Unicomb, L., Johnston, R. B., Halder, A. K., Sharker, M. A. Y. & Luby, S. P. 2012 Interim evaluation of a large-scale sanitation, hygiene and water improvement programme on childhood diarrhea and respiratory disease in rural Bangladesh. *Social Science & Medicine* **75** (4), 604–611.

Hulland, K., Martin, N., Dreibelbis, R., Valliant, J. D. & Winch, P. 2015 *What Factors Affect Sustained Adoption of Safe Water, Hygiene and Sanitation Technologies? A Systematic Review of Literature*. EPPI-Centre, Social Science Research Unit, UCL Institute of Education, University College London, London.

Kalam, M. A., Davis Jr., T. P., Shano, S., Uddin, M. N., Islam, M. A., Kanwagi, R., Islam, A., Hassan, M. M. & Larson, H. J. 2021 Exploring the behavioral determinants of COVID-19 vaccine acceptance among an urban population in Bangladesh: implications for behavior change interventions. *PloS one* **16** (8), e0256496.

Parker, A. A., Stephenson, R., Riley, P. L., Ombeki, S., Komolleh, C., Sibley, L. & Quick, R. 2006 Sustained high levels of stored drinking water treatment and retention of hand-washing knowledge in rural Kenyan households following a clinic-based intervention. *Epidemiology & Infection* **134** (5), 1029–1036.

Perspectivity n.d. Available from: https://sprockler.com/help#faq.

Pickering, A. J., Djebbari, H., Lopez, C., Coulibaly, M. & Alzua, M. L. 2015 Effect of a community-led sanitation intervention on child diarrhoea and child growth in rural Mali: a cluster-randomised controlled trial. *The Lancet Global Health* **3** (11), e701–e711.

Prochaska, J. O. & DiClemente, C. C. 1983 Stages and processes of self-change of smoking: toward an integrative model of change. *Journal of Consulting and Clinical Psychology* **51** (3), 390.

Prüss-Üstün, A., Wolf, J., Corvalán, C., Bos, R. & Neira, M. 2016 *Preventing Disease Through Healthy Environments: A Global Assessment of the Burden of Disease From Environmental Risks*. World Health Organization, Geneva.

Taylor, D. L., Kahawita, T. M., Cairncross, S. & Ensink, J. H. 2015 The impact of water, sanitation and hygiene interventions to control cholera: a systematic review. *PLoS one* **10** (8), e0135676.

White, S., Thorseth, A. H., Dreibelbis, R. & Curtis, V. 2020 The determinants of handwashing behaviour in domestic settings: an integrative systematic review. *International Journal of Hygiene and Environmental Health* **227**, 113512.

WHO 2020 *Global Report on the Epidemiology and Burden of Sepsis: Current Evidence, Identifying Gaps and Future Directions*. Geneva.

WHO and UNICEF 2018 *Drinking Water, Sanitation and Hygiene in Schools: Global Baseline Report 2018*. World Health Organization and the United Nations Children's Fund, New York.

WHO and UNICEF 2019 *WASH in Health Care Facilities: Global Baseline Report 2019*. World Health Organization and the United Nations Children's Fund, Geneva.

WHO and UNICEF 2021 *Progress on Household Drinking Water, Sanitation and Hygiene 2000–2020: Five Years into the SDGs*. World Health Organization (WHO) and the United Nations Children's Fund, Geneva. Available from: https://washdata.org/ (accessed 30 August 2021).

First received 29 January 2021; accepted in revised form 28 April 2022. Available online 16 May 2022

doi: 10.2166/h2oj.2022.070

Scorecard diagnostic tool for the enabling regulatory environment of faecal sludge management

H. N. Mbogo [a,*], P. G. Home[a], J. M. Raude[a] and R. W. Mwangi[b]

[a] Department of Soil, Water and Environmental Engineering, Jomo Kenyatta University of Agriculture and Technology, P.O. Box 62 000-00200 Nairobi, Kenya

[b] School of Computing and Information Technology, Jomo Kenyatta University of Agriculture and Technology, P.O. Box 62 000-00200, Nairobi, Kenya

*Corresponding author. E-mail: hngugi@jkuat.ac.ke

HNM, 0000-0002-2715-6158

ABSTRACT

Expeditious situational analysis of the enabling regulatory environment (ERE) of faecal sludge management (FSM) is vital for strengthening sanitation systems. However, existing diagnostic tools employ broad indicators, neglecting the detailed assessment of the policy, legal and institutional frameworks for each step along the FSM chain. This paper presents a web-based integrated diagnostic tool for evaluating the quality and adequacy of policy in guiding equity, targeting of resources, quality of service, financial considerations and institutional roles and responsibilities. The tool evaluates the legal framework based on laws, standards and regulations emphasizing on means of enforcement. It evaluates the institutional framework based on investors, service providers, regulators and consumers as guided by the literature. Each indicator is assigned a score of 1: green (effective), 0.5: yellow (limited) and 0: red (poor) for the user interface, containment, emptying, transport, treatment, disposal and reuse of faecal sludge. Built on Laravel Framework Hypertext Preprocessor (PHP), the tool links the scores into a single index averaged into terciles as 0–0.33 (poor), 0.34–0.66 (limited) and 0.67–1 (effective). This helps to identify areas of priority in a given context. The tool successfully facilitated a participatory pilot study in Kenya based on individual stakeholders' opinion. However, the tool does not provide the specific details leading to a given score. Therefore, its application should precede a detailed evaluation of each indicator in order to generate specific details per indicator per step of FSM.

Key words: effectiveness, enabling regulatory environment, scoring criteria, FSM chain, single index

HIGHLIGHTS

- An open-access web-based scorecard diagnostic tool for a participatory situational analysis of the enabling regulatory environment of faecal sludge management.
- Employs key performance indicators and scoring criteria to facilitate a detailed assessment of the policy, legal and institutional frameworks.
- Specifically assesses the quality of the policy frameworks.
- Speeds up the identification of areas of priority.

GRAPHICAL ABSTRACT

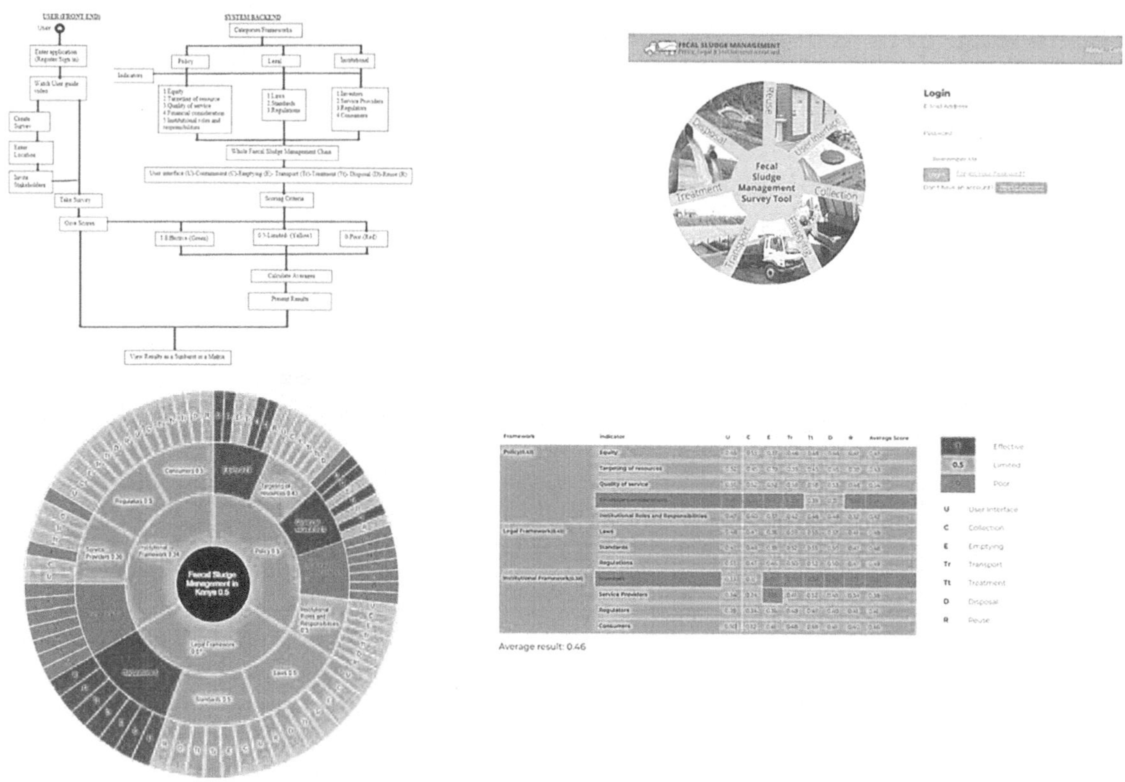

INTRODUCTION

Achievement of the sustainable development goal (SDG), target 6.2.1, on safely managed sanitation services for all by 2030 requires an all-inclusive sanitation approach that combines both sewered and non-sewered sanitation services (UNICEF & WHO 2020). However, 46% (3.6 billion) of the global population lacked access to these services by 2020 and at the current rate of acceleration, more than 2.8 billion people might miss this target (WHO | UNICEF JMP 2021). Nevertheless, with more than 3.1 billion people using onsite sanitation technologies worldwide, non-sewered systems are driving considerable progress in achieving this goal especially in developing countries. The number of people using onsite sanitation technologies is expected to grow to 5 billion by 2030 (Strande *et al.* 2014).

However, poorly managed non-sewered sanitation facilities pose great human and environmental health risks along the whole faecal sludge management (FSM) chain. The FSM chain consists of the user interface (toilet), containment, emptying, transport, treatment, reuse and/or disposal of faecal sludge (Strande *et al.* 2014). Therefore, the measures needed to protect human and environmental health should be clearly laid out in an effective regulatory framework and firmly enforced along the whole FSM chain (Bassan 2014). Most developing countries lack such clear regulatory frameworks to address each step along the FSM chain. Others lack specific guidelines, regulations or national standards to specifically address FSM (ESAWAS Regulators Association 2019). In other countries where multisectoral-based regulatory frameworks exist, standards and the regulations are often ignored in practice due to unclear mandates among the several regulatory agencies (Burr & Waititu 2018). Thus, FSM services end up neglected both in delivery and in regulation along the entire chain. In addition, lack of rules, reporting mechanism and enforcement procedures have led to difficulties in tracing the past trends and the current outcomes of FSM in most developing countries (Jayathilake *et al.* 2019).

An enabling regulatory environment outlines the who, the how and the what to regulate for inclusive service provision. It consists of the policy, legal and institutional arrangements including the regulatory tools and instruments required (Coglianese 2012; ESAWAS Regulators Association 2019). For an enabling regulatory environment to successfully promote sustainable FSM services, it needs to be effective to achieve its planned goals (Kirkpatrick & Parker 2004). The World Health Organization (WHO) recommends a situational analysis of existing policies, legislation and practice as the first step towards achieving safely managed sanitation services for all (WHO 2018). The policy, the legal and the institutional considerations should be evaluated holistically for the entire FSM chain but singly and exhaustively for each step (Bassan 2014). In addition, rapid assessment of sanitation policies is recommended for every country in order to assess their adequacy and effectiveness in responding to the prevailing sanitation situations (African Ministers' Council on Water (AMCOW) 2021).

Unfortunately, only a few developing countries apply some form of assessment on the enabling regulatory environment and where it is applied, the methods adopted are usually partial and unsystematic (Kirkpatrick & Parker 2004). Measuring effectiveness involves comparing a given phenomenon to the best practice or standard of high performance (Stake & Schwandt 2011). Diagnostic tools facilitate the comparison between current and preferred status in order to identify areas of improvement. They utilize a set of indicators to measure the relevant outcomes of concern and a research design to support the conclusions (Olivier 2017). There exist innumerable diagnostic tools for evaluating different aspects of the enabling environment of water, sanitation and hygiene (WASH). Some are described in the sanitation tool compendium (Dey *et al.* 2016) and others in the mapping of WASH sustainability tools (Schweitzer *et al.* 2014). However, there isn't a sanitation sector-wide consensus on a common framework or the number or scope of aspects to be evaluated (Gensch & Tillett 2019).

Some of the existing tools such as World Bank's City Services Delivery Assessment (CSDA) (Ross *et al.* 2016) and Sector Functionality Framework (SFF) (WSUP 2019) employ broad indicators such as enabling, developing and policy/mandates, which subsume the critical elements of the enabling regulatory environment into the broader indicators. Others such as the Global Analysis and Assessment of Sanitation and Drinking Water (UN-Water (GLAAS) and World Health Organization 2019) and the WASH Bottlenecks Analysis Tool (UNICEF 2020) integrate FSM into the general WASH provision. This neglects the detailed assessment of the policy, legal and institutional frameworks, particularly the quality and adequacy of the policy framework as guided by the Guidelines for the Assessment of National Sanitation Policies (Elledge *et al.* 2002; Water Supply and Sanitation Collaborative Council and World Health Organization 2005; ESAWAS Regulators Association 2019; AMCOW 2021). Other tools fail to include all the steps of FSM from the user interface to reuse. Others combine disposal and reuse as a single step overlooking the fact that disposal and reuse of faecal sludge and its products should be guided and regulated differently since they pose differing human and environment health risks (Strande *et al.* 2014; WHO 2018).

Therefore, this paper presents the contents, development and web-testing process of Faecal Sludge Management; Policy, Legal and Institutional (FSMPLI) Scorecard tool. This is an integrated web-based diagnostic tool for evaluating the effectiveness of the three critical elements of the enabling regulatory environment along the entire FSM chain. The paper also presents the results of a pilot study in Kenya where the scorecard tool was tested for applicability.

METHODS

Defining the indicators, conditions and criteria

Indicators are the factors used to denote the degree to which the three elements of the enabling regulatory environment meet their objectives based on their existing state or output. The factors were defined by identifying the suitable attributes and the expected standards of each element based on the literature (Coglianese 2012). A three-point Likert-type of scale was applied where the ideal condition of each indicator formed the stem statement, while the three-response scale formed the scoring criteria (Johns 2010). Each criteria was assigned a value and colour code as 1 (green) for effective, 0.5 (yellow) for limited and 0 (red) for poor (Ross *et al.* 2016) as in Table 1.

The policy framework

A policy is a statement of intent, a commitment, a principle or a rule that guide decision making to achieve a given outcome. It serves as key stimulus for local action and create conditions within which FSM stakeholders operate (Elledge *et al.* 2002; AMCOW 2021). The policy framework was assessed through five of the seven key elements applicable for the entire sanitation service chain (Water Supply and Sanitation Collaborative Council and World Health Organization 2005; ESAWAS Regulators Association 2019).The five indicators included equity, targeting of resources, quality of service, financial considerations and institutional roles and responsibilities (Elledge *et al.* 2002). Health and environmental considerations were incorporated into quality of service because quality of any sanitation service affects human and environmental health (Bassan 2014). The regulation

Table 1 | Indicators, conditions and criteria

Indicator	Conditions	Criteria
Policy framework		
Equity	Guidance for steering resources and service provision to all citizens and support for inclusive participation provided	■ Clear and specific guidance for inclusive service provision and inclusive participation supported (0.5) Guidance for inclusive service provision is provided but not clear or specific and inclusive participation somehow supported (0) Little or No guidance for inclusive service provision and/or inclusive participation is not supported
Targeting of resources	Specific FSM service need and its impact outlined with targets and timelines specified	■ Service need clearly stated and Targets clearly specified (0.5) Service need and targets stated but not clear or specific (0) No/unclear service need and no/unspecific targets stated
Quality of service	Minimum acceptable level of service in terms of health, safety and environmental standards and how they would be achieved, regulated and monitored specified	■ Comprehensive Requirements for specified health, safety and environmental risks provided, interventions on how to achieve all the standards provided and means and procedures for monitoring specified (0.5) Some requirements to address health, safety and environmental risks provided, some interventions on how to achieve some of the standards provided and some means and procedures for monitoring specified but not comprehensively (0) No Requirements to address any health, safety and environmental risks provided, no interventions on how to achieve the standards provided and no means or procedures for monitoring specified
Financial considerations	FSM need costed, incorporated in investment plans and payment for services guided	■ Need costed, clearly outlined in public financial plans and payment for services clearly guided (0.5) Need not costed adequately, partially included in budget/financial plans and payment for services somewhat guided (0) Need not costed nor outlined in budget/plans and payment for services not guided
Institutional roles and responsibilities	Roles and responsibilities of public, private and non-state actors defined, capacity development addressed and consultations among stakeholders supported	■ Roles and responsibilities clearly defined, capacity development clearly addressed and stakeholder consultations supported (0.5) Roles and responsibilities not clearly defined/overlap, capacity development not clearly addressed and stakeholder consultations not clearly supported (0) Roles and responsibilities not defined, capacity development not addressed and stakeholder consultations not supported
Legal framework		
Laws	Laws exists specifying the incentives, Sanctions (Penalties) and means to enforce accountability provided	■ Up to date Laws exists and enforcement tools clearly defined (0.5) Laws and enforcement tools exists but not quite clear/inadequate/outdated (0) No laws exists nor any enforcement tools defined

(Continued.)

Table 1 | Continued

Indicator	Conditions	Criteria
Standards	Norms, standards and guidelines exist, published and applied	■ Specific and clear standards and guidelines exist, are published and applied (0.5) Some standards and guidelines exist but not adequately applied (0) No standards and guidelines exist or applied
Regulations	Regulatory framework exists incorporating the set standards, guides payment of services and specifies means of enforcement	■ Clear, updated regulations exist, including the standards, tariff guidelines and clear means of enforcement (0.5) Some regulations exist but without clear standards or tariff guidelines and means for enforcement not very clear or are outdated (0) No regulations at all or if they exist have not included standards or tariff guidelines and no means of enforcement specified
Institutional framework		
Investors	FSM need is funded and progress monitored adequately	■ Adequate investment to meet FSM need to achieve the set target and progress is monitored regularly (0.5) Some investment to meet FSM need but not adequate to achieve the set target and progress is monitored but not regularly (0) No or very little investment to meet FSM need to achieve the set target and no monitoring is done
Service providers	Service providers have the technical, operational and strategic capacity, meets demand and targets and compliance to standards, protecting public and environmental health	■ Adequate capacity, meets all demand and targets and compliance to standards (0.5) Inadequate capacity, meets some quantified demand and compliance to standards (0) No/very little capacity, meets no/unquantifiable demand and compliance to Standards unverified/unknown/none
Regulators	Independent regulator(s), regulates quality and payment of services, monitors service provision and enforces the legal requirements effectively	■ Regulator(s) independent, regulates quality and payment for service, monitors service provision and enforces adequately effectively (0.5) Regulator(s) independent, regulates quality and/or payment of services, monitors and enforces for compliance but not adequately (0) No regulator or regulator(s) not independent or regulator offers low or no regulation of quality or payment of service, little/no monitoring or enforcement for compliance
Consumers	Consumers are available, are willing to pay and have a means to seek redress	■ Consumers are available, are willing and can afford to pay and means to seek redress is/are available and efficient (0.5) Consumers are available, are willing but not all can afford to pay and means to seek redress is/are available but not quite efficient (0) Very few if any Consumers are available and/or willing to pay; means to seek redress is/are unavailable/inefficient

and monitoring aspects of the policy were also incorporated into the quality of service since monitoring assesses progress while regulation ensures quality and efficiency of sanitation services. The capacity development aspects were incorporated into institutional roles and responsibilities since capacity development involve improving the ability of different institutions to perform their assigned roles (AMCOW 2021).

The legal framework

The legal framework is a broad set of constitutional, legislative, regulation, jurisprudential and managerial rules that facilitate the implementation and the enforcement of policies. For it to be effective, it should include clear roles and responsibilities for each stakeholder (ACE Project 2021). It should also incorporate requirements for compliance and means for enforcement. These include licensing procedures, monitoring and reporting mechanisms, performance indicators, penalties and sanctioning procedures for non-compliance as well as incentive for outstanding performance (Kameri-Mbote & Akech 2011). The legal framework was evaluated by the existence and operationalization of laws, standards and regulations with an emphasis on incentives, penalties and sanctioning procedures employed (SNV and ISF-UTS 2016; WHO 2018; ESAWAS Regulators Association 2019).

The institutional framework

The institutional framework comprises the different organizations and other stakeholders involved in FSM services provision. Effective institutional frameworks facilitate the design, delivery, regulation and consumption of FSM services (Mumssen *et al.* 2018). It was evaluated based on operationalization of the roles and responsibilities of the actors of FSM (investors, service providers, regulators and consumers) in implementing the policy framework and enforcing the legal framework (Chua *et al.* 2005).

Tool development

The web-based scorecard diagnostic tool was built using the Laravel Framework Hypertext Preprocessor (PHP) for server-side processing. Laravel (PHP) is one of the most popular open-source web development frameworks. It is flexible, secure and it supports rapid application development (McCool 2012). The data were stored in MySQL which is also a popular open-source Structured Query Language database management system. MySQL allowed addition, access and processing of the data stored in it, while offering a flexible programming environment (Christudas 2019). The tool was organized into models as described in Table 2.

The results in FSMPLI-Scorecard were displayed in either a sunburst or matrix chart developed on the Chart.Js library. Chart.Js is also a free open-source data visualization library used to create a wide range of responsive web-based visualization (Da Rocha 2019). The tool was hosted on an Apache web server with security-enhanced

Table 2 | Models of FSMPLI-Scorecard in Laravel PHP

Model	Attributes	Relationships in Laravel	Description
Category	name, description	'HasMany' ->sections	Specification of the elements and indicators of evaluation
Location	name, description		Specification of the geographical location under study
Result	user_id, date, survey_id, status, date_started, date_completed	'hasMany'->SurveyTaking, 'hasOne->Survey, 'hasOne'->User	Provision for graphical representation of the results in a sunburst or in matrix format
Section	name, description, category_id, A, B, C	'hasMany'->SurveyTaking, 'hasOne'->Category	Description of the three response levels of the scoring criteria for effective, limited and poor
Survey	title, description, location_id, status	'hasOne'->Location	Description of the title of the survey being undertaken
Surveytaking	user_id, U, C, E, Tr, Tt, D, R, remarks, survey_id, section_id, category_id, result_id	'hasOne'->Category, 'hasOne'->Section, 'hasOne'->Survey	Specification of the scores per indicator per element per step of the FSM chain from User interface (U), Containment (C), Emptying(E), Transport (Tr) Treatment (Tt), Disposal(D) and Reuse (R).
User	name, email, phone, password, is_admin		Point for user and the administrator's authentication

Linux which enhances the mandatory access control policy. The mandatory access control denies unnecessary access to the web application but allows required access. This provides the web server functionality while limiting potential damage to it (Gosselin & Schommer 2001). A user is required to register as an administrator to set up a diagnostic survey for any predefined area of study. The administrator then invites the relevant stakeholders to give their opinion on the effectiveness for each indicator for each step along the FSM service chain. The tool has a user guide video developed at fiverr.com: a freelance platform offering digital services (Green *et al.* 2018). Figure 1 shows a flow chart for FSMPLI-Scorecard.

Web testing

FSMPLI-Scorecard was tested for functionality using the PHP unit. Test protocols were created for each application unit including the category, location, result, section, survey and user. This was done by creating scenarios such as insert data, validation, update data, view data and delete data in each unit (Alrashed 2018). It was tested for browser compatibility with Internet Explorer, Edge, Firefox, Safari, Opera, Chrome, iOS and Android using power mapper (Mesbah 2014). The server IP address for FSMPLI-Scorecard was tested for loading and stress using artillery (Howe *et al.* 2021). It was also tested for security in terms of package vulnerabilities, Secure Socket Layer (SSL) access and directory access using a Laravel package to assess the ease of access by unauthorized users (Dahle 2020).

Pilot study

Purposively selected stakeholders and FSM experts were invited to test the functionality of tool in application through a pilot study of the effectiveness of the regulatory environment of FSM services in Kenya. They included officials from government ministries, regulators, academia, non-state actors and service providers, as shown in

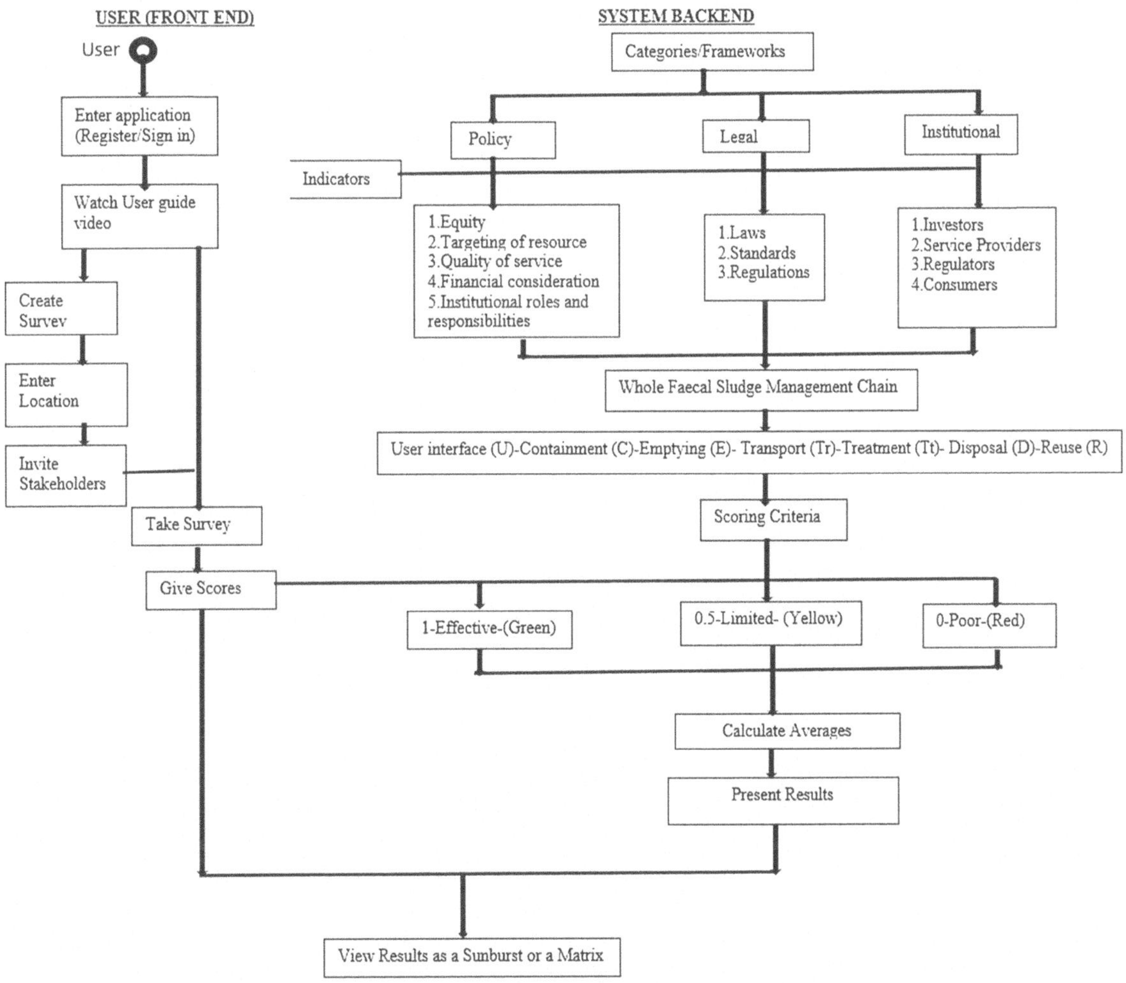

Figure 1 | Flow chart for the FSMPLI-Scorecard diagnostic tool.

Table 3 | Participants of the pilot study in Kenya

Category	No. of respondents	Organization
National Government	1	Ministry of Health
	1	Ministry of Water & Sanitation and Irrigation
	1	National Irrigation Authority
Regulator	1	National Environment and Management Authority
	1	Water Services Regulatory Board
Standardization Agency	1	Kenya Bureau of Standards
Service provider	1	Nairobi City Water and Sewerage Company
Non-state actors	2	Water and Sanitation for the Urban Poor
	1	Sanergy
Private sector	1	Jiji Water Works Ltd
	1	Kenya Septage Emptiers Association
Capacity building network	1	Water Capacity Network
Academia	3	Jomo Kenyatta University of Agriculture and Technology
	1	University of Nairobi

Table 3. After an exposure to the tool in a stakeholders' workshop, the participants assigned each indicator a score of 1 for effective, 0.5 for limited or 0 for poor for each step along the FSM service chain based on their individual experiences and best of knowledge. The tool calculated the average score for each indicator, the average score for each element and the overall score as a single index.

RESULTS AND DISCUSSION

The content

The FSMPLI-Scorecard numerically evaluated the effectiveness of the three elements of the enabling regulatory environment to show areas of strength and weakness as well as identify areas of priority for each indicator along the FSM chain (Ross *et al.* 2016). It considered each step individually from the user interface, containment, emptying, transport, treatment, disposal and reuse (Strande *et al.* 2014). Considering each step ensures that the different actors, infrastructure and procedures applied per step are adequately addressed in the policy, legal and institutional frameworks (Bassan 2014). This consideration expanded the SFF (WSUP 2019) which evaluates the sanitation sector in general without narrowing down to the specific steps of FSM. It also expanded the World Bank's CSDA (Peal *et al.* 2014; Ross *et al.* 2016; Scott *et al.* 2019) which leaves out the user interface stage and also considers disposal and reuse as a single step.

FSMPLI-Scorecard not only examined the existence of a policy framework but assessed its quality based on the predefined key elements of sanitation policies (Elledge *et al.* 2002) and the African guidelines for sanitation policy (AMCOW 2021). By employing predefined and already published indicators, the tool allows local and contextualized interpretation of the indicators while maintaining their universal applicability. They are narrow in scope compared to other existing diagnostic tools (Schweitzer *et al.* 2014; Dey *et al.* 2016), to effectively facilitate a detailed assessment of each element. Table 4 shows the alignment and differences between FSMPLI-Scorecard and similar tools.

FSMPLI-Scorecard tool

The open-access web-based FSMPLI-Scorecard diagnostic tool found at http://fsmpli-scorecard.com/ facilitated an integrated participatory evaluation of the policy, legal and institutional frameworks of the FSM service chain based on predefined indicators and scoring criteria (Mansourian *et al.* 2011). It covered all the steps of FSM from the user interface (U), Collection (C), Emptying (E), Transport (Tr), Treatment (Tt), Disposal (D) and Reuse (R). The tool's front page is shown in Figure 2.

The tool can be used by any registered administrator at any internet supplied geographical location across the globe (Rogers *et al.* 2020). By integrating the different aspects of measurement, the tool facilitates a comprehensive situational analysis of the enabling regulatory environment. It rates the quality of each element according to

Table 4 | The alignment and differences between FSMPLI-Scorecard and similar tools

Tool	Objective	Alignment	Difference/Improvement
City Services Delivery Assessment (CSDA)	• Qualitative analysis of the quality of the enabling environment for FSM-based on nine indicators organized in three broad pillars: enabling, developing and sustaining. • (attempts to cover all the six dimensions of governance however scattered and not organized as government support, legal and regulatory frameworks, institutional arrangement, skills and capacity, financial arrangements and social and cultural acceptance (Lüthi *et al.* 2011)).	• Focus on assessment of the quality of the enabling environment of FSM • Based on a three-scale Likert-type of scoring criteria	• Narrows down to ERE of FSM • Includes the user interface and separates Disposal and Reuse • Improved and expanded the indicator on the availability of policy to focus on its quality and adequacy as guided by the Guidelines for the Assessment of National Sanitation Policies • Assesses the institutional framework in terms of operationalized roles by service providers, investors, customers and regulators • Assesses the legal framework by the existence and operationalization of laws, standards and regulations with an emphasis on incentives, penalties and sanctioning procedures employed • Facilitates an individual stakeholder-based assessment but also integrates the scores to a participatory evaluation of ERE. It calculates the average score per indicator along the FSM Chain per participant, Average score per indicator across the participants and overall average score for the ERE per participant and among the participants.
• Sector Functionality Framework (SFF)	• Assessment of the genera sanitation Sector Functionality. • Uses 21 broad qualitative indicators using four-level Likert-type scale: high, medium, low and zero • Does not zero down to individual steps of the FSM chain	• Focus on assessment of the quality of the enabling environment of the general sanitation sector • Based on Likert-type scale of scoring criteria	• Narrows down to the enabling regulatory environment of FSM • Focusses on each step of FSM • Expands the broad indicator on policy/mandates and regulatory effectiveness • Gives a quantitative score to the indicators for easier assessment • Facilitates individualized yet participatory evaluation of ERE
• Governance Capacity Framework (GCF) for Cities	• Assessment and comparison of governance capacity for cities concerning water scarcity, flood risks, waste water and solid waste treatment and urban heat islands. Utilizes nine governance condition each defined by three indicators using afive-level Likert-type of scoring scale ranging from very encouraging (++) to very limiting (–).	• Focus on assessment of the quality of the enabling environment of various environmental issues • Based on Likert-type scale of scoring criteria	• Narrows down to the enabling regulatory environment of FSM • Focusses on each step of FSM • Gives a quantitative score to the indicators for easier assessment • Facilitates individualized yet participatory evaluation of ERE of FSM

(*Continued.*)

Table 4 | Continued

Tool	Objective	Alignment	Difference/Improvement
• Global Analysis and Assessment of Sanitation and Drinking Water (GLAAS) • WASH Bottlenecks Analysis Tool (United Nations Childrens Fund (UNICEF) & The World Bank Group 2014)	• Assessment of the enabling/ Limiting conditions on general Water, Sanitation and Hygiene (WASH)	• Focus on assessment of the quality of the enabling environment of WASH • Based on Likert-type scale of scoring criteria	• Narrows down to the enabling regulatory environment of FSM • Focusses on each step of FSM • Gives a quantitative score to the indicators for easier assessment • Facilitates individualized but participatory evaluation of ERE

Governance Capacity Framework (GCF) for Cities (Koop *et al.* 2017).

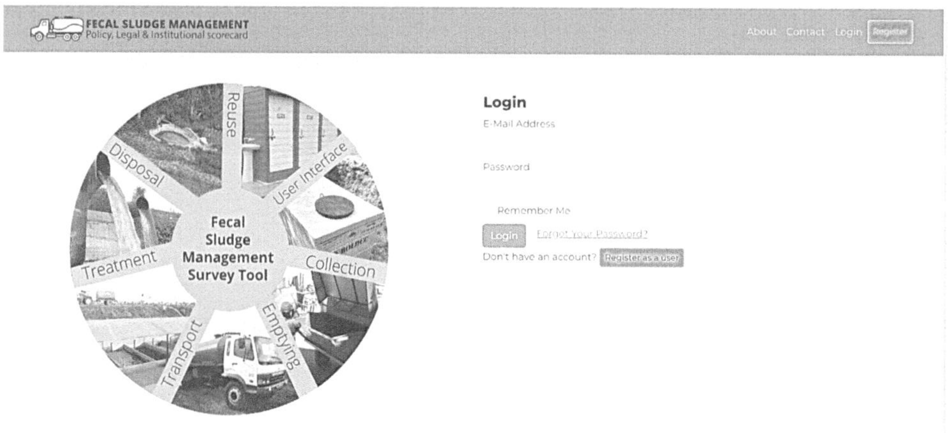

Figure 2 | FSMPLI-Scorecard web-based front page.

the predefined indicators and scoring criteria along the entire FSM service chain. This holistic diagnosis could reduce the time, resources and expertise required for the evaluation (Ross *et al.* 2016).

Presentation of results

The three Likert-type scale applied in FSMPLI-Scorecard sufficiently identified the indicators in need of improvement for each step, without necessarily breaking them into finer categories (Johns 2010). The predefined scoring criteria standardizes the meaning of the scores thus reducing ambiguity by providing a common understating of the scores among diverse stakeholders with diverse interest across the sector (Rogers *et al.* 2020). The tool calculates the average score for each indicator, then the average score for each element under diagnosis and the overall score as a single index. The overall score indicates the effectiveness of the enabling regulatory environment categorized in terciles as the following: 0–0.33 (poor), 0.34–0.66 (limited) and 0.67–1 (effective) (Meyer *et al.* 2019). This three-level numerical scale aims to qualitatively indicate areas of priority to stakeholders and not necessarily a quantitative evaluation (Ross *et al.* 2016).

The results are presented in charts using different colour codes of green for effective, yellow for limited and red for poor. This speeds up the visualization and interpretation of the outcome for each indicator. The presentation helps to establish the probable cause–effect relations for the score attained by a given indicator to the specific step of FSM (Burr & Waititu 2018). The tool is balanced in the sense that, the overall score is an average score of each element per indicator per step of FSM. This denotes an even distribution of weight among the elements and across the range of indicators (Kaplan & Norton 1992). The sunburst presentation of the results enhances the perception of each element and each indicator as important and the interrelationship between the elements is

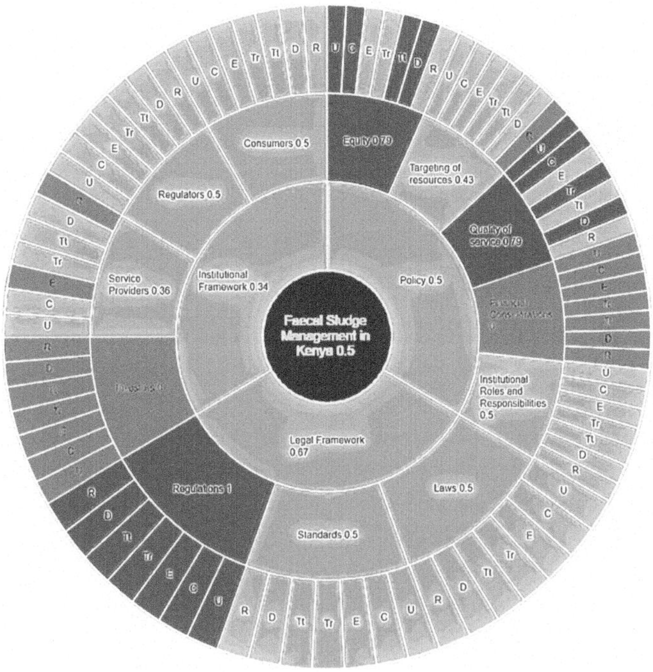

Figure 3 | An example of a sunburst presentation of results in the FSMPLI-Scorecard tool.

implicitly understood (Schulz *et al.* 2011). Figure 3 shows an example of a sunburst presentation of results in FSMPLI-Scorecard tool. However, due to the chain nature of FSM services, the results can be presented in a matrix format as shown in Figure 4 depending on user preference.

Tool web testing

All the units in the tool passed for functionality in all the scenarios tested. Therefore, the tool is expected to perform its functions such as inserting data, validation of data, updating data, viewing data and deleting data as it was designed to (Alrashed 2018). The tool was found compatible with all the browsers tested as shown in Table 5. Therefore, it can be accessed and applied through any of the stated browsers.

For the loading test, a ping to the server address retuned with 0% loss of data in an approximate average round-trip of 158 ms. It was able to support 496 concurrent virtual users in 820.7 ms in the stress testing. Therefore, it was found reliable and capable of performing its functions satisfactorily regardless of a high number of concurrent users (Subraya & Subrahmanya 2000) In the security test, none of packages were found vulnerable to external attack and only the public directory is accessible from the browser since the projects code is stored in in accessible a directory. Therefore, the tool is secure in terms of access from unauthorized users and external attacks (Kundu 2012).

Framework	Indicator	U	C	E	Tr	Tt	D	R	Average Score
Policy(0.43)	Equity	0.65	0.53	0.37	0.46	0.46	0.46	0.41	0.47
	Targeting of resources	0.52	0.45	0.39	0.38	0.45	0.45	0.38	0.43
	Quality of service	0.55	0.52	0.52	0.58	0.58	0.53	0.48	0.54
	Financial Considerations	0.21	0.21	0.16	0.23	0.39	0.31	0.24	[illegible]
	Institutional Roles and Responsibilities	0.47	0.40	0.37	0.42	0.48	0.48	0.32	0.42
Legal Framework(0.49)	Laws	0.48	0.47	0.36	0.59	0.55	0.57	0.41	0.49
	Standards	0.47	0.48	0.38	0.52	0.55	0.50	0.47	0.48
	Regulations	0.53	0.47	0.45	0.50	0.52	0.50	0.47	0.49
Institutional Framework(0.38)	Investors	0.33	0.31	[illegible]	[illegible]	[illegible]	[illegible]	[illegible]	[illegible]
	Service Providers	0.34	0.34	[illegible]	0.41	0.52	0.45	0.34	0.38
	Regulators	0.38	0.34	0.36	0.48	0.47	0.40	0.41	0.41
	Consumers	0.50	0.52	0.41	0.48	0.48	0.41	0.40	0.46

Average result: 0.46

1 Effective
0.5 Limited
0 Poor

U User Interface
C Collection
E Emptying
Tr Transport
Tt Treatment
D Disposal
R Reuse

Figure 4 | Results of the pilot study on the enabling regulatory environment by FSM stakeholders in Kenya.

Table 5 | FSMPLI-Scorecard browser compatibility

Browser	IE	Edge	Firefox	Safari	Opera	Chrome	iOS		Android
Version	11	95	93	15	80	95	≤ 14	15	95
Critical Issues	✓	✓	✓	✓	✓	✓	✓	✓	✓
Major Issues	✓	✓	✓	✓	✓	✓	✓	✓	✓
Minor Issues	✓	✓	✓	✓	✓	✓	✓	✓	✓

Pilot testing

The enabling regulatory environment of FSM in Kenya received an average score of 0.47 through the multisectoral FSM stakeholders' opinion survey. The policy framework was rated as limited with an average score of 0.45, across the five indicators. However, financial considerations were found poor for most FSM services with an average score of 0.28. The legal framework was also rated as limited with an average score of 0.5 across the three indicators. The institutional framework was also found limited with a low score of 0.39. It was particularly found poor in terms of the investors for most FSM services as well as service providers for emptying services. The results as shown in Figure 4 are in a matrix format.

The survey was carried out based on each participant's inherent knowledge and experience soon after the initial exposure to the scoring framework. It is therefore likely that they all did not have the necessary in-depth knowledge or prior experience with each element or similar understanding to give identical scores for each indicator per step of FSM services. They also probably did not have time to access and assess available evidence to score the indicators patently. However, the tool linked each stakeholder's opinion to give an average score per indicator per step and also an overall score for the enabling regulatory environment in Kenya. By linking these opinions together, the tool would improve the quality of the diagnosis and also avoid biased conclusions (Rogers *et al.* 2020).

However, there was no moderation for consensus building since it was a pilot test on the tool, but stakeholders' participation formed a basis for a sector-wide dialogue towards consensus building (Rogers *et al.* 2020). Therefore, the participants requested for a subsequent all-inclusive stakeholders' forum to discuss the outcome of the diagnosis towards a common ownership of the results. They also suggested to be provided with the scoring criteria prior to the survey, so as to consider available evidence for each indicator and to allow consultations among colleagues so as to present institutional-based perspectives.

CONCLUSION AND RECOMMENDATIONS

FSMPLI-Scorecard evaluated the effectiveness of the enabling regulatory environment of FSM based on the indicators described in the Regulation Strategy and Framework for Inclusive Urban Sanitation Service Provision Incorporating Non-Sewered Sanitation Services developed by the Eastern and Southern African Water and Sanitation (ESAWAS) Regulators Association. However, it adapted, improved and expanded some broad indicators from several existing diagnostic tools such as the World Bank's CSDA and WSUP's SFF in order to specifically assess the quality of the policy framework according to the Guidelines for the Assessment of National Sanitation as provided by the Water Supply and Sanitation Collaborative Council and WHO. It also considered the African Sanitation Policy Guidelines provided by the African Ministers' Council on Water.

The tool was found functional in its application by successfully facilitating a participatory diagnosis of the effectiveness of the policy, legal and institutional frameworks of FSM services along the entire service chain. It proved effective in identifying areas of improvement across all the indicators along the entire chain. For instance, the financial considerations in the policy framework, the investors and service providers in the institutional framework were identified as the main priority areas of consideration in the Kenyan pilot study. However, the tool does not provide the specific details leading to a given score assigned to each indicator per step. Therefore, its application should precede a detailed evaluation of each indicator in a given context in order generate specific details per indicator per step of FSM. This would lead to development of evidence-based priorities towards an effective enabling regulatory environment of FSM services.

The open-access web-based FSMPLI-Scorecard can be reliably and safely applied to diagnose the enabling regulatory environment of any predefined study area such as a region, a country, a city or even a smaller geographical extent. However, it should be evaluated against a user's opinion and preferences.

Participation of the sector stakeholders, revealed diverse perspective and opinions on the current status of each indicator. This initiated the identification of priorities towards a joint commitment among the stakeholders for accelerated safely managed sanitation services for all. However, the tool does not dictate the specific remedial actions to be taken to improve the low scoring elements but it establishes the expected goal such that each respective arm or actor should adopt the most appropriate action plan to arrive to those goals.

To draw a more comprehensive and conclusive diagnosis, more and representative stakeholders should be invited to participate in the opinion survey. An all-inclusive stakeholders' forum should be held to discuss the outcome of the diagnosis towards a consensus building and a common ownership of the results. This would be essential for the development of coherent framework of actions for strengthening the policy, legal and institutional framework of FSM services in a given context.

Ethical considerations and approval

Ethical aspects considered in this research included the following: informed voluntary participation, right to refuse or withdraw, confidentiality and privacy, risks and benefits and payment for participating in the study. They were presented in consent forms that were signed by each participant prior to data collection: The research was approved by the Jomo Kenyatta University of Agriculture and Technology Institutional Ethics Review Committee (JKUAT IERC); Ref. No. JKU/2/4/896B and was licensed by the National Commission for Science and Technology and Innovation (NACOSTI) Ref. No. 932560.

ACKNOWLEDGEMENTS

FSMPLI-Scorecard was developed under a research (partially) funded by the African Doctoral Dissertation Research Fellowship (ADDRF) award offered by the African Population and Health Research Centre (APHRC) and funded by the Bill and Melinda Gates Foundations through a project grant to APHRC. We wish to appreciate our colleagues at the Jomo Kenyatta University of Agriculture and technology for their positive critique and suggestions throughout the study. We appreciate all the key interview informants, the validation workshop participants and their institutions for their invaluable contributions towards the development and testing of this tool. We acknowledge the contributory comments and suggestions of Rémi Kaupp of WaterAID UK during the development of the indicators and scoring criteria. We also appreciate, Mbitrix Technologies Limited for their partnership in the tool development.

DATA AVAILABILITY STATEMENT

All relevant data are included in the paper or its Supplementary Information.

REFERENCES

ACE Project 2012 Legal Framework. Retrieved June 13, 2021, from https://aceproject.org/ace-en/topics/lf/default.

African Ministers' Council on Water (AMCOW) 2021 *African Sanitation Policy Guidelines*.

Alrashed, W. A. 2018 *Evaluating the e Effectiveness of Web Application Testing Techniques Using Automated Tools*. McMaster University Hamilton, Ontario, Canada.

Bassan, M. 2014 Institutional frameworks for faecal sludge management: faecal sludge management: systems approach for implementation and operation. In: L. Strande, M. Ronteltap & D. Brdjanovic (eds). *Faecal Sludge Management: Systems Approach for Implementation and Operation*. IWA Publishing, London, pp. 255-272.

Burr, P. & Waititu, A. 2018 *Evaluation of Urban WASH Sector Functionality Kenya 2018 Baseline Report*.

Christudas, B. 2019 Practical microservices architectural patterns. In: *Practical Microservices Architectural Patterns*. pp. 877–884. https://doi.org/10.1007/978-1-4842-4501-9

Chua, C. E. H., Khoo, H. M., Straub, D. W., Kadiyala, S. & Kuechler, D. 2005 The evolution of e-commerce research: a stakeholder perspective. *Electronic Commerce Research and Applications* **6** (February), 262–280.

Coglianese, C. 2012 *Measuring Regulatory Performance – Evaluating the Impact of Regulation and Regulatory Policy. Organisation for Economic Co-operation and Development* (Vol. August).

Dahle, T. 2020 *Large Scale Vulnerability Scanning Development of a large-scale web scanner for detecting vulnerabilities*. Thesis from Department of Informatics, Faculty of Mathematics and Natural Science, University of Oslo, Oslo, Norway.

Da Rocha, H. 2019 *Learn Chart.js: Create interactive visualizations for the Web with Chart.js 2, 1st edn*. Packt Publishing Ltd, Birmingham, UK.

Dey, S., Patro, S. A. & Rathi, S. 2016 *Sanitation Tool Compendium*, vol. 53. Bengaluru, Karnataka, India.
Elledge, M. F., Rosensweig, F., Warner, D. B., Austin, J. H. & Perez, E. A. 2002 *Guidelines for the Assessment of National Sanitation Policies*. Washington, DC.
ESAWAS Regulators Association 2019 *Regulation Strategy and Framework for Inclusive Urban Sanitation Service Provision Incorporating Non-Sewered Sanitation Services.*
Gensch, R. & Tillett, W. 2019 *Strengthening Sanitation and Hygiene in the Wash Systems Conceptual Framework Strengthening Wash Systems: Tools for Practitioners, (August)*. pp. 1–32.
Gosselin, M. J. & Schommer, J. 2001 *Confining the Apache Web Server with Security-Enhanced Linux.*
Green, D. D., McCann, J., Thao, V., Lopez, N. & Ouattara, S. 2018 Case study gig economy and the future of work. *Management and Economics Research Journal* **4**, 281–288.
Howe, C., Kraisingkorn, A. T. & Light-Hillary, N. 2021 *Artillery*. Available from: https://www.npmjs.com/package/artillery (accessed 18 November 2021)
Jayathilake, N., Drechsel, P., Keraita, B., Fernando, S. & Hanjra, M. A. 2019 Guidelines and regulations for fecal sludge management from on-site sanitation facilities. Resource Recovery and Reuse (Vol. 2019).
Johns, R. 2010 SQB methods fact sheet 1: Likert items and scales. *Survey Research Network* **1** (March), 1–11.
Kameri-Mbote, P. & Akech, J. M. M. 2011 *Kenya: Justice Sector and the Rule of law: A Review by AfriMAP and the Open Society Initiative for Eastern Africa.*
Kaplan, R. S. & Norton, D. P. 1992 The balanced scorecard - measures that drive performance. *Harvard Business Review* **70** (1), 71–79.
Kirkpatrick, C. & Parker, D. 2004 Regulatory impact assessment and regulatory governance in developing countries. *Public Administration and Development* **344**, 333–344.
Koop, S. H. A., Koetsier, L. & Doornhof, A. 2017 Assessing the governance capacity of cities to address challenges of water, waste, and climate change. *Water Resources Management* **31**, 3427–3443. https://doi.org/10.1007/s11269-017-1677-7.
Kundu, S. 2012 Web testing: tool, challenges and methods. *International Journal of Computer Science Issues* **9** (2), 481–486.
Lüthi, C., Morel, A., Tilley, E. & Ulrich, L. 2011 *Community-Led Urban Environmental Sanitation Planning: CLUES – Complete Guidelines for Decision-Makers with 30 Tools.*
Mansourian, A., Taleai, M. & Fasihi, A. 2011 A web-based spatial decision support system to enhance public participation in urban planning processes, 8596. https://doi.org/10.1080/14498596.2011.623347.
McCool, S. 2012 *Laravel Starter. Packtlib.*
Mesbah, A. 2014 Automated cross-browser compatibility testing Automated Cross-Browser Compatibility Testing, (May), 561–570.
Meyer, A. M., Siri, G, Becker, I., Betz, T., Bödecker, A. W., Robertz, J. W., Krause, O., Benzing, T., Pilotto, A. & Polidori, M. C. 2019 The Multidimensional Prognostic Index in general practice: One-year follow-up study. *International Journal of Clinical Practice* **73** (12), 1–9. https://doi.org/10.1111/ijcp.13403.
Mumssen, Y., Saltiel, G. & Kingdom, B. 2018 Aligning Institutions and Incentives for Sustainable Water Supply and Sanitation Services. *Aligning Institutions and Incentives for Sustainable Water Supply and Sanitation Services*, (May). https://doi.org/10.1596/29795.
Olivier, B. H. 2017 The use of mixed-methods research to diagnose the organisational performance of a local government. *SA Journal of Industrial Psychology* **43** (1), 1–14. https://doi.org/10.4102/sajip.v43i0.1453.
Peal, A., Evans, B., Blackett, I., Hawkins, P. & Heymans, C. 2014 Fecal sludge management (FSM): analytical tools for assessing FSM in cities. *Journal of Water, Sanitation and Hygiene for Development* **4** (3), 371. https://doi.org/10.2166/washdev.2014.139.
Rogers, B. C., Dunn, G., Hammer, K., Novalia, W., de Haan, F. J., Brown, L., Brown, R. R., Lloyd, S., Urich, C., Wong, T. H. F. & Chesterfield, C. 2020 Water Sensitive Cities Index: A diagnostic tool to assess water sensitivity and guide management actions. *Water Research* **186**, 116411. https://doi.org/10.1016/j.watres.2020.116411.
Ross, I., Scott, R., Mujica, A., Smith, M. & White, Z. 2016 *Fecal Sludge Management: Diagnostics for Service Delivery in Urban Areas Tools and Guidelines.*
Schulz, H. J., Hadlak, S. & Schumann, H. 2011 The design space of implicit hierarchy visualization: a survey. *IEEE Transactions on Visualization and Computer Graphics* **17** (4), 393–411. https://doi.org/10.1109/TVCG.2010.79.
Schweitzer, R., Grayson, C. & Lockwood, H. 2014 Mapping of water, sanitation and hygiene sustainability tools. *Triple-S Working Papers*, *10*(May), 43 p.; 7 fig.
Scott, R. E., Ross, I., Hawkins, P., Blackett, I. & Smith, M. D. 2019 *Diagnostics for Assessing City-Wide Sanitation Services*. pp. 111–118. https://doi.org/10.2166/washdev.2018.113
SNV and ISF-UTS 2016 *Exploring Legal and Policy Aspects of Urban Sanitation and Hygiene.*
Stake, R. E. & Schwandt, T. A. 2011 On discerning quality in evaluation. In: *The SAGE Handbook of Evaluation*. pp. 405–418. https://doi.org/10.4135/9781848608078.n18
Strande, L., Ronteltap, M. & Brdjanovic, D. 2014 *Faecal Sludge Management* (Vol. 9781780404).
Subraya, B. M. & Subrahmanya, S. V. 2000 Object driven performance testing of Web applications. In: *Proceedings – 1st Asia-Pacific Conference on Quality Software, APAQS 2000*. pp. 17–26. https://doi.org/10.1109/APAQ.2000.883774.
UNICEF 2020 *Review of the WASH Bottleneck Analysis Tool (BAT): Improving the WASH BAT as A Tool for Planning and Partnering for Sustainability.*
UNICEF & WHO 2020 *State of the World's Sanitation: An Urgent Call to Transform Sanitation for Better Health, Environments, Economies and Societies.*

United Nations Childrens Fund (UNICEF), & The World Bank Group 2014 Water, Sanitation and Hygiene Bottleneck Analysis Tool. *Online*, (May), 2014.

UN-Water (GLAAS) and World Health Organisation 2019 *National Systems to Support Drinking-Water, Sanitation and Hygiene: Global Status Report 2019.*

Water Supply and Sanitation Collaborative Council and World Health Organization 2005 *Sanitation and Hygiene Promotion: Programming Guidance.*

WHO 2018 *Guidelines on Sanitation and Health.*

WHO | UNICEF JMP 2021 *Progress on Household Drinking Water, Sanitation and Hygiene 2000–2020.*

WSUP 2019 *Strengthening the Enabling Environment for Citywide Inclusive Water and Sanitation: an Evaluative Framework.*

First received 1 February 2021; accepted in revised form 15 May 2022. Available online 30 May 2022

doi: 10.2166/h2oj.2022.028

Strengthening country-led water and sanitation services monitoring and data use for decision-making: lessons from WaterAid experience in four countries

Ceaser Kimbugwe[a], Tim Davis[b], Fraser Goff[b], Ellen Greggio [c,*], Sokhadeva Chanthet[d] and Benjamin Kiap[e]

[a] WaterAid Uganda, Uganda
[b] WaterAid Australia, Australia
[c] WaterAid UK, UK
[d] Ministry of Rural Development, Cambodia
[e] Department of National Planning and Monitoring, Government of Papua New Guinea, Papua New Guinea
*Corresponding author. E-mail: ellengreggio@wateraid.com

EG, 0000-0003-2575-6552

ABSTRACT

Water, sanitation, and hygiene (WASH) are fundamental human rights, of critical importance to health, education, wellbeing, and economic prosperity. To fulfil these human rights and drive progress towards universal and sustainable access to WASH services, government service-level monitoring processes and data use are vital for effective decision-making and accountability. Despite increasing sector efforts to improve WASH data access, there is limited evidence of this translating into effective data use to inform effective planning for equitable access and budgeting and of the factors affecting this. Four case studies where WaterAid has worked with national government and sector stakeholders to strengthen WASH monitoring processes in Uganda, Cambodia, Papua New Guinea (PNG), and Myanmar were analysed through an analytical framework to understand the impact of different factors and related system-strengthening activities towards outcomes of increased data coordination, timely and relevant data availability and data use to inform decision-making in WASH service delivery. The analysis highlighted that strengthening activities aiming at improving indicators, data collection and analysis, and the type of data collection and visualisation technology have a direct impact on improving WASH sector coordination and timely data availability. However, to ensure strengthening activities support data use for decision-making, they need to be developed from within and adapt to the on-going wider political economy systems evolution, including formal processes such as decentralisation and evolving informal political drivers.

Key words: data, decision-making, MIS, monitoring, system strengthening, WASH

HIGHLIGHTS

- Analysis of factors leading to improvements to WASH data availability and data use.
- Indicators, data analysis, data needs mapping, ICT, sector coordination platforms are strategic entry points for strengthening country-led monitoring system.
- Approach to strengthening WASH monitoring requires recognition of the wider WASH system and political economy.

INTRODUCTION

There is a growing recognition that universal access to water, sanitation, and hygiene (WASH) services will only be realised through a system-wide approach (Hollander *et al.* 2020; Valcourt *et al.* 2020) to drive progress in WASH systems and ensure a sustainable realisation of the human rights to water and sanitation. The human rights to water and sanitation state that these services should be available and accessible to all, be safe, of acceptable quality and affordable, and offer privacy and dignity to users (UN 2010). Underpinning their achievement are the principles of non-discrimination and equality, access to information and transparency, participation, accountability, and sustainability. Governments have an obligation to progressively realise these rights, underpinned by human rights principles. Monitoring processes play a key role in the progressive realisation of rights by ensuring the availability of information and data use for effective and evidence-based planning, budgeting, service improvement, and accountability. Timely availability of reliable WASH data is essential to proper service planning and decision-making; 'with accurate data on who has access to water and sanitation, and at what level of service, States can prioritise the provision of services to the people who need them the most' (de Albuquerque 2014, p. 22). At the same time, rights holder participation in monitoring and data use is key to improving accountability between service users and service providers, and cycles of continuous learning, adaptation, and improvement (Da Silva Wells *et al.* 2013; Kempster 2020).

In WASH, as well as the provision of many other public services, there is a trend in developing countries towards greater decentralisation of power and responsibilities for planning, budgeting, and service provision (Faguet 2014; Gadenne & Singhal 2014). In theory, making decisions about services closer to the local level increases accountability and greater responsiveness (Faguet 2014; Gadenne & Singhal 2014). While the literature on decentralisation and governance highlights the political nature of decision-making (Giné Garriga *et al.* 2015; Kempster 2020), it seldom refers to the availability and use of data to ensure well-informed and evidence-based decision-making by local leaders or for rights holders and civil society to hold local leaders to account for their decisions.

Efforts and investments to improve and strengthen national WASH monitoring systems and data availability have become increasingly prevalent in the sector (Schouten & Smits 2015), accelerated by the increasing availability and use of Information Communication Technologies (ICTs) and the increased possibility of data generation at all levels (World Bank 2018). Much of the literature on WASH monitoring systems focus on such technical aspects of data production as robustness, timeliness, scale, communication, and participation (Da Silva Wells *et al.* 2013; Cronk *et al.* 2015; Giné Garriga *et al.* 2015). One practical framework which sets out guidance for WASH agencies to assess WASH Monitoring and Evaluation (M&E) systems and design interventions to strengthen them is the '12 components of the National WASH M&E system' (Dickinson 2016). This framework outlines factors and conditions required for an effective WASH monitoring system, based on UNAIDS organising framework for a functional national HIV M&E system (UNAIDS 2008). The 12 components represent a useful checklist of components, and helpful framing of desirable performance results in a well-functioning national WASH M&E system. Less analysis exists of factors that lead to better data use in the sector. Research reviewing the factors for successful ICT-based water supply services reporting (Welle *et al.* 2015; Williams *et al.* 2016) identified some of the key conditions necessary for the successful use of data, including service providers' leadership of the initiative, clarity of service provider responsibility, and available financing to respond to water supply data; and overall, the need to integrate data generation initiatives as part of larger service delivery reforms to improve service providers accountability. Kempster (2020) presents a WASH data-informed decision-making framework guidance that takes into consideration local political economy and behavioural factors.

Despite the more extensive literature on the technical aspects of WASH monitoring systems, there is limited documented evidence of success in WASH data generation and use to support decision-making, and consequently, there is limited analysis and literature of the main driving factors/conditions that can lead to successful improvements to monitoring processes and data use. This paper aims to respond to this literature gap, drawing on experiences from WaterAid's approach and work to strengthen WASH M&E systems in four countries: Uganda, Myanmar, Papua New Guinea, and Cambodia, and focusing on the analysis of driving factors and activities that gradually lead to transformations in WASH monitoring and data use. While the ultimate goal of WaterAid's work to strengthen government-led WASH M&E systems is to achieve strong, transparent, and data-driven decision-making, it should be acknowledged that the strengthening process is not linear and the interventions are based on different starting points for each country case study.

BACKGROUND

WaterAid's approach to strengthening sector monitoring acknowledges the need to go beyond the production of data and its analysis and instead work within the wider system to strengthen institutions, processes, behaviours (Hallsworth *et al.* 2018), factors, and actors, including the complex interactions required to enhance the use of data for decision-making (Kempster 2020). To identify key driving factors impacting improvements to data availability and use, WaterAid established a review team that had worked on strengthening WASH sector monitoring in systems in four countries. The country case studies were selected to present geographic and intervention type diversity. A summary of the background, context, and activities is shown in Table 1.

Table 1 | Description of the four case studies analysed in this paper: background and activities done

Case study	Background and description of strengthening monitoring intervention
Cambodia	The right to water was enshrined in Cambodia's Law on Water Resources Management in 2007. The Cambodian WASH sector's efforts to realise universal access identified the need for a comprehensive system to collect, manage and share information about the progress of WASH in the rural Cambodian context in 2013. An MIS was needed to enable the sector to effectively collect data, monitor progress, and evaluate activities undertaken; to monitor rural WASH performance against the existing targets agreed upon in the National Action Plan (NAP) for rural WASH; and to share information to relevant stakeholders and to provide transparency in WASH progress to the public. The Ministry of Rural Development (MRD) is responsible for coordinating local authorities on rural WASH and ensuring WASH services are available to all Cambodians living in rural areas. A phased MIS development was led by MRD through a sector working group comprising government and non-government actors, including WaterAid, UNICEF, and Plan International. Engagement and ownership of the MIS were underpinned by a Ministry regulation that established institutionalised management structures and roles at the national and sub-national level. In the first phase of the MIS (2018), nationwide data were collected by provincial MIS committees on six key indicators from the NAP. MIS phases 2 (2019) and 3 (2020) developed the annual data collection and validation process through the development of standard data entry templates, capacity development, coordination structures, the annual collection of secondary data from government and non-government actors, data validation and review processes with other sector actors, data analysis, mapping, and reflection. Indicators were expanded so that by the MIS phase 3 more than half the NAP indicators were being measured. The MIS provides a means to measure the progress towards the target outcomes and outputs in the NAP, thereby improving the visibility of WASH sector performance and improving accountability of WASH programme implementation.
Myanmar	In Myanmar, WASH sector policies had been newly developed in 2016 with new standards and targets, in line with SDGs and progressive realisation of rights to water and sanitation. The water supply sector had historically been strongly centralised and investment in new infrastructure was donor-led with limited autonomy to the local government level. This was reflected in water services data availability, which was limited to donor reports of new water infrastructure. In 2018, the Department for Rural Development (DRD), in response to new policies implementation, identified the need for improvements to their rural water supply monitoring systems, particularly the need to track progress toward SDG 6.1 water service levels at the national level, while also support planning and budgeting processes at the local level. Digitalisation and routine monitoring process creation were identified as initial steps to increase the availability of quality and timely data. WaterAid partnered with DRD to provide technical support to assess rural water monitoring gaps, then co-develop, pilot, and adapt a model for routine monitoring of rural water supply service levels. The process included indicator review and harmonisation through sector coordination and consultation, which lead to the adoption of sector best practice indicators (SDG6 JMP questions including accessibility, availability, quality, and quantity of water, World Bank Rural Water Supply Sustainability Matrix) (Requejo-Castro *et al.* 2017). This was followed by piloting the proposed new monitoring model and tool (ICT-based data collection and analysis through the use of the freely available mWater platform) in three districts representing a variety of water supply contexts. Over 200 government staff were involved, trained, and participated in the process, and informed the regular review and adaptations of the proposed model. This learning phase was followed by the development of a fully locally customised rural water supply MIS to inform local planning (identifying communities with lower services, planning, and budgeting maintenance) and also by aggregating data, to track national progress. This was scaled up country-wide with leadership from DRD and by the creation of a routine monitoring framework with defined responsibilities. The process has been hindered and slowed down by political changes in 2021.
Papua New Guinea	In Papua New Guinea, effective processes for subnational WASH planning, budgeting and service delivery are weak and emerging. The Department of National Planning and Monitoring (DNPM), developed PNG's first WASH Policy in 2015. The WASH policy recognises that access to safe, convenient and sustainable water, sanitation and hygiene are basic human rights and provides a framework for sector reform through action at District level. The lack of quality and timely data on WASH services is a barrier for government and the broader WASH sector, hampering efforts to increase access and progressively achieve the rights to water and sanitation. To establish PNG's first WASH monitoring system WaterAid partnered with UNICEF and the European Union to build a project MIS hosted on the mWater system, The project utilised common indicators and data collection forms and was populated with data from five pilot districts across the country, with each district formulating their own 5-year WASH investment plans based on the data. The project MIS demonstrated to DNPM the potential to customise, institutionalise and scale the MIS

(*Continued.*)

Table 1 | Continued

Case study	Background and description of strengthening monitoring intervention
	and the system was formally handed over to DNPM in August 2019. Based on experiences from the project MIS, WaterAid was engaged by DNPM through a formal Technical Advisor contract to support DNPM to further build, enhance, customise, institutionalise and scale PNG's first WASH Monitoring system, this process has been ongoing since August 2019. The first phases of the work and focused on development of the national WASH indicator framework (through national level technical working group), customising the WASH Management Information System (MIS), development and piloting of training materials and programmes, establishing subnational monitoring processes linked to the development of WASH investment plans. Significant work is ongoing to effectively achieve ongoing monitoring and recurrent finance at subnational level with the aim of driving improvements to planning, accountability and realisation of the PNG WASH policies aim to ensure the human rights to safe, convenient and sustainable water and sanitation. The establishment of the monitoring system was driven by national leadership and the desire to obtain, quality and timely data to track progress against policy targets, improve coordination and drive evidence-based planning and finance. In effect, quality data have provided an example of a viable approach to subnational planning of WASH service provision.
Uganda	In Uganda the presence of multiple WASH government agencies and institutions sharing the WASH mandate led to inconsistent and incomplete WASH statistics, hindering coherent cross-sectoral WASH planning and coordination between line ministries of water, health and education. To address this, WaterAid Uganda in collaboration with other actors, worked with different government authorities to support a process of indicators harmonisation to improve existing monitoring processes and data use to inform local and national planning. In particular, with the leadership of the Health Ministry, addressed national reforms of the Health Management Information System (HMIS) and included WASH in health care facilities indicators and household onsite sanitation data, in line with global monitoring best practices by JMP. The national changes to the HMIS cascaded down to local authorities, mandated onsite sanitation data collection. While some local authorities such as the Kampala City Council Authority (KCCA) had collected comprehensive sanitation data from schools, households, and public facilities, the reliability and use of this data in the planning processes were yet to be appreciated. WaterAid supported improvements at the local level by demonstrating the use of ICT-based data collection to reduce time and efforts spent on paper-based data collection and to establish clear roles and incentivise Village Health Teams (VHTs) supporting data collection.

METHODS

To conduct the review of each country's case study, the WaterAid review team developed a structured analytical framework based on the 'From Data to Decisions' policy brief and data use planning guide developed by ODI (Overseas Development Institute) and WaterAid (Kempster 2020) and the 12 components framework (Dickinson 2016). The structured analytical framework (Figure 1) was designed to review two main aspects of the monitoring system interventions:

- *Intervention activities for each factor:* Key strengthening activities were identified for each factor (see Supplementary Material, Appendix 1), drawing on areas and activities included in the 12-component framework (Dickinson 2016).
- *Contribution of factors towards continuous government monitoring and data use outcomes*: The perceived contribution of each factor in driving success towards the following four key outcomes of successful WASH monitoring and data use at a national level:

1. Coordinated monitoring processes and tools are used in the sector.
2. Relevant, timely, and quality data are available.
3. Decision-makers use data for tracking progress, planning, and budgeting.
4. Informed decision-making impacts WASH services provision, sustainability.

Within the framework, the driving factors were grouped into four categories, in line with categories from Kempster (2020), and defined as follows:

a. Data needs (Data)
b. Processes and technology (Processes)

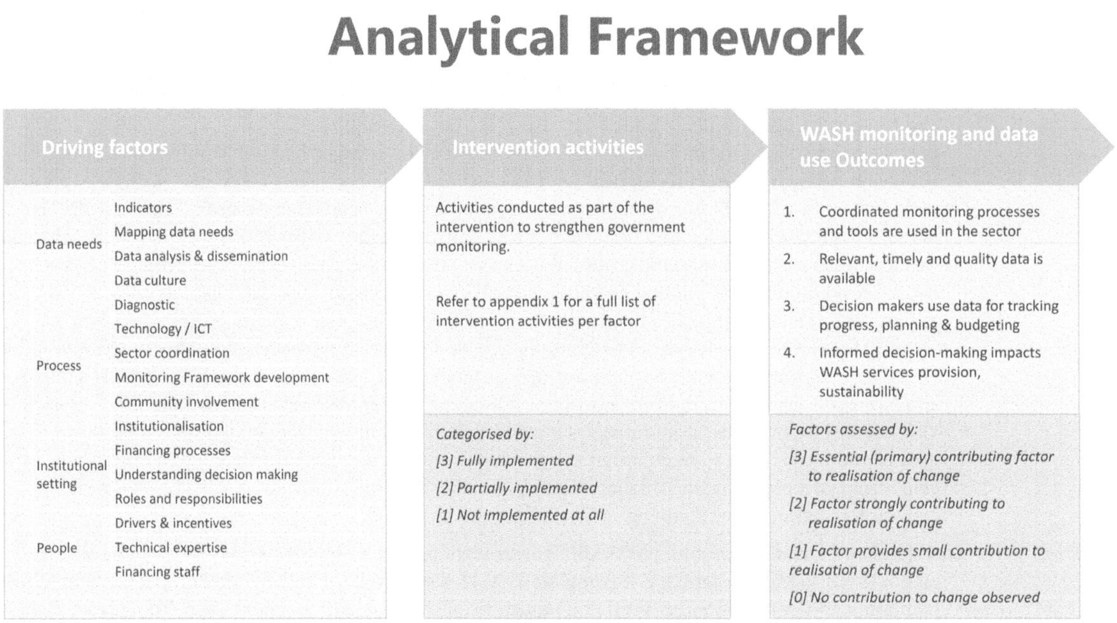

Figure 1 | The analytical framework based on the 12-component framework (Dickinson 2016) and Data 2 Decisions Planning Guide (Kempster 2020).

c. Institutional settings (Context)
d. People and Drivers (People)

A rating scale assessed the extent to which each factor was perceived to contribute to the successful achievement of each of the outcomes of continuous monitoring and data use listed above (including four scorings: no contribution to change observed (0), factor provides a small contribution to realisation of change (1), factor strongly contributes to realisation of change (2), and essential (primary) contributing factor to realisation of change (3).

Quantitative responses to each of these rating scales have been complemented by qualitative explanations to justify the rating selected. While it is acknowledged that other systematic factors also influence realisation of outcomes, particularly for data use, the scoring developed aimed at identifying which strengthening activity has had a larger impact on different outcomes within each country's context, to inform future intervention design and prioritisation.

The method to review the four case study interventions included:

- Development of the analytical framework (summarised in Figure 1 and more detail in Supplementary Material, Appendix 1).
- Testing and refinement of the analytical framework based on review team's own experiences and existing documentation of the monitoring and data use interventions.
- Population of the framework analysis templates through in-depth interviews in July 2021 with at least one government counterpart from each of the four-country contexts who were selected based on their role as focal point or coordinator for the government WASH monitoring system. These interviews involved:
  1. The review team explains the analytical framework, and terms and definitions.
  2. Discussion of each of the activities completed for each factor as part of the monitoring system intervention; quantitative assessment using a rating scale of the extent to which each activity was fully, partially, or not at all included in the intervention to strengthen the WASH M&E system; and a qualitative explanation of the rating.
  3. Discussion of each of the four monitoring and data use outcomes; quantitative assessment using a rating scale of the extent to which each factor was perceived to contribute to the successful achievement of each of the outcomes; and a qualitative explanation of each rating.
- Normalisation of the ratings across the four case studies to facilitate cross-country comparison and facilitate cross-country comparison.

FINDINGS

The results from the analysis in each of the four case studies are as follows:

Table 2 presents the relative contribution of each of the factors (from the sum of the contribution of three factor activities) towards the overall four outcomes. This supports identifying the key factors and associated strengthening activities that had a larger impact on the overall outcome of improved data production and use.

Table 3 presents the cumulative assessment of all factors towards the monitoring and data use outcomes for each of the case studies. This supports the identification of what outcome was mostly impacted by the strengthening activities analysed.

Cambodia

In Cambodia, standardisation of monitoring processes and availability of data through the Management Information System (MIS) has facilitated a culture of regular upwards reporting and aggregation of data which has

Table 2 | Contribution of factors towards WASH M&E system outcomes in each of the four case studies

Factors categories	Factors	Assessment score per factor for four assessable outcomes Green = stronger contributing factor, Orange = lower contributing factor (maximum score 12 for each factor)				
		Cambodia	Myanmar	PNG	Uganda	Average
Data needs	Indicators	58% (7)	41% (5)	100% (12)	42% (5)	60%
	Mapping data needs	25% (3)	42% (5)	67% (8)	33% (4)	42%
	Data analysis & dissemination	67% (8)	42% (5)	75% (9)	58% (7)	60%
	Data culture	58% (7)	33% (4)	67% (8)	17% (2)	44%
Process	Diagnostic	67% (8)	25% (3)	58% (7)	33% (4)	31%
	Technology/ICT	33% (4)	42% (5)	75% (9)	33% (4)	46%
	Sector coordination groups	42% (5)	17% (2)	75% (9)	50% (6)	46%
	Monitoring framework development	58% (7)	8% (1)	75% (9)	25% (3)	42%
	Community involvement	0% (0)	0% (0)	0% (0)	17% (2)	4%
	Institutionalisation	67% (8)	17% (2)	42% (5)	8% (1)	33%
	Financing	0% (0)	25% (3)	0% (0)	0% (0)	6%
Institutional settings	Understanding decision-making	0% (0)	25% (3)	58% (7)	33% (4)	29%
	Roles and Responsibilities	50% (6)	8% (1)	50% (6)	50% (6)	40%
People	Divers and incentives	50% (6)	25% (3)	33% (4)	33% (4)	35%
	Technical expertise	33% (4)	33% (4)	67% (8)	58% (7)	48%
	Financing staff	33% (4)	25% (3)	33% (4)	8% (1)	25%

Table 3 | Relative contribution of monitoring intervention factors towards M&E system outcomes in each of four case studies

	Outcome No progress observed Factors contributed to only limited / early progress towards outcome Factors contributed to a strong progress towards outcome Factors contributed to realised outcome	Extent of progress towards outcomes			
		Cambodia	Myanmar	PNG	Uganda
1	Coordinated monitoring processes and tools are used in the sector				
2	Relevant, timely and quality data is available				
3	Decision makers use data for tracking progress, planning & budgeting				
4	Informed decision-making impacts WASH services provision, sustainability				

significantly improved progress tracking against sector plans and enabled inter-province comparison. However, there is still limited evidence of the MIS data's use for provincial or district WASH service planning and budgeting.

The following factors have contributed significantly to the successful establishment of the Cambodia MIS, and were among the highest scores in Table 2.

- *Institutionalisation:* The Cambodian government, through the Ministry of Rural Development (MRD), demonstrated strong ownership and leadership of the MIS process from the beginning. The leadership of the MIS development was consultative with the wider sector but was embedded within existing government roles and responsibilities at national and provincial levels through an official *Prakas* (Regulation) which encouraged upward accountability. In addition to formalised responsibilities, the MIS working group developed clearly defined annual work plans, flow charts, and procedures to operationalise the MIS within the mandate of the MRD.
- *Data analysis and dissemination:* Annual analysis of data, sharing of summaries, and creation of dashboards for sub-national government to extract relevant indicators helped to deepen the understanding and engagement of MIS stakeholders and increased the utility of the data for progress monitoring at national and sub-national levels. Annual reflection workshops were held to disseminate the results and collectively reflect on the sector's progress, as well as gather feedback to refine the MIS itself.
- *Monitoring framework development:* The MIS is the system for monitoring progress against the indicators already identified in the National Action Plan (NAP) for Rural WASH. The need to report progress against the NAP's monitoring framework was a strong motivation for establishing the MIS. The second phase of the NAP was developed concurrently with the MIS, and lessons learned from using the MIS to monitor NAP progress informed the selection and short-listing of indicators in the second NAP.
- *Indicators:* The process of harmonising indicators helped bring together government and non-government actors to build sector coordination and consensus. Prioritising a manageable list of indicators also helped to encourage timeliness in the data gathering processes and build trust in the data quality.
- *Data culture:* The Minister of Rural Development's support for the MIS and creation of a government core group to lead the MIS contributed to prioritisation of this intervention and gave weight to coordination and tool development. Annual training and orientation on indicators, data collection, and interpretation of results increased understanding of data management and use at national and sub-national levels and helped planners and implementers to see how the MIS could help monitor progress against the NAP.

Three factors/activities received particularly low scores in Table 2, and represent significant risks to the continued sustainability of the MIS in Cambodia:

- *Process financing:* While government staff, employed by MRD, form the backbone of the MIS, the costs associated with travel, workshops, consultants for developing ICT dashboards and printing of documents have been largely borne by non-governmental organisations (NGOs) and development partners, and advocacy for allocation of government funding to cover these costs has been limited.
- *Understanding of decision-making:* the absence of structured or systematic assessment of fiscal, administrative and political data use is likely a key factor for the continued limited use of MIS data for (service) planning and budgeting. Provincial WASH Working Groups, established at the same time as the MIS and led by provincial political leaders, are not yet routinely drawing on MIS data for their annual planning and budgeting, likely because of the service level data they need for such processes is not yet available within the MIS. Significantly, the decentralisation of responsibilities for rural WASH in Cambodia occurred during the establishment of the MIS. District administrations, who have not been involved in the development or use of the MIS to date, now have additional decision-making responsibilities for WASH and the MIS may not be meeting their data needs.
- *Community involvement:* The MIS development has initially focused on how data can be made available for WASH planning and performance monitoring but with minimal consideration or involvement of community and service users in the process. This represents a risk by not considering the data needs of users or data use for accountability from the system's inception.

The *understanding of the decision-making:* factor mentioned above is one reason for the relatively limited progress observed towards the outcome of *data use for WASH service improvement* among the four outcomes. Without a strong understanding of how data are, and could be, used for decision-making and integrated with planning and budgeting, several other factors which would be expected to contribute strongly to data use also rated poorly on contribution to this outcome such as understanding the individual motivations and incentives for data use, and allocation of funding and technical resources to the monitoring focal points. In addition, the indicators being collected through the Cambodia MIS until now have mostly focused on performance measurement (e.g. number of villages receiving Community-Led Total Sanitation (CLTS), number of MRD staff participating in training) and while they provide a good summary at national level of progress across the NAP indicators, data on service access or quality indicators which would be most useful for service provision and sustainability decision-making (e.g. household access to sanitation, or water supply functionality) are still being integrated in the current phase of the MIS.

As in the other case studies, the strongest outcome progress in the Cambodia case study was towards the *coordination of sector monitoring tools and processes* and *relevant, timely availability of quality data* (Table 3). The factor ratings (Table 2) indicate this is predominantly for the following two reasons:

(a) The MIS was primarily established to fill a gap identified in the review of the first National Action Plan for rural WASH, in consistent processes and mechanisms for data collection, analysis, and reporting between different provinces or at the national level.
(b) The mechanism through which the MIS was developed and refined, a government-led working group with the active participation and buy-in of key non-government sector actors had the effect of strengthening coordination and sharing knowledge between sector actors.

Myanmar

The Department of Rural Development from the Government of Myanmar identified the need to strengthen the national rural water supply monitoring as a key step toward progress monitoring and informed planning towards the achievement of Sustainable Development Goal (SDG)6. Key drivers for change included the department leadership and clear progress vision, a keen technical savvy government team driven by career progression, and a goal to adopt technology innovation to improve data availability, access, and sharing. The process included co-designing with the government at the local and national level, and adapting regularly to feedback and changes required.

Through the analysis of the interventions that led to the strengthening of the monitoring system and data use (Table 2 and Supplementary Material, Appendix 2), we can identify the key factors and associated strengthening activities leading to progress and successful outcomes, in particular, the outcomes related to coordinated monitoring and availability of relevant, timely and quality data;

- *Indicators definition and harmonisation:* the collaborative process of introducing, testing, and adapting new sector best practice indicators for rural water supply monitoring, including Joint Monitoring Programme (JMP) SGD6 monitoring indicators, and World Bank rural water supply sustainability indicators (World Bank 2017) led to a step change into governmental and sectoral shifting from infrastructure and service delivery reporting to water supply service levels monitoring in line with SDGs. Furthermore, the sector-wide consultation for indicators reviews and harmonisation with other WASH stakeholders was key to engaging stakeholders in the use of a common monitoring indicators framework (for example, UNICEF fully adopted the government led rural water supply indicators and data collection tool).
- *Data analysis and dissemination:* WaterAid co-designed with government staff the data analysis approach through the selection of appropriate data visualisations (maps, charts, tables) summarising key indicators data to respond to different information needs and capacity for data interpretation at local and national levels.
- *Introduction, piloting, and adapting of ICT and locally appropriate technology:* in Myanmar, the adoption of ICT-based technology had progressed rapidly, however mostly for private use rather than to support government-led processes. By introducing and piloting a freely available ICT platform that was accessible and usable with local government's own phones without any modification or investment, it was possible to demonstrate to national and local governments how to leverage existing technology to accelerate data collection and real-time analysis. Access to real-time data and its analysis was seen as an essential step change, particularly by the national government.
- *Mapping data needs* at different levels and for different uses at the beginning of strengthening activities supporting the *generation of a data culture* (through regular trainings, demonstrations) were also a key driver for achievement of outcomes.

It is instead observed that some factors and strengthening activities, such as the creation of *roles and responsibilities* and routine monitoring framework, which we would expect to strongly contribute to outcomes such as 'decision-makers use data for tracking progress, planning and budgeting' did not yet (in the period analysed) contribute to the outcomes, particularly 3 and 4. This is however due to the short period of intervention analysed (2 years) and progress slowed down the process due to political changes.

Furthermore, the scale-up and adoption of the indicators and ICT-based data collection approach introduced was facilitated by the top-down political model, while the limited outcomes related to data used to support local level planning and budgeting were also limited by the weak decentralisation of WASH planning and budgeting processes in Myanmar. This is due to the historical politically centralised management and weak decentralisation reforms. Quality data access improvement was acting as an entry point for a wider reform towards decentralised planning and budgeting processes. While a data culture has been growing within the DRD teams, additional work is required to *shift the culture of 'reporting upwards' to data use for improvements to planning and budgeting* particularly at subnational level.

Identification of both political champions and technical experts within government and the co-designing and adaptive approach (including regular reviews and changes to the system proposed) together with the generation of a *common vision*, with government ownership but also encouraging *sector coordination, discussion and engagement in the process* have been key to ensure suitability of processes and development of a system that responded to data needs at local and national level.

Papua New Guinea

The lack of effective government monitoring in Papua New Guinea (PNG) is a significant barrier to the WASH sector. Until recently, the Department of National Planning and Monitoring (DNPM) did not have nationally agreed to WASH indicators with definitions. Both at the national and subnational level the roles and responsibilities for data collection, processing, management, and use were poorly defined. Recently, progress towards establishing an effective national WASH monitoring system has been driven by national level leadership and the requirement for tracking process towards the National WASH Policy (DNPM 2015) and Medium-Term Development Plan 3 (MTDP3) (DNPM 2018) targets. A national level technical working group on WASH monitoring was established. Leading a co-design and consultation process to map and agree with data and decision-making needs of government, development partners and NGOs. A WaterAid staff member was formally engaged through a consultant contract as a National M&E advisor within DNPM to document data needs, indicators, and definitions through the development of the DNPM National WASH M&E framework, monitoring manual, development and enhancement of the WASH MIS and supporting subnational data use through District planning processes and the development of 5-year WASH investment plans.

Through the analysis of the interventions that led to the strengthening of the PNG monitoring system and data use (Tables 2 and 3), we can identify key factors and associated strengthening activities that led to progress towards all four outcomes areas. In PNG, several factors were the key to the enhancement of WASH M&E systems. These gains were driven primarily by five key areas:

- *Sector coordination groups:* the formation of a national level technical working group chaired by DNPM and brought together the key government departments, development partners, and NGOs to map data needs and decision-making processes of government and non-government stakeholders. While experiences of the working group at a subnational level were considered in the process consultation on data needs with the District government was only possible in eight pilot districts out of 89 total districts across the country.
- *Indicator definitions, harmonisation:* WaterAid with the Government of PNG and the TWG undertook a collaborative process to document key and non-key indicators, definitions, and their intended uses. DNPM adopted SDG indicators and extended indicators for subnational planning and finance processes and formalised the indicators with the National WASH M&E framework (DNPM 2020).
- *Monitoring framework development*: a collaborative process was led by WaterAid to establish PNG's first national WASH M&E framework, outlining key indications, definitions and standard data collection forms. The development of the M&E framework led to enhancement of sector coordination, notably key WASH sector stakeholders adopting standardised data collection formats aligned to national WASH policy, MTDP3 and providing effective data for subnational planning and budgetary processes.
- *Demonstration of appropriate ICT systems to establish national level WASH MIS:* the PNG WASH policy provides a framework for action at the District level, through key development partner programmes, several

INGOs partnered with District Development Authorities across the country to systematically gather District-wide WASH data. Three key development partner programmes (World Bank, DFAT, and EU/UNICEF) were used to pilot national data collection forms. Data were collected by government and INGOs via the mWater system from 12 districts across the country with data and indicators visualised on a common dashboard at the national level. Based on the pilot, DNPM formally adopted the system and assigned staff to progressively administer and manage the system. The DNPM National WASH MIS has driven progress towards the availability of relevant quality and timely data.

- *Data Analysis and Dissemination*: data analysis and dissemination can be problematic in low resource settings such as rural PNG. The WASH MIS automatically processes analysis based on the national indicators and has been key to achieving data use by District level government. The WASH MIS data have been used by decision makers at District level to formulate 5-year WASH investment plans highlighting targets, required financing and roles and responsibilities for service improvement.

The analysis of the outcomes (Table 3) indicates that the intervention of strengthening WASH monitoring in PNG resulted in notable gains in all four outcome areas while noting that most support has been provided in outcome areas 1 and 2. While 'coordinated monitoring processes and tools' has the most impact, it should be noted that the most energy and resources were focused on this area. Efforts to enhance coordination for development partners to engage in collective monitoring processes, develop and formalise indicators and data collection forms and use a common system were found to be a viable pathway to accelerate progress.

While efforts have resulted in progressive improvement to relevant, timely and quality data and the formation of an MIS to visualise and use incoming data, the approaches have drawn largely on 12 districts where INGOs are closely partnered with District Development Authorities. While DNPM have plans to scale the approaches across the country, analysis is based on experiences from the 12 districts and significant efforts are needed to institutionalise monitoring processes, particularly in districts with less development partner support.

The PNG National WASH policy provides a framework for the development of 5-year WASH investment plans at the district level. The key to success has been linking data collection processes within the National M&E system to District level planning and budgeting processes. The resulting approach has encouraged subnational leadership and fostered transparent decision-making through the formation of district WASH committees for planning and budgeting processes. This has been observed in eight of the 12 districts that have contributed data to the WASH MIS.

It is noted that WASH monitoring finance has scored poorly in PNG. It should be noted that within the phasing for M&E system rollout DNPM is working to strengthen WASH MIS systems and processes and rollout the system throughout the country. This includes supporting the subnational government to allocate financial and human resources for recurrent monitoring.

To date, the government of PNG has not established community or citizen reporting mechanisms within the national WASH MIS. Due to limited resources, existing efforts have focused on institutionalising the system within the public service. Access to WASH services is low in many parts of the country and service providers such as District Development Authorities are poorly resourced to respond to community demands, therefore there can be a reluctance to prioritise community reporting. This may be prioritised in the future as the MIS continues to evolve and strong systems and processes are built.

Uganda

In Uganda, the *coordinated monitoring processes and tools are used in the sector* and *relevant, timely and quality data are available outcomes* were achieved more consistently across different interventions but also contribute to *Informed decision-making*. WaterAid, in partnership with the Ministry of Health (MoH), Ministry of Water and Environment, WHO, UNICEF, Makerere University, and Emory University among other stakeholders, facilitated the development of a national measurement framework for WASH in healthcare facilities (HCFs) and in communities. JMP service level indicators were localised for communities and HCFs and these have been integrated in the MoH-HMIS revised in 2018. A national data collection tool was also developed by contextualising the Emory University's WASH-Con Tool which included systems strengthening indicators around planning and budgeting for WASH in HCFs. The standardised indicators have since been widely adopted nationally and used by MoH, UNICEF, WaterAid, Kampala Capital City Authority (KCCA), and the Ministry of Water and Environment to collect national data on the status of WASH in HCFs. These data have then resulted in the development of national guidelines for WASH in HCFs. Furthermore, WaterAid Uganda has continued to support the KCCA indicators harmonisation process for WASH in schools, by integrating JMP WASH in school core questions and supporting updating the city-wide online database for WASH in public primary and secondary schools covering 79 primary schools and 22 secondary schools in Kampala. This capacity support initiative has helped KCCA map out schools with different WASH needs and is informing financial investment from both public funds and allocation of development partner support including NGOs. It also has improved information sharing and use among the different stakeholders in terms of timely data generation and reporting.

Based on Table 2, in Uganda, the factors that have had largest impact on outcomes included:

- *Sector coordination groups:* The MoH has continued to convene the national sanitation working group that provides a platform for discussing cross-sectoral coordination and integration of WASH data for planning and decision-making. By partnering with the Water Resources Institute of the Ministry of Water and Environment WaterAid has drafted the development of a Training manual on SDG6 to among other objectives define roles and responsibilities in monitoring SDG6 targets and indicators.
- *Roles and responsibilities:* the definition of clear roles at different levels has been key for the routine generation of data and its use. From data collection at community level, data management and analysis at health centre level and aggregation at national level where data are used to inform policies and planning processes.
- *Data analysis and dissemination:* the support towards improved data analysis and dissemination led to increased data access and use outcomes. WaterAid supported the building capacity of civil society organisations coordinated under the Uganda Water and Sanitation NGO Network to proactively contribute to data availability compiled and disseminated during the annual Joint Sector Reviews (JSR) as part of the Sector Performance Report. The JSR platform has been used to disseminate data and inform data-based cross-sectoral policy discourse and decision-making on priority undertakings for government and non-state actors to focus on.
- *Technical expertise:* the technical support in introducing and demonstrating best practice indicators in line with SDG service levels indicators has led to an increased understanding of WASH service levels and the use of this data to inform planning processes, moving from coverage data use only to consideration for service levels (reliability, accessibility, quality, etc). This was also complemented by support on use of digital tools for data collection and analysis which were rapidly adopted by government staff.
- *Indicators definition and harmonisation:* There process of sector wide indicators harmonisation across line ministries aimed at alignment with the global SDG commitment on universal access and JMP indicators, particular with HMIS indicators on health care facilities and sanitation, has led to increased coordination across the sector and increased availability of consistent WASH data.

LIMITATIONS OF THE ANALYSIS

- Each case study presented had different starting points, in terms of existing monitoring processes and frameworks, drivers, data needs, and data use scope limiting the full comparability. In Myanmar, there was no previous presence of a rural water supply service levels monitoring system with only upwards donor-led or large investment reporting processes in place. Understanding of service level monitoring beyond functionality monitoring was low. In Cambodia, a clear monitoring framework had been established through the sector-agreed National Action Plan for rural WASH however there was no consistent monitoring system prior to the MIS; each province had developed its own templates, procedures, and indicators for monitoring, often aligned with the data requests of NGOs and development agencies. This variety of existing monitoring and data culture and processes leads to different approaches and impacts of strengthening activities and factors in the different case studies. For example, PNG MIS development was focused in 12 pilot districts while Cambodia was nation-wide, but the data gathered were more limited in scope compared to PNG.
- The analysis is limited to assessing the strengthening interventions and their impact. While the analysis is grounded in the political economy context of each of the case studies, it does not aim to analyse other possible processes and activities occurring outside the interventions which might have influenced the outcomes.

DISCUSSION

From the analysis of the four WaterAid's country experiences of strengthening government national monitoring systems, it was possible to identify activities (and associated factors) that have the largest impact across all case studies. These included the following (in order of score):

- Indicators: introduction of best practice indicators, including SDG6 JMP indicators, clear definition setting and support for the sector wide indicators harmonisation process.
- Data analysis and dissemination: development of targeted and locally contextualised and appropriate data analysis generation and sharing at different government levels and sector.
- Technical expertise: provision of technical training, and expertise on WASH services monitoring to national and subnational government staff.
- Sector coordination groups: developing and facilitating sector working group/committee to harmonise indicators and coordinate WASH monitoring processes.
- Technology (ICT) demonstration, piloting, and adapting to local context needs for data collection, management, and analysis.
- Data use culture: understanding existing data use, data needs at a different level and regular demonstration and advocacy on data use for planning, progress tracking, and policy adaptations.
- Mapping data needs: Mapping pre-intervention/previous data use, mapping data needs and gaps at different levels, identifying and aligning with sector priorities, policies, targets, and related indicators.
- Monitoring framework development: development of government WASH monitoring frameworks at the national level help to formalise and embed indicators within the public service. This provides consistency and lends support to the subnational rollout and institutionalisation of the monitoring system.

This leads to the interpretation that the *above factors are viable and strategic entry points that contribute to accelerating progress towards strengthening and institutionalising coordinated government-led WASH monitoring processes.* Efforts of development agencies like WaterAid to adapt and improve suitable data collection and analysis processes, introduce best practice indicators and support indicators harmonisation, map data needs, and demonstrate ICT systems through co-design processes with the government can have a direct impact and influence progressive improvement of government monitoring at scale. In particular, it can lead to improvements to sector coordination processes and data availability.

The WASH sector places a high emphasis on putting decision-makers at the forefront of data needs (Da Silave Wells *et al.* 2013; Dickinson 2016; Kempster 2020). Logically decision makers need quality, and timely data to effect evidence-based decisions, however, *data use and decision-making processes are not always clear cut:* data needs, decision-making processes, and responsibilities by government and service providers are diverse and evolve based on changes in context, political influence, personal incentives, and overall political economy. Priorities and data needs of national level stakeholders may not align with their subnational counterparts and therefore data analysis and visualisation need to reflect these different needs, even if based on the same dataset. In PNG, district-level government staff were found to have low policy awareness with very minimal awareness of SDG indicators and service levels. District level decision-makers such as local level government managers do not have formal processes or criteria for how a WASH project should be selected and to what standard of services should be delivered. Lacking this, priorities commonly align with the provision of basic services (for example, if a community had an existing rainwater tank or not). While the PNG WASH MIS displays SDG service levels, useful for national level policy tracking and reporting, this had little bearing on decision-making processes at the subnational level. Similarly, in Myanmar, national decision-makers identified SDG indicators and water service levels tracking as a key data need for progress comparability at an international level, while local government staff only identified the need to track key sub-indicators mostly associated with coverage and functionality of water supply, as this information aligned more clearly with their core responsibilities and targets.

From the analysis, we have observed that activities associated with strengthening *institutional settings and people factors,* including understanding and leveraging current data use, identification of stakeholders' drivers, development of monitoring frameworks, have had a limited direct or immediate impact the four outcomes. On one hand, this is unsurprising as these represent the factors that are intrinsically linked to government functioning and therefore less likely to be influenced over short-term intervention by external agencies like WaterAid. However, these had an indirect impact on developing data and data use culture, for example, in the PNG case study, which made significant progress towards the intended outcome of data impacting WASH service provision and sustainability. Despite the limited direct impact, the case study narratives demonstrate that the successful strengthening of indicators, data collection, and analysis could only be effective if developed *with an understanding and recognition of the wider system, its factors, formal and informal processes and with regular reviews and adaptations of approaches propose to the regularly changing political economy change.* This is in line with sector recognition of the need to work within the wider system to strengthen institutions, processes, behaviours (Hallsworth *et al.* 2018) and that improvements to monitoring systems or introduction of ICT can only support improved services and accountability if coupled with wider reforms (Williams *et al.* 2016) linked to regulatory processes, management models, sector financing that impact how data can be used.

The culture and attitudes around data among WASH system actors, including community and civil society, are central to the use of data for improved service planning and budgeting, but also in making the step to fulfilling the human rights principle of accountability. In an effort to shift from a 'reporting culture', *activities that aim to strengthen monitoring processes represent entry points towards the first step of the creation of a 'data use culture'* in which data are not just reported but used for evidence-based decision-making, course correction and reducing inequality in the provision of services (Figure 2). As such, a sector culture of joint reflections, formal and informal networking and social learning encourage continuous learning, and use of data to manage WASH services adaptively (Da Silva Wells *et al.* 2013). The review team observed some progress in moving towards a 'data use culture' in the WASH systems in each of the four case studies. The *progressive realisation from a 'data use culture' to a 'data-based accountability culture'* in WASH is a longer process, but an essential shift through which monitoring and data availability contribute to the accountability mechanisms through which rights holders hold duty bearers to account for fulfilment of the rights to water and sanitation (de Albuquerque 2014). The shift from a 'data use culture' to a 'data-based accountability culture' is often dependent on the realisation of other institutional reforms, such as the implementation of regulatory processes, policy targets and requirements, leadership, clarity on management models and responsibilities, and decentralisation of planning and budgeting processes to the local level. It also requires strengthening of citizen capacity to understand monitoring indicators and their rights in relation to WASH, as well as avenues to integrate monitoring systems with mechanisms for accountability such as involving civil society rights groups in the development and strengthening of the monitoring system and facilitating data analysis and interpretation (Da Silva Wells *et al.* 2013; de Albuquerque 2014). Nevertheless, improved monitoring processes, quality and timely data and capacity to interpret and use data are necessary steps towards the greater accountability assumed by WASH decentralisation processes (Faguet 2014; Gadenne& Singhal 2014). Without improved monitoring and availability of data for both decision-makers, civil society, and rights holders, local leaders may use any conveniently available WASH data which may serve their political purposes rather than being accountable for service users' satisfaction and realisation of human rights to water, sanitation, and hygiene without discrimination on the grounds of wealth, race, class, sex, religion or geography.

Often WASH monitoring literature advocates for the creation of a uniform national monitoring system, (Schouten & Smits 2015; Dickinson 2016). Through our case studies, we have found that the creation of uniform national monitoring systems needs to be balanced with *co-designing appropriate approaches and phasing to strengthening WASH Monitoring with government stakeholders*, reviewing indicator appropriateness, and understanding and responding to different data needs and feedback processes at the local level in order to avoid reinforcing a 'reporting culture' with top-down centralised monitoring processes reducing local level data

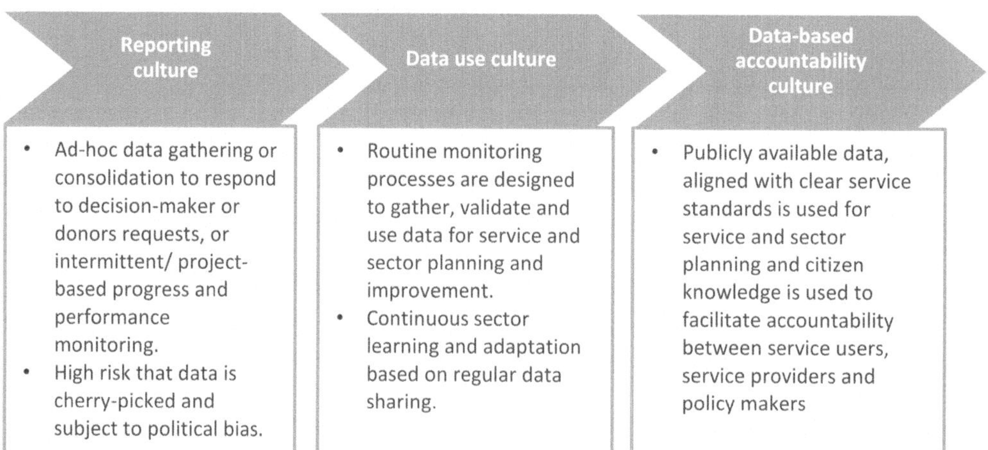

Figure 2 | Progressive realisation of 'data-based accountability culture' based on review team's interpretation of the review findings.

ownership. In Cambodia, we observed that the creation of a national MIS to standardise and aggregate data for progress monitoring reinforced reporting culture from the provincial level to the national level. This system adequately met the progress-tracking needs of the national level (and likely provincial level), however, it did not necessarily provide the WASH data needed by district decision-makers to plan, prioritise, and budget for local WASH services once they received these responsibilities through decentralisation. By contrast, in Uganda the collaborative, inter-ministerial co-design of the national monitoring framework and data collection tool for WASH in HCFs prioritised application and data use for local-level planning and budgeting of WASH in HCFs, and influenced the updating of new standards, leading to its increased coordinated adoption and utilisation at all levels. In contrast, we observed that who collects data impacts how the data are used and interpreted; therefore, it is important to involve decision-makers (including from the national level) during data collection and validation processes, rather than fully delegating to local level staff. In Uganda, without government participation in data collection, data analysis findings cannot be officially reported or referenced.

Identification of *political and technical champions* within government staff at the national and sub-national levels has been key to driving change and progress towards improvements towards harmonisation of indicators and data availability in line the with literature (Da Silva Wells *et al.* 2013; Kempster 2020). In Myanmar, high level political leadership with a clear vision and top-down culture, combined with the presence of technical savvy young engineers driven by career progress and technical expertise, accelerated progress and scale-up of the newly introduced monitoring system (Figure 3). Likewise in Cambodia, strong government ownership and institutionalisation of the MIS benefited from the active engagement of political champions and their official allocation of responsibility for the MIS through a government regulation, combined with dedicated technical staff resources within the MRD.

Reflections on the low-rated factors:

- *Community involvement* has generally had a low impact, probably because all country case studies represented early stages of establishing Monitoring systems. As these are government owned monitoring systems, there needs to be a fully established data use by the government and sufficient resourcing and capacity to respond to low services and community demands. Seeking community review and feedback without government or service providers' ability to respond can undermine the trust and people's willingness to continue supporting and operationalising the monitoring system.
- *Financing:* Again, as the case studies analysed present evidence from early-stage monitoring system development, effective long-term financing of monitoring systems was yet to be fully established. While in all contexts it was recognised as an essential step for the sustainability of the monitoring system, there was a recognition that the value of the monitoring system as an essential support to evidence-based decision-making still needed to be demonstrated in order to attract budget allocation funding.

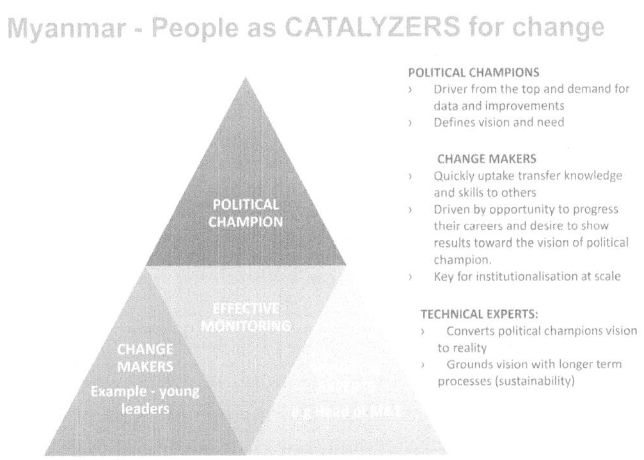

Figure 3 | Identification of three key groups of catalyser for government-led rural water monitoring change in Myanmar.

IMPLICATIONS AND LEARNING

Based on the experiences from the four case studies presented, we suggest the following implications and lessons learned for future efforts aiming to strengthen WASH monitoring systems:

1. *Approach monitoring systems as part of the broader political economy:* accurate, relevant and timely data are critical to effective decision-making and the performance of the WASH sector, however other factors influence and drive decision-making and data use. Decision-making processes are inherently political and vulnerable to political change; power dynamics, interests, and counterproductive incentives can trigger over/under reporting, while biased and selective data use, in the absence of rigorous data validation or standardised indicators, can undermine monitoring processes by eroding confidence and trust in the quality of data and its representativeness. Wider institutional arrangements, such as regulatory settings, decentralisation, and cross-ministerial coordination, can either promote or inhibit data-informed decision-making. Regular political economy analysis and mapping of informal and formal decision-making processes before and during activities to strengthen monitoring systems contribute to the ongoing evolution and long-term sustainability.
2. *Co-designing with data users:* user-centred co-design monitoring systems support meeting the data needs and technical capacity of formal and informal decision-makers, sector stakeholders, and rights holders themselves. Upfront identification of technical and political government champions is key to supporting this while adapting to continuous changes.
3. *Engage with monitoring processes strategically:* based on the analysis of the four case studies, strategic key entry points for agencies/actors aiming to strengthen government-led WASH monitoring processes includes efforts to strengthen:

   Mapping data needs: Mapping pre-intervention/previous data use, mapping data needs at different levels, identifying and aligning with sector priorities and related indicators.

   Indicators: introduction of best practice indicators, clear definition setting, and harmonisation process.

   Data analysis: development of targeted and locally contextualised data analysis generation and sharing at different government levels and sector.

   Technical expertise: provision of technical training and expertise to national and subnational government staff.

   Technology (ICT) demonstration, piloting, and adapting to local context need for data collection, management, and analysis.

   Sector coordination groups: Developing and facilitating sector working group/committee to coordinate WASH monitoring processes.
4. *Data availability and use for planning and budgeting, while critical, will not by itself guarantee evidence-based accountability:* the availability of relevant and timely data access can lead to improved WASH sector coordination processes (through forums, sector working group who use data to increase investment coordination, progress monitoring, etc) and lead to improved 'data use culture'; however, it is not sufficient to achieve a 'data-based accountability culture' which also requires realisation of other institutional reforms in the sector, particularly related to regulatory processes and clear responsibilities.
5. *Strengthen a sector culture of reflection, learning and adaptation:* to ensure an effective shift of 'data use culture' into WASH accountability, activities to strengthen monitoring need to respond and adapt to changes in the dynamic political and institutional contexts. Fostering a culture of continuous learning, reflection and adaptation can help stakeholders respond to the inevitable changes in political drivers that define sector targets (and therefore indicators), institutional settings such as decentralisation processes (therefore responsibilities for decision-making) and key actors (e.g. turnover in political or technical government staff who are acting as catalysers). Engaging civil society rights holder groups early in the monitoring discussions might ensure data culture does not stop evolving once data are available for service providers' use but continue to build momentum towards a culture of data-based accountability and rights holders using monitoring systems to hold duty bearers to account.
6. *Frameworks developed to support planning and targeting of WASH monitoring strengthening activities can be useful to identify weaknesses and entry points for WASH monitoring strengthening:* however, these would need to be complemented with political economy analysis that include analysis of roles and responsibility, drivers for data use, champions, etc. Furthermore, co-designing with government is key to develop effective monitoring processes.

7. *Ensure recurrent financing for monitoring:* financing of monitoring processes by government is still limited, with high dependency on supporting agencies. Sustainability of stronger WASH monitoring systems requires sectoral monitoring data to be integrated in core government public financial management functions. Significant advocacy is needed, based on demonstrated value of the monitoring system, to promote recurrent financial resource allocations by government.

ACKNOWLEDGEMENTS

Funding for WaterAid to support the monitoring systems interventions described in this paper was provided by the Water Supply and Sanitation Development Programme (a World Bank concession loan) and EU/UNICEF-funded KlinpelaKomunitiProjek (PNG), WaterAid UK (Myanmar, Uganda), and the H&M Foundation through the SusWASH Programme (Cambodia, Uganda).

DATA AVAILABILITY STATEMENT

All relevant data are included in the paper or its Supplementary Information.

REFERENCES

Cronk, R., Slaymaker, T. & Bartram, J. 2015 Monitoring drinking water, sanitation and hygiene in non-household settings: priorities for policy and practice. *International Journal of Hygiene and Environmental Health* **218** (8), 694–703. doi:10.1016/j.ijheh.2015.03.003.

Da Silva Wells, C., van Lieshout, R. & Uytewaal, E. 2013 Monitoring for learning and developing capacities in the WASH sector. *Water Policy* **15** (52), 206–225. doi:10.2166/wp.2013.120.

De Albuquerque, C. 2014 *Realising the Human Rights to Water and Sanitation: A Handbook by the UN Special Rapporteur*. United Nations Office of the High Commissioner for Human Rights, Lisbon, Portugal.

Department of National Planning and Monitoring 2015 *PNG National Water, Sanitation and Hygiene (WaSH) Policy 2015–2030*. Government of the Independent State of Papua New Guinea. DNPM.

Department of National Planning and Monitoring 2018 *Medium Term Development Plan III 2018–2022. Development Planning Framework and Strategic Priorities*. Government of the Independent State of Papua New Guinea. DNPM.

Department of National Planning and Monitoring 2020 *Monitoring and Evaluation Framework for Water, Sanitation and Hygiene*. Government of the Independent State of Papua New Guinea. DNPM.

Dickinson, N. 2016 *Organising Framework for Functional National WASH Monitoring and Evaluation Systems*. UNICEF, IRC and Akvo. Available from: https://www.ircwash.org/sites/default/files/20160714_organizing_framework.pdf.

Faguet, J. P. 2014 Decentralization and governance. *World Development* **53**, 2–13. http://dx.doi.org/10.1016/j.worlddev.2013.01.002.

Gadenne, L. & Singhal, M. 2014 Decentralization in developing economies. *Annual Review of Economics* **6**, 581–604. http://dx.doi.org/10.1146/annurev-economics-080213-040833.

Giné Garriga, R., Jiménez Fdez. de Palencia, A. & Pérez Foguet, A. 2015 Improved monitoring framework for local planning in the water, sanitation and hygiene sector: from data to decision-making. *Science of the Total Environment* **526**, 204–214. http://dx.doi.org/10.1016/j.scitotenv.2015.04.078.

Hallsworth, M., Egan, M., Rutter, J. & McCrae, J. 2018 *Behavioural Government: Using Behavioural Science to Improve How Governments Make Decisions*. The Behavioural Insights Team, London.

Hollander, D., Ajroud, B., Thomas, E., Peabody, S., Jordan, E., Javernick-Will, A. & Linden, K. 2020 Monitoring methods for systems-Strengthening activities toward sustainable water and sanitation services in Low-Income settings. *Sustainability* **12**, 7044–7060. doi:10.3390/su12177044.

Kempster, S. 2020 *From Data to Decision, Developing User-Centred Monitoring Programmes for Water, Sanitation and Hygiene Policy Brief*. WaterAid.

Requejo-Castro, D., Giné-Garriga, R., Flores-Baquero, Ó., Martínez, G., Rodríguez, A., Jiménez Fdez. de Palencia, A. & Pérez-Foguet, A. 2017 SIASAR: A country-led indicator framework for monitoring the rural water and sanitation sector in Latin America and the Caribbean. *Water Practice and Technology* **12** (2), 372–385. doi:10.2166/wpt.2017.041.

Schouten, T. & Smits, S. 2015 *From Infrastructure to Services: Trends in Monitoring Sustainable Water, Sanitation and Hygiene Services*. IRC and Practical Action. Available from: https://www.ircwash.org/resources/infrastructure-services-trends-monitoring-sustainable-water-sanitation-and-hygiene.

UNAIDS 2008 *Organizing Framework for a Functional National HIV Monitoring and Evaluation System*. UNAIDS Monitoring & Evaluation Reference Group. Available from: https://www.unaids.org/sites/default/files/sub_landing/files/20080430_JC1769_Organizing_Framework_Functional_v2_en.pdf.

UN General Assembly 2010 *The Human Right to Water and Sanitation: Resolution/Adopted by The General Assembly*. A/RES/64/292r. Available from: https://www.refworld.org/docid/4cc926b02.html.

Valcourt, N., Javernick-Will, A., Walters, J. & Linden, K. 2020 System approaches to water, sanitation, and hygiene: a systematic literature review. *International Journal of Environmental Research and Public Health* **17**, 702–720. Available at: doi:10.3390/ijerph17030702.

Welle, K., Williams, J., Pearce, J. & Befani, B. 2015 *Testing the Waters: A Qualitative Comparative Analysis of the Factors Affecting Success in Rendering Water Services Sustainable Based on ICT Reporting*. Institute of Development Studies. Available from: https://opendocs.ids.ac.uk/opendocs/handle/20.500.12413/7099.

Williams, J., Welle, K. & Pearce, J. 2016 *How can ICT Initiatives be Designed to Improve Rural Water Supply?* IDS and WaterAid, Brighton.

World Bank 2018 *Data for Development. An Evaluation of World Bank Support for Data and Statistical Capacity*. Independent Evaluation Group, Washington, DC. World Bank. Available from: https://documents.worldbank.org/en/publication/documents-reports/documentdetail/705581506620506441/data-for-development-an-evaluation-of-world-bank-support-for-data-and-statistical-capacity.

World Bank Group 2017 *Toward a Universal Measure of What Works on Rural Water Supply: Rural Water Metrics Global Framework*.World Bank, Washington, DC. © World Bank. Available from: https://openknowledge.worldbank.org/handle/10986/27950License: CC BY 3.0 IGO.

First received 29 January 2021; accepted in revised form 11 May 2022. Available online 25 May 2022

doi: 10.2166/h2oj.2022.066

E-learning for WASH systems strengthening: lessons from a capacity-building platform

A. Leal [a,*], A. Saleh[b] and J. Verhoeven[c]

[a] IRC, Morelia, Michoacan, Mexico
[b] IRC, Lozells, Birmingham, UK
[c] IRC, Utrecht, The Netherlands
*Corresponding author. E-mail: ee.analeal@gmail.com, leal@ircwash.org

AL, 0000-0002-0620-8561

ABSTRACT

Research was undertaken to evaluate the performance of the Water, Sanitation, and Hygiene (WASH) Systems Academy, a free e-learning platform. The course, 'WASH systems strengthening: The basics', was assessed in three different learning arrangements: Case 1: MOOC (Massive Open Online Course, self-paced and self-guided), Case 2: MOOC with structured support, and Case 3: MOOC with blended learning. In each of these cases, the extent to which e-learning can contribute to improved understanding and application of WASH systems strengthening was investigated. The research was conducted between October 2020 and July 2021 using a modified Kirkpatrick model supplemented by the Outcome Harvesting methodology to assess which MOOC format was successful in performance and application. The study analysed data from the e-learning platform, user surveys, and interviews with water and sanitation e-learning experts and selected users. A comparison of the cases demonstrated that overall, the three learning arrangements were effective based on the reactions of the users, with completion rates of 21, 65, and 100% for MOOC, MOOC with structured support, and MOOC with blended learning, respectively. The MOOC with a blended learning approach was the most effective in terms of performance. This approach allowed participants to have a better understanding of the content with extra face-to-face support while completing the course.

Key words: blended learning, capacity building, MOOC, online learning, systems strengthening, WASH systems

HIGHLIGHTS

- The research assesses how a Massive Open Online Course (MOOC) could build capacities for the Water, Sanitation, and Hygiene (WASH) systems strengthening by analysing the performance and application of its content in three different learning arrangements (self-paced, structured support, and blended learning).
- In comparison, all cases have high completion rates of 21, 65, and 100%, respectively.
- The MOOC contributes to reach a wider audience but tailor-made approaches to have contextualised content.

GRAPHICAL ABSTRACT

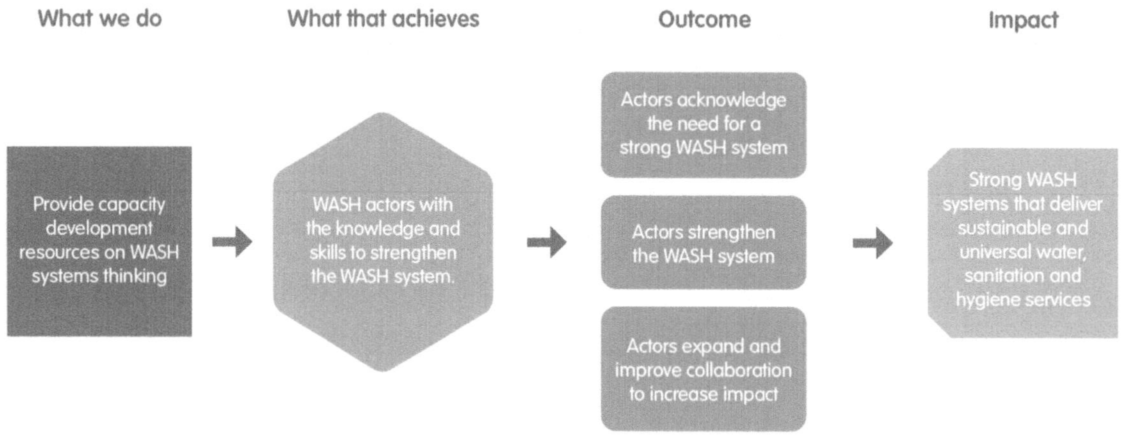

CASES	REACTION & LEARNING	BEHAVIOUR CHANGE
Case 1 **MASSIVE OPEN ONLINE COURSE (MOOC)** 94 countries, 1512 users	Case 1 MASSIVE OPEN ONLINE COURSE (MOOC) Completion rate: 21% User satisfaction score: 4.5/5.0	RELATIONS • Meetings with various key stakeholders PRACTICES • Shift in monitoring and evaluation focus from collecting data at country level, to monitoring advocacy at community level
Case 2 **MOOC WITH STRUCTURED SUPPORT** India, 17 users	Case 2 MOOC WITH STRUCTURED SUPPORT Completion rate: 65% User satisfaction score: 4.5/5.0	ACTIONS AND ACTIVITIES • Modification of monitoring questionnaires at local level • Capacity building to do WASH context assessment with specific tools • More confidence with WASH jargon • Critical thinking in career
Case 3 **MOOC WITH BLENDED LEARNING** Ethiopia, 63 users	Case 3 MOOC WITH BLENDED LEARNING Completion rate: 100% User satisfaction score: 4.4/5.0	

INTRODUCTION

Gaps in human resource capacity in the water and sanitation sector have been highlighted for at least the last three decades (Suter & Lüthi 2021) hindering progress in providing adequate water and sanitation services. A 2014 study by the International Water Association (IWA) highlighted that there can be no sustained progress in achieving Sustainable Development Goal 6 (SDG 6) – ensure the availability and sustainable management of water and sanitation for all – without effective capacity building and training of an adequate number of people (IWA 2014). The United Nations GLAAS report of 2019 (UN-Water 2019) showed that even where most countries report having policies for drinking water (94%), sanitation (94%), and hygiene (79%), less than 14% have sufficient human resources to implement these policies.

Water, sanitation, and hygiene (WASH) service delivery are provided through a series of interactions between social and ecological systems (Da Silva Wells *et al.* 2013). Given the complexity and socio-technical nature of WASH service delivery, systems thinking is a core competency (Gabrielsson *et al.* 2020). The capacity to understand such systems in a holistic manner is fundamental for engaging with the actors involved (Hollander *et al.* 2020; IHE-Delft 2020). Systems thinking helps to understand water, hygiene, and sanitation as a system, however, conventional training usually focuses on specific disciplines rather than multidisciplines, hence a holistic approach is required (Wiek *et al.* 2011). Enrolments in Massive Open Online Courses (MOOCs) globally have recently increased, making these an increasingly popular vehicle for building capacities.[1] As in other sectors,

[1] For example, enrolment at Coursera, an online platform that offers MOOCs, was 640% higher from mid-March to mid-April 2020 than during the same period last year, growing from 1.6 to 10.3 million. This surge corresponds to lockdowns across the world due to the COVID-19 pandemic. See https://www.classcentral.com/report/mooc-providers-response-to-the-pandemic/.

the COVID-19 pandemic has led to increased interest in online training in the water and sanitation sectors. MOOCs have the potential to make a substantial contribution to capacity development in the water and sanitation sectors, reaching larger audiences than other training modalities (Otto *et al.* 2019). They provide an unrestricted educational platform that offers free knowledge, tools, and experiences to learners globally with internet access (Johnson 2021; Siemens 2013).

This paper aims to evaluate the performance and application of a MOOC with the core topic of the Water, Sanitation, and Hygiene (WASH)[2] systems strengthening[3] in three different learning arrangements. It explores performance in terms of user reactions and learning, and actual application (behaviour change) and follow-up actions that might contribute to the strengthening of water and sanitation systems. The three cases are:

- Case 1: MOOC as self-paced and self-guided,
- Case 2: MOOC with structured support (India), and
- Case 3: MOOC with blended learning (Ethiopia).

The MOOC used was 'WASH systems strengthening: The basics'[4], which forms the basis of all three cases. This MOOC has been accessible for free since September 2019 via the e-learning platform the WASH Systems Academy[5], developed by IRC WASH (IRC 2021). The study was conducted between October 2020 and July 2021.

A modified Kirkpatrick model (Kirkpatrick 1994) was used to evaluate the performance of the MOOC and explore the application of its content at three levels, namely *reaction*, *learning*, and *behaviour change*. The research is based on quantitative and qualitative data analysis using data extracted from the e-learning platform, surveys among participants, individual interviews, and focus group discussions with practitioners in the field who have used the platform and to some extent applied its content. Additionally, an Outcome Harvesting methodology (Wilson-Grau 2018) was used to analyse the behaviour changes.

WASH Systems Academy

IRC designed the MOOC 'WASH systems strengthening: The basics' as the basis for its free e-learning platform the WASH Systems Academy. The Academy's ambition is to target a large audience by offering high-quality and strategic content for promoting sustainable, universal water and sanitation, and hygiene service delivery through continuous systems strengthening towards SDG 6. The Academy was launched in September 2019 with the course 'WASH systems strengthening: The basics'. After 1 year, it reached 2,000+ users from more than 40 different countries with two additional MOOCs on offer. By July 2021, the Academy had 3,100+ users from 110 countries with seven MOOCs available.

The parameters of success of a MOOC to potentially contribute to systems strengthening are reflected in the ability of MOOCs to attract a large audience from diverse backgrounds who share a joint interest to gain knowledge and skills on water and sanitation service delivery. Other parameters of success are positive user evaluations and high course completion rates. Acknowledging that such parameters could be a first step in the long process related to recognising the complexity and intertwined nature of the different components (or building blocks) of the WASH systems.

The WASH Systems Academy's Theory of Change (ToC) 2019–2030 (Figure 1) describes the vision and actions for contributing to achieving SDG 6 and highlights how WASH education interacts with and potentially contributes to systems strengthening. The ToC maps out the intermediate outcomes considered most critical for the sector to deliver universal and sustainable WASH services and the WASH Systems Academy's proposed contribution to those outcomes.

---

[2] Water, Sanitation, and Hygiene (WASH) Systems refers to all the social, technical, institutional, environmental, and financial factors, actors, motivations, and interactions that influence WASH service delivery in a given context (Huston & Moriarty 2018). In other words, it is a network of diverse people and organisations working towards delivering water, sanitation, and hygiene services (IRC 2021).

[3] Systems strengthening is an approach that focuses on recognising the complexity and fundamentally inter-linked nature of the different aspects of real world, in this case of WASH systems. Instead of trying to ignore complexity – for example, by focusing on a specific, time-limited project – a systems approach engages with it in the belief that doing so will lead to solutions that are more meaningful and more sustainable (Huston & Moriarty 2018).

[4] See WASH systems strengthening: The basics: IRC (www.ircwash.org).

[5] Available at www.washsystemsacademy.org.

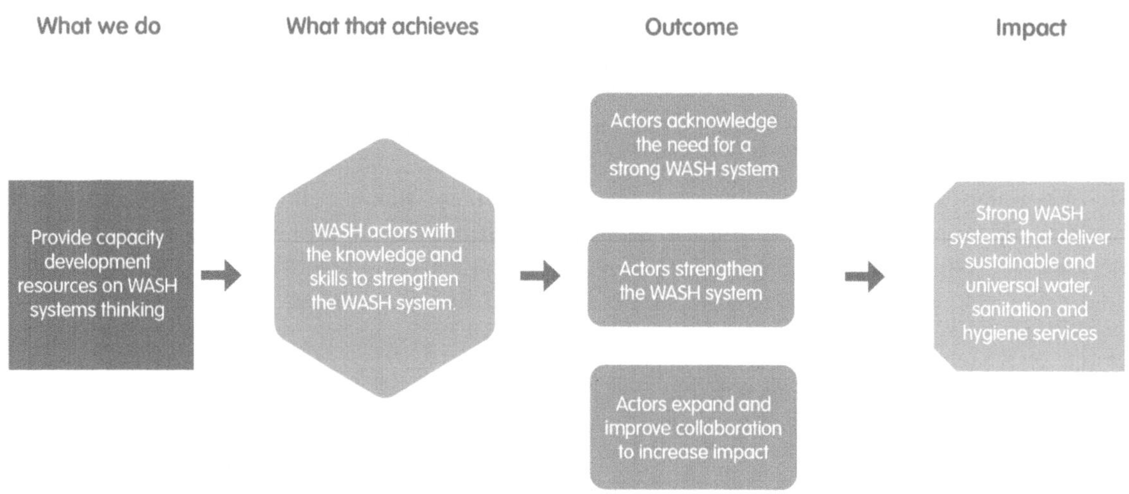

Figure 1 | ToC of the WASH Systems Academy.

The MOOC: 'WASH systems strengthening: The basics'

This MOOC course covers the basic concepts of WASH systems strengthening, with an estimated average duration of 16 h. The MOOC aims to explain the systems strengthening approach, how and why it was developed, and how to start applying it. Its content is divided into nine sessions that need to be completed in consecutive order with three multiple choice tests:

- Session 1: Introducing the WASH Systems Academy
- Session 2: From broken pumps to lasting systems
- Session 3: What is WASH systems strengthening?
- **Multiple choice test on sessions 1–3**
- Session 4: WASH is a service
- Session 5: Sanitation systems strengthening
- Session 6: Hygiene promotion for WASH systems strengthening
- **Multiple choice test on sessions 4–6**
- Session 7: WASH systems strengthening away from home
- Session 8: Leaving no one behind
- Session 9: The roadmap to sustainable WASH services
- **Multiple choice test on sessions 7–9**

Each session contains a combination of activities including podcasts with WASH system experts, animations, forum discussions, quizzes, and exercises to create one's own materials to promote WASH systems strengthening.

RESEARCH DESIGN AND METHODS

The research compared three different learning arrangements within particular contexts and selected through an emergent sampling, particularly suitable when the research being conducted is exploratory in nature (Cohen & Crabtree 2006). The three cases with their key features are shown in Figure 2.

A modified Kirkpatrick model (Kirkpatrick 1994) was used to evaluate the performance of the MOOC and explore the application of its content (see Figure 3). The Kirkpatrick model consists of four levels: *reaction, learning, behaviour change*, and *results* (Kirkpatrick 1994). To categorise the findings, we have defined the performance as *reaction* and *learning* (levels 1 and 2) and the application as *behaviour change* (level 3). Part of the performance is also a comparison of costs per user for the three learning arrangements in terms of capital expenditure (CapEx) and Operational and minor maintenance expenditure (OpEx).

This study did not consider the fourth level *results*, as this requires organisational change (i.e., changes of more than one person), and users of Case 1 (i.e., the MOOC) participate as individuals. Hence, this level does not apply. Achieving results in level 4 of systems change is also a long-term process of multiple years and the

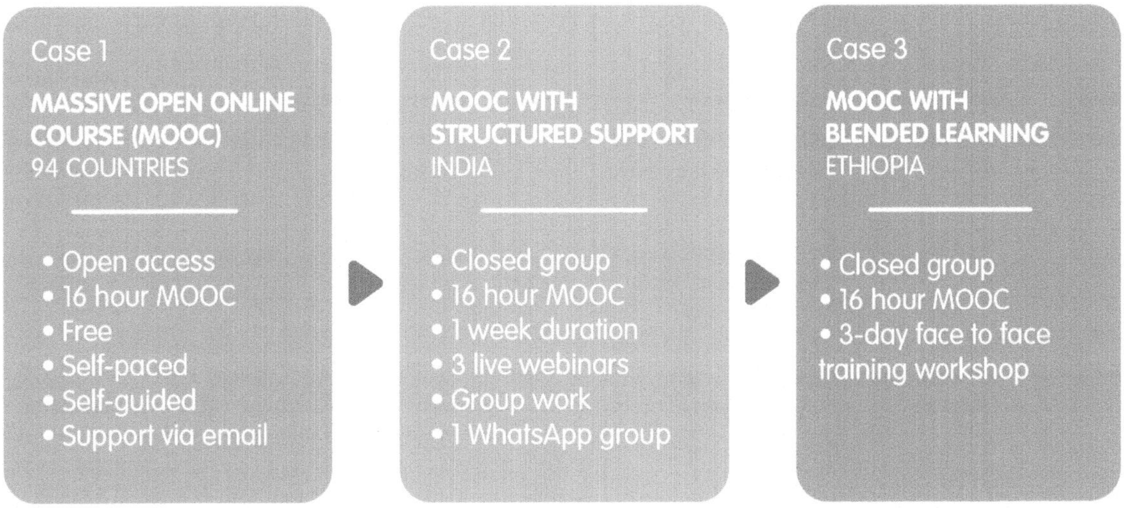

Figure 2 | Key features of the three cases.

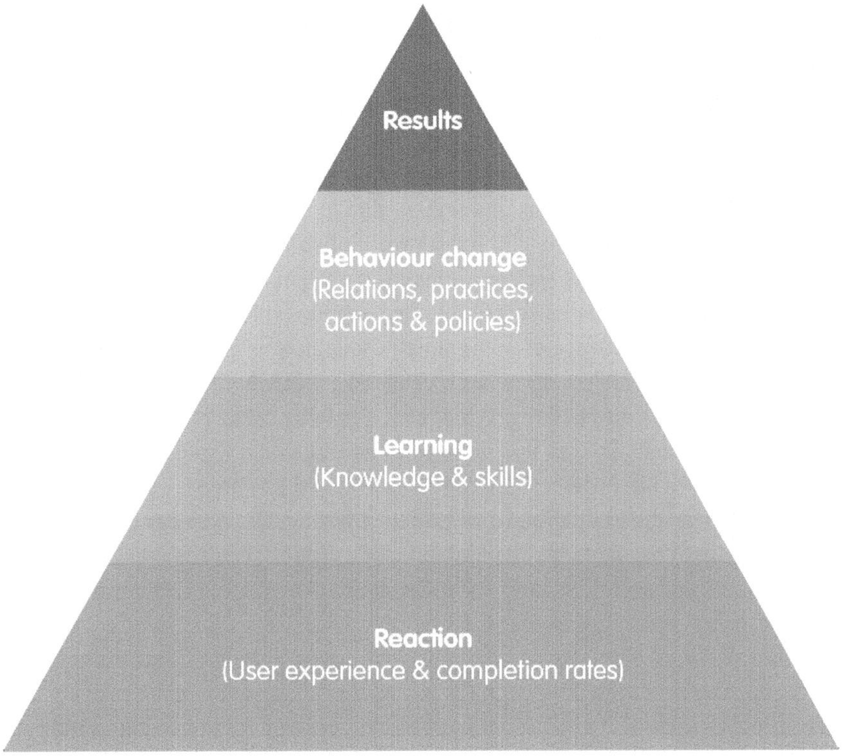

Figure 3 | Kirkpatrick model applied to the WASH Systems Academy.

scope of the study is short-term (October 2020 and July 2021). It is, therefore, too early to draw conclusions regarding changes at the *results* level.

Level 1 '*Reaction*' was assessed in terms of completion rates and user experience, and level 2 '*Learning*' was evaluated in terms of knowledge and skills. Level 1 '*Reaction*', level 2 '*Learning*', and partly level 3 '*behaviour change*' were assessed with quantitative and qualitative data analysis on user engagement on the E-learning Platform and through anonymous user surveys. The user survey includes four open-ended questions and seven statements on content, navigation, and user experience that users rate with a Likert Scale of 1–5, with 1 being 'strongly disagree' and a score of 5 'strongly agree'. Level 3 '*behaviour change*' on applying elements of WASH systems strengthening was explored through in-depth interviews with selected users, through open-ended questions in the user survey and open response sections of the MOOC, and in Cases 2 and 3 through group discussions. Reported changes were categorised using the Outcome Harvesting methodology (Wilson-

Table 1 | Assessment of performance and application with the modified Kirkpatrick model and data collection methods

Indicator	Level	What it represents	How it is measured	Data collection method
Performance[a]	1 – Reaction	Performance in terms of user response to the course	Completion rates and user experience	• E-learning platform • User surveys
	2 – Learning	Performance in terms of the acquisition of new knowledge or skills by a user	Extent of new knowledge and skills gained	• E-learning platform • User surveys
Application	3 – Behaviour change	Application in terms of modifications in observable events by user	Changes in actions, activities, relations, and practices (i.e., outcome harvesting)	• E-learning platform • User surveys • Interviews • Group discussions

[a]Performance was supplemented by the LCCA to understand the investment required to developed each case.

Grau 2018) in terms of changes in relations, practices, actions, and activities. For an overview of data collection methods, see Table 1.

Outcome Harvesting is a method that enables monitoring and evaluation of specific initiatives. It has been widely used by networks, non-governmental organisations, and research centres among others (Wilson-Grau & Britt 2012). The method contributes to the collection of evidence on what has been achieved (the outcome) whether good or bad, planned, or unplanned, and works in a reverse direction to verify and explore how the initiative contributed to that specific change by using specific questions. Unlike common evaluation methods that measure progress against pre-set outcomes, this method highlights how a given initiative has contributed to outcomes, in the sense of behavioural changes in relationships, actions, activities, practices, or policies. That is why it was chosen as it can be used to collect evidence of achieved effects and systems change, (Wilson-Grau 2018; Bori & Huston 2020). The methodology facilitates rigorous investigation into whether and how the MOOC contributed to behavioural changes in relationships, actions, activities, or practices. No policy changes were expected to be found as they require organisational changes that go beyond an individual's sphere of influence.

Although the Kirkpatrick model has been criticised (Bates 2004), it has been used for decades and remains one of the most widely used tools for the evaluation of workplace training due to its systematic approach (Paula *et al.* 2016). The modified Kirkpatrick model helps to gain an insight from a user's perspective. It also gives ideas for improvements in user knowledge and skills after completing the course and on the user's application of knowledge which makes clear if the course was successful or not.

To examine the costs per user associated with the three cases, linked to Performance, the life-cycle cost approach (LCCA) was applied (see Table 2). An LCCA is a methodology to comprehensively identify and analyse the full costs of delivering WASH services, including infrastructure and both direct and indirect support. Of the six life-cycle cost categories, only CapEx and OpEx are applicable to the three cases. The first Capital maintenance expenditure (CapManEx) for the studied MOOC is planned for 2022 as it is expected that every MOOC needs a major overhaul or a replacement every 3–4 years.

Data collection

The study was conducted between October 2020 and July 2021 with:

- Case 1: MOOC was open to all free of charge between September 2019 and October 2020.
- Case 2: MOOC with structured support to selected users from India in February 2021.
- Case 3: MOOC with blended learning for selected users from Ethiopia in three pilots between April and June 2021.

Case 1: MOOC

For Level 1, Reactions, two user surveys were sent out to participants (July 2020, $n = 201$, and July 2021 $n = 423$) to solicit feedback on the overall performance and user experience. For Level 2, Learning, each user answered a

Table 2 | The six life-cycle cost categories applied to the three studied cases

Life-cycle cost category	Application to the WASH Systems Academy	Application to the three studied cases
CapEx	The initial investments in the construction of the E-learning platform and the development of each course.	The costs of customisation of the training materials[a] for each case are part of this category.
OpEx	The recurrent costs of operating the E-learning platform, including continuous updating of the courses and the platform. It also includes the costs of hosting the E-learning platform on the IRC (cloud-based) server and administration of the platform and all courses. Answering questions from users via one centred e-mail address training@ircwash.org is also part of this category.	All costs to run the training are part of this category.
CapManEx	This includes expenditure on major maintenance, repairs, and replacement of (parts of) the E-learning platform and each course.	Not Applicable (NA).
Expenditure on direct support (ExpDS)	The costs incurred by IRC in planning, coordinating, and monitoring of the E-learning platform.	NA
Expenditure on indirect support (ExpIDS)	The costs of IRC on business development, yearly planning, and coordination of the WASH Systems Academy.	NA
Cost of capital (CoC)	The cost of accessing finance as interest rates on loans is not applicable to the WASH Systems Academy.	NA

[a]The training materials include the MOOC, PowerPoint Presentations used in the webinars and training workshop, and group work.

feedback survey after completing the course. For Level 3, Behaviour Change, exploratory interviews with users were held. The interviewees were randomly selected from a user survey sent out to users who had taken or completed the MOOC. Individuals who mentioned they had or were planning to apply the knowledge in their work were identified and those who gave their e-mail addresses were contacted.[6] With those who responded, interviews were conducted to find out more about how the Academy has helped them strengthen WASH systems. Additionally, some users who could not be part of the interviews self-reported, via e-mail, on the behavioural changes that occurred after taking the course. This was incorporated into the data analysis by looking at the topic and actions of the behaviour changes.

Case 2: MOOC with structured support (India)

For Levels 1 and 2 'Reactions and Learning', participants took part in a survey after finishing the training. Furthermore, facilitator feedback was collected through interviews, and participants' grades from the online assessments (three tests) taken during the course.

Case 3: MOOC with blended learning (Ethiopia)

For Levels 1 and 2 'Reactions and Learning', participants took part in a survey after finishing the training. Furthermore, informal interviews with the governmental agencies that participated were held. For Level 3, Behaviour Change, focus group discussions with participants and key governmental staff were held.

With respect to the limitations of this study, we acknowledge a potential bias in our user surveys since the options offered were positively oriented. The responses were simplified to encourage participants to answer them. However, a Likert scale was used, so that participants could express the extent to which they agree or disagree with the statements while also giving space for more feedback.

Moreover, a self-completion survey itself can be considered as biased as users were willing to take time out to report back, and you may assume that they only do that if they enjoyed the course. Participants with opposing views are not likely to take part, so no negative feedback could be gathered this way.

Another limitation is that the user surveys were answered by a very specific audience, 65% of the users who had completed the survey in June 2020 were from NGOs. This type of audience may have already been familiar with

[6] Refer to 'Text 1' in the Supplementary Material for copy of e-mail sent to interviewees.

the content. To create long-lasting systems change, stakeholders with different backgrounds need to take the courses and not just NGOs. Actors such as governments (national and subnational), private sector, donors, civil society organisations (CSOs), and community-based organisations (CBOs) are all key stakeholders in systems strengthening.

RESULTS

This section presents the findings on the performance and application for all three cases – Case 1: MOOC, Case 2: MOOC with structured support, and Case 3: MOOC with blended learning. Together the three cases reached a global audience of 1,594 users from 47 countries as shown in Figure 4. The largest representation of users was in Africa followed by Asia, Europe, and North America.

As Table 3 shows, the audience reached by Case 1: MOOC is different from Cases 2 and 3. Users of the standalone MOOC have relatively fewer years of work experience related to WASH and are predominantly working for an NGO. In Cases 2 and 3, users are predominantly male and have more than 5 years of work experience

Figure 4 | Country representation of the three WASH Systems Academy cases (as of October 2021).User representation of Case 1 MOOC = mint green circles, Case 2 MOOC with structured support (India) = light blue, and Case 3: MOOC with blended learning (Ethiopia) = dark blue. The size of the mint green circles represents the size of the number of users per country in comparison. Please refer to the online version of this paper to see this figure in colour: http://dx.doi.org/10.2166/h2oj.2022.066.

Table 3 | User data for the three WASH Systems Academy cases

			Years of work experience related to WASH in %			Gender		
	Number of users	Number of countries reached	<5	5–10	>10	Male	Female	Working for NGO in %
Case 1[a]: MOOC	1,512	47	47	23	30			65
Case 2: MOOC with structured support	17	1 (India)	25	31	44	71	29	88
Case 3: MOOC with blended learning	63	1 (Ethiopia)	8	22	70	92	8	0
Total	1,594	47	100	100	100	100	100	

[a]For Case 1, data is used for the period September 2019 to October 2020. Data for gender for Case 1 is only systematically collected from June 2021.

related to WASH. For Case 1, data on gender was only systematically collected from June 2021 and not for the period of this study (i.e., September 2019 to October 2020). However, the user data from June 2021 to September 2021 for Case 1: MOOC also shows that 94% of the users are male ($N = 308$).

Table 4 shows that the performance of all three cases in terms of user *learning* and *reaction* is comparably positive. In all three cases, users strongly recommend the course, scoring the statement with weighted average scores of 4.6, 4.7, and 4.6 (for MOOC, MOOC with structured support, and MOOC with blended learning, respectively) on a scale of 1–5 with 1 being 'strongly disagree' and a score of 5 'strongly agree'.

Case 1: self-paced MOOC

In total 1,512 users registered for 'WASH systems strengthening: The basics' between September 2019 and October 2020. Figure 5 indicates the 47 countries where users were located who complete the user survey sent out in June 2020 ($N = 202$). The two regions with the highest representation were Africa (48%) and Asia (22%), followed by Europe (16%) and North America (12%).

Table 4 | Performance (simple average scores) in terms of user *learning* and *reaction* for the three WASH Systems Academy cases

Level	Indicator[a]	Case 1: MOOC	Case 2: MOOC with structured support (India)	Case 3: MOOC with blended learning (Ethiopia)
Learning	I learned something new.	4.6	4.7	4.5
Reaction	Course completion rate.	21%	65%	100%
	I would recommend this course.	4.7	4.9	4.5
	I found the content interesting.	4.6	4.4	4.6
	The content motivated me to continue with the course.	4.5	4.5	4.6
	The content was easy to understand.	4.4	4.4	4.2
	The platform is easy to use.	4.5	4.8	4.3
	I did not have any technical problems.	4.2	4.1	3.9
Reaction and learning	Weighted average course user satisfaction score	**4.5**	**4.5**	**4.4**
Number of users feedback surveys received (*N*)		470	10	52
LCCA – cost per user range		€12–24	€176–470	€220–480

[a]Scale of 1–5 with 1 being 'strongly disagree' and a score of 5 'strongly agree'. Scores are weighted average scores.

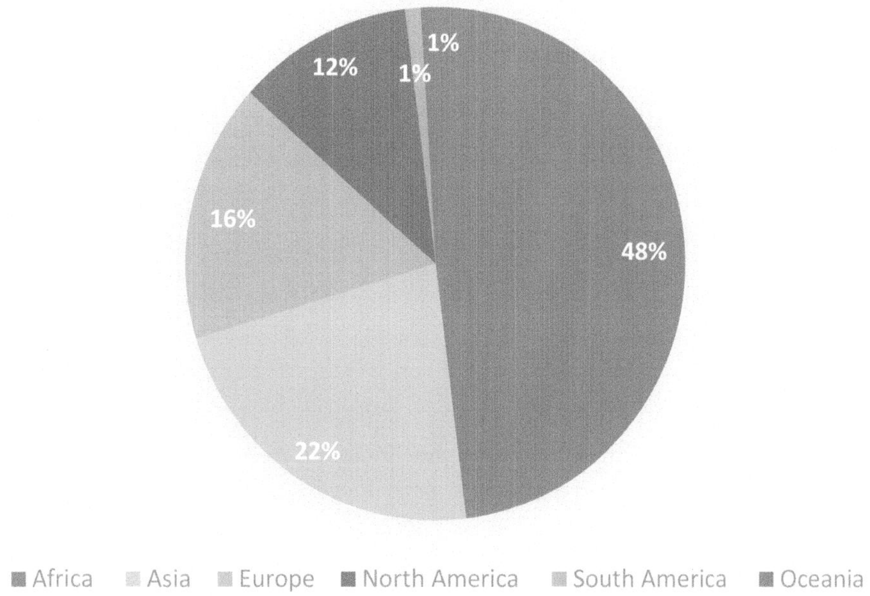

Figure 5 | Location of the user of the MOOC 'WASH systems strengthening: The basics' in percentage (%).

The users of the MOOC who had completed the survey in June 2020 ($N = 203$) were predominantly from NGOs (i.e., 65%) and government (national) at just over 7%, closely followed by research and education (6%), consultants (6%), private sector (5%), and government (subnational) at only 3%. At the bottom are donors (2%) and CBOs or CSOs (1%).

The costs associated with Case 1: Self-paced MOOC range between 1,500 and 3,000 Euro per month (Table 5). In Case 1, no CapEx is incurred as the same MOOC is continuously made available for free between September 2019 and October 2020. The operational expenditure (OpEx), given per month, mainly relates to time spent by IRC to answer questions from users via e-mail on the content of the MOOC or technical questions on the platform and costs to host the platform (and all MOOCs) on the IRC server. When new materials such as guidelines, tools, etc. became available, this was added to the MOOC, and time spent by IRC to update the MOOC is included in the OpEx.

The cost range is given per month as the MOOC has continuously run since September 2019.

Application: behaviour change

In the user survey sent out in June 2020, 59% indicated ($N = 195$) that they found opportunities to apply content from the MOOC in their work on elements of WASH systems of learning and adaptation, planning and budgeting, monitoring, and water resource management. In the open-ended section of the survey, interviewees described ways in which the topics of the course had been applied in their work, below are two examples.

Interviewees mentioned examples (Box 1) of how they have applied knowledge and in particular how they have changed the way they work, also with others. As well as their priorities in different elements of their work, for example, the idea of working and implementing change from the local level up to the national level. Most changes occurred in actions and activities (see Figure 6).

Performance: reaction and learning

Performance of the MOOC is measured by the *'reaction'* of users in terms of completion rates and user experience. The completion rate of the MOOC between September 2019 and October 2020 was 21% ($N = 341$ users).

In terms of user experience, the MOOC scored an average of 4.5 (July 2021, $n = 423$) on a scale of 1–5 with 1 being 'strongly disagree' and 5 'strongly agree'. Users learned something new (Box 2) and would recommend the course, and the average scores were 4.6 and 4.7 (see Table 4).

Case 2: MOOC with structured support

In February 2021, the IRC India programme was approached by Tata Trusts to develop a structured and supported 1-week online learning programme for 17 staff members of Tata Trusts' implementing partner organisations (17–24 February). Due to the COVID-19 pandemic, using traditional training methods (classroom) was not possible in India. Therefore, a structured and supported online learning programme was chosen. The approach included:

Table 5 | Costs associated with Case 1: self-paced MOOC per month in Euro (€)

Life-cycle cost category	Costs Case 1: self-paced MOOC per month in Euro (€)
CapEx	No costs are incurred in this category as no customisation of the MOOC is in need.
OpEx	€1,500–3,000
Total costs	**€1,500–3,000**
Costs per user (range; group size averages 126)	**€12–24**

Box 1 | Quote from interviews that highlight the outcomes after taking the course (Application).

Interviewee 1: Data manager (Environmental health practitioner), Liberia.

'Interviewee 1 mentioned how they can now help *improve the reporting systems* to detect communities with a high risk of returning to open defecation. They have shifted from collecting quality data at the district level and are starting to *focus more on monitoring and advocacy at the community level*. Their future plan is to work from the bottom up to enforce change.'

Figure 6 | Changes detected in relations, practices and actions, and activities via outcome harvesting regarding the application of the MOOC.

Box 2 | Quote from interviews that highlight outcomes after taking the course (Performance).

Interviewee 2: Budget Specialist Officer, Uganda.

'Interviewee 2 mentions the MOOC helped them understand *how governments use financial flows and budgets for WASH*. Also, it allowed them to *better understand the WASH jargon*.'

- A self-paced online MOOC (i.e., 'WASH systems strengthening: The basics')
- Three live webinars of 2 h each
- Group work in each webinar

This was complemented by a WhatsApp group for participants and course facilitators, where participants could ask questions and connect outside of the live webinars.

In terms of the user profile, most participants were male implementing partners of Tata Trusts with more than 5 years of work experience related to WASH. One participant worked for subnational government and one for a consultancy company.

The costs associated with Case 2: MOOC with structured support range between 3,000 and 8,000 Euros (see Table 6). For Case 2, CapEx to customise the MOOC, the webinars, and group work for the target audience is minimal. CapEx includes mainly providing the MOOC that is branded with the logo of the client to a closed group of selected users and adding examples related to the target audience to the webinars. The OpEx relates to running the course, facilitating the webinars, and supporting users via e-mail and WhatsApp. Monitoring the behaviour of users on the MOOC and reporting to the client are also included in OpEx.

Performance: reaction and learning

Performance in terms of completion rates was 65% (11 out of 17 users). Performance in terms of *reaction and learning* scored well with users rating the course with an average score of 4.5 ($n = 10$) on a scale of 1–5 (see Table 4).

In open-ended response sections of the user survey and in group discussions of the webinars, users described the course as 'interesting', 'relevant for their context', and 'very informative'. They thought that the webinars,

Table 6 | Costs associated with MOOC with structured support in Euro (€)

Life-cycle cost category	Costs Case 2: MOOC with structured support in Euro (€)
CapEx	€500–2,000
OpEx	€2,500–6,000
Total costs	**€3,000–8,000**
Costs per participant (with a group size of 17 participants)	**€176–470**

group work, and WhatsApp group were useful additions to the course. Also, participants mentioned that without the facilitated face-to-face group sessions, they would never have started an online course or completed one.

Some participants offered suggestions for improving the format of the MOOC, webinars, and coursework. These suggestions included: conducting the training in local languages and translating the MOOC into local languages; increasing one-to-one interaction between participants and facilitators; providing more books, articles, and other supplementary reference materials; and accommodating more diverse settings and contexts of participants.

The participants' learning was evaluated by assessing the three multiple choice tests in the MOOC. The course grades showed promising results. The average course grade for participants who completed all five of the assessments was 94%, indicating a good understanding of the course content. The group work and discussions that were part of the live webinars also reflected this.

Application: behaviour changes

Although the performance is looking promising, it is still too early to see behaviour changes. The application of new knowledge and skills by participants will be further investigated in a new study from 2022.

Case 3: MOOC with blended learning (Ethiopia)

Between April and June 2021, the MOOC 'WASH systems strengthening: The basics' was used for a blended learning approach in three pilots in Ethiopia on request of the Ethiopian Water Technology Institute (EWTI).[7] The blended learning pilots combined the MOOC with group work and discussions held in an in-person training setting. The pilots took place in Bishoftu town and Bahir Dar, Ethiopia:

- Pilot 1 (26 April to 8 May): Training of Trainers (ToT) of 26 EWTI trainers.
- Pilot 2 (7–9 June): Regional training in the Amhara region of 18 participants.
- Pilot 3 (17–19 June): Regional training in the Oromia region of 19 participants.

In terms of user profile, Case 3 was aimed at targeting government staff (see Table 7) as governments are ultimately responsible for service provision in their country. Pilot 1 reached government staff at the national level, while pilots 2 and 3 predominantly reached subnational governments at regional and district levels and utilities.

The costs associated with Case 3: MOOC with blended learning range between 5,500 and 12,000 Euros (see Table 8). In Case 3, CapEx consists of time to customise the MOOC and the 3-day training workshop for the

Table 7 | User profile of Case 3 MOOC with blended learning in Ethiopia

	Years of work experience			Gender		Source of employment		
Pilot	**<5**	**5–10**	**>10**	**Male**	**Female**	**Government (national)**	**Government (subnational)**	**Utility**
1. ToT	5	7	14	23	3	26		
2. Amhara		3	15	18			14	4
3. Oromia		4	15	17	2		17	2
Total	**5**	**14**	**44**	**58**	**5**	**26**	**31**	**6**

Table 8 | Costs associated with MOOC with blended in Euro (€)

Life-cycle cost category	**Costs Case 2: MOOC with blended learning in Euro (€)**
CapEx	€1,500–3,000
OpEx	€4,000–9,000
Total costs	**€5,500–12,000**
Costs per participant (with a group size of 25 participants)	**€220–480**

[7] For more information see https://www.eweti-ethiopia.com/.

target audience. The CapEx includes providing the MOOC that is branded with the logo of the client to a closed group of selected users. The OpEx mainly relates to running the workshop. Monitoring the behaviour of users on the MOOC and reporting to the client are also included in the OpEx.

Performance: reaction and learning

The performance of the MOOC with a blended learning approach proved successful in terms of completion rates. All 65 people, 28 at the national and 37 at the regional level (i.e., Amhara and Oromia regions), completed the course. Performance in terms of user satisfaction and learning indicated that users were positive about the approach, rating it with an average score of 4.4 (see Table 4).

In the group discussions part of the course, users mentioned that the concept of WASH systems strengthening and partaking in a MOOC was completely new to them. Before the COVID pandemic, EWTI provided only face-to-face courses and seminars. Users indicated in the discussions not to have had much interest in MOOCs because the options offered by EWTI were more appealing. The issue of the English language was another reason why some users did not use MOOCs. In the evaluation group discussion and the user survey, they indicated that they enjoyed the online e-learning platform and quickly felt comfortable using it (see Table 2). Participants wrote in the open-ended questions in the survey that the course made them feel 'happy and confident'.

In terms of learning, through the group discussions and reflections, participants showed a good understanding of the key concepts underlying WASH systems thinking, i.e., how all actors and factors contribute to systems strengthening. This was reflected in the three multiple choice tests in the MOOC which had a total average score of 96% correct ($n = 40$).

Application: behaviour change

It is too early to assess signs of behaviour change on WASH systems strengthening in Case 3. However, in the preliminary results, it is possible to identify some changes in terms of actions and activities. For instance, based on the initial pilot in April, EWTI decided to (1) include 'WASH systems strengthening: The basics' in its training package as a mandatory course for all its trainees and (2) cascade and pilot the course in the Amhara and Oromia regions in June. Based on feedback received in a group discussion with participants and key EWTI staff after the initial pilot, IRC and EWTI identified necessary adaptations for the course to be cascaded to the regional level.

The initial pilot ToT resulted in a new course titled *WASH Systems Strengthening for Ethiopia*. The adaptation focused on relating all the key concepts, such as the SDGs, service delivery models, WASH service levels, and the right to water and sanitation, to the context of Ethiopia. For example, after discussing the right to water and sanitation, information was added on how these rights are embedded in the Ethiopian constitution with a link to the constitution as part of the course resources. Key sector documents were also added to the course resources such as Ethiopia's 2017 Voluntary National Review of the SDGs and Ethiopia's Growth and Transformation Plan II (GTP II). The course resources were adapted to the local context, including images, to improve further uptake. Terms and definitions for Ethiopia were added to the course glossary on service authorities (i.e., woreda water office) and service providers (i.e., WASH Committees or WASHCOs, rural water board, rural utility, and urban utility).

The new course 'WASH Systems Strengthening for Ethiopia' was piloted in two regions (i.e., Amhara and Oromia regions) in June. In the group discussion part of the course, participants indicated that they wanted to provide the training to their district stakeholders. They felt confident that they could take this on themselves. However, for this to be successful, the MOOC would first need to be translated into the local language(s). The face-to-face discussions in the pilots were conducted in three local languages and the MOOC in English.

Other first indications of application were found in the open-ended part of the user survey (June 2021, $n = 51$) where users indicated to be 'confident and striving to teach or give trainings about WASH systems to appropriate audiences throughout the country'.

OBSERVATIONS AND CONCLUSION

MOOCs are often criticised for their high incompletion rates.[8] Large MOOC providers generally have completion rates of less than 10% (Diver & Martinez 2015). Davis *et al.* (2017) also report low course completion

[8] Defined as the percentage of enrolled students who completed the course (Jordan 2015).

rates ranging between 5 and 10%, while Jordan (2015) informs a median value of 12.6%. In all our three cases, completion rates were above average ranging from 21% in Case 1 MOOC to 65% in Case 2 MOOC with structured support and 100% in Case 3 MOOC with blended learning.

However, Case 2, MOOC with structured support, also showed the difficulty of sustaining participation among working professionals who have limited time for additional activities outside of their work. This is compared to Case 1 where working professionals have the flexibility of 1-year free access to the WASH Systems Academy to complete the MOOC. It took users of the stand-alone MOOC (i.e., Case 1) on average 40 days to complete the MOOC spread over 1–8 weeks. This falls in the range provided in industry reports that find that MOOCs range in length from 1 to 16 weeks, although there may be significant variations (Bowden 2020).

In general, MOOCs tend to dive into topics that only specialists in the field are familiar with and are not effective in promoting hands-on skills (Hall & Villareal 2015; Cundell & Sheepy 2018). To overcome these common obstacles faced by MOOCs, next to high incompletion rates, blended instruction is reported to be more effective than purely online learning or purely face-to-face (Bornstein 2012). Blended learning can result in high levels of student achievement and can be more effective in creating sustainable behaviour changes than just face-to-face learning (Saritepeci *et al.* 2015).

In the case of blended learning in all three pilots in Ethiopia, users felt that this approach created more active user engagement than a traditional face-to-face training. They found the self-study through the online course to be motivating because it visualised their progress. The self-study and activities offered on the MOOC – such as exercises, quizzes, and forum discussions – enabled participants to self-assess their understanding of the content. The contextualisation of the MOOC through group discussions and face-to-face interaction with an instructor, which was part of Case 3, created a learning environment where users could better relate to and understand the content. However, to further scale up the MOOC with blended learning at the subnational level, the MOOC needs to be translated into local languages to utilise the full potential of this approach.

The case of the MOOC with blended learning in Ethiopia was also successful at reaching government actors at national and subnational levels (i.e., regional and district) and to some extent service providers (i.e., utility). Users of the MOOC and the MOOC with structured support (i.e., Cases 1 and 2) were predominantly working at NGOs. The stand-alone MOOC (i.e., Case 1) attracted an audience with relatively few years of work experience related to WASH than the structured support and blended learning (i.e., Cases 1 and 2). Together the three approaches were able to reach a varied audience in terms of work experience related to WASH and in terms of stakeholders. However, the audience in all cases was predominantly male. More investigation is planned to identify ways to attract more female trainees.

The parameters of success that potentially contribute to systems strengthening are not just shown in the large audience that was attracted, the positive user evaluations and high completion rates (i.e., performance of the MOOC in terms of user reaction and learning). First indications of intermediate outcomes in application regarding modifications in observable events by users in changes in actions, activities, relations, and practices have also been found (i.e., behaviour change). Users reported opportunities to apply content from the MOOC in their work on elements of WASH systems of learning and adaptation, planning and budgeting, monitoring, and water resource management.

In this regard, the MOOC showcases to some extent that the expected intermediate outcomes from the ToC (Figure 2) are feasible, hence, moving a step closer to reaching impact. At the same time, the MOOC helps as a recognition that capacity development could be a first step to realising behavioural changes and results that feed into strengthening the WASH systems.

CONCLUSION

The experience of the WASH Systems Academy shows that MOOCs can be a versatile and effective capacity development tool. The performance of all three cases was successful in terms of completion rates and user satisfaction. Users in all three cases also strongly recommend the course and have found opportunities to apply content from the MOOC in their work.

From this short assessment, we can conclude that the first arrangement (Case 1: MOOC) is best suited for reaching a large number of people via an e-learning platform and has the lowest cost of all three cases. The second set-up (Case 2: MOOC with structured support) is good for encouraging the completion rate by providing live support and contextualisation of the content in webinars and via WhatsApp. However, this was at a much higher cost than Case 1. The third arrangement (Case 3: MOOC with blended learning) is successful regarding completion rates, attracting a different audience of national and subnational government actors and service providers, and creates more active user engagement but with significant higher costs than Cases 1 and 2.

ACKNOWLEDGEMENTS

The authors would like to acknowledge the following people for their contributions to this paper; Dechan Dalrymple for designing all images, Gezahegn Lemecha Boru and Zewdu Assefa for the case of Ethiopia, Ruchika Shiva and Shiny Saha for the case of India, Samantha Wind for the section on methods and the case of India, Angela Huston, John Butterworth, and Camilo Benitez for their support in reviewing this paper, and Tettje van Daalen for editing this paper. The WASH Systems Academy and the MOOC 'WASH Systems Strengthening: The basics' discussed in this paper were developed with funding from the Conard N. Hilton Foundation and the Ministry of Foreign Affairs of the Netherlands (DGIS).

DATA AVAILABILITY STATEMENT

All relevant data are included in the paper or its Supplementary Information.

REFERENCES

Bates, R. 2004 A critical analysis of evaluation practice: the Kirkpatrick model and the principle of beneficence. *Evaluation and Program Planning* **27**(3), 341–347. https://doi.org/10.1016/j.evalprogplan.2004.04.011.

Bori, S. & Huston, A. 2020 Collective action for the WASH SDGs: lessons from using a learning alliance approach in three countries. In: *6th International Symposium Knowledge and Capacity Development for Water Management, 5*.

Bornstein, D. 2012 *Open Education for A Global Economy*. New York Times. Available from: https://opinionator.blogs.nytimes.com/2012/07/11/open-education-for-a-global-economy/ Accessed 27 July 2021.

Bowden, P. 2020 *Beginners Guide to Massive Open Online Courses (MOOCs)*. Class Central. Available from: https://www.classcentral.com/help/moocs#:~:text=How%20long%20to%20complete%20a,from%20one%20student%20to%20another Accessed May 2021.

Cohen, D. & Crabtree, B. 2006 *Qualitative Research Guidelines Project*. Available from: http://www.qualres.org/HomeOppo-3815.html Accessed May 2021.

Crichton-Smith, H. 2020 *Sustainable Development Goals*. Accelerating progress towards SDG 6: a system strengthening approach for water, sanitation and hygiene that leaves no one behind. Available from: https://sustainabledevelopment.un.org/partnership/?p=30208 Accessed May 2021.

Cundell, A. & Sheepy, E. 2018 Student Perceptions of the Most Effective and Engaging Online Learning Activities in a Blended Graduate Seminar. Available on https://www.semanticscholar.org/paper/Student-Perceptions-of-the-Most-Effective-and-in-a-Cundell-Sheepy./8be65e4f7b83184cc7f369ef10090409d55c21ea#citing-papers.

Da Silva Wells, C., Van Lieshout, R. & Uytewaal, E. 2013 Monitoring for learning and developing capacities in the WASH sector. *Water Policy* **15**(Suppl. 2), 206–225. https://doi.org/10.2166/wp.2013.120.

Davis, D., Chen, G., Jivet, I., Hauff, C., Kizilcec, R. F. & Houben, G. J. 2017 Follow the successful crowd: raising MOOC completion rates through social comparison at scale? In: *ACM International Conference Proceeding Series*, pp. 454–463. https://doi.org/10.1145/3027385.3027411.

Diver, P. & Martinez, I. 2015 MOOCs as a massive research laboratory: opportunities and challenges. *Distance Education* **36**(1), 5–25. https://doi.org/10.1080/01587919.2015.1019968.

Gabrielsson, S., Huston, A. & Gaski, S. 2020 *Reframing the Challenges and Opportunities for Improved Sanitation Services in Eastern Africa through Sustainability Science*. Springer Link. Available from: https://link.springer.com/chapter/10.1007/978-981-15-5358-5_4 Accessed July 2021.

Hall, S. & Villareal, D. 2015 The hybrid advantage: graduate student perspectives of hybrid education courses. *International Journal of Teaching and Learning in Higher Education* **27**(1), 69–80. Available from: https://eric.ed.gov/?id=EJ1069791.

Hollander, D., Ajroud, B., Thomas, E., Peabody, S., Jordan, E., Javernick-Will, A. & Linden, K. 2020 Monitoring methods for systems-strengthening activities toward sustainable water and sanitation services in low-income settings. *Sustainability (Switzerland)* **12**(17). https://doi.org/10.3390/su12177044.

Huston, A. & Moriarty, P. 2018 *Understanding the WASH System and its Building Blocks: Building Strong Wash Systems for the SDGs*. IRC WASH Systems Series, The Hague, Netherlands.

IHE-Delft 2020 Delft agenda for action on knowledge and capacity for the water sector achieving. In *6th International Symposium Knowledge and Capacity Development for Water Management*, pp. 3–5. Available from: https://www.un-ihe.org/sites/default/files/2020_delft_agenda_2.pdf Accessed July 2021.

IRC 2021 *IRCWASH*. Available from https://www.ircwash.org.

IWA 2014 *WASH Human Resource Capacity Gaps in 15 Developing Economies – An Avoidable Crisis*, p. 56. Available from: https://iwa-network.org/wp-content/uploads/2016/03/1422745887-an-avoidable-crisis-wash-gaps.pdf. Accessed July 2021.

Johnson, J. 2021 *Statista*. Digital Populations Worldwide. Available from: https://www.statista.com/statistics/617136/digital-population-worldwide/. Accessed July 2021.

Jordan, K. 2015 Massive open online course completion rates revisited: assessment, length and attrition. *International Review of Research in Open and Distributed Learning* **15**(6), 341–358.

Kirkpatrick, D. L. 1994 *Evaluating Training Programs: The Four Levels*. Berrett-Koehler, San Francisco.

Otto, D., Caeiro, S., Nicolau, P., Disterheft, A., Teixeira, A., Becker, S., Bollmann, A. & Sander, K. 2019 Can MOOCs empower people to critically think about climate change? A learning outcome based comparison of two MOOCs. *Journal of Cleaner Production* **222**, 12–21. https://doi.org/10.1016/j.jclepro.2019.02.190.

Paula, M., Whitsed, C. & Giradi, A. 2016 Applying the Kirkpatrick model: Evaluating an interaction for learning framework curriculum intervention, issues. *Educational Research* **26**(3), Murdoch University, Australia. Available from https://www.iier.org.au/iier26/paull.pdf.

Saritepeci, M. & Çakır, H. 2015 The effect of blended learning environments on student motivation and student engagement: A study on social studies course. *Education and Science* **40**(177), 203–216.

Siemens, G. 2013 *MOOCS are Really a Platform*. Elearnspace. Archived from the original on 21 January 2013. Retrieved 9 August 2021.

Suter, F. & Lüthi, C. 2021 Delivering WASH education at scale: evidence from a global MOOC series. *Environment and Urbanization* **33**(1), 99–116. https://doi.org/10.1177/0956247820987759.

UN-Water 2019 *National Systems to Support Drinking-Water, Sanitation and Hygiene: Global Status Report 2019*. UN-Water Global Analysis and Assessment of Sanitation and Drinking-Water. Available from: https://www.unwater.org/publications/un-water-glaas-2019-national-systems-to-support-drinking-water-sanitation-and-hygiene-global-status-report-2019/. Accessed February 2021.

Wiek, A., Withycombe, L. & Redman, C. L. 2011 *Key Competencies in Sustainability: A Reference Framework*. Springer. Available from: https://www.netzwerk-n.org/wp-content/uploads/2017/01/Wiek-et-al.-2011-Key-competencies-in-sustainability.pdf. Accessed July 2021.

Wilson-Grau, R. 2018 *Outcome Harvesting: Principles, Steps, and Evaluation Applications*. IAP Information Age Publishing, Charlotte, NC.

Wilson Grau, R. & Britt, H. 2012 Outcome Harvesting, MENA Office, Ford Foundation. Retrieved from http://www.managingforimpact.org/sites/default/files/resource/outome_harvesting_brief_final_2012-05-2-1.pdf.

First received 31 January 2021; accepted in revised form 23 March 2022. Available online 5 April 2022

doi: 10.2166/h2oj.2022.044

A systems approach to improving access to water, sanitation, and hygiene (WASH) in schools in Odisha, India

Narendra Singh Chouhan [a,*], Monika Oledzka Nielsen[a], Pratibha Singh[b], Swathi Manchikanti[b], Vivek Pandey[a], Jeffrey P. Walters[c] and Karishma Kadyan [b]

[a] UNICEF India, 44 Surya Nagar, Bhubaneswar, Odisha 751003, India
[b] UNICEF India, Lodhi Rd, Near 73, Lodhi Gardens, Delhi 110003, India
[c] School of Engineering and Technology, University of Washington Tacoma, 1900 Commerce Street, Tacoma, WA 98402-311414, USA
*Corresponding author. E-mail: nschouhan@unicef.org

NSC, 0000-0003-3231-7785; KK, 0000-0001-5100-1810

ABSTRACT

As part of the humanitarian response to create an enabling environment for children at schools, and to contribute to achieving Sustainable Development Goals (SDGs) 1, 3, 4, and 6, UNICEF India envisages to improve water, sanitation, and hygiene (WASH) facilities in schools. In 2019, UNICEF in partnership with the Government of Odisha rolled out a systems approach to WASH for 58,000 schools to improve equitable access to safe WASH facilities in schools and to address any gaps that may exist. Schools were assessed through a digital monitoring application and ranked on 39 benchmarks for WASH facilities and practices. Approximately 13,000 schools were ranked as 1 and 2 stars on a 5-star scale (with 5-star being the best), which necessitated structuring schedules for improvement throughout the year. The systems approach to strengthening key building blocks was followed through including advocacy for policy and planning, infrastructure improvement, leveraging finances, capacity building, behavior change, institutional strengthening, accountability, and monitoring. School Swachhata (cleanliness) Action Plans (SAPs) were developed for more than 5,000 schools, and necessary behavior change was encouraged by involving school management committees and by instituting child cabinets of student leaders as role models to promote positive WASH behaviors among others. The complex interconnection between stakeholders was assessed using the social network analysis to highlight the level of interaction and stakeholder roles that guided the success of the Odisha WASH program to guide future WASH in school programs in Odisha and other states in India. Overall, the study findings suggest that the Odisha WASH program's success is owed, in part, to a diverse and multi-layered coordination structure between the district, block, and community-level stakeholders.

Key words: India, Odisha, social network analysis, systems approach, WASH in schools, UNICEF

HIGHLIGHTS

- A systems-level approach with evidence based planning was adopted to inform improvement plans for WASH in schools in Odisha, India.
- A core focus of the program's success was capacity building and system strengthening at various levels.
- Existing resources were also leveraged towards improvement in WASH access, practices, and functionality.
- Social network analysis was used to study how stakeholder coordination influenced program success.
- Findings revealed program success was bolstered by diverse and multilayered stakeholder coordination.

GRAPHICAL ABSTRACT

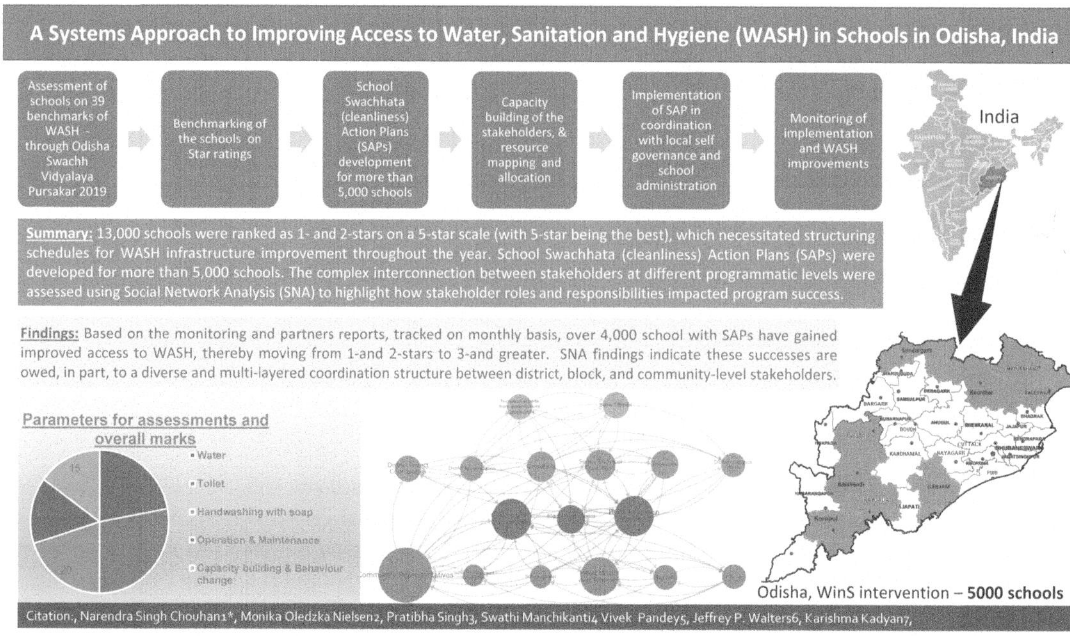

INTRODUCTION

Water, sanitation, and hygiene (WASH) in schools aims to make a visible impact on the health of children through the improvement in their hygiene practices and those of their families and communities. It also aims to improve curriculum and teaching methods while promoting hygiene practices and community ownership of water and sanitation facilities in schools. It improves children's health, school enrollment, attendance, and retention and paves the way for a new generation of healthy children. Indeed, the societal implications of successful (or unsuccessful) provision of sanitation are far-reaching, and the detrimental impact of poor water and sanitation, especially on childhood health and the development of human capital, is well-documented (Chatterley *et al.* 2013; Spears *et al.* 2013; Dreibelbis *et al.* 2014). Interventions to alleviate issues with global access to WASH services are often inhibited by a complex tapestry of dynamic and interconnected factors and stakeholders engaging in service delivery programs ((Walters & Javernick-Will 2015; Liddle & Fenner 2017). Thus, the WASH sector is increasingly conferring on the notion that service provision is a complex or wicked problem (Casella *et al.* 2015; Walters *et al.* 2017) that requires the use of systems thinking and modeling to navigate the so-called 'WASH systems approach' (Zhang *et al.* 2016; Cannon *et al.* 2022; Walters *et al.* 2022).

India has had its fair share of challenges and accomplishments in providing WASH in schools. The Government of India's strong commitment to providing schools with adequate WASH facilities is supported by legislation and by the Right to Education Act (2009), which necessitates ensuring drinking water and sanitation facilities in schools (Right to Education Act India 2009). In the recent past, various national flagship programs – which often further translate to state flagship programs launched by Government to support this quest for safe and adequate water and sanitation for all. The most noteworthy of these is perhaps the Swachh Bharat Mission (SBM) or the Clean India Mission, which was launched in 2014 by the Government to support this quest for safe and adequate WASH for all, with government data indicating that over 110 million household toilets were constructed during the 2014–2019 period and all villages across the country declared themselves open-defecation free (ODF) (SBMG 2020). The Indian state of Odisha, which is home to a substantial proportion of the population belonging to socially marginalized groups, with 17% of its population belonging to scheduled castes (SCs) and 23% belonging to scheduled tribes (STs) (Roy 2018) achieved its ODF status in September 2019 (SBM 2019). However, the ODF status achieved was based only on the access to a toilet at the household level, but not the access to toilets in institutions such as schools, preschool centers, and health care facilities.

WASH improvement in any segment requires a systems approach. A systems approach is based on the concept that sustainable development is a result of interrelated and multi-dimensional technical and non-technical factors and associated complex network of stakeholders that engage within development programming (Hjorth &

Bagheri 2006; Amadei 2016). A WASH systems approach requires curating an enabling environment for different building blocks (IRC 2018). As summarized by IRC (2018), these building blocks include advocacy, policy, and planning from the highest level to the lowest governance level, the financing mechanism for infrastructure improvement, access to WASH infrastructure (new construction, repair, and maintenance of existing infrastructure), capacity building and behavior change promotion, partnerships, and institutional accountability (to ensure role and participation of governing body/local self-governance, education and district administration, civil society organizations (CSOs), student cabinet, and school management committee (SMC) in the process) and monitoring on key aspects including planning, resource allocation, implementation, and quality assurance. We summarize these building blocks in more detail in Figure 1.

A systems approach for WASH in schools, therefore, is composed of related and dependent WASH components, which in interaction in schools, form a unitary whole (Casella *et al.* 2015). The systems approach for strengthening WASH facilities in schools in Odisha started with a baseline assessment covering the key elements like water supply, sanitation, hand hygiene, operation and maintenance, the Capacity building of key stakeholders, and learning and behavior change/practices. The systems, building blocks from policy advocacy, planning, capacity building, financing, monitoring, and institutional accountability were prioritized as part of the systemic approach. Participatory planning requires the involvement of concerned stakeholders. This includes identifying public concerns and values and developing a broad consensus on planned initiatives. It is also about utilizing the vast amount of information and knowledge that stakeholders hold to find workable, efficient, and sustainable solutions (CAP-NET 2008).

This paper highlights a state-level programming approach where an approach: 'Swachhata Action Plans' (SAPs) was applied, a key part of The Odisha Swachh Vidayalaya Puraskar (Odisha Clean School Award) program, in Odisha, India. Within this study, we sought to gain insight into the interconnected and complementary aspects of the WASH program through a systems lens to learn about the aspects of the program, and the stakeholders that supported and executed this successful program.

Strengthening WASH facilities in schools has always been a cross-cutting subject that involves various institutions and diversified stakeholders. The level of interaction among the stakeholders is an essential parameter to influence the outcome. Each stakeholder has an important and distinguished role that affects the outcome of the program.

This study utilized our team's intimate case knowledge of the Odisha WASH program to perform a social network analysis (SNA) to gain insight into the complex underlying elements of successful (or unsuccessful)

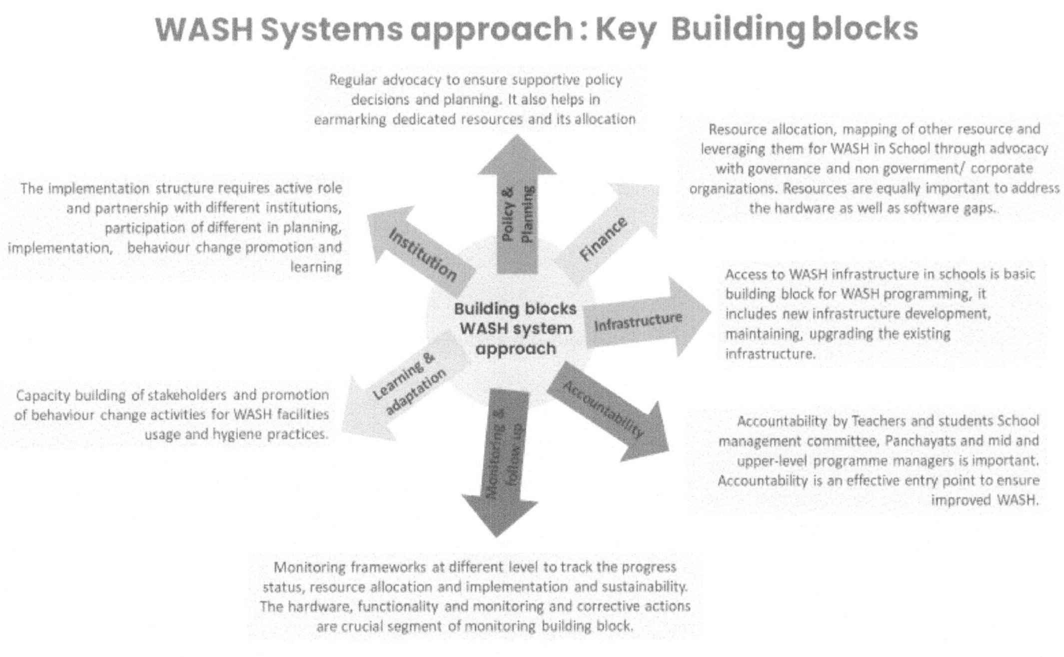

Figure 1 | WASH systems approach – key building blocks (adapted from IRC 2018).

planning, stakeholder coordination, and action for sustainable WASH school programs within the Odisha WASH program in India.

The overarching research questions (RQs) that guided this study are:

RQ1: How did the Odisha WASH program structure and the interaction of stakeholders within this structure affect the success of WASH in school programs in Odisha, India?

RQ2: What do the beneficial and suboptimal aspects of the program and the supporting stakeholder network inform for future WASH programming in Odisha in particular and India in general?

In the next section, we provide a detailed summary of recent national efforts for WASH in schools in India with a focus on the Odisha WASH program in the State of Odisha. Then, we describe the mixed-method research approach, centered on SNA, that was used to explore the systemic aspects of the Odisha WASH program and program stakeholders, focusing on the structural aspects of the social network that appeared to contribute to the Odisha WASH program outcomes, whether positive or negative. We use the findings from these analyses to point to future recommendations for future research and practice in India and beyond focusing on WASH in schools.

ODISHA CASE STUDY CONTEXT

WASH in schools refers to a combination of technical and human development components that are necessary to produce a healthy school environment and to develop or support appropriate health and hygiene behaviors. The technical components include facilities for drinking water, hand-washing, toilet, and soap in the school compound for use by children and teachers. Access to adequate toilets not only reduced the rate of school dropout, especially for girls but also the chances of the transmission of diseases (UNICEF 2012a). The provision of WASH facilities in school secures a healthy school environment and protects children from illness and exclusion. It is the first step towards a healthy physical learning environment, benefiting both learning and health. Children who are healthy and well-nourished can fully participate in school and get the most from their education. Hygiene education in schools helps promote those practices that would prevent water and sanitation-related diseases as well as encourage healthy behavior in future generations of adults. Facilities for menstrual hygiene management at the school level ensure reduced absenteeism, especially for adolescent girls (SBSV 2015). In 2014, the Ministry of Human Resource Development (now the Ministry of Education), the Government of India launched 'Swachh Bharat Swachh Vidyalaya' (SBSV), or *Clean India*, *Clean School*, complementing the ongoing national SBM, to promote improved and better access to water and sanitation facilities in all government schools of India. A key feature of the campaign was to ensure that every school has a set of functioning and well-maintained WASH facilities.

To further motivate well performing schools across the country, the Government of India, in the 2016–2017 financial year, launched a national initiative called the *Swachh Vidyalaya Puraskar* (Clean School Award). The explicit purpose of the award was to honor schools that have undertaken significant steps towards fulfilling the mandate of the Swachh Bharat Swachh Vidyalaya (Clean India, Clean School) program. The key indicators for the selection were based on school scores obtained in five subcategories shown in Table 1 *Swachh Vidyalaya Puraskar* (Clean School Award). In 2018, 52 schools across the nation were awarded nationally based on a series of validation exercises at various levels viz District, State and National level (through internal as well as external

Table 1 | National guideline for school scoring for Swachh Vidyalaya Puraskar Award (Ministry of Human resource Development 2017)

Parameters of the assessments	Maximum score
Water	22
Toilet	28
Handwashing with soap	20
Operation and maintenance	15
Capacity building and behavior change	15
Total	100

agencies). National awardees were selected among the schools, which secured 90–100% of the total scores, using the weighted criterion shown in Table 1.

Within Odisha, only one school qualified for the national level award from the 21,511 schools who have participated, only 197 out of the 21,511 of the state's schools assessed were even moderately ranked. According to the Odisha School Education Program Authority (a state-level government body), there are over 66,765 schools in Odisha, of which 54,911 are government-run (OSEPA 2019). While the Government of Odisha has a well-functioning state-level educational dashboard to monitor the education-related activities for all schools, the dashboard does not adequately track access to gender-segregated water and sanitation services, hand hygiene, and operation and maintenance, and behavior change components. Access to gender-segregated WASH facilities is important to ensure safety and privacy and make a crucial difference in the continued education of girls. Girls need clean water to wash themselves and/or their menstrual cloths and a place to dispose of their menstrual pads. If girls do not have access to these facilities at school, they often stay at home while they are menstruating. In fact, gender-segregated toilets that are located in convenient, safe locations at school can protect girls from violence and assault (Levy & Houston 2017). Well-maintained and gender-segregated toilets not only ensure a reduction in dropouts particularly for girls from schools but also create a better learning environment (UNICEF 2012b).

Sustainability of WASH in school was a big factor to try systems approach for WASH in school, the government has invested a lot in the infrastructure, capacity building and behavior change promotion, despite investment, poor operation and maintenance and other factors has led to slippage in WASH compliance in schools. It led to exploring the systems thinking approach for WASH in school.

There was a clear need to improve the existing WASH facilities to create an enabling environment for school children in Odisha. The motive was not just to construct toilets, but to create child-friendly and gender-segregated sanitation facilities and strengthen the system for regular upkeep to sustain WASH services.

In 2019, to accelerate the progress of WASH compliance in schools, the state launched the 'Odisha Swachh Vidayalaya Puraskar' (Odisha Clean School Award) *to benchmark all schools on WASH through* the School WASH rating tool – 39-point checklist – (Ministry of Human Resource Development 2017) (Supplementary Material, Information-SI-1). UNICEF supported the government as a technical partner in this initiative. The benchmarking allowed to assess schools on the existing WASH access and in addressing the existing gaps. The application of this tool effectively demonstrated how evidence-based data can improve service delivery and budget allocations across the schools.

With support from UNICEF India, the state department overseeing education, adapted from the School WASH rating tool in the state context that covered the following five main thematic areas of WASH relevant to schools: water supply, toilets, handwashing with soap, operations and maintenance, and behavior change and capacity building. Application of this checklist took place within the five-step process as shown in Figure 2. The last thematic area also assessed awareness of menstrual hygiene. The responses against the weighted indicators ranked from 1 star to 5 stars on the basis of WASH facilities available in the schools, where 5 stars were granted to schools scoring 90% or above (Table 2).

The benchmarking for the Odisha Swachh Vidyalaya Puraskar in 2019 was done in approximately 58,000 schools across the state and from all 30 districts. This benchmarking and stratification of schools on a scale of 1 star- to 5 stars was done for all government and government-aided schools. This assessment process was used within a larger program structure as shown in Figure 2.

There was effective utilization of information technology to capture the real-time information related to WASH infrastructure. Data for each school was collected through a website and an android-based mobile application called *Odisha Swachh Vidyalaya Puraskar app*, developed with technical assistance and expertise from UNICEF (GoO letter 6829 4.9.2019 and OSVP SOP). The authorized users, the school-level stakeholders, had access to this android-based application and website, wherein they could key in the status of WASH compliance of the allotted schools. The schools registered themselves against their unique UDISE (Unified district information system for education) code and submitted the information as per the prescribed format (See Supplementary Material, SI-2), covering information on key parameters as categorized in Table 1.

On the basis of the score, schools were ranked on a scale of 1–5 stars as mentioned in Table 1. After the benchmarking was done, 3,296 schools were ranked in the category of 1 star, which meant that 3,296 schools were ranked poor in terms of WASH compliance and needed significant improvement. Additionally, 10,173 schools were ranked under the category of 2-star schools, which meant they had fair compliance with the WASH facilities but still needed significant improvements. The schools with the most need or those with the most significant gaps

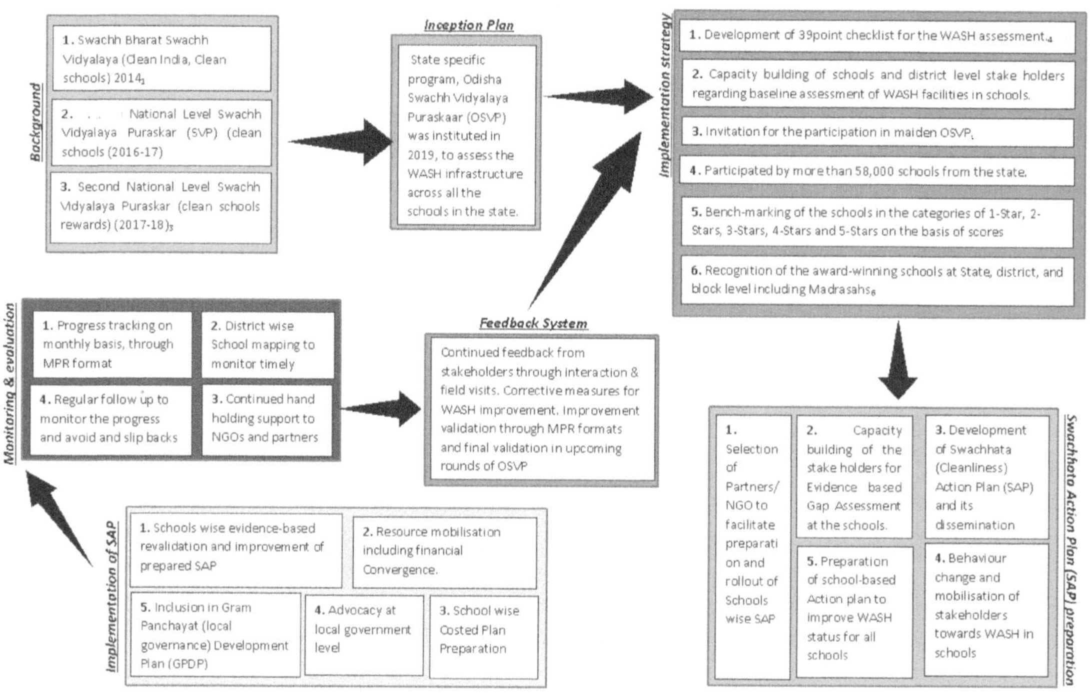

Figure 2 | Odisha WASH in school programs process diagram.

Table 2 | Swachh Vidyalaya Puraskar benchmarking scoring scale against 39 assessed indicators

Score	Star rating	Remarks
90–100% of the norms	*****	Excellent Keep it up !!!
75–89% adherence to the norms	****	Very Good
51–74% adherence to the norms	***	Good, but there is a scope of improvement
35–50% adherence to the norms	**	Fair, needs improvement
Below 35% adherence to the norms	*	Poor: needs considerable improvement

(ranked under the 1- and 2-star category) were shortlisted and funded within SAPs (Cleanliness Action Plan) (Supplementary Material, SI-3). The administration of these identified schools was facilitated in preparing their action plans, which now also sought to bridge the existing gaps in WASH facilities. These action plans also helped schools in exploring different sources of funds mobilization and resource mapping.

To sensitize key stakeholders, a series of capacity-building activities were organized at different levels, namely state, district, block, and school levels. Teachers, head teachers, and School Management Committee (SMC) members were capacitated on WASH, and on developing evidence-based school WASH improvement plans, i.e., School SAP, and regular school-based activities were initiated to promote the WASH behavior among children. Special focus was given to the operation and maintenance of WASH facilities for sustainability and long-term usage.

UNICEF and state-level monitoring partners – civil cell (a civil engineering wing supporting infrastructural development within the Education department) at state, and District administration and CSOs – implementation partners at districts, then instituted a real-time progress reporting mechanism on a monthly basis, to which assigned focal points for each school had to target, focal points include CSO volunteers, cluster coordinators, and technical consultants. This included regularly assessing both infrastructural investments, as well as behavioral practices on the school campus and updating the monthly status on WASH improvement on key indicators as per the WASH benchmarking.

The implementation of SAP at the school level was continuously monitored and evaluated against a monthly progress tracking format. The progress of each NGO partner against the implementation of SAP was tracked on a regular basis through monitoring reports. Continuous feedback (shown in Figure 2) on the implementation of SAP was ensured through regular interactions with stakeholders and field visits to schools. The feedback collected was effectively utilized in the improvement of the SAP process and issuing guidance for corrective action in planning from the state level. The feedback and observations from practitioners and NGOs were also used in establishing linkages on the influence of key stakeholders in the process through SNA.

To encourage long-term safe sanitation practices, SMCs, comprising of parents, teachers, and child role models were educated on the importance of WASH practices and were assigned responsibilities to take care of WASH infrastructures and create awareness about the same. Complementarily, during the SAP preparation process, teachers, headmasters, and government education officers were trained on how to maintain sanitation facilities. Furthermore, to ensure the effectiveness in the implementation of action plans prepared, a cohort of stakeholders was further trained on the implementation part. Local champions, including School Cabinet, (SMC), Mid-Day Meal Cook cum helpers and janitors, etc. were identified and trained at each school as well, with a total of 9,655 stakeholders trained, including government officials. The capacity-building efforts were done in a cascading manner, with master trainers further training grassroot-level implementers.

Panchayat Raj Institutions (PRIs) members (Elected community representatives) are the nodal points through which the local self-government of rural areas is realized. It is a three-tier system of rural governance via PRIs at the district, block, and village levels. The involvement of PRIs is important to ensure the equal distribution of different services and resources for rural development and to maintain sustainability. In the SAP process, the role of PRIs is highly realized as a key stakeholder leveraging resources.

While the SAPs were based on credible analysis of the existing situation, gap assessment, and prioritization of interventions for better outcomes, mapping and leveraging of resources, a limited share from Annual Composite School Grant was available to the schools as a resource, schools were provided only a 10% earmarked grant from the Annual Composite School Grant (between 25,000 and 100,000 INR, based on the number of students in school) for WASH activities under the Samagra Shiksha Abhiyan (A National Education Program in India). This amount would not be sufficient to ensure WASH improvement in the case of 1- and 2-star category schools where the WASH infrastructure gaps were larger and required resources to ensure safe WASH conditions of schools.

Given this gap in funding, the tied and untied grants available in Panchayats were identified as an opportunity to address the gaps. The 15th Finance Commission (FC) India has recommended that 60% of rural local bodies' grants from the central divisive tax pool be spent on water conservation, drinking water provisions, and managing household waste and excreta management, including maintaining the ODF status achieved under the SBM in 2019. The grant has both tied and untied components. The tied grant – 60% of the total – has to be spent on water and sanitation-related activities. The remaining 40% is untied: Local bodies are free to choose how to spend it, as per the priorities and Panchayat responsibilities mandated in schedule 11 of the Constitution of India. (Government of India 2020).

The planning and utilization of the grants are dependent on the decision at the Gram Sabha/Local self-governance level. For inclusion of school-level WASH improvement action points in Panchayat planning, the SAP process was followed with situational analysis, priorities, and plan with the estimate, and then it was presented to the local self-governance body – i.e., Panchayat or Gram Panchayat (GP) to consider and include in Gram Panchayat Development Plans (GPDP). ('Panchayat' means an institution of self-government constituted under article 243B (Article 243(b) in The Constitution of India 1949), for the rural areas – it is the lowest governing institution for rural areas in India), it acts as the executive committee of the village general body (i.e., Gram Sabha). The Gram Sabha means a body consisting of persons registered in the electoral rolls relating to a village comprised within the area of Panchayat at the village level. The Gram Sabha is the fulcrum and general body of the Panchayati Raj and village development. People use the forum of the Gram Sabha to discuss local governance and development and make need-based plans for the village.

The Panchayat implements development programs under the overarching mandate, supervision, and monitoring of the Gram Sabha. All decisions of the Panchayat are taken through the Gram Sabha (Village Meeting) and no decision is official and valid without the consent of the Gram Sabha. The members of the GP are elected by the Gram Sabha. The School SAP were placed before GP to consider and include in GPDP. Furthermore, UNICEF helped in mobilizing funds for the SAPs from various government partners, CSOs, and through corporate social responsibility (CSR). Government funding, such as that for local community schemes example Tribal area development funds, District Mineral funds, etc., was also leveraged. The latter funding mechanism, overseen by rural governance bodies, was often allocated for operations and maintenance of school WASH structures in each school, in order to make it a priority for communities.

Overall, school-based WASH interventions variously aim to: (i) reduce the incidence of diarrhea and other hygiene-related diseases; (ii) improve school enrollment, school performance, and attendance; and (iii) influence hygiene practices of parents and siblings whereby children act as agents of change in their households and communities (McMichael 2019). With similar objectives, and to provide an impetus to schools to fortify WASH facilities, the Government of Odisha announced the Odisha Swachh Vidyalaya Puraskar to acknowledge and support WASH compliance in schools. After the state announced the award initiative in mid-2019, altogether, 58,553 schools enrolled themselves in the self-reported assessment. As per baseline findings, 13,491 schools earned 1- and 2-stars (OSEPA 2019). Five thousand of these 1- and 2-star schools were supported in developing SAPs for the year, with the aim of having each school meet higher benchmarks by the end-of-year assessment.

To ensure the continuous and effective monitoring for the implementation of SAPs by the field implementation partners like CSO and schools, a monthly monitoring tool was developed to track the progress of action plans. The partners have used the school network of teachers, and cluster coordinators and coordinated with block officials to track the periodic progress. The SNA analysis of key stakeholders has highlighted that community representatives and school-level stakeholders have a prime role in the success of the entire process followed by support from the cluster, block, and district-level stakeholders in the overall planning, Implementation, and monitoring of the program. Based on the partner's report, tracked on a monthly basis, over 4,000 schools with SAPs have gained improved access to WASH, thereby moving from 1- and 2-star to 3-star and greater (Figure 3). One thousand schools were unable to leverage necessary resources to progress forward in the ranking system, even with SAPs in place, as limited resource availability at local self-governance level on other development priorities.

The pace of the progress was also depending on the perception and interest of the local community and school staff. Financing was an important factor, the short term, and midterm improvements were managed by leveraging resources available within schools and Panchayats, but for major infrastructure upgrades (long-term plan) the budget proposal was proposed in the annual state education plan for approval from state/national government and it took time in approval as a one-time annual event. GP-level stakeholder's role and action was also an underlying cause in funding allocation, which was guided by the local priorities, if community-level water supply issues were pending then Panchayat's first priority was fund allocation and improving water supply, the institutional WASH issues moved to secondary priority list on other priorities due to overall limitation of funds.

Currently, regular support is being ensured through cluster resource coordinators, SMCs, and technical field partners, to support planning at the local self-governance and district levels to leverage resources locally.

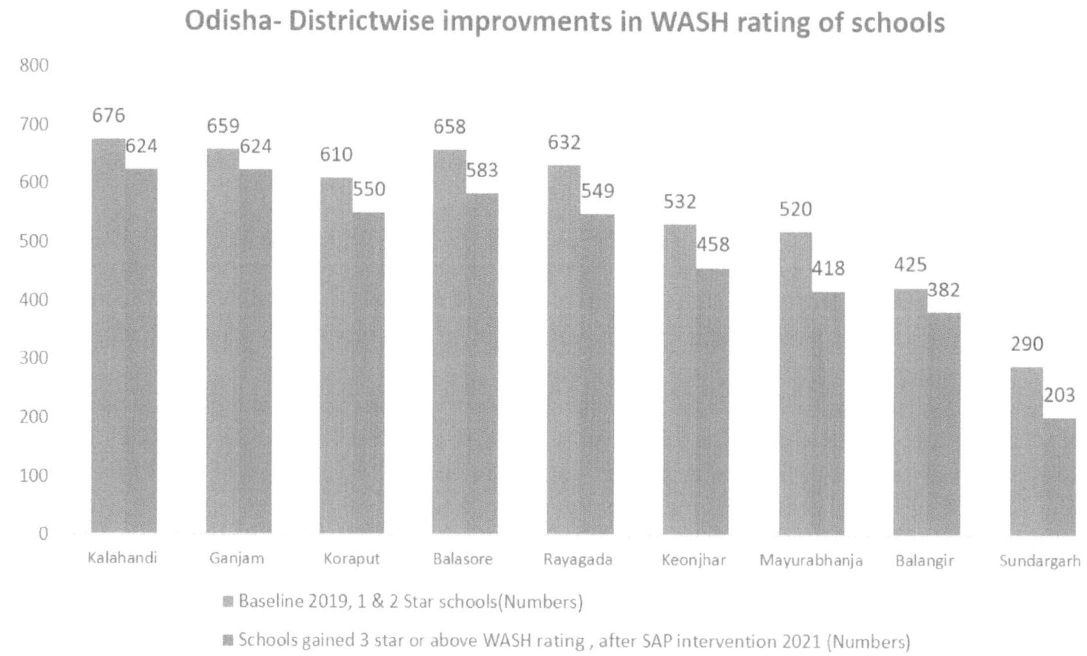

Figure 3 | WASH improvements 2019–2021 in 1- and 2-star category schools.

METHODS

In this section, we describe the SNA approach used to identify stakeholders (also called 'nodes' in the social network), to create the social network within which these stakeholders are interconnected (called 'edges' in the social network), and the metrics used to structurally evaluate the core stakeholders influencing Odisha WASH program success. We begin by describing how the data were collected, followed by the SNA metrics used to measure salient aspects of stakeholder interconnectivity (addressing RQ1). We then provide an overview of the Odisha WASH program stakeholders considered in this analysis. Finally, we conclude with a description of the qualitative rationale through which the SNA metrics are evaluated to highlight key stakeholders in the Odisha WASH program network (addressing RQ2).

Data collection

The identification of stakeholders to include in the SNA, as well as the connections between these stakeholders, were derived based on the cognitive social structures perceived by the members of the research team (Krackhardt 1987). Many of the research team members currently work directly within the Odisha WASH program to support the entire process of intervention and reporting of case studies for evaluating and disseminating success and lessons learned throughout the program. This approach to data collection was used in light of the challenges in performing a survey – the process typically used to collect social network data from a larger pool of stakeholders (Wasserman & Galaskiewicz 1994). The process for identifying stakeholders and their interaction based on the cognitive social structures of the research team entailed a robust and lengthy synchronous and asynchronous discussion, whereby key stakeholders were systematically and iteratively identified, and the directed pairwise connections (from → to) were used to develop a network edge list (Table A1 in the Supplementary Material) Odisha WASH program.

Identification of key program stakeholders entailed a lengthy discussion on how these stakeholders communicate and coordinate within the hierarchical structures present in the Odisha WASH program. Proper execution of the Odisha WASH program relied on the communication and coordination between stakeholders at the state-, district-, block-, and community levels. State-level stakeholders are the policy and decision makers who drive the entire program, including the planning, resource allocation by the state government and monitoring, and administration. State-level stakeholders that were included in the SNA were the Secretaries, Project Director, and State focal points from development partners like UNICEF. District-level stakeholders are executive managers of the program for the district and ensure the execution of the program in line with state/national flagship programs and circulars. District-level stakeholders included in the SNA were District Collectors, District Education Officers, and District Project Coordinators. Consultants, Senior Technical Consultants, and Engineers, Block-level stakeholders are executive managers to ensure quality program implementation including basic infrastructure development and ensuring academic works. Block-level stakeholders included in the SNA were Cluster-level Coordinators, Block Education Officers, and Block Development Officers. Finally, community-level stakeholders are key implementers and frontline for ensuring service delivery to the community and children. Community-level stakeholders included in the SNA were SMCs, School Cabinet, Community Representatives, Parents, Students, and Head Master and Teachers. A summary of these stakeholders and their roles and responsibilities is shown in Table 3. The stakeholder groups were identified by the authors of this paper, having extensive experience and connection with Odisha WASH program activities.

Given the service-level at which the research team engage within the overall SAP system context, communication and coordination between the different stakeholders at this level were prioritized highly to map the key agents within the Odisha WASH program in order to learn about the structural characteristics driving coordination and communication of the many stakeholders involved towards sustainable WASH service provision in Odisha schools. Thus, the connections (edges) between stakeholders in the resulting social network were considered to qualitatively represent the presence (or in the case of no connection, absence) of communication and/or coordination between stakeholders to execute Odisha WASH program activities.

SNA metrics – nodal centrality

SNA is a diagramming methodology based on graph theory used to understand structural interaction and process-based relationships between network nodes (Freeman 1978; Wasserman & Galaskiewicz 1994; Borgatti 2005). Nodes represent a stakeholder in the network, and 'edges' represent links or relationships between nodes (Wasserman & Galaskiewicz 1994). Node centrality analysis is a quantitative process used to evaluate key players in

Table 3 | Summary of SNA stakeholders and their roles and responsibilities, as identified and defined by the authors

Level	Stakeholder	Roles & Responsibilities	Relevant WASH systems building block
State	State Officials	Policy level formulation; Implementation planning and management; Financial planning and management; Resource leveraging at state level; Monitoring	Policy financing, planning, and monitoring
	Technical experts from development agencies/UN	Technical support to government; Technical and knowledge partner; Advocacy for policy reforms; State–district-level capacity building, demonstration of different program/approaches; System strengthening by providing technical experts and partners to support the program	Planning, institution strengthening, learning and adaptation, and monitoring
District	District Administration	Administrative support and overall monitoring of program execution; Coordination with the state department; Resource mobilization	Planning, financing, infrastructure development, and monitoring
	District Education Officers	Administrative management, planning and resource allocation, and monitoring of education programs; Evidence-based assessments; Reporting to district level; Implementation of SAPs; Roll out of capacity building plan	
	District Project Coordinators	Administrative management, planning and resource allocation, and monitoring of education programs; Evidence-based assessments; Reporting to district level	
	Senior Technical Consultants	Stakeholder mapping, technical guidance; Capacity building; Mapping of schools	Infrastructure development, handholding, and quality assurance monitoring
	Engineers	Technical supervision, including development of estimates for infrastructural improvement and supportive supervision to SMC and schools; Monitoring and quality assurance of civil works	
	Consultants	Stakeholder mapping; Capacity building; Mapping of schools; Infrastructure development monitoring; School SAP development	Planning, learning and behavior change promotion, and monitoring
Block	Cluster-level Coordinators	Supportive supervision to schools; Monitoring and follow-up; Reporting to block and district level	Institution strengthening, monitoring
	Block Education Officers	Block-level administration of education program; Planning and monitoring; Reporting to district level	Planning, financing, and monitoring
	Block Development Officers	Planning and allocating resources for all development works for under different block-level development schemes; Monitoring of development works executed at GP level	Infrastructure improvement
Community	Head Master and Teachers	Sensitization of students; School management; Planning O&M of the infrastructure; Supportive supervision	Planning, implementation, accountability, and institutional strengthening
	SMCs	Planning and execution of school-level development works; Creating an enabling environment for learning; Development of school development plan	Planning, implementation, monitoring, learning, and behavior change promotion
	School Cabinet	Management of student cabinet; Behavior change promotion among students; Support in O&M and monitoring	Planning, learning and behavior change

(*Continued.*)

Table 3 | Continued

Level	Stakeholder	Roles & Responsibilities	Relevant WASH systems building block
	Community Representatives/ PRIs	Planning; Resources allocation implementation and monitoring of development works in their GP area	promotion, and implementation
	Parents	Support in school development; Help in behavior change promotion among children; Presenting school-level issues in GPs planning meetings	
	Students	Agent of change; School-level monitoring; Sensitization	

communication and coordination. Three distinct node centrality evaluations were conducted in this study to provide unique structural insights into the key stakeholders within the SAP network: (i) degree centrality, (ii) betweenness centrality, and (iii) eigenvector centrality. Degree centrality measures the number of incident edges that pass into (degree-in) or out (degree-out) of a node. It measures how directly connected a particular stakeholder is with other stakeholders, providing insight into which stakeholders are local hubs for communication and coordination. For the purposes of evaluating key actors in the Odisha WASH program network, degree centrality was calculated as a combination (sum) of degree-in and degree-out scores. Betweenness centrality measures a stakeholder's ability to 'bridge', within shortest (or geodesic) paths, between other stakeholders and thus measures a stakeholder's control of information and coordination through a network, and their ability to efficiently share this information (Freeman 1978; Borgatti 2005). Finally, eigenvector centrality is a modified measurement of degree centrality that considers a stakeholder's direct connection with other highly connected nodes. In so doing, eigenvector centrality measures a stakeholder's overall influence or leadership within the entire social network. Calculation of eigenvector centrality, especially for large networks, is a complex and iterative process, whereby solutions on centrality scores are arrived at using eigenvectors, and often, an associated eigenvector algorithm is used to find the dominant eigenvector solution that represents the eigenvector centrality scores for each node. Readers who are interested in knowing more about this metric can refer to Wasserman & Galaskiewicz (1994). All centrality analyses and SNA visualizations presented in this paper were performed in the online and open-source Kumu platform (Kumu.oi).

Interpretation of findings

Degree, betweenness, and eigenvector centrality scores each tell unique stories on how social network structure, and the relative location of stakeholders within that structure, influence communication, and information transfer. These SNA metrics were compared and contrasted for the Odisha WASH program stakeholder, both visually and quantitatively, to guide discussion on the structural characteristics of stakeholder interaction and the influence on SAP process outcomes. Specifically, degree centrality was used to inform which stakeholders serve as communication and coordination hubs within their local networks. Betweenness centrality was used to identify the stakeholders that facilitate information transfer and program coordination across the state, district, block, and community levels. Finally, eigenvector centrality informed which stakeholders are the overall leader's information transfer and coordination. Stakeholders with the highest degree, betweenness, and eigenvector centrality were considered the greatest overall influencers on SAP planning and execution. This discussion is supported by authors' intimate knowledge of the WASH and SAP process and is used to inform recommendations for future WASH in school programming using the SAP process in Odisha in particular and in other states in general. The SNA results and discussion are presented in the subsequent section.

RESULTS AND DISCUSSION

This section presents results obtained through the SNA of Odisha WASH program stakeholders and discusses the core structural aspects – as analyzed using the aforementioned nodal centrality measures – of the SAP social network driving service delivery strategies. We conclude this section with a summary of the study learning that can be applied to the WASH in the school sector in the other Indian States. Figure 4 shows the SAP social network derived based on our group's combined cognitive social structure, with nodes divided by the four Odisha WASH

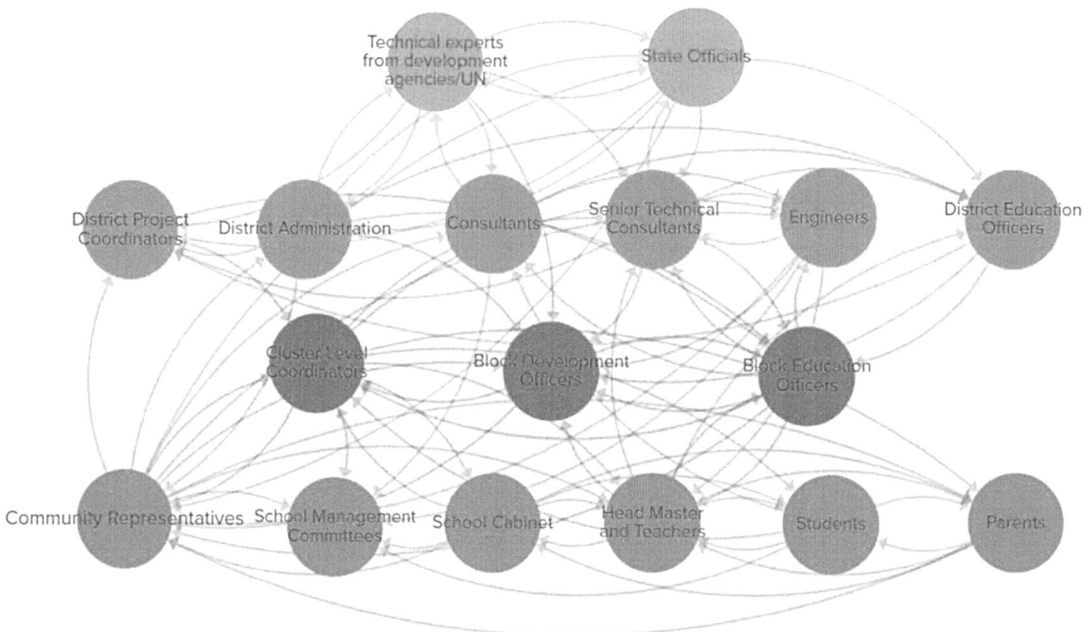

Figure 4 | The Odisha WASH program social network in Odisha. Blue nodes (circles) are state-level stakeholders, yellow nodes are district-level stakeholders, green nodes are block-level stakeholders, and orange circles are community-level stakeholders.

program institutional levels. Table 4 summarizes the degree, betweenness and eigenvector scores of SAP Odisha program stakeholders.

Findings from degree centrality

Degree centrality evaluates the top local influencers. Figure 5 shows the SAP Odisha program social network with node sizing based on degree centrality scores. The local influencers within the Odisha WASH program, based on degree centrality scoring were Community Representatives (21), Senior Technical Consultants (14), Headmaster & Teachers (14), and Cluster-level Coordinators (14). These results make sense, as the SAP approach is formed around evidence-based planning at the community (school) level for WASH improvement through for better learning environment led by the school and community-level stakeholders. These evidence-based planning activities are

Table 4 | Degree, betweenness, and eigenvector scores and rankings for SAP Odisha program stakeholders, with top-three of each metric in 'bolded underlined' text

Level	Stakeholder	Degree	Betweenness	Eigenvector
State	State Officials	12	0.028	0.049
	Technical experts from development agencies/UN	9	0.024	0.022
District	District Administration	10	0.016	0.029C_D
District Block	District Education Officers	8	0.015	0.051
	District Project Coordinators	8	0.026	0.058
	Senior Technical Consultants	14	0.052	0.080
	Engineers	10	0.039	0.064
	Consultants	8	0.051	0.041
	Block Development Officers	11	0.040	0.089
Block Community	Block Education Officers	13	0.110	0.071
	Cluster-level Coordinators	14	0.112	0.071
	SMCs	6	0	0.064
Community	Head Master and Teachers	14	0.100	0.080
	School Cabinet	9	0.008	0.034
	Community Representatives/PRIs	21	0.197	0.086
	Parents	9	0.019	0.059
	Students	9	0.016	0.046

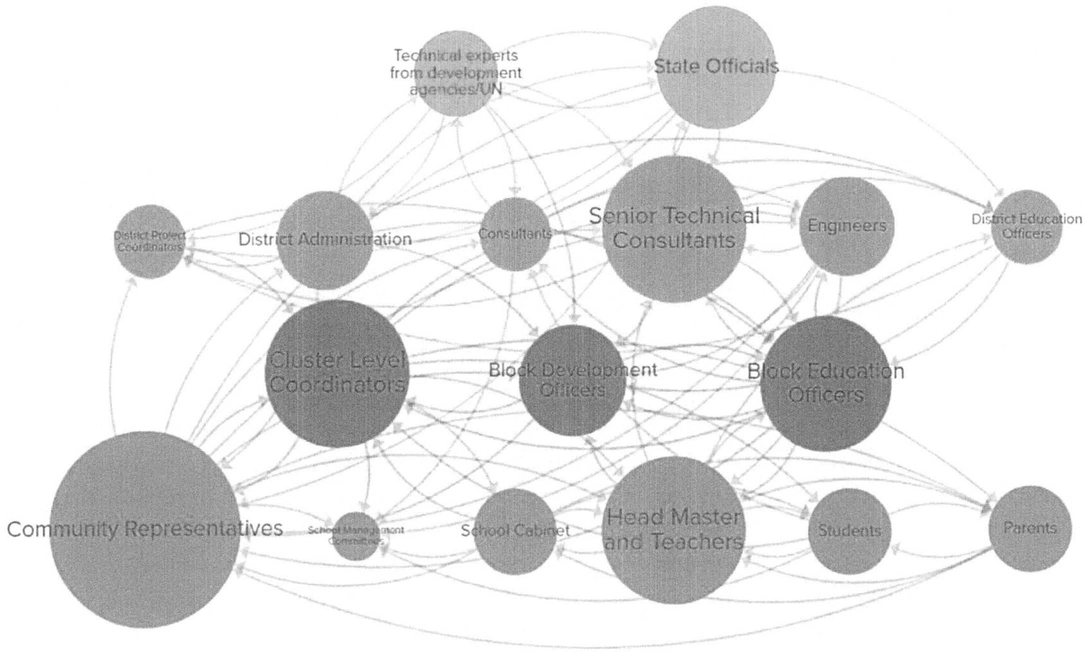

Figure 5 | Odisha SAP social network, with nodes (and node fonts) sized by degree centrality scores; larger nodes and font sizes indicate a higher degree centrality score.

the responsibilities of the Head Master & Teachers, who are supported and supervised by the Cluster Coordinator periodically. Among other important tasks, Community Representative (PRI institutions) also engage in the infrastructural and operation and maintenance action and development plan with technical guidance from Engineers within the wing of civil construction led by Senior Technical Consultants at the district level.

Findings from betweenness centrality

Betweenness centrality identifies the stakeholders who serve as brokers for information and coordination between the state, district, block, and community levels. Figure 6 shows the SAP Odisha program social network with node sizing based on betweenness centrality scores. The top-three most central stakeholders based on

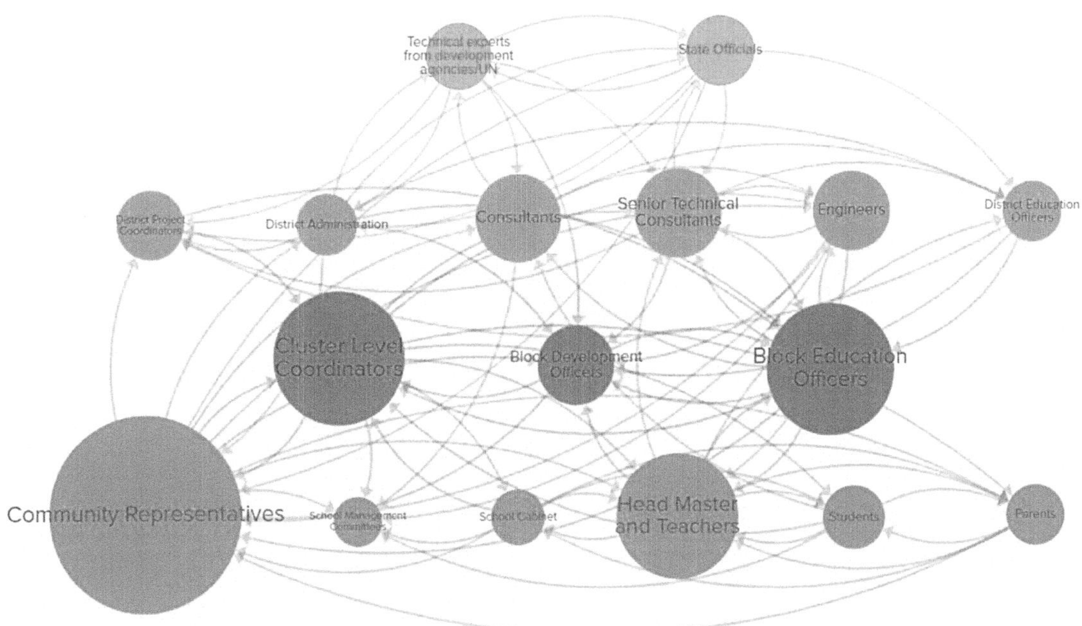

Figure 6 | Odisha SAP social network, with nodes (and node fonts) sized by betweenness centrality scores; larger nodes and font sizes indicate a higher betweenness centrality score.

betweenness centrality scoring were Community Representatives (0.197), Cluster-level Coordinators (0.112), and Block Education Officers (0.110). Indeed, block-level officials support the allocation of resources and approval from the Block development office for GP and the timely planning of the capacity building for better execution of the SAP approach in their blocks. Community Representatives are deeply engaged in school plans in GPDP. Block Education Officers have an important role in issuing approval for development works and resource allocation. In the SAP approach, all of these stakeholders prioritize WASH actions guided by the sensitization created by the state- and district-level stakeholders for the implementation of the SAP approach.

Findings from eigenvector centrality

Eigenvector centrality reveals the leaders of the network, regardless of their level of influence. Figure 7 shows the SAP Odisha program social network with node sizing based on eigenvector centrality scores. The top-three leaders within the Odisha WASH program network in Odisha based on eigenvector centrality scoring were Block Development Officers (0.089), Community Representatives (0.086), and Head Masters and Teachers (0.080). Within the Odisha WASH program, each school is a separate unit and many aspects of WASH programming depend on the motivation of teachers and school administration to prioritize and sustain efforts. The motivation of Head Masters and Teachers has helped in the development of evidence-based action plans, and their interaction with Community Representatives (PRIs) and Block-level stakeholder, which has helped in timely approvals and implementation. While these three stakeholders are highlighted as the leader of information flow and coordination within the Odisha WASH program, it is important to note that the role of each stakeholder is equally important in the SAP approach. Overall, eigenvector centrality indicates that the leaders of the Odisha WASH program coordination network are district, block, and community-level stakeholders.

Overall, the structural analyses of the SAP social network, using degree, betweenness, and eigenvector centrality, suggest that the success of the Odisha WASH program success are owed, in part, to a diverse and multilayered coordination structure between the district, block, and community-level stakeholders. Indeed, the SAP school initiative clearly motivated a state-wide movement for achieving greater WASH access and coverage in schools. State, district officials, and school-level stakeholders felt motivated and empowered for action. The greater potential was captured to translate the systems approach inputs into results for better WASH programming for schools. The school-level stakeholders' involvement and accountability ensured the effective SAP process and its role in WASH improvement and practices promotion and O&M. These interactions with community representatives helped in leveraging resources and improved monitoring of WinS.

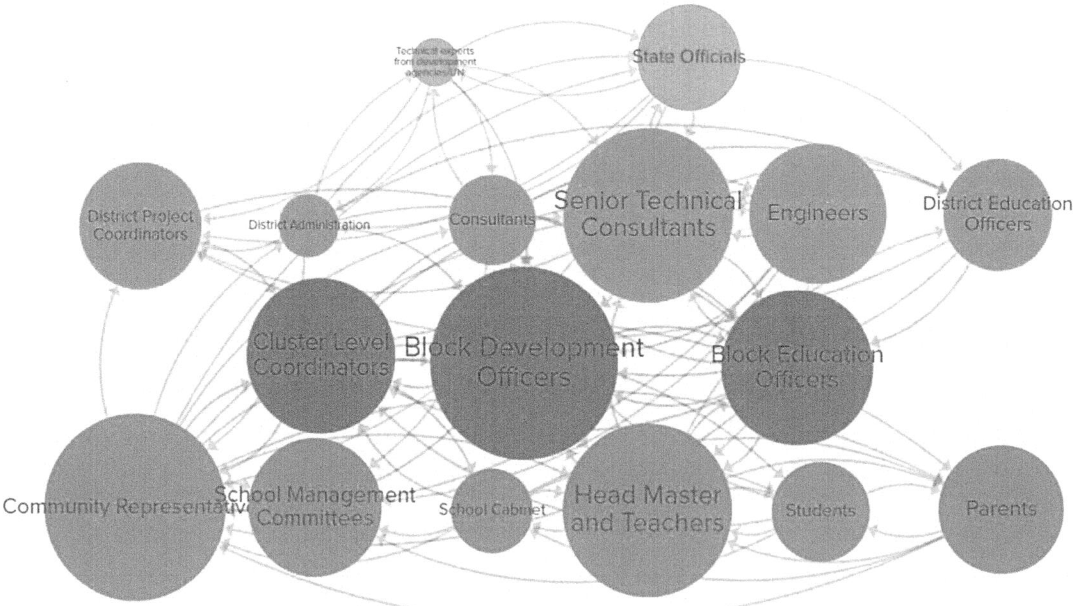

Figure 7 | Odisha SAP social network, with nodes (and node fonts) sized by eigenvector centrality scores; larger nodes and font sizes indicate a higher eigenvector centrality score.

Wider implications for WASH systems

The Odisha WASH program involves various stakeholders interacting in complex networks across different institutional levels, as shown in the SNA analysis. As per SNA analysis for Degree, Betweenness, and Eigenvector, we see that community representative have the highest overall influence on network flow and coordination activities followed by Teachers & Headmasters and then Cluster-Level Coordinators. These findings highlight the decentralized bottom-up planning with supportive supervision and monitoring of cluster coordinators. As the process is related to infrastructural development, the role of Senior Technical Consultants (Engineers) also emerged as integral to Odisha WASH programming by providing important technical guidance in estimation planning and implementation. The continued interaction and follow-up with different stakeholders WASH in school helped in increasing the motivation level of schoolteachers and students to prioritize WASH-related actions, and the ongoing dialogue with community representatives helped in leveraging the resources available under different GP-level schemes. It also helped in resource allocation for the operation and maintenance of WASH. The SAP approach included interaction and involvement of stakeholders from state, district, block, and GP level (community-level). The SNA showed that policy level stakeholders in the state-level played an inceptual and supporting role in implementing the approach in different districts with the administrative decision and resource allocation. The community-level stakeholders played a vital role in the entire process as a bottom approach starting with the school level and promoting evidence-based planning, mapping, and leveraging resources and implementation. The monitoring and feedback and actions from the district and block-level stakeholders with systematic and timely coordination for leveraging resources and technical expertise emerged within the SNA analysis as ways these stakeholders can address the gaps related to capacity and resources.

The year-long monitoring exercise within SAP was effectively able to demonstrate how fundamental planning efforts, such as making budgets based on evidence-based infrastructural assessments, can significantly improve facilities at schools. Evidence-based budgets are based on validated gaps and priority action as identified by the SMC and GPs. In addition, the SAPs were able to demonstrate to the government the importance of prioritizing and converging resources for WASH. However, the SAPs would not have been as effective without significant investment in the capacity building of key role models – from students in child cabinets to teacher and parent champions, all found to be central to the network of key stakeholders. Regular monitoring and follow-up through cluster resource coordinators also served as consistent reminders to teachers and other champions in schools. In the end, the critical activity that supported the Odisha WASH program's success was data collected through the monthly progress reports from CSO partners, which were used to maintain track of the status of the key WASH indicators in schools and identify weak areas to shore up with community action.

Study limitations

This study, like any study of this kind, has limitations the reader must be aware of. First, the small sample size of individuals (i.e., six of the seven authors) providing their cognitive interpretation of the social network, potentially limits the generalizability of the study findings. Second, it is not unlikely that the position of authors in the SAP network potentially influences their perception of the network – imposing or omitting stakeholders or links between stakeholders (Krackhardt 1987). However, despite these study limitations, we believe the application of the SNA approach, based on the intimate case knowledge of the authors, improves understanding of the core players and roles within the Odisha WASH program and offers salient insights for how to continue promoting the program's success, and offers insights for similar programs in India and beyond.

CONCLUSION AND FUTURE WORK

This paper aimed to highlight a systems approach to WASH in schools performed in the Odisha state through a process and systems-centric evaluation of stakeholder communication and engagement. SNA enabled us to evaluate the core stakeholders driving SAP approach operations, and demonstrated, through various social network metrics, the importance of a diverse, bottom-up structure for providing access to WASH infrastructure in schools.

Making schools WASH compliant is not a one-time activity. Not all the Odisha schools were found to move forward at the same pace and scope. Constant monitoring, evaluation, and overall, advocacy, is necessary to ensure all schools have adequate WASH services. For example, the 1,000 schools in Odisha state that have not progressed are now being prioritized for allocation of resources through including in the Annual Budget plan of state to ensure meeting the resource gaps, further they are prioritized for the next round of assessments and refresher training. Constant follow-up meetings to encourage GP planning is being done to ensure resource mobilization to address the priority areas. The tapping of resources available from different corpora like mineral funds, CSR, and tribal area development funds are being advocated. Regular operations and maintenance investment remain an important focus area to avoid any slip-backs. The CSR funds were largely leveraged for the construction of toilet facilities with running water supply provision. For other WASH infrastructure improvements largely government resources were used from GP or other schemes. Continued advocacy for dedicated resources for operation and maintenance is required. The evidence-based planning at the lowest level and support from program structure on key building blocks of the systems approach can ensure the success and sustainability of the WinS program. The decentralized planning and interaction with key stakeholders created ownership for WASH improvement.

The SAP school initiative clearly motivated a state-wide movement for achieving greater WASH access and coverage in schools. Through regular reporting on progress, state and district officials, and school-level stakeholders felt motivated. There is now greater potential than ever to capture that momentum and translate this learning into sustained investment in WASH programming for schools going forward.

The monitoring exercise is ongoing; progress in and of schools is being tracked on the key WASH parameters and rankings are currently being generated to provide constant feedback and guidance through the help of technical partners supporting the intervention in the field. Overall, the key to these efforts was, and will continue to be, a data-driven systems approach to target entrenched and complex issues to WASH service provision in schools; an effort that is only possible through a well-coordinated network of stakeholders at various institutional levels.

DATA AVAILABILITY STATEMENT

Data cannot be made publicly available; readers should contact the corresponding author for details.

CONFLICT OF INTEREST STATEMENT

The authors declare there is no conflict.

REFERENCES

Amadei, B. 2016 *A Systems Approach to Modeling Community Development Project.* Momentum Press LLC, New York, NY.

Article 243(b) in The Constitution of India 1949 Available from: https://legislative.gov.in/sites/default/files/COI-updated.pdf.

Borgatti, S. P. 2005 Centrality and network flow. *Social Networks* **27** (1), 55–71. https://doi.org/10.1016/j.socnet.2004.11.008.

Cannon, R. A., Mihelcic, J. R., Ghebremichael, K. & Zhang, Q. 2022 Strategies to improve performance of community-managed water systems with system dynamics modeling. *Journal of Environmental Engineering* **148** (2), 04021079. https://doi.org/10.1061/(ASCE)EE.1943-7870.0001958.

CAP-NET 2008 *Integrated Water Resource Management for River Basin Organisations. Training Manual.* CAP-NET, Pretoria.

Casella, D., Tongeren, S. & Nikolic, I. 2015 *Change in Complex Adaptive Systems: A Review of Concepts, Theory and Approaches for Tackling 'Wicked' Problems in Achieving Sustainable Rural Water Services.* IRC Working Paper.

Chatterley, C., Linden, K. G. & Javernick-Will, A. 2013 Identifying pathways to continued maintenance of school sanitation in Belize. *Journal of Water, Sanitation and Hygiene for Development* **3** (3), 411–422. https://doi.org/10.2166/washdev.2013.128.

Dreibelbis, R., Freeman, M. C., Greene, L. E., Saboori, S. & Rheingans, R. 2014 The impact of school water, sanitation, and hygiene interventions on the health of younger siblings of pupils: a cluster-randomized trial in Kenya. *American Journal of Public Health* **104** (1), e91–e97. https://doi.org/10.2105/AJPH.2013.301412.

Freeman, L. C. 1978 Centrality in social networks conceptual clarification. *Social Networks* **1** (3), 215–239. https://doi.org/10.1016/0378-8733(78)90021-7.

Government of India 2020 *Water Sanitation Hygiene Tied Funding Provision in Local Self Governance – Panchayat Level.* Government of India. Available from: https://www.panchayat.gov.in/documents/20126/0/Joint+letter+dt+17-3-20.pdf/40bdc8c4-0aa6-49b8-08ab-db5ee05a1097?t=1599722365604. Accessed 26 July 2021.

Hjorth, P. & Bagheri, A. 2006 Navigating towards sustainable development: a system dynamics approach. *Futures* **38** (1), 74–92. https://doi.org/10.1016/j.futures.2005.04.005.

IRC 2018 *Building Strong Systems for SDGs, Understanding the WASH Systems and its Building Blocks.* Available from: https://www.ircwash.org/sites/default/files/084-201813wp_buildingblocksdef_newweb.pdf. Accessed 15 May 2022.

Krackhardt, D. 1987 Cognitive social structures. *Social Networks* **9**, 109–134. https://doi.org/10.1016/0378-8733(87)90009-8.

Levy, J. K. & Houston, A. 2017 *Gender, Education, and Global Health: What's the big Deal About Toilets?* Available from: https://publichealth.wustl.edu/gender-education-global-health-whats-big-deal-toilets/. Accessed 29 July 2021.

Liddle, E. S. & Fenner, R. 2017 Water point failure in sub-Saharan Africa: the value of a systems thinking approach. *Waterlines* **36** (2), 140–166. https://doi.org/10.3362/1756-3488.16-00022.

McMichael, C. 2019 Water, sanitation and hygiene (WASH) in schools in low-income countries: a review of evidence of impact. *International Journal of Environmental Research and Public Health* **16** (3), 359. https://doi.org/10.3390/ijerph16030359.

Ministry of Human Resource Development 2017 *Swachh Vidyala Puraskar 2017–2018. Ministry of Human Resource Development*. Government of India. Available from: https://www.education.gov.in/sites/upload_files/mhrd/files/Guidelines%20SVP%202017-18_0.pdf. Accessed 16 July 2021.

OSEPA 2019 *School Scenario 2018–19 Odisha Schools Education Program Authority Dashboard*. Government of Odisha. Available from: http://osepa.odisha.gov.in/.

Right of Children to Free and Compulsory Education Act 2009 *Indian Legislation*. Available from: http://www.indianlegislation.in/BA/BaActToc.aspx?actid=2933. Accessed 11 August 2021.

Roy, S. N. 2018 *School Sanitation in Odisha*. Centre for Policy Research.

SBGM 2020 *Swachh Bharat Mission Grameen Phase II Operational Guidelines*. Available from: https://swachhbharatmission.gov.in/SBMCMS/guidelines.htm. Accessed 21 July 2021.

SBM 2019 *Household Toilet Coverage Across India*. Swachh Bharat Mission. Available from: https://sbm.gov.in/sbmdashboard/IHHL.aspx. Accessed 22 March 2021.

SBSV 2015 *Clean India: Clean Schools – A Handbook*. Government of India. Swachh Baharat Swachh Vidyalaya: A national Mission Clean India Clean School Handbook. Available from: http://14.139.60.153/bitstream/123456789/10502/1/Swachch_Bharat_Swachch_Vidhalaya.pdf. Accessed 15 July 2021.

Spears, D., Ghosh, A. & Cumming, O. 2013 Open defecation and childhood stunting in India: an ecological analysis of new data from 112 districts. *PLoS ONE* **8** (9), e73784. https://doi.org/10.1371/journal.pone.0073784.

UNICEF 2012a *Child Friendly Schools*. UNICEF. Available from: https://www.unicef.org/media/66486/file/Child-Friendly-Schools-Manual.pdf. Accessed 10 August 2021.

UNICEF 2012b *WASH for School Children South Asia Report*. Available from: https://www.ircwash.org/sites/default/files/unicef_wash_for_school_children_south_asia_report.pdf. Accessed 10 August 2021.

Walters, J. & Javernick-Will, A. 2015 Long-term functionality of rural water services in developing countries: a system dynamics approach to understanding the dynamic interaction of factors. *Environmental Science & Technology* **49** (8), 5035–5043. https://doi.org/10.1021/es505975 h.

Walters, J. P., Neely, K. & Pozo, K. 2017 Working with complexity: a participatory systems-based process for planning and evaluating rural water, sanitation and hygiene services. *Journal of Water Sanitation and Hygiene for Development* **7** (3), 426–435. https://doi.org/10.2166/washdev.2017.009.

Walters, J. P., Valcourt, N., Javernick-Will, A. & Linden, K. 2022 Sector perspectives on the attributes of system approaches to water, sanitation, and hygiene service delivery. *Journal of Environmental Engineering* **148** (6), 13. https://doi.org/10.1061/(ASCE)EE.1943-7870.0002010.

Wasserman, S. & Galaskiewicz, J. 1994 *Advances in Social Network Analysis: Research in the Social and Behavioral Sciences*. SAGE Publications Inc. Cambridge University Press, Cambridge, UK.

Zhang, Q., Prouty, C., Zimmerman, J. B. & Mihelcic, J. R. 2016 More than target 6.3: a systems approach to rethinking sustainable development goals in a resource-scarce world. *Engineering* **2** (4), 481–489. https://doi.org/10.1016/J.ENG.2016.04.010.

First received 31 January 2021; accepted in revised form 2 June 2022. Available online 15 June 2022

PUBLISHING

doi: 10.2166/h2oj.2022.061

Working with local governments to increase access to WASH services: a case of WaterAid's participatory approaches in Nigeria

Tariya Sarauta Yusuf [a,*], Anna Murray[b] and Chukwumerije Okereke[c]

[a] WaterAid UK, 6th floor, 20 Canada Square, London E14 5NN, UK
[b] The Aquaya Institute, P.O. Box 1603, San Anselmo, CA 94979, USA
[c] Centre for Climate and Development, Alex-Ekueme Federal University, Ndufu-Alike (AEFUNAI), Ikwo, Ebonyi State, Nigeria
*Corresponding author. E-mail: tariyayusuf@wateraid.org

TSY, 0000-0002-1333-8292

ABSTRACT

Attempts to implement participatory processes are often complicated by realities which make objectives of participation unattainable. Hence, if participation is to mean more than good intentions, it is vital to understand the strategies deployed by international nongovernmental organisations (INGOs) in participatory development. This study employed a qualitative case-study approach to evaluate the participatory processes employed in two WASH projects implemented by WaterAid Nigeria attempting work with Local Government Areas: the Sustainable Total Sanitation project and the HSBC Water programme. WaterAid Nigeria's participatory development approach recognises the responsibility of the various tiers of government for its citizens' wellbeing, especially the need for partnership with the Local Government Area (LGA) towards the delivery of WASH services. The state, as the 'duty bearer', is responsible for the provision of WASH services. Thus, both projects centred on government capacity building and community empowerment through the establishment and capacity development of WASH units in LGAs, insistence on counterpart funding and the establishment of water, sanitation and hygiene committees (WASHCOMs). Despite these attempts, findings highlighted the persistence of inefficiencies within the LGA WASH system that rendered partnerships ineffective. At the community level, WASHCOMs were not sustainable. Despite the challenges of implementing partnerships between INGOs and LGAs, such partnership remains a way to improve the provision of WASH services in Nigeria and other low- and middle-income countries. To better work within the participatory model, WASH INGOs should explore how to better use existing LGA arrangements for service delivery. Additionally, the LGA, as a duty bearer, could foster genuine, non-discriminatory spaces for the participation of communities by taking political, socio-economic and cultural differences into account.

Key words: capacity building, counterpart funding, local governance, participatory development, partnerships

HIGHLIGHTS

- In the WASH sector, a growing feature of participatory development is the building of partnerships between different tiers of government and international nongovernmental organisations (INGOs) to deliver development projects.
- In two case study projects in Nigeria, WaterAid elicited partnerships with Local Government Areas (LGAs) in WASH service delivery through the formation of WASH units, capacity building, insistence on counterpart funding and formation of WASH committees.
- Various frustrations existed with implementing LGA-INGO partnerships. Despite capacity building attempts, there was still inefficiency, bureaucracy and aid dependency within the LGA WASH system that served as impediments to effective partnership. Institutions and structures established to facilitate partnership with LGAs were not sustainable.
- To more effectively work within the participatory model, WASH INGOs could seek to understand the power dynamics in the LGA system and explore how to better use existing LGA arrangements.
- Additionally, the LGA, as duty bearer responsible for WASH service provision, could foster genuine, non-discriminatory spaces for participation of communities by taking political, socioeconomic and cultural differences into account.

INTRODUCTION

United Nations Sustainable Development Goal 6 seeks to ensure safe drinking water and sanitation for all by 2030. As of 2020, 2 billion people globally lack safely managed drinking water services, 3.6 billion people lack safely managed sanitation services, and 2.3 billion people lack basic hygiene services (WHO/UNICEF 2021). Globally, the areas with largest needs for water, sanitation, and hygiene (WASH) services are rural areas or fragile contexts of low- and middle-income countries (LMICs) like Nigeria. It is estimated that about 39 million people in Nigeria, representing about 19% of the country's population, lack access to improved sanitation facilities, while 45 million inhabitants do not have access to safe water services (ibid). The provision of WASH services in these contexts has been historically funded or implemented by governments, multilateral agencies, or international or local NGOs working with a top-down approach, where programmes are designed and implemented without the participation of stakeholders (Kamruzzaman 2013). A human rights approach to WASH, with its emphasis on accountability, access to information, non-discrimination, attention to vulnerable groups and participation, is a crucial aspect of achieving the SDGs (COHRE AAS SDC and UN-HABITAT 2007). Increasingly, participatory development approaches have been applied to WASH programmes in LMICs (Wamuchiru 2017).

In the participatory development, local actors are engaged in the development process. Participation by the poor and marginalised is expected to lead to a greater achievement of development goals and agendas (Ribot 2011). This approach was born out of the presumed failure of the 'top-down' development approach (Chambers 2008; Kamruzzaman 2013); it, therefore, rejects centralised decision-making processes in favour of an inclusive 'bottom-up' approach (Kapoor 2005).

Participatory development has several aims in promoting bottom-up planning and inclusion. First, participatory development seeks to promote local community empowerment (Kapoor 2005). Empowerment here means the pooling of financial and human resources to achieve collective strength and countervail power. It also means the enhancement of technical skills, planning and managerial competence, and analytical and reflective abilities (Kapoor 2005; Tsang *et al.* 2009). Second, participatory development seeks to ensure the ownership of development programmes (Ndabaga *et al.* 2015), where people take the initiative in establishing project activities. Lastly, there is a wide, if contested, assumption that stakeholder participation will enhance the sustainability of development projects by creating project ownership leading to willingness to assume responsibility relevant to project success (Gleitsmann *et al.* 2007).

In the WASH sector, a growing feature of participatory development is the building of partnerships between relevant governments and international nongovernmental organisations (INGOs) to deliver development projects in LMICs (Sharma 2015).

This approach takes different forms but primarily involves INGOs working with private and public institutions to encourage beneficiary participation in WASH projects (Trinborg 2007). Instead of replacing the state (Rose & Miller 1992), INGOs work through state institutions to deliver services.

Attempts to implement participatory processes are often complicated by realities which make the goals of participation unattainable (Cohen & Uphoff 1980). Hence, if participation is to mean more than good intentions, it is vital to understand the strategies deployed by INGOs in participatory development and how these interact with the existing structural and political dynamics to either facilitate or stifle promises of participation.

This paper employs a qualitative case-study approach to evaluate participatory processes employed in two WASH projects in Nigeria. The Sustainable Total Sanitation (STS) project and HSBC Water programme (HWP) projects were both implemented by WaterAid Nigeria in partnership with Local Government Area (LGA) institutions (Table 1). The aim was to document the approach and effectiveness of INGOs working in partnership with local government institutions to deliver WASH services. Specifically, the paper outlines and explores (i) the strategies and procedures adopted by WaterAid in promoting local participation and empowerment; (ii) the effectiveness of these strategies; (iii) their effect on existing administrative structures, inclusion and power relations, and (iv) the broader implications for the long-term sustainability of WASH projects.

METHODS

The case-study organisation is WaterAid Nigeria. The specific projects are the STS, a large programme part-funded by the Bill and Melinda Gates Foundation, and the HWP funded by the HSBC Bank in six countries including Nigeria. The case-study projects were selected for this research based on their avowed commitment to participation as a strategy for governing WASH in communities.

Data collection involved three phases: a pilot phase to plan the data collection and pilot the tools; a field-based visit for initial data collection through interviews, observations and document review; and follow-up interviews with the respondents.

Table 1 | Descriptions of two case-study projects implemented by WaterAid in Nigeria

	STS Project	HWP – Phase 1
Project aim	The project aims to improve the effectiveness, efficiency, inclusion and sustainability of all sanitation approaches in three states of Nigeria	The programme aims to significantly improve access to safe water, sanitation and hygiene for people living in the poorest communities in Nigeria
Project partners	Bill and Melinda Gates Foundation, Institute of Fiscal Studies, Federal, State and LGAs, and Beneficiary Communities	The Hongkong and Shanghai Banking Corporation Limited (HSBC) water programme, Earth Watch, WaterAid, World Wildlife Fund, federal, state, and LGAs, and beneficiary communities
Years implemented	2012–2016	2012–2017
States covered	Jigawa, Ekiti and Enugu	Benue, Plateau and Bauchi
LGAs	**Jigawa:** Gumel, Kaugama Sule Tankarkar and Maigatari **Ekiti:** Ikole, Moba, Ilejemeje and Ekiti South West **Enugu:** Nkanu East, Igbo-Etiti and Udenu	**Bauchi**: Dass, Bogoro, Ningi and Ganjuwa **Plateau:** Pankshin, Kanke, Langtang North and Bokkos **Benue**: Logo and Ado
Target beneficiaries	625,000	Water: 173,000 Sanitation and hygiene: 316,000

States, LGAs and communities participating in the STS and HWP projects were used as the sampling frame. For the STS, fieldwork was carried out in two project states: Enugu and Ekiti. Two LGAs in Enugu and four LGAs in Ekiti were purposively selected. It was not possible to visit communities in Jigawa state because of the Boko Haram insurgency in Northern Nigeria, but remote interviews were conducted. For the HWP, all three project states of Bauchi, Plateau and Benue were included. Three LGAs from Bauchi and Plateau states and two from Benue were purposively selected from a list of all LGAs where the projects were implemented. For each LGA, two communities, one with the highest rate of open defaecation and another with the lowest rate of open defaecation, were purposively selected for water, sanitation and hygiene committees (WASHCOM) interviews.

A pilot trip was conducted in 2013 to determine the categories of interview respondents and pilot complete interview guidelines and questionnaires. Open structured key informant interviews (KIIs) were held with two staff from the STS project and three staff from the HWP project using a snowballing technique.

In 2015, data were collected through KIIs, focus group discussions (FGD), the observation of village sanitation planning meetings, and project document review. Interviews were conducted with LGA WASH staff (9), national NGO staff (4), WaterAid staff (6) and WASH consultants from the private sector (5). KIIs were conducted by the researcher in English using semi-structured and open-ended questionnaires. Interviews with WaterAid staff and WASH consultants were recorded. A total of seven FGDs were held with 11 WASHCOMs. FGDs were conducted with members of WASHCOM. Discussions with WASHCOMs and interviews with LGA staff were not recorded because respondents did not grant consent, but written notes were taken. Both discussions and interviews were conducted in person, except for LGA staff in Jigawa, which were carried out using Skype.

For the STS project, two community mobilisation meetings were observed in Enugu state, and four sanitation planning meetings were observed in Ekiti state. All community mobilisation and sanitation planning meetings were recorded. Additionally, a checklist was used to collect information on attendance and participation during meetings. Finally, 69 project and WaterAid strategy documents were reviewed.

In 2016, all key informants (LGA staff, WASH consultants and national NGO staff) were contacted again for follow-up discussions. Interviews were open-ended discussions to understand if there were changes in the project environment since the last interviews were conducted. Interviews were conducted through telephone, Skype or WhatsApp and lasted an average of 2 h. A group chat room was established on WhatsApp with the five WASH consultants. The chat room allowed for continuous interaction about the interview questions for 1 month.

Interview documents were analysed qualitatively in Nvivo analytical software (QSI International, Doncaster, Australia) by indexing, coding, summarising, identifying and mapping patterns.

RESULTS AND DISCUSSION

Responsibility for the provision of WASH services in Nigeria

The Nigeria National Water and Sanitation policy drafted in 2004 defines the responsibility of WASH service provision as being shared across federal, state and local governments. The policy was developed by the federal government of Nigeria with support from INGOs. The federal government of Nigeria is the primary custodian of the human rights to water and sanitation by being the signatory to relevant global conventions. The Federal Ministry of Water Resources is responsible for national policy formulation, data collection, monitoring and mobilising funds. They provide support, coordinate and regulate the other tiers of government and the private sector and in the fulfilment of their responsibilities. State governments, through water boards and rural water supply and sanitation agencies, are responsible for supplying potable water to their states, including the 'establishment, operation, quality control and maintenance of water supply schemes' (FMWR 2004: 10). State governments also provide technical support to local governments. Local governments take responsibility for the 'establishment, operation and maintenance of rural water supply schemes, in conjunction with the benefiting communities. This shall be through the establishment of a unit that shall be technically equipped, adequately funded and manned' (FMWR 2004: 10).

In addition to government responsibility, the policy defines the importance of non-state actor participation – including the private sector, NGOs and communities. Specifically, the role of NGOs is specified as providing technical expertise and additional finance.

Project descriptions

WaterAid, founded in 1981 and working in 27 countries, is an INGO focused on improving access to WASH in the world's poorest communities. WaterAid is committed to service delivery to governments and the communities where it implements its programmes. In addition to providing services, WaterAid influences the various tiers of government to respond to the need to provide WASH services.

The STS, implemented from 2012 to 2016, aimed to end open defaecation through a total sanitation approach for over 500 communities in 12 LGAs in three states. The HWP, implemented from 2012 to 2017, aimed to provide water access to 173,000 beneficiaries and improved sanitation and hygiene services to 316,000 beneficiaries in 10 LGAs within three states (Table 1).

First, participating states and LGAs were selected through a vulnerability assessment. Then, WaterAid established WASH units in all participating LGAs lacking them and provided office equipment and established project management systems. Staff were deployed to WASH units by LGAs. WaterAid strengthened the capacity of WASH units in project management and technical components of the projects. LGA WASH units then selected project communities, collected baseline WASH access data and established WASHCOMs in all project communities to support the implementation of WASH activities. WaterAid also trained LGA WASH unit staff to serve as community-led total sanitation (CLTS) facilitators for the project. The facilitators then conducted community sensitisation or 'triggering' meetings, where community members were persuaded to commit to end open defaecation. During triggering meetings, CLTS facilitators selected volunteer 'natural leaders' and village health promoters in each community to support hygiene behaviour change, toilet construction and an end to open defaecation. Following the triggering meetings, LGA WASH unit staff, natural leaders and WASHCOMs promoted latrine construction through ongoing household visits and community meetings. Communities were also encouraged to construct public latrines with funds provided by WaterAid through LGA WASH units. LGA WASH unit staff regularly monitored communities against national requirements for open defaecation free (ODF) certification, which is granted by the National Task Group on Sanitation after inspection by LGA, state and/or national sanitation groups. In addition to the sanitation work, under the HWP project, water points such as boreholes with hand pumps and motorised boreholes were also constructed in selected communities. Finally, an end line assessment was carried out in project communities to measure changes in WASH access at the end of each year.

Participatory development approach

WaterAid delivers its programmes through a rights-based approach. The rights-based approach perceives the State as the primary custodian (duty bearer) of the rights to water and sanitation, and therefore aims to build accountability into the WASH service sector by stressing to citizens their rights to WASH services and empowering citizens through partnership and participation to claim their rights from the state. The WaterAid Nigeria Country Strategy for 2011–2015 summarises how its interventions are to be carried out in Nigeria.

'The strategy is to maintain a clear focus. We will advocate the rights-based approach with emphasis on sustainability and collaboration with others. We will engage in direct partnerships and investments in local organizations and governments to both provide services and to drive demand and accountability through empowered citizens claiming their rights and exercising choice. We will also focus on equity and inclusion so that WASH reaches those who are often marginalised and missed out.' (WaterAid 2010: 4)

The implementation of both WASH projects by WaterAid included several key components used with the aim of increasing project participation, ownership, empowerment and sustainability. These include the formation of WASH units at the local government level, the establishment of WASHCOMs at the community level, requirements for counterpart funding and capacity building.

LGA WASH unit formation

The HSBC and STS projects were implemented through WASH units, which are the lowest state organisation involved in delivering WASH services to communities in Nigeria. The establishment of WASH units in LGAs is required by the draft Nigeria National WASH policy. The primary responsibility of WASH units is the service delivery and maintenance of WASH facilities, including procurement and construction, as well as sanitation and hygiene behaviour change campaigns and community trainings.

The rationale for WASH units in WaterAid projects is illustrated by the following:

'WaterAid believes that local government authorities, who find themselves at the frontline of basic service provision, are key to the achievement of the water and sanitation Millennium Development Goal (MDGs no 7).' (WaterAid 2008: 3)

'WaterAid in Nigeria focuses its efforts on facilitating the establishment of functional systems such as the WASH units in the Local Government system, WASHCOMs at the community level, and the establishment of Water Consumer Associations as owners and managers of small town water schemes.' (WaterAid 2010: 7)

Prior to the establishment of WASH units, various departments within the LGA were responsible for carrying out water and sanitation functions. The works departments were mainly responsible for the construction of boreholes and latrines, while the health departments were responsible for hygiene and health education. The WASH unit as a sub-department attempted to bring together all LGA WASH functions under the same roof. Other LGAs where INGOs have not implemented WASH projects were not likely to have WASH units.

The WASH unit was headed by a WASH unit coordinator who was a staff of the LGA. The WASH unit coordinator related directly with WaterAid on project management issues. The LGA unit coordinator was directly responsible for implementing activities of the WASH unit, reporting daily to a programme manager, a staff of WaterAid. Other WASH unit staff included a monitoring and evaluation officer, hygiene officer, sanitation officer, water officer and a finance officer. Salary of all WASH unit staff was paid by the LGA. The LGA was also expected to pay the unit a monthly imprest towards office administration. A management committee made up of state and non-state actors like traditional rulers, community leaders, LGA senior staff and members of civil society provided an oversight function to the WASH unit.

Administration of finances for the two projects was managed directly by WaterAid through the LGA WASH units. Funds for project implementation were transferred from WaterAid to the LGA WASH units, which operated separate accounts for the project in their name. The LGA unit coordinator administered the delivery and use of funds with the supervision of WaterAid and the management committee. The signatories to the WASH unit account are, however, two members of the management committee who were not staff of the LGA. The LGA executive committee has no direct control over project funds. This is to ensure that non-state actors participate in decisions on how funds are utilised and that funds are not diverted to other LGA activities.

Community WASHCOM formation

WASHCOMs were the primary institutions through which communities were expected to participate in WASH projects. A WaterAid Nigeria internal document, in describing the rationality for WASHCOMs, states:

'The overall ownership and sustainability of projects...will primarily be attained through the community-led processes. WASHCOMs will be formed and involved in the project implementation cycle. The processes will ensure that communities are at the centre of their development taking the lead and responsibility at every level of project implementation. This will ensure social sustainability of projects and promote acceptance at the community level.' (WaterAid 2015: 4)

WASHCOMs in these two projects had 12–15 members and were responsible for deciding where WASH infrastructure like water points and communal toilets will be sited, supporting construction and providing operation and maintenance. They also support the construction of household toilets and community hygiene behaviour change activities. WASHCOMs were expected to meet once a month and keep minutes of meetings. WASHCOMs were also expected to keep financial records, equipment and labour costs, community maps and community action plans.

The community general assembly, a community decision-making forum made up of all the residents of a community, was responsible for selecting members of WASHCOMs. Members of WASHCOMs were selected to represent all genders, ages, social classes and zones in the community. To carry out their responsibilities, WASHCOMs worked through additional volunteers. Members of WASHCOM and other volunteers were unpaid.

Counterpart funding requirements

In the STS and HWP projects, WaterAid required the payment of counterpart funds, or cost share, between WaterAid, government and beneficiaries in line with the Nigeria National WASH Policy. A WaterAid report on the sustainability of water supply in Tanzania links attaining sustainability in WASH services to the financial commitment of project beneficiaries.

'Sustainability today invariably depends upon communities taking financial responsibility for their schemes, which, if achieved, will enable scarce resources from government and donors to be targeted specifically on areas where there is no improved water supply.' (WaterAid 2015: 11)

Three tiers of government were expected to make financial contributions towards the provision of WASH services in Nigeria (FMWR 2004). The federal government had the major responsibility for capital investment in the construction of rural water facilities (50%), while state and LGA governments were to contribute 25 and 20%, respectively. The remaining 5% was to be contributed by communities. In addition to contributing to the cost of constructing new water facilities, the various tiers of government and communities were also expected to share financial responsibility for the operation and maintenance of water schemes. Communities were responsible for 70% of the cost of operations and maintenance of rural WASH facilities, and LGA units were responsible for 30%. To raise funds for operation and maintenance, some communities charge a token for water collection. Communities that contribute counterpart funds for the construction of WASH facilities were expected to show ownership and establish systems for maintenance.

For the STS and HWP projects, the financial contribution from state agencies and other project beneficiaries known as the 'counterpart fund' was deposited into the LGA WASH unit account. This fund was then utilised for project implementation together with funds deposited by WaterAid. In addition, the LGA was expected to provide the WASH unit with weekly funds for project consumables and transport.

Capacity building of partners and beneficiaries

The STS and HWP projects both employed the capacity building of both government and community institutions, as WaterAid believes the lack of strong institutions and systems has led to poor WASH access in rural communities.

The WaterAid country strategy, in discussing the need for capacity building of partners at the local level, states:

'Capacity building is expected to lead to inclusive and effective participation of citizens in demanding accountability from government at the local level. The aim of strengthening capacity is to raise awareness of citizens on their rights to water, sanitation and hygiene services.' (WaterAid 2010: 9)

Capacity building was deployed in two stages. First, WaterAid staff and consultants strengthened the capacity of LGA WASH units, and then the WASH units were expected to build capacity in the communities. Capacity building of the WASH units focused on enhancing logistical support, such as the provision of office equipment, as well as technical support. Technical capacity was strengthened through training and workshops, including an annual partner round table where all project stakeholders came together to plan implementation. Other training, such as participatory budget tracking training delivered to WASH unit staff, allowed them to strengthen the capacity of communities to demand services from government in keeping with a rights-based approach. In addition to the trainings described above, each year LGA WASH units were also given an opportunity to identify additional capacity building needs.

The capacity building of WASHCOMs was facilitated by LGA WASH unit staff using WASHCOM mentoring plans. The plans included sessions on documentation, resource mobilisation, operations and maintenance of WASH infrastructure, developing community action plans and monitoring.

Effectiveness of the participatory development approach

Each component of the participatory development approach described above had its own successes and challenges within the two WASH projects studied.

LGA WASH units

In establishing WASH units and departments within the LGAs, WaterAid believes that governance and decision-making powers will be brought closer to the people, resulting in more effective community participation and empowerment. Narratives from interviews with WASH consultants and WaterAid staff echoed this vision and described the WASH units as leading to more effective WASH programmes:

> *'Before the creation of WASH units, interventions in the WASH sector at LGA level were scattered. The creation of WASH units helps bring all WASH activities in the LGA under one roof.' (LGA WASH unit staff)*

However, not everybody shares the view that the creation of WASH units in participating local governments is a necessary condition for successful implementation. Rather, some critical voices opined that WASH units are created by INGOs primarily to enable them to ensure the accountability of project funds and deliver services within LGAs without getting caught in the LGA bureaucratic bottleneck:

> *'We do not need new units. What is the works department doing? What is the health department doing? WASH units are just duplication of responsibilities.' (LGA WASH unit Coordinator)*

> *'WASH units are created to ensure the accountability of donor funds. It is not about sustainability and ownership.' (WASH Consultant)*

According to WaterAid, the creation of WASH units for these two projects was successful at reducing bureaucracy in terms of turnaround time for project implementation. Furthermore, the creation of WASH units increased the attention paid to WASH and budgetary allocations in the project LGAs.

> *'Our observation is that LGAs that do not have WASH units do not have a strong WASH activity. We believe that having WASH units is an important for implementing WASH programmes.' (WaterAid Staff)*

However, the WASH units are still part of the local government system, where a poor attitude to work and complacency were major issues affecting partnership and relationships between WaterAid and its LGA partners:

> *'The civil service mentality to work is very visible in the way [the LGA WASH unit staff] work. All the efforts we make to change orientation to project management has met a lot of difficulty.' (WaterAid Staff, STS project)*

Community WASHCOMs

Among the two projects, WASHCOMs were created in all project communities. WASHCOMs provided a platform for communities to discuss and reach decisions on WASH challenges. They also set user fees and collected them for the operations and maintenance funds.

However, the requirements of monthly meeting and record-keeping were not well followed in the STS and HWP projects. In WASHCOM observations, most had notebooks for keeping meeting minutes, but only 23% had records of meetings that were dated within a month of the fieldwork, and minutes were non-existent in several WASHCOM notebooks.

WaterAid Staff explained that WASHCOMs commonly do not meet regularly or keep records:

'Minutes of meetings is the immediate evidence of activity; when you visit WASHCOMs you will be told that the last time they met was six months ago. It should be fortnightly. After a project life span, you don't find any WASHCOM that will feed into the design I have described.' (WaterAid STS project staff)

However, a WASHCOM member explained that the regular frequency of meetings specified by the WASH Policy was not seen as necessary for the WASHCOM to effectively complete its mandate:

'We do not need to meet all the time. We all have lot of things we are busy with. When there is an issue, we resolve it wherever we are. Moreover, the cost of hosting the meetings is expensive. Whoever is hosting has to prepare food for those attending from his pocket.' (WASHCOM member)

Interviews showed that the vision of WASHCOMs transcending beyond the project cycle to define decision-making and the way activities are conducted in communities is seldom realised.

'In a field interview of existing structures, most communities cannot remember the names of members or the number of members in the WASHCOM. Functionality is directly proportional to inducement of funding and technical support. I have not seen any WASHCOM functioning actively without support.' (WaterAid staff)

Counterpart funding requirements

Despite the counterpart funding requirement, LGAs have been often unable to make direct financial contributions towards projects. In the 2015 WaterAid annual round table report, 95% of LGAs mentioned unwillingness of LGAs to pay counterpart funding as a major factor affecting the implementation of the projects in their respective LGAs. According to a WaterAid staff member:

'We have not had success with getting communities and LGAs to contribute counterpart funding. They do not take it seriously. They sign agreements, but nothing happens.'

Several reasons were presented for why LGAs and communities were unable to pay counterpart funds in these two projects. A major reason was the lack of financial autonomy of LGAs:

'LGAs in Nigeria run joint accounts with state governments. Projects are implemented jointly with the state government. LGAs do not therefore have autonomy over their funds. This makes payments of counterpart funds difficult.' (National NGO staff)

Another reason was not related to the availability of funds, but instead the structure of the LGA and decisions about where funds are spent. LGA staff reported instances in which LGAs drill boreholes through other departments instead of giving the money to the WASH unit:

'...the only funds that come to our department is the funds from WaterAid. The LGA budgets for boreholes and other WASH facilities. This money is given to the works department instead of the department for environment where the WASH unit is.' (WASH unit Coordinator)

The LGA is responsible for WASH financing with other stakeholder funds supplementing. The consistent failure of LGAs and communities to meet their counterpart funding obligations, as in these two projects, may lead to unequal partnerships as one group may feel obligated to the other.

Capacity building

Capacity building increased WASH units' abilities to implement project activities. Staff obtained project equipment and developed competence in various aspects of project implementation, including facilitation and monitoring.

Despite capacity building attempts, there still was inefficiency, bureaucracy and aid dependency within the LGA WASH system in these projects. Frequent transfers, poor attitudes to work, complacency and allegiance to civil service structures were impediments to effective partnership between WaterAid and LGA WASH units. This was also not unique to these projects. In a 2013 annual round table report, all 16 LGAs in attendance mentioned the frequent transfer of WASH unit staff as an impediment to project implementation.

Challenges also existed for capacity building the community level. WASHCOMs were expected to understand how to promote good hygiene behaviour, mobilise resources, monitor progress and carry out basic maintenance of WASH infrastructure. Discussions with WASHCOM members, however, revealed that the opportunity to attend WaterAid training meetings was considered a privilege because of the monetary incentives provided to enable project beneficiaries to attend.

> *'Community members and WASHCOMs compete to be nominated to attend training and consultative meetings. (WASHCOM member)'*

Conflicts sometimes arose where members of WASHCOMs thought they should be nominated to participate in training activities instead of other members. The motivation for rural community members therefore was not the knowledge to be acquired from attending training sessions or the desire to put that knowledge to use, but the material incentives such as training allowances. Such conflicts can affect the effective functioning of WASHCOMs and the potential for collective action towards WASH service delivery.

Administrative structures and power relations

The establishment of LGA WASH units and WASHCOMs as part of the two projects affected existing administrative structures and power relations at both the local government and community levels.

WASH units and new power relationships

WASH units were created to sit as part of a larger department within the LGA, yet they were also formally tied to WaterAid as the manager of project finances. This administrative arrangement led to confusing reporting arrangements for some WASH unit staff:

> *'I report to two people, my director at the LGA and WaterAid STS project coordinator.' (WASH unit coordinator)*

Reporting arrangements like this, where an LGA coordinator reports directly to a project coordinator at WaterAid, sets up new hierarchies thereby potentially creating conflicts between coordinators and their directors. The coordinators, who are officers junior to the directors, in this arrangement have autonomous power to deal directly with WaterAid and manage funds that are meant for the WASH unit on a day-to-day basis without involving the department directors. A national NGO staff explains:

> *'Some decisions in the WASH units are taken without the knowledge of the Directors. A good example is during the construction of institutional latrines. The directors in some LGAs were not aware of the cost of most items procured by the WASH coordinators.' (National NGO staff)*

However, WaterAid staff interviewed provided narratives contrary to the accounts described above. WaterAid staff stated that Directors who supervised WASH units were fully aware of the daily interactions within the WASH units and received information on the activities of the unit, especially during the WaterAid annual round table meeting.

> *'We relate directly with WASH coordinators to reduce bureaucracy and speed up programme implementation. We hold several other interactive sessions with departmental directors.' (WaterAid staff)*

This structure meant the WASH units, while reporting and existing within the local government system, and as part of a wider department, operated as autonomous units within the LGAs. The creation of self-functioning autonomous units within existing departments in the local government enabled timely project implementation. However, such arrangements created new power relationships within the LGA structure, which may have led to conflicts among staff members and therefore could have lowered the effectiveness of the unit at achieving its programmatic goal.

Furthermore, attempts to enhance the performance of WASH units also enabled WaterAid to implement its own bureaucratic processes and directly manage resources provided to LGAs. This was expected to increase project efficiency and reduce misappropriation of project resources, but in practice, WASH units unintentionally served as instruments of control, where WaterAid managed project implementation without having to relinquish power or funds to the LGA.

WASHCOMs and community power relationships

Under the STS and HWP projects, pre-existing associations in project communities were asked to take on WASH responsibilities instead of creating completely new WASHCOMs. It was expected that having existing committees to perform the role of WASHCOMs would maintain local power structures and ensure project sustainability. A WaterAid staff explained this reasoning:

> *'Instead of having separate committees which run parallel to community development associations, we have decided to work with existing committees while defining roles and responsibilities.' (WaterAid STS project staff)*

In adopting existing community institutions, the projects took several steps to ensure that such institutions conformed to the requirements for the formation of WASHCOMs:

> *'...the existing associations are examined and made to meet the criteria set by the project. This may mean increasing the number of women on the executive committee of the associations or expanding it to accommodate other aspects of equality.' (WaterAid Programme Support Manger)*

While the idea of working with existing community organisations intended to prevent duplication and lead to sustainability, the process of making local organisations conform to project standards had the unintended consequence of affecting community structures by altering existing power relationships. Some WASHCOMs in project communities reportedly became unintentionally authoritarian, taking over responsibilities originally intended for traditional rulers and other leaders in communities. WASH consultants revealed:

> *'...WASHCOMs are forming parallel government to that of traditional rulers in communities. They usurp powers and impose fines on households and use their members to execute that.' (WASH consultant)*
>
> *'We have had cases where chairmen of WASHCOMs will ask a town crier to call for a meeting by ringing the village bell. This is the purview of the village chief and in this case the chief did not take it lightly.' (WASH consultant)*

Decentralisation and the effect of new power relations on participation

By creating LGA WASH units and WASHCOMs, WaterAid attempted to create various decision-making centres to promote government and community participation in WASH service delivery, thus allowing governance to be decentralised and occur at lower levels. The theory was that decentralised institutions such as WASH units would be closer to WASH beneficiaries and provide opportunities for beneficiaries to participate in the management of WASH resources (Babu 2009). A number of challenges, however, limited the ability of WASH units to take responsibility within the projects and therefore limited the ability of beneficiaries to effectively participate in the management of WASH resources. Findings show that WASH units did not fit well into existing government systems, and LGAs were unable to meet project engagement requirements even despite capacity building attempts.

Some participatory approaches intentionally disrupt the order of hierarchal institutions, creating new and different spaces in which different rules of the game apply (Cooke & Kothari 2001); however, this was not the intent here. With these projects, WASH units unintentionally set up new dynamics and power relationships between the state, civil society and beneficiary communities.

These new dynamics limited the participation of the decentralised institutions in the delivery of WASH services. The rules and structures that define the involvement of decentralised institutions, while seeking to ensure efficiency, may serve to prevent the empowerment of communities.

Weak or unsuitable institutional arrangements have been blamed for poor performance of decentralised systems worldwide (Willetts *et al.* 2020). Universal decentralisation of water services has been found to not only *not improve* participatory access to WASH services but may instead undermine effective governance (Marcus 2007).

In trying to implement LGA–INGO partnerships, INGOs often overlook an institutional capacity of local government agencies and instead give currency to a simplified view of strengthening capacity (Mosse 2004). Attempting to facilitate partnerships in the WASH sector through projects like the STS and HWP, which have clear timelines and strict targets, questions the promise of empowerment through participation.

Inequalities

The projects highlighted a few areas of deepening inequalities through the establishment of LGA WASH units and WASHCOMs.

Participatory approaches and the inclusion of the marginalised

First, the requirement for WASHCOMs to keep records results in targeting educated members of communities to become members and officers. While formal education is not an official guideline, there is a preference for educated people. These requirements may unintentionally exclude vulnerable and marginalised beneficiaries who may not have the educational level or be able to afford the time and money required to meet responsibilities.

Cooke & Kothari (2001) report that development institutions prefer, and require, various forms of participation and reporting techniques that are sometimes beyond the scope of marginalised people. Truly marginalised people and the oppressed are sometimes not welcomed into formal institutions and do not usually have the capacity to organise themselves into formal institutions like WASHCOMs, either due to lack of adequate skills or financial constraints. Consequently, they are left out of development. Cleaver (2001) observed that the mere setting up of formal organisations and the specification of their membership does not necessarily overcome exclusion because the wider structural factors that shape such conditions are often left untouched. Codifying the rights of the vulnerable must involve far wider reaching measures than the requirement that they sit on a committee, or individually speak at meetings.

Second, as previously discussed, the failure of both LGAs and communities to meet counterpart funding obligations may lead to their exclusion from WASH services. Wealthy community members or politicians may contribute their own funds if the community is unable to meet counterpart funding requirements, and may influence the outcome of the project to his or her personal advantage, as described by a project WASH consultant:

> *'...Decisions on where to site WASH facilities may be made solely by the person that has paid the fund on behalf of the community. This may mean that marginalised members of the community are denied access to WASH services.' (WASH consultant)*

These projects recognised the diversity within project communities and made deliberate efforts to ensure the representation of women, people living with disability and children. In some cases, separate meetings were held for different gender categories to encourage participation. There was, however, a lack of sensitivity by the projects to political and cultural differences within project communities. The public decision-making meetings such as CLTS triggering and community council meetings assumed an absence of such differences in project communities.

White (1996) observes that a quota for the inclusion of women, and women simply being there, as is the practice in WASHCOMs, does not mean that women eventually have a real say or that they will be able to represent others and speak for them. Such measures can only be facilitated by fuller participation which may be difficult to measure. These accounts may imply that women do not influence decision-making within the projects. The influence exerted by women may not lie within the technical institutional arrangement prescribed by the project. For instance, WASH Consultants reported cases where women mobilised informally to boycott the use of certain water points because they were unhappy about where the water point was sited.

Genuine, non-discriminatory spaces for participation could be fostered by taking the diversity in communities into account in organising participatory processes.

Participatory approaches and shared responsibility

Attempts to implement participatory programmes by the projects through WASHCOMs illustrate an attempt to implement a neoliberal governance regime which looks to shift responsibility away from states, relies on multi-stakeholder networks independent of the state to provide WASH services normally associated with the state, discards centralised decision-making processes and attempts to get rural dwellers to bear some responsibilities for the provision of WASH services. The LGA-copying techniques of the private sector by 'contracting out' and 'outsourcing' responsibilities to WASHCOMs (Kapoor 2005; Fraser 2018). WaterAid performing the duties of the state seeks to encourage villagers as responsible citizens to play an active role in service delivery which is the responsibility of the traditional state (Rose 2000; Dean 2010). WASHCOMs as representative grassroot organisations are expected to strengthen local participation in the provision of WASH services by increasing the levels of democracy, transparency and accountability (Hoogesteger 2016).

Several factors hinder the implementation of this regime of neoliberalism characterised by shared responsibilities between the state and citizens (Sangameswaran 2010). WASH units within the LGAs do not fully capture or control government-related functions and capacities related to water provision, WASHCOMs find it difficult to keep up with their obligations and duties assigned to them, as was seen in these case-study projects. In some cases, elite power is reproduced through recentralisation and privileged access to decision-making (Taylor 2007), which was also described in these projects. The dynamics of communities and their real-life situations are sometimes not considered in designing participatory programmes. This limits the empowerment of communities and places a burden on communities.

Participation as an approach to sustainability and ownership

WaterAid deploys participatory methods as a strategy towards promoting the ownership and sustainability of WASH services. Sustainability, which forms the bedrock for the engagement of communities in WASH services, stems not only from a neoliberal rationality but also an economic rationality, in which donors seek to reduce expenditure on WASH services by ensuring the maintenance and functionality of WASH facilities through WASHCOMs (Kleemeier 2000).

The non-sustainability of WASHCOMs and other structures created by the projects calls into question the effectiveness of participatory processes in promoting the ownership of WASH services. Despite the promising theoretical framework, the vision of community project structures transcending beyond the project cycle to define decision-making and WASH service delivery in communities was not realised.

CONCLUSIONS

This article explores the strategies and procedures adopted by WaterAid to promote the participation of local government institutions and communities in WASH projects through the review of two WASH participatory development projects implemented by WaterAid in Nigeria. WaterAid emphasises the need for the state as a duty bearer to play a leading role in the provision of WASH services in Nigeria and for partnership and participation between states and citizens. WaterAid elicits the participation of LGAs and communities in the provision of WASH services through the establishment of WASH units in LGAs, the establishment of WASHCOMs, counterpart funding, capacity building and promotion of a rights-based approach, in which an attempt is made to use training and mentoring as a strategy to guide rural dwellers to demand for their rights from the state as a duty bearer. While well-intentioned, various frustrations existed with the participatory development processes, the capacity of LGA staff was limited, LGAs and communities were unable to contribute counterpart funds, and WASH units and WASHCOMs unintentionally created new power relationships which created inequalities and may affect sustainability. Despite these challenges, the creation of WASH units and WASHCOMs has increased coordination in LGAs and provided structures for decision at lower levels. Capacity building increased the ability of WASH units to participate in project implementation. Despite the frustrations of implementing partnerships between the LGA and INGOs, partnerships with the state remain a way to improve the provision of WASH services in Nigeria and other LMICs. To more effectively work with this model, WASH INGOs could seek to understand the power dynamics in the LGA system and explore how to better use existing LGA arrangements in setting up participatory processes for the provision of sustainable WASH services. To ensure the payment of counterpart funds, one approach could be for donor agencies to place the responsibility for the payment of LGA counterpart funds on state governments since LGAs may lack financial autonomy. Additionally, the LGA, as a duty bearer responsible for WASH service provision, could foster genuine, non-discriminatory spaces for the participation of communities by taking political, socio-economic and cultural differences into account in organising participatory processes.

DATA AVAILABILITY STATEMENT

All relevant data are included in the paper or its Supplementary Information.

CONFLICTS OF INTEREST

The authors declare there is no conflict.

REFERENCES

Babu, A. 2009 Governmentality, active citizenship and marginalisation: the case of rural drinking water supply in Kerala, India. *Asian Social Science* **5** (11), 89–98.

Chambers, R. 2008 *Revolutions in Development Inquiry*. Institute of Development Studies, London.

Cleaver, F., 2001 Institutions, agency and the limitations of participatory approaches to development. In: *participation: The new Tyranny?* (Cooke, B. & Kothari, U., eds). Zed Books, London, pp. 102–121.

Cohen, J. & Uphoff, N. 1980 Participation's place in rural development: Seeking clarity through specificity. *World Development* **8** (3), 213–235.

COHRE, AAAS, SDC and UN-HABITAT 2007 *Manual on the Right to Water and Sanitation*. Available from: https://unhabitat.org/manual-on-the-right-to-water-and-sanitation.

Cooke, B. & Kothari, U. 2001 *Participation: The New Tyranny?* Zed Books, London.

Dean, M. 2010 *Governmentality: Power and Rule in Modern Society*, 2nd edn. Sage Publications Ltd, London.

Federal Ministry of Water Resources 2004 'National Water Sanitation Policy'. Department of Water Supply and Quality Control. Draft Final. Available from: https://water resources.gov.ng

Fraser, G. 2018 Foucault, governmentality theory and 'Neoliberal Community Development'. *Community Development Journal* **55** (3), 437–451.

Gleitsmann, B., Kroma, M. & Steenhuis, T. 2007 Analysis of a rural water supply project in three communities in Mali: participation and sustainability. *Natural Resources Forum* **31** (2), 142–150.

Hoogesteger, J. 2016 The politics of water democracy: insights from grass root struggles in the Ecuadorian Highlands. *Asia Pacific View Point* **58** (1), 74–85.

Kamruzzaman, P. 2013 Civil society or 'comprador class', participation or parroting? *Progress in Development Studies* **13** (1), 31–49.

Kapoor, I. 2005 Participatory development, complicity and desire. *Third World Quarterly* **26** (8), 1203–1220.

Kleemeier, E. 2000 The impact of participation on sustainability: an analysis of the Malawi Rural Piped Scheme Program. *World Development* **28** (5), 929–944.

Marcus, R. R. 2007 Where community-based water resource management has gone too far: poverty and disempowerment in Southern Madagascar. *Conservation and Society* **5** (2), 202–231.

Mosse, D. 2004 *'Cultivating Development. An Ethnography of Aid Policy and Practice'*. Pluto Press, London.

Ndabaga, E., Oduor, J., Iribagiza, M. & Iravo 2015 A Stakeholders' participation in planning and designing of locally based UNDP projects: a case of ministry of disasters and refugees affairs (midimar). *European Journal of Business and Social Sciences* **4** (8), 141–153.

Ribot, J., 2011 Participation without representation: Chiefs, Council and Forestry law in the West African Sahel. In: *The Participation Reader* (Cornwall, A., ed.). Zed Books, London.

Rose, N. 2000 Community, citizenship, and the third way. *American Behavioral Scientist* **43** (9), 1395–1411.

Rose, N. & Miller, P. 1992 Political power beyond the state: problematics of government. *The British Journal of Sociology* **43** (2), 173.

Sangameswaran, P. 2010 Rural water reforms in Maharashtra: the role of neoliberalism. *Economic and Political Weekly* **45** (4), 62–69.

Sharma, A. 2015 Sustainable and socially inclusive development of urban water provisioning: a case of Patna. *Environment and Urbanization ASIA* **6** (1), 28–40.

Taylor, M. 2007 Community participation in the real world: opportunities and pitfalls in new governance spaces. *Urban Studies* **44** (2), 297–317.

Trinborg, K. J. 2007 *Performing Participation Stakeholders, Translations & Power in the World Bank Participation Source Book*. Master's Thesis *Submitted to the Department of Social Anthropology*, University of Oslo.

Tsang, S., Burnett, M., Hills, P. & Welford, R. 2009 Trust, public participation and environmental governance in Hong Kong. *Environmental Policy and Governance* **19** (2), 99–114.
UNICEF/WHO 2021 *Progress on Drinking Water, Sanitation and Hygiene in Africa 2000–2020*. Available from: https://washdata.org/ (accessed 4 May 2022).
Wamuchiru, E. 2017 Beyond the networked city: situated practices of citizenship and grassroots agency in water infrastructure provision in the Chamazi settlement, Dar es Salaam. *Environment and Urbanization* **29** (2), 551–566.
WaterAid 2008 *Think Local, Act Local: Effective Financing of Local Governments to Provide Water and Sanitation Services*. Available from: www.wateraid.org.
WaterAid 2010 *Country Strategy 2010–2015*. Available from: www.wateraid.org (accessed 4 January 2022).
WaterAid 2015 *Undoing Inequality: Inclusive Water, Sanitation and Hygiene Programmes That Deliver for all in Zambia. A Midterm Review*. Available from: www.wateraid.org.
White, S. 1996 Depoliticising development: the uses and abuses of participation. *Development in Practice* **6** (1), 6–15.
Willetts, J., Mills, F. & Al'Afghani, M. 2020 Sustaining community-scale sanitation services: co-management by local government and low-income communities in Indonesia. *Frontiers in Environmental Science* **8**, 98. doi:10.3389/fenvs.2020.00098.

First received 31 January 2021; accepted in revised form 29 June 2022. Available online 14 July 2022

© 2022 The Authors

doi: 10.2166/h2oj.2022.054

Bridging the rural–urban divide in sanitation with a cluster-based approach to faecal sludge management: a case study from Dhenkanal district in Odisha, India

Shipra Saxena [a,*], Narendra Singh Chouhan [a], Sujoy Mojumdar[b], Monika Oledzka Nielsen[a], Swathi Manchikanti[b], Anindita Mukherjee [c], Neha Agarwal [c], Megha Rani [a], Anwesa Dutta[a], Rajendran Vignesh [d] and Jeffrey P. Walters [e]

[a] UNICEF India, 44 Surya Nagar, Bhubaneswar, Odisha 751003, India
[b] UNICEF India, Lodhi Rd, Near 73, Lodhi Gardens, Delhi 110003, India
[c] Center for Policy Research, Dharam Marg, Diplomatic Enclave, Delhi 110021, India
[d] Sree Dattha Institute of Engineering and Science, Nagarjuna Sagar Road, Sheriguda (V), Ibrahimpatnam (M), Hyderabad 501510, India
[e] Civil Engineering, School of Engineering and Technology, University of Washington Tacoma, 1900 Commerce St, Tacoma, WA 98402, USA
*Corresponding author. E-mail: ssaxena@unicef.org

SS, 0000-0002-7070-0606; NSC, 0000-0003-3231-7785; AM, 0000-0001-8819-2403; NA, 0000-0002-5866-9976; MR, 0000-0001-6515-8974; RV, 0000-0002-6515-2553; JPW, 0000-0001-6655-1233

ABSTRACT

Halfway into the sustainable development goal (SDG) period, the rural and urban divide in sanitation persists. As of 2020, less than half of the global rural population has access to safely managed sanitation. In India, the Swachh Bharat (Clean India) Mission – Rural helped over 100 million rural households to construct individual toilets and access at least basic sanitation during 2014–2019. Expectedly, the increase in toilet usage has led to an urgent need for faecal sludge management (FSM). The present paper describes a novel model, rooted in an urban–rural partnership, to increase access to FSM services among rural households. In 2020–2021, we piloted the model in the Dhenkanal district in Odisha, which had a functional urban faecal sludge treatment plant (FSTP) and publicly run desludging trucks. The model adopted a five-step approach that included a data-led situational assessment, model development, stakeholder consultation, legal formalization of urban–rural partnership, and capacity building. Upon its implementation, the partnership transformed the rural sanitation service chain and resulted in the safe collection, conveyance, and treatment of 278 kL of faecal sludge from rural households within the first 5 months of implementation. As rural governments in India and other developing countries strive to achieve safely managed sanitation by 2030, the urban–rural partnership model discussed in the paper can present a viable pathway for rapidly scaling-up FSM services.

Key words: cluster-based planning approach, faecal sludge management (FSM), rural governance, rural–urban convergence, safely managed sanitation, sustainable development goals (SDGs)

HIGHLIGHTS

- The increased access to toilet in Dhenkanal district has produced a high dependence on single pits and septic tanks and consequently, heightened a need for FSM.
- A novel model based on urban-rural partnerships is presented for scaling up FSM services among rural households.
- Under the partnership, the urban local government has formally expanded the service catchment area beyond its boundary to include 17 neighbouring rural settlements This novel arrangement has been able to treat 278 KL of sludge or 93 truckloads from rural areas leveraging urban FSTP infrastructure.

GRAPHICAL ABSTRACT

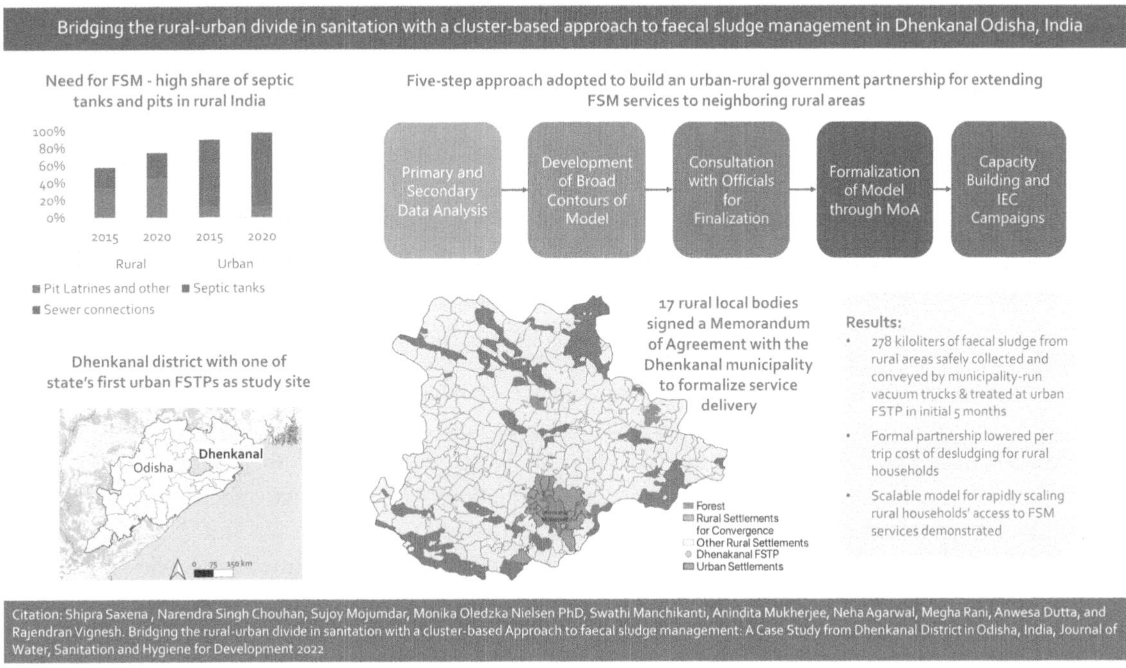

1. INTRODUCTION

As we approach 2030, rapid progress is required to meet the sustainable development goal (SDG) of universal 'safely managed sanitation', which includes better sanitation facilities (designed to hygienically separate excreta from human contact), facilities that are not shared with other households, and where excreta are safely disposed of *in situ* or removed and treated off-site (WHO-UNICEF JMP 2018). Over the past 6 years, in large part due to the drive initiated by the Swachh Bharat Mission, India has made significant strides in furthering access to individual household toilets for over 100 million households, mainly in rural communities where coverage gaps were larger (Ministry of Jal Shakti 2022). In 2019, 4,360 statutory towns and 262,769 gram panchayats, or rural local bodies, had declared themselves open defaecation free (ODF), indicating that all households within their census territory had toilets (Ministry of Housing and Urban Affairs (MoHUA) 2022).

The emerging evidence on the impact of unprecedented gains in access to individual household toilets among rural households hints at its positive correlation with water pollution abatement and improvement in the overall quality of life. For instance, according to a study by UNICEF India and the Department of Drinking Water and Sanitation, the Ministry of Jal Shakti (the erstwhile Ministry of Drinking Water and Sanitation) in 2019, villages that could not attain ODF status were 10% more likely to have bacterial contamination of their soil and over 12 times more likely to have contamination in their water sources (Ministry of Jal Shakti, UNICEF India, and SRI Delhi 2019). An economic assessment by Hutton *et al.* (2020) found that households with toilets in India, living in ODF settlements, saved approximately USD 727, primarily owing to costs saved from averting health and costs. Furthermore, Jadhav *et al.* (2016) have shown that access to household toilets is positively correlated with a reduction in the instances of violence against women and an increased sense of dignity and safety. While access to improved toilet facilities is associated with these multidimensional benefits, it is still only the first step towards ensuring universal access to safely managed sanitation. Moreover, the continuous usage of these newly constructed toilets itself is contingent on the household's ability to keep the toilets functional through the timely evacuation and management of the faecal sludge collected from the households via on-site sanitation (OSS) systems like septic tanks and pits.

A proper faecal sludge management (FSM) system provides for the safe desludging, conveyance, and treatment of faecal waste before it is reused or discharged in the open environment (Harada *et al.* 2016). However, until recent times, centralized sewerage systems comprising vast networks of underground pipes to convey wastewater to large-scale sewage treatment plants (STPs) had been considered the gold standard in wastewater and sludge management (Prasad & Ray 2019). However, these systems are resource-intensive to build, operate, and maintain. They require an uninterrupted supply of electricity, a vast amount of water for conveyance and treatment, and skilled operators and managers for the operation and maintenance (O&M) of the network and the plants (Dodane *et al.* 2012). Moreover, given the typical complexity and fragmentation endemic to local governance and funding mechanisms in India, sewage development projects come with unpredictably long gestation periods. Resultantly, as per the latest data from the National Sample Survey Organisation (NSSO) of India 2018, sewage systems serve only 17% of the total households and 39% of the total urban households. Therefore, the on-site containment of wastewater combined with FSM emerges as an indispensable solution to ensure ODF sustainability and safely managed sanitation, especially in rural areas. Owing to the diversity of India's settlements, underpinned by varying trends in urbanization, hydrogeological structures, and unregulated procurement of the various sludge containment options, there is a need to create a diverse portfolio of strategies and models for implementing an all-inclusive FSM strategy.

Currently, the National Policy on Faecal Sludge and Septage Management (FSSM) issued by the Ministry of Housing and Urban Affairs (MoHUA 2017) (2017) in India, evinces an institutional commitment to achieve the goal. However, its scope is limited to cities and towns. Similarly, other national programmatic investments, such as the Atal Mission for Urban Rejuvenation and Transformation (AMRUT), have underwritten FSM infrastructure in the 500 most populous cities of India (Ministry of Urban Development 2015). On the other hand, rural sanitation programmes since the 1980s have confined themselves to the construction of twin leaching pit toilets for providing the complete faecal sludge treatment at the household level itself (Dasgupta & Agarwal 2021). This is because unlike septic tanks and single pits, twin pits allow for near-continuous use without the need for mechanized or semi-mechanized desludging and off-site treatment of evacuated faecal sludge. After one pit fills up, it is sealed, and the wastewater is diverted to the second pit using a simple valve mechanism. In the time that it takes the second pit to fill up (usually 2–3 years), the contents of the first pit are expected to have dewatered and decomposed to a humus-like material that is safe for manual removal and ready for reuse as a soil conditioner (Tilley *et al.* 2014).

However, despite their continued promotion under government programmes, twin pits continue to have limited adoption in rural areas in favour of septic tanks and single pits. Resultantly, under the Swachch Bharat Mission – Gramin (SBM-G) Phase II, it is recommended to retrofit single pits to twin pits for obviating the need for FSM (Bhol *et al.* 2020). While twin leaching pits are an efficient system for achieving safely managed sanitation, due to their inherent operating principle, they are unlikely to meet the objective of complete household-level treatment in low-density and low groundwater table settings. Moreover, though limited, existing research on the techno-economic feasibility of retrofitting single pits to twin pits shows that households exhibit low willingness to pay for the implementation of such a strategy (Agarwal *et al.* 2020).

Therefore, a retrofitting-focused strategy in rural areas and the confinement of FSM to only urban areas will impede India's trajectory towards universal safely managed sanitation, especially among its rural population, which accounts for a majority of India's total population, at 898 million residents as of 2020 estimates (World Bank 2020). If India is to speed up its progress towards SDG 6, it is important to include a focus towards including rural communities as part of the target population for FSM too.

This paper presents a novel service delivery model for achieving universal, safely managed sanitation based on urban–rural convergence and transcending the false dichotomy typical of programmatic investments. The model was piloted in the Dhenkanal district of Odisha and provides a rapidly scalable pathway to increase access to safely managed sanitation among rural households in other Indian districts as well. While the model was implemented in an existing, well-functioning urban FSM system, planners and practitioners designing FSM systems in the future could incorporate the principle of urban–rural convergence from the onset and plan investments for a larger service area of urban–rural continuum.

The development and operationalization of the model have benefitted from strong political will, the lack of which could present a challenge in settings with limited institutional buy-in. The model also presupposes a specific pattern of socio-spatial transformations common in India, which leads to a preference for urban-like infrastructure among rural households and proximity between urban and rural settlements (Dasgupta *et al.* 2017). Therefore, the model may hold applicability in global contexts where rural and urban settlements are governed independently but share a common need for FSM and spatial continuity.

2. LITERATURE REVIEW AND STUDY CONTEXT

2.1. Global FSM

Managing faecal waste systems and practices differs from region to region based on underlying problems. Knowledge on the operational factors of faecal sludge is complemented by different procedural aspects like the quantification of sludge, the characterization of faecal sludge, features of collection, transport, and appropriate treatment. The evolution of sanitation practices shifted enormously with the advancement of treatment technologies. Some of the important studies carried out globally would provide the current scenario of the challenges faced, improvement in managing the challenges faced, changes in policies from countries, etc.

OSS technologies simplify the management of faecal waste to a large extent, and these systems are largely prevalent in the urban regions of sub-Saharan Africa ranging from 65 to 100% (Strauss *et al.* 2000). It is important to note that these on-site technologies are found in low- and middle-income countries and could result in the accumulation of faecal sludge. The existence of sewer systems and faecal sludge systems were also tried in unison in some countries, especially in Japan providing some useful insights into studying the FSM (Gaulke 2006). Apart from this study, there exists a variety of different technologies and a wide range of septic tanks used in Bangkok, Dakar, and Buenos Aires, whereas various types of pit latrines are predominant in Kampala, Nairobi, Kenya, and some other cities (Strande *et al.* 2014). A Water Sanitation Programme (WSP) (2014) study analysed the systems used in FSM in 12 cities representing different regions, sizes, and levels of service delivery. In these studies, dependence on OSS technologies, minimum percentage on-site management, unsystematic, unhygienic manual emptying, and unplanned regulations in treatment facilities of the waste are some of the common issues identified. A persistent challenge across these study contexts in particular and in the sanitation sector in general has been finding an adequate funding for sanitation development. Investment and cost are significant factors in implementing FSM technologies (Dodane *et al.* 2012).

Changing perceptions of how sanitation should be implemented, through networked or non-networked approaches, with innovation towards more flexible approaches, greater technical know-how, new strategies, combined with sound financial mechanisms will allow innovations to prosper. Governance and institutional structures for sanitation projects are also problematic. Up until recently, public organizations provided the most basic consumer utilities, such as water, sanitation, and transportation. Many non-state actors, especially in developing countries like India, Bangladesh, Nepal, and some African countries, have invested heavily for the betterment of the sanitation sector in recent years. Scholars and analysts have yet to acknowledge the combination of state and non-state providers prevalent in the sector and define their role through regulation. Where decentralized approaches are being implemented, non-state actors are already involved (Kelkar & Seetharam 2019). Therefore, it is significant that developing countries are already using hybrid unregulated systems for service delivery. However, the provision of water and sanitation services has largely been an activity of the public and heavily affected by the politics of user charges and the fate of regimes, where the general sentiment has been protective. Due to the emergence of NGOs, private companies, and social enterprises in this sector, service provision has been highly fragmented and unregulated, leading to challenges in replication and scalability. Sector-wise, these structures have helped to resolve issues in some areas but fail to address the larger sanitation problem.

2.2. Current status of FSM in India

In 2017, around the time of MoHUA's release of its FSSM policy, many states began conducting FSM pilots in cities and smaller towns. As of July 2020, more than 19 Indian states have issued state-level FSM policies (FSSM Policy 2017), and 13 have constructed 109 faecal sludge treatment plants (FSTPs) with another 193 underway (National FSM PMU GoI 2020). Although cities adopted varying business models, the policy mandated the formation of a City Sanitation Task Force (MoHUA 2017) to coordinate the planning and implementation of the FSM-inclusive city sanitation plan to ensure streamlined support and inclusion of overlooked communities.

As per the Joint Monitoring Programme (JMP) ladder, in India (Table 1), 67% of rural households have access to at least basic sanitation facilities, 8% with limited or shared, and 2% have access to unimproved sanitation facilities, whereas 22% households are practising open defaecation, which has improved from 40% in 2015 to 22% by 2022 (WHO & UNICEF 2021).

2.3. FSM in rural India

As of the 2011 Census, only 17% of India's total population was served by the centralized sewerage system with the majority of the coverage being restricted to urban areas. The rest of the population currently depends on OSS

Table 1 | Status of the management of faecal sludge and wastewater in India

		Sanitation status (%)						
		Proportion of the population using improved sanitation facilities (excluding shared)				Proportions of the population using improved sanitation facilities (including shared)		
Year	Location	Safely managed	Disposed *in situ*	Emptied and treated	Wastewater treated	Pit latrines and other	Septic tanks	Sewer connections
2015	Total	36	33	<1	3	27	30	11
	Rural	37	37	<1	<1	34	23	<1
	Urban	33	24	<1	9	13	46	31
2020	Total	46	42	<1	4	35	36	13
	Rural	51	50	<1	<1	46	28	1
	Urban	37	27	<1	11	13	51	34

systems, such as leach pits and septic tanks, for containing and partially treating faecal waste on-site. Generally, OSS systems (MoHUA 2017) are considered relatively easy to build and maintain in rural areas, given that there are several safe, low-cost, and low-effort options. These include septic systems, single pits, composting toilets, and twin leaching pits, among others. These OSS systems, however, come with their own challenges for maintenance, continued functionality, and performance. There needs to be periodic removal and conveyance of faecal waste to an off-site treatment facility like an FSTP, where it can be scientifically treated before disposal.

Twin leach pits were widely promoted during SBM-G Phase I as a sustainable self-contained system. However, in rock soil or high water-table areas, the risk of overflowing or contamination of groundwater was higher. In addition to the twin pits in unsuitable terrains, the toilets constructed with a single pit or septic tanks also present a need for FSM to ensure the safe management of effluent once filled. The single leaching pit remains one of the most widely used sanitation technologies – especially in rural India – that stores the faecal matter over time. Once the single pit is filled, it requires a second pit to empty the contents of a single pit with the safe management of sludge to protect the environment.

Prior to the large-scale construction of subsidized toilets, over 40% of rural households (38.5 million) had toilets connected to septic tanks as per Census 2011. The continued preference for septic tanks is highlighted in a 2017 study by WaterAid India (WaterAid India 2019), which revealed that rural households prefer septic tanks despite the government efforts to promote twin leach pits, especially as their socioeconomic status improved. The study also found that many septic tanks that were constructed did not comply with all the standards set by the Bureau of Indian Standards, making them more susceptible to system failure and contamination. According to the NARS 18–19, 34% of all rural toilets at the national level are connected to septic tanks without soak pits, while only 3% are connected to tanks with soak pits. Nearly 19% of the rural toilets are connected to single pits.

The differential prevalence of the three main types of OSS systems in rural India is also determined by the diverse characteristics of the rural settlements. Administrative boundaries of 'urban' and 'rural' notwithstanding, the Census of India 2011 identified census towns, as rurally governed, urban-like settlements having more than 5,000 residents and a population density of 400 people per square kilometre with 75% or more of the workforce employed in non-agricultural activities. As of 2011, there were 3,894 of them compared to 1,362 in 2001. Similarly, there are large and densely populated villages (LDVs) which like census towns are administered as rural and have a population of at least 1,000 and a density of 400 persons per square kilometre (Dasgupta *et al.* 2017). Through this study, Dasgupta *et al.* (2017) show that the dependence on septic tanks steadily increases on going from villages to largely and densely populated villages further to census towns and finally, urban statutory towns. Therefore, the demand for FSM shows a clear pattern in relation to the rural settlement's demographic characteristics and its proximity to urban areas – creating an opportunity for graded FSM interventions in rural areas.

2.4. FSM in the state of Odisha, India

The state of Odisha has been at the forefront of FSM and urban sanitation in India and issued the Odisha Urban Sanitation Policy and Odisha Urban Sanitation Strategy in 2017. The state also established two of the lighthouse urban pilots on FSM in smaller towns and nine FSM pilots in bigger AMRUT cities. Following their success, the state is currently scaling-up FSM to all its 114 urban local bodies and poised to emerge as a pioneer in rural sanitation with the release of the Odisha Rural Sanitation Policy, 2020. The policy provides a systematic framework to understand and address the gaps in rural sanitation and covers several aspects like access to toilets, behavioural change for elimination of open defaecation, FSM, greywater management, solid waste management, and menstrual hygiene management.

The administrative structure for sanitation and FSM is splintered across the urban–rural divide in the state, with two distinct arms that govern urban and rural developments in Odisha. At the state level, while the Panchayati Raj and Drinking Water Department (PR&DWD) governs rural administrations, the Housing and Urban Development Department (H&UD) performs a similar function for urban local bodies also called municipal authorities. Following the enactment of the Orissa Grama Panchayat Act, 1948, the state introduced a three-tier system of rural governance via Panchayat Raj Institutions (PRIs) at the district, block, and village levels. Additionally, in 1992, the 73rd amendment of the Constitution of India further strengthened the role of PRIs – repositioning gram panchayats as institutions of rural local governance. Therefore, at the local level, gram panchayats, with the support of their constituent committees (e.g. Village Water and Sanitation Committee (VWSC)), are responsible for water and sanitation service delivery. Urban local bodies, municipal corporations, and municipalities fulfil the function in urban settlements.

2.5. Study context

Indian states are divided into smaller governance units called districts that comprise multiple urban (municipalities) and rural local bodies (gram panchayats). The state of Odisha is divided into 30 districts including Dhenkanal (Figure 1). It is located in the central part of the state and is spread out over 4,452 km^2. As per the Census of India 2011, the district had a total population of 1.2 million with about 270,000 urban and rural households residing in 3 statutory towns and 199 gram panchayats. The rural areas in the district comprise more than 90% of its total population.

The majority of the blocks in the Dhenkanal district are considered to be at a safe level of groundwater development that is measured as a ratio of annual groundwater draft and net annual groundwater availability in percentage by the Central Ground Water Board (CGWB). The groundwater table varies between 2 and 10 metres below the ground level (mbgl) annually (Figure 2).

Located near the coastal belt of the state, about three-fourths of the district's land area falls in the 'very high damage risk zone' and the remaining in the 'high damage risk zone with respect to cyclones'. In addition to cyclones, the district also suffers from moderate risk to earthquakes. The Dhenkanal Municipality in the district is one of the first small towns in the country to have introduced a city-wide FSM system. The municipality commissioned a nature-based FSTP capable of treating 27 kL per day (KLD) of sludge in 2018. Given the district's

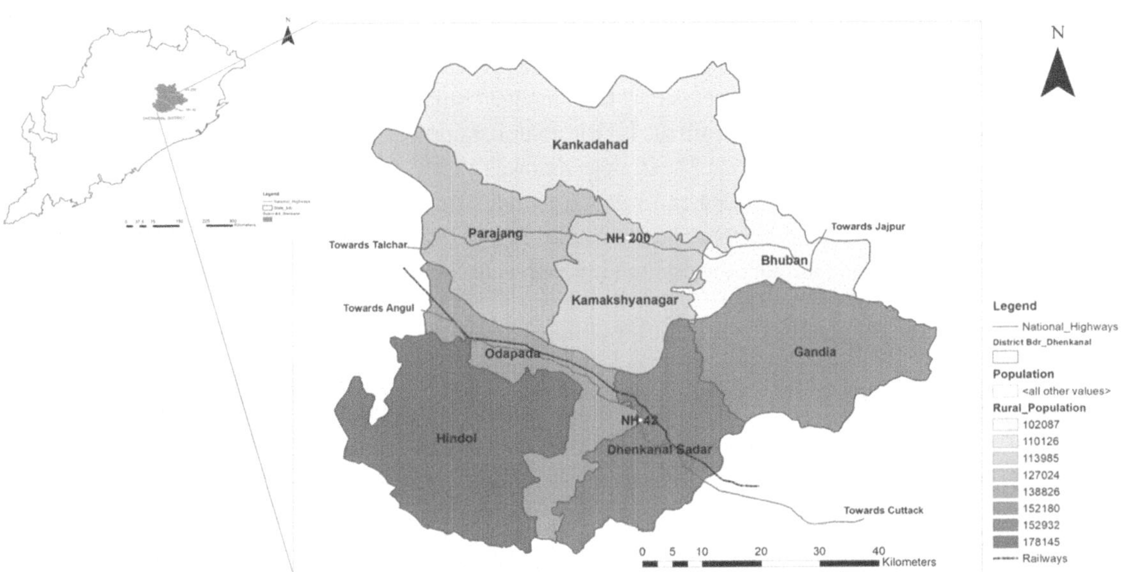

Figure 1 | Location of the district and population in different blocks of the district.

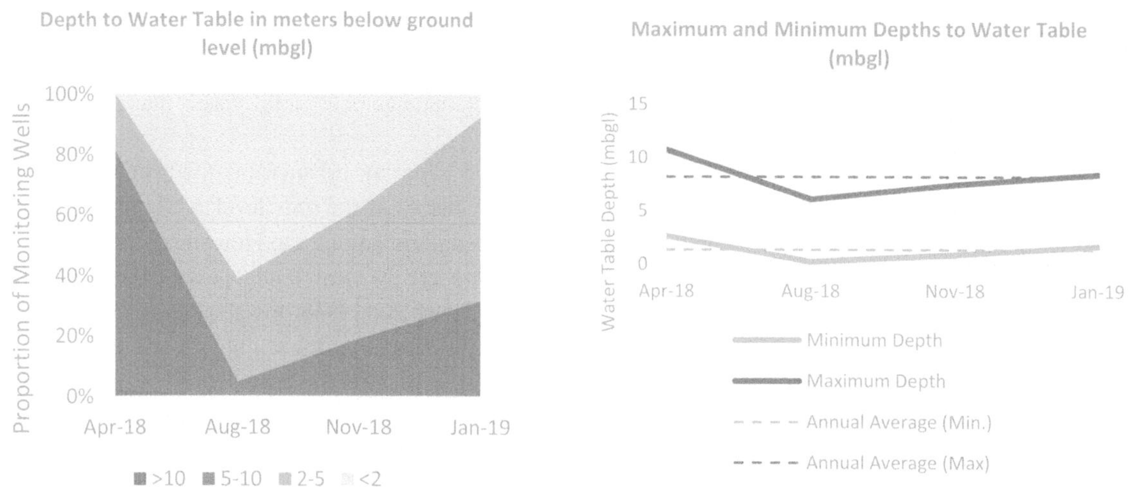

Figure 2 | Depth to the groundwater table in the Dhenkanal district.

attainment of ODF status across its rural and urban areas in 2019, the next step was to institute district-wide safely managed sanitation.

As of late, 17 gram panchayats and the Dhenkanal Municipality signed a one-on-one Memorandum of Agreements (MoAs) on 28 December 2020. The MoAs set the price of desludging at INR 1,250 (USD 17) and INR 750 (USD 10) per trip for vehicles of 3 and 1 KL sizes, respectively, which are lower than the *ad hoc* price that rural households were paying for the services before the operationalization of the model. Moreover, to bring down the cost of desludging services per household further, the request allocation allows for the pooling of multiple requests in one trip. During the first 5 months (Feb–Jun 2021) of its operations, 278 KL of sludge or 93 truckloads from rural areas have been treated in the FSTP, thereby preventing its indiscriminate disposal from polluting the open environment. While the average fees paid by rural households has decreased following the formalization of the plug-in model, total revenue amounting to INR 140,000 or USD 1882 has been collected during this period, thus supporting and ensuring service sustainability.

Currently, the state government is testing the approach in six additional districts to further explore the adaptability required for the model to be replicable state-wide. The national government has also paved the way for the institutionalization of the approach by espousing it in an official letter issued in September 2021.

3. METHODOLOGY AND APPROACH

In the last couple of years, the state of Odisha has witnessed a rise in the requirement for FSM services in its rural areas and a parallel scale-up of urban FSM systems across all its cities and towns. The phenomenon presented a unique opportunity to rapidly increase access to safely managed sanitation in rural Odisha by creating a partnership between urban and rural local governments for the co-utilization of FSM systems. Motivated by it, we sought to pilot and study a model of urban–rural FSM strategy that was effective, scalable, and built on an understanding of the mismatch between supply and demand relating to infrastructure, awareness, governance capacity, and regulation. Figure 3 presents the process through which this progressive FSM strategy was realized. In this process, study results, as informed through the preliminary and secondary data analysis (Step 1 of this process shown

Figure 3 | Process adopted for developing and institutionalizing the urban–rural convergence model in Dhenkanal, Odisha.

in Figure 3), were used to pilot the implementation of the urban–rural FSM strategy. The subsequent sections in this paper present these steps and associated outputs in more detail.

3.1. Model development

Given the Dhenkanal FSTP was the only available functional facility in the district, we considered the available spare capacity of the FSTP as a given boundary condition. Therefore, we undertook an assessment of FSTP's capacity utilization as the first step, which we discuss in Section 5. As a logical next step, we aimed to estimate the size of the rural population that the FSTP could comfortably serve over the next 3 years. A practical estimate of the current and future infrastructure demand would require knowing the average:

- share of the rural population that depends on single pits and septic tanks and would therefore require FSM services,
- the sizes of the single pits and septic tanks,
- the number of users,
- the hydraulic loading of single pits and septic tanks, and
- the average desludging frequencies of the systems.

Owing to the paucity of universal benchmarks for these parameters in the rural context, we had to determine their values using a primary household survey as discussed in Section 4. Moreover, given the novelty of toilets and the deficiencies in behavioural change towards regular and consistent toilet usage, we were interested in understanding how well the relatively hypothetical infrastructure demand would translate to the uptake of services among rural households. Therefore, the survey was also used to ascertain the rural households' willingness to pay for the two mainstream alternatives to safely managed sanitation, for example, by retrofitting existing single pits to twin pits and FSM. Interviews with government officials, conducted as part of the survey helped to define the need for capacity building and to assess the role that Information, Education, Communication (IEC) campaigns, could play in shifting household behaviour towards safer sanitation outcomes.

Following data collection and analysis, a list of gram panchayats was identified in the vicinity of the Dhenkanal municipality that the urban FSTP could serve by matching supply against demand. As a part of the evaluation, we determined if the municipality's existing fleet of desludging vehicles (DVs) would suffice to cater to both urban and rural households with requisite promptness on the receipt of service requests. The findings were utilized to build the urban–rural convergence model in consultation with urban and rural local body officials and the district administration. For its institutionalization, the rural and urban local governments were supported to formalize the terms of the engagement in an MoA. The operationalization of the model and service delivery was followed by the capacity building of gram panchayat officials and IEC campaigns for raising awareness in households about the need for and access to FSM services.

3.2. Data collection and analysis

Assessing the gap towards safely managed sanitation in a given setting requires building a holistic understanding of the existing infrastructure, behaviour, and capacity. Data on the prevalent OSS systems typology, their characteristics, and local hydrogeology are key factors for estimating the need for FSM in any given region. Government datasets like the Census of India, National Sample Survey, and National Annual Rural Sanitation Survey, while providing adequate information on the OSS systems in use (including at the settlement level), do not provide detailed information on the size of OSS systems, toilet usage, and hydraulic loading, among others, that are crucial for infrastructural planning. Therefore, a district-wide sample survey of households led to a deeper understanding of the sanitation facilities in use in rural areas of the Dhenkanal district.

The survey included a sample of 1,000 households in the district. This allowed us to report the results with a margin of error of less than 5% and a confidence level of 99% according to the Cochran formula and achieve representativeness of the findings at the district level. The next steps, as discussed below, focused on determining how to allocate these 1,000 households across the district:

- *Gram Panchayats*: Since gram panchayats form the primary local governance unit in rural areas, it became the first-stage sampling unit. In order to approximate a normal probability distribution at the gram panchayat level, a sample size of 30 households was adopted per panchayat (Kwak & Kim 2017). Therefore, considering the total sample size of 1,000 households and a gram panchayat sample of 30 households, the sample was distributed across 33 gram panchayats (or 1,000/30). Probability proportionate to size sampling was deployed to select the 33 gram panchayats out of 199 in the district.

- *Villages*: Since gram panchayats are typically comprised of one or more villages often with varying socioeconomic characteristics, the gram panchayat sample was allocated across three villages in each panchayat. Here, simple random sampling was used to select the three villages of the total. In the case where the gram panchayat had three or fewer villages, all the villages were selected for surveying. In both cases, the sample was distributed equally across the selected villages.
- *Households*: Within each village, the households were randomly selected to ensure the maximum geographical coverage by the field enumerators.

Overall, the household survey covered 97 villages in 33 gram panchayats. The questionnaire for the household survey sought information on the household's socioeconomic characteristics, access to the toilet and its usage, type, and characteristics of the OSS system, access to desludging services, willingness to pay for services, and other aspects related to sanitation such as water supply and drainage. It was administered in the computer-assisted personal interviewing mode using the Survey123 application provided by the Environmental Systems Research Institute (ESRI). A quantitative analysis was conducted using Microsoft Excel 16.0 and Stata Standard Edition 13.0.

In addition to the household survey, we interviewed key stakeholders including gram panchayat officials, members of the VWSCs, local masons, and desludging service providers. Understanding the governance side of the service delivery enabled us to further contextualize and triangulate the quantitative data collected through household surveys. These interviews were conducted in 8 out of the 33 GPs selected for the survey.

4. RESULTS AND DISCUSSION

This section presents the results from the primary survey and secondary data analyses and discusses how these data-informed programmatic actions took place within the urban–rural FSM model in the Dhenkanal district in Odisha. We conclude this section with a higher-level reflection on study insights and their broader application to global FSM programmes.

4.1. Survey results on FSM technology use, demand, and feasibility

The data sample (Table 2) from the primary survey consisted of 60% male respondents and 40% female respondents. Thirty percent of the sample was from the general category, 34% from other backward classes (OBCs), 24% from scheduled castes (SCs) and 12% from scheduled tribes (STs). Most of the respondents are either self-

Table 2 | Demographic profile of the sample

Socioeconomic variable		Percentage of responses
1. Gender	Female	60
	Male	40
2. Social category	General	29
	OBC	34
	SC	24
	ST	13
3. Income source	Self-employed in agriculture	35
	Self-employed in non-agriculture	17
	Casual labour in agriculture	4
	Casual labour in non-agricultural	31
	Salaried – public/government	3
	Salaried – private	9
	No source of income	2
4. Monthly per capita expenditure (MPCE)	≤INR 500	14
	INR 500–1,000	37
	INR 1,000–2,000	42
	>INR 2,000	7
5. House type	Kutcha	28
	Pucca	72

employed in agriculture (35%) or casual labourers in the non-agricultural sector (31%), and 80% of the sample has monthly per capita expenditure (MPCE) between INR 500 and 2,000 (USD 7–29). It was found that the SC and ST category households were concentrated in the lower consumption categories. Overall, nearly all households report residing in self-owned houses, with 28% of the sampled households residing in *kutcha* or hut-like dwellings, and 72% in *pucca* or brick and cement dwellings. Subsequently, we present salient findings from this survey including (i) an increased dependence on single pits and septic tanks, (ii) low techno-economic feasibility for retrofitting, and (iii) unsafe manual desludging services due to high on-ground demand for FSM.

4.1.1. Increased dependence on single pits and septic tanks

A substantiative increase in the accessibility to toilet facilities was seen in the data, a likely result from the progress under SBM-G, between 2014 and 2019 (Figure 4). The toilets that have been constructed under the scheme are predominantly connected to a single pit (84%). All the eight-gram panchayat *sarpanches* that were interviewed and confirmed that single leaching pits were the predominant system constructed under SBM-G in their villages. The prevalence of single pit increases as one moves from higher-to-lower consumption quintiles. From the stakeholder interviews, it was found that the primary reasons for preferring single pits include low cost for household, rapid construction targets for contractors, and social norms of purity–impurity with the Y-junction. Richer households preferred septic tanks, which show a statistically significant correlation ($p < 0.001$) with non-SBM or privately constructed toilets.

4.1.2. Low techno-economic feasibility for retrofitting

The data highlighted a low preference for retrofitting a single-pit toilet to a twin-pit one, as it involves both financial and technical challenges. Although 25% of the households believed that the addition of a second pit would make toilet use more convenient and maintenance easier and sustainable, a lower proportion reported that they would be willing to pay for such an intervention. Nearly 73% of the households who expressed the willingness to pay reported that they would be willing to pay up to INR 500 (USD 7) for retrofitting, a fraction of the anticipated retrofitting costs. *Swachhagrahis*, or community sanitation volunteers, interviewed for the study also reaffirmed that financial support or subsidy would motivate households for retrofitting (UNICEF-CPR 2020).

Research shows that less than 2% of the reported single pits have a junction chamber for connecting a second pit, thereby further complicating the retrofitting process. A district-wide retrofitting drive would also entail a comprehensive capacity-building programme with only one out of five masons interviewed reporting any prior experience in retrofitting. Therefore, the need arises to facilitate beneficiary-led retrofitting of all dysfunctional or ill-constructed toilets and OSS systems to ensure their continued functionality and usage through market linkages, creating awareness among communities, building capacities of masons, other key stakeholders, etc. Even if these challenges are overcome, the high groundwater table underlying the district restricts the strategy's hydrogeological suitability. Figure 5 shows a map of groundwater availability (and feasibility) within the Dhenkanal district. The map is created by using district and GP/village boundaries provided by the Census 2011 District handbook and overlaying the CGWB groundwater data on this base to assess the feasibility of retrofitting strategy in the district.

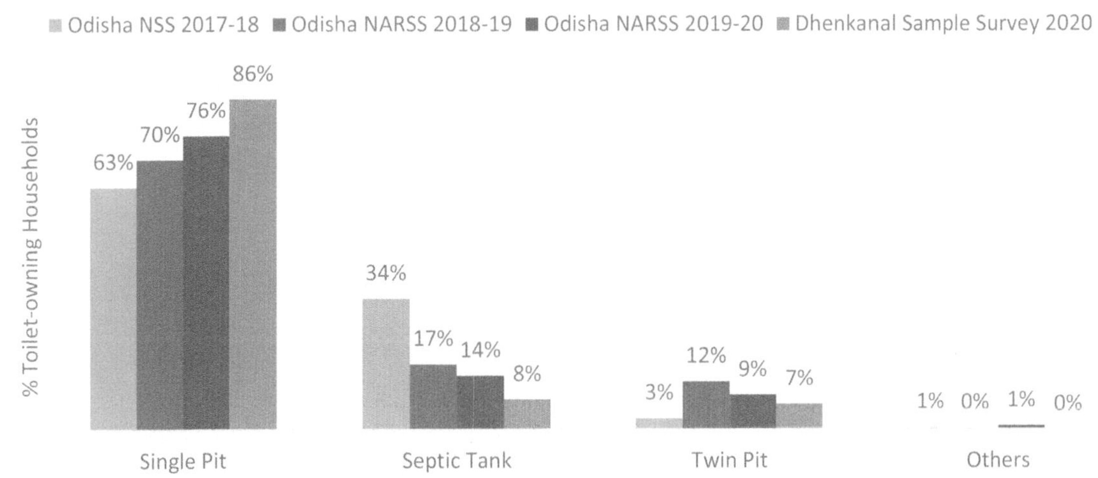

Figure 4 | Different types of OSS systems in use in Odisha.

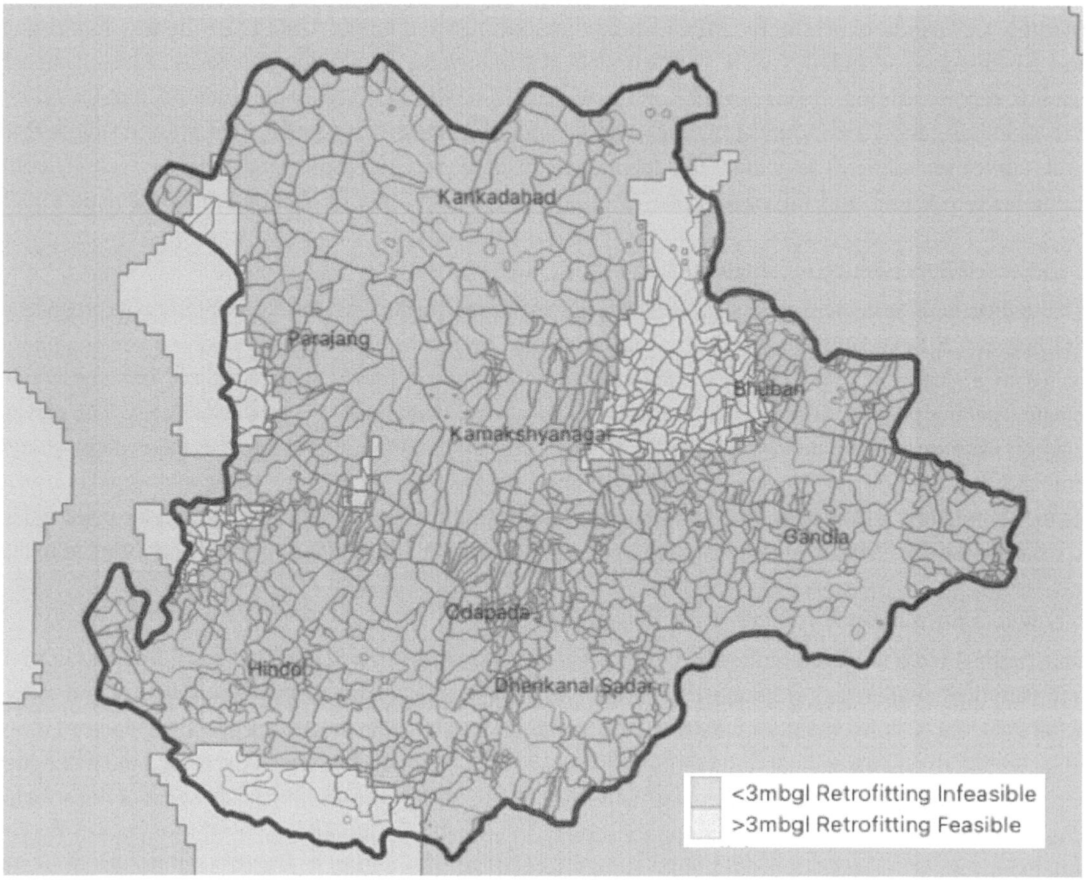

Figure 5 | Retrofitting feasibility in the Dhenkanal district guided by groundwater criteria based on groundwater data (mbgl) published by [CGWB 2019–Central Ground Water Board. Ground Water Year Book 2019-2020] 2020.

4.1.3. Unsafe manual desludging services due to high on-ground demand for FSM

In view of the above situation, with large numbers of septic tanks and single-pit toilets, both of which require emptying, the institution of FSM services would be critical towards preventing risks to public health and the environment but also for the safety and dignity of those providing manual desludging services (Figure 6). The data show that a high share of households engage in manual labour for desludging their OSS systems in the absence of widely available, affordable, safer mechanized desludging services. As the demand for desludging services increases concomitant to increased toilet usage, the inaccessibility of safe desludging services could further entrench the dependence on manual labour. Therefore, it is imperative to establish proper FSM services and safely managed sanitation in rural areas.

4.2. Results from secondary and primary data to evaluate household demand and implementation strategies

Subsequently, we provide a summary of the results obtained from the analysis of the primary household survey, operational data from the Dhenkanal FSTP, and other operational data to evaluate the available spare capacity of

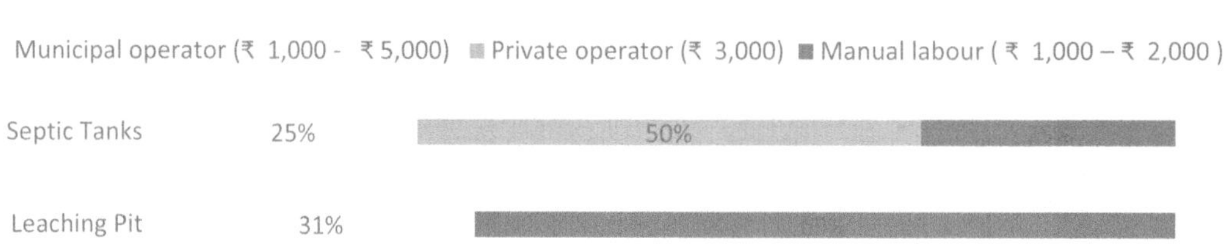

Figure 6 | Prevalence of different types of desludging service providers in Dhenkanal.

FSTP, to estimate the demand for services from rural households, and to estimate the associated implications this demand has on the need for additional DVs. These results, combined with the other insights from the primary survey, informed the piloting, formalization, and future monitoring and evaluation of an urban–rural FSM programme, as well as a monitoring and evaluation strategy for future programme implementation.

4.2.1. Estimating the available spare capacity of FSTP

An analysis of the capacity utilization of the urban FSTP over a 2-year period, 2019–2021, was deployed to estimate the spare capacity and ascertain the maximum range of the rural demand for FSM that could be served. We observed that although the utilization fluctuated between months and across individual days, on average, the capacity utilization was consistent at 50% or about 13 KLD for both years (Figure 7). While data from more years of operation are necessary to understand and identify any cyclical patterns in demand for services, the trend for the first 2 years suggests overall constancy in the amount of sludge arriving at the FSTP for treatment each year.

4.2.2. Estimating demand from rural households

Once the available capacity was established, we estimated the hypothetical demand from rural households over a 3-year period based on average values of parameters including household size, septic tank and single-pit sizes, and hydraulic loading that we assessed using the survey data. Due to the significant difference in the average sizes of septic tanks and single pits reported by households, 16,990 and 600 L, respectively, and the difference in the benchmarked sludge accumulation rates (Bureau of Indian Standards 1985), the demand for the two types of systems were estimated separately. To arrive at the total demand, we weighted the estimated demand per septic tank and single pit by the respective number of households dependent on them. We used a relative share reported by the survey and the total number of households from governmental records to calculate the latter. The estimation considered two scenarios – first, where erstwhile and ongoing IEC campaigns focused on toilet usage would result in 100% usage of toilets in the next 3 years and second, where the campaigns meet limited success and toilet usage, and resultingly, the demand for FSM services would ramp up slowly.

We started with an initial radius of 10 km for the extended rural service area and listed all the gram panchayats falling within it (Figure 8). Since the constituent villages of a gram panchayat may fall outside or within the cut-off radii independently, an inclusive criterion was adopted while preparing the list of gram panchayats. Based on the aforementioned 2 scenarios and the list of 17 gram panchayats, the estimated rural demand was settled in a range of 9–12 KLD, which is within the available spare capacity of the FSTP calculated in the previous step.

4.2.3. Estimating requirement for additional vehicles

The Dhenkanal municipality owns two DVs of 3,000 KL and is the primary and sole provider of demand-based desludging services in the town. It has contracted out the day-to-day management and operations for desludging service delivery to a women self-help group (WSHG). Like many other small towns and cities in Odisha, it lacks private service providers. Noting the inaccessibility of DVs in settings with high density and narrow approach roads over 2 years of operation, the Urban Local Body (ULB) procured an additional smaller-sized vehicle of 1 KLD, which has been under operation since October 2020. Considering that the rural demand for FSM will ramp up slowly, it was concluded that for the next 2 years the existing fleet will suffice. In the medium term, an additional DV of 3 KLD – allowing the pooling of desludging requests – will be required to cover the full-service area.

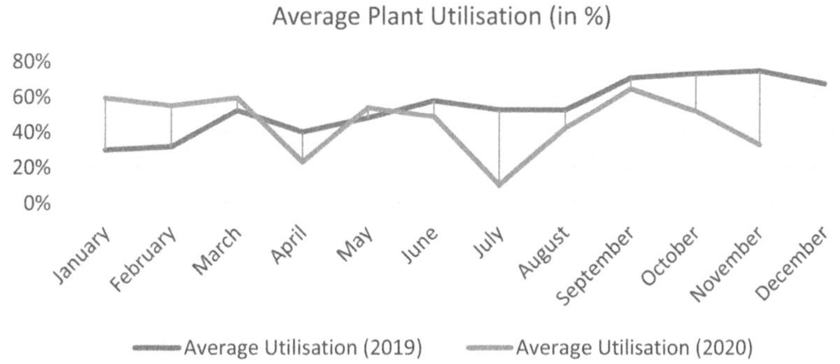

Figure 7 | Capacity utilization at FSTP.

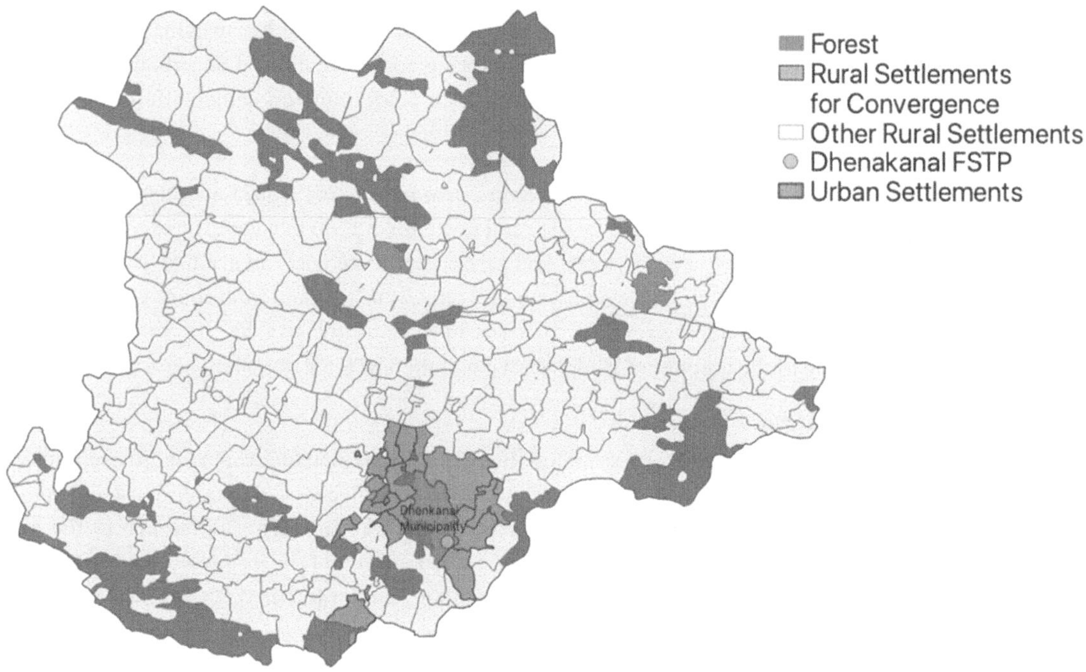

Figure 8 | Location of the Dhenkanal municipality and gram panchayats identified for urban–rural convergence.

4.2.4. Formalizing governance and regulation

The governance and coordination mechanisms between the rural and urban local bodies for the delivery of demand-based desludging services to rural households and the treatment of the collected faecal sludge at the urban FSTP have been articulated in an MoA. Given that the plug-in model is based on the rural–urban and rural–rural coordination of multiple local bodies, it was critical to set out a clear framework that specifies the terms of their cooperation. It helps define the roles and responsibilities of the district/block/local governments and provides a clear framework for the day-to-day engagement between the urban and rural local bodies, as applicable, on issues of invoice generation, payment transfers, monitoring, and recordkeeping, among others.

Specific to the municipality, the MoA requires it to amend its contract with the WSHG to ensure that the extended service area is included under the purview of the group. Other major responsibilities of the municipality include:

- Ensuring the safe treatment of faecal sludge collected from OSS systems,
- Extending the services of the call centre/toll-free helpline number/mobile application currently available to urban households for requesting services to rural households,
- Providing the necessary equipment and personal protective gear to the sanitation workers and ensuring their safety,
- Maintaining digital records of complaints and their redressal, and
- Developing desludging tariff in consultation with the gram panchayat and officially notifying it.

Under the MoA, the gram panchayat is similarly responsible for:

- Maintaining a database of OSS systems and
- Undertaking IEC campaigns to build awareness of households about the availability of services and the importance of timely desludging and FSM.

Most importantly, the MoA provides systematic review and monitoring of the model, under the guidance of the district administration, to facilitate continued and smooth service delivery.

4.2.5. Future progress, monitoring, and future adjustments

The 17 gram panchayats and the Dhenkanal Municipality signed one-on-one MoAs on 28 December 2020. The MoAs set the price of desludging at INR 1,250 (USD 17) and INR 750 (USD 10) per trip for vehicles of 3 and 1 KL sizes, respectively, which are lower than the *ad hoc* price that rural households were paying for the services before the operationalization of the model. Moreover, to bring down the cost of desludging services per household further, the request allocation allows for the pooling of multiple requests in one trip. During the first 5 months (Feb–Jun 2021) of its operations, 278 KL of sludge or 93 truckloads from rural areas have been treated in the FSTP, thereby preventing its indiscriminate disposal from polluting the open environment (Figure 9). While the average fees paid by rural households has decreased following the formalization of the plug-in model, total revenue amounting to INR 140,000 or USD 1882 has been collected during this period thus supporting and ensuring service sustainability.

The state government is testing the approach in six additional districts to further explore the adaptability required for the model to be replicable state-wide. The national government has also paved the way for the institutionalization of the approach by espousing it in an official letter issued in September 2021.

The formalized urban–rural convergence model for FSM and safely managed sanitation is the first of its kind in India. To ensure its operational success and efficiency, the team worked with the urban local body to modify its Management Information System (MIS) to track service delivery to rural areas. The MIS-tracked parameters include the date on which desludging was requested, the time taken to fulfil the request, the fees charged, the type of OSS system, the name of the gram panchayat, the roundtrip distance covered, etc.

As the model matures, the time may come for a dynamic repository of data and more on-ground consultations to answer more questions and ensure that community access to safely managed sanitation is sustained and equitable. Some of these could be:

- Is the service delivery to rural households timely? Does it differ from the average amount of time taken to fulfil requests from urban households? How does it impact the sustainability of the ODF status?
- Are the DVs running at optimal capacity? Does the municipality need more DVs to deliver timely services to a larger service area?
- Is the FSTP beginning to near the designed capacity? Is the demand from rural households exceeding its estimate? Could the municipality augment the FSTP? Could the reduction of service area and the development of greenfield FSM systems for the areas left out work as a more realistic solution?
- Is pooling of requests, especially for the low-volume single pits, in a single trip of the DV (to and from the FSTP) a viable and efficient mechanism to bring down costs for desludging services per household?
- Is the desludging tariff applicable to rural areas adequately covering the cost of service delivery? Does the gram panchayat need to subsidize services to balance affordability with the financial sustainability of services?

4.3. Higher-level insights for global FSM programmes

Our results indicate that the household preference for different sanitation systems has transcended the rural–urban divide (Dasgupta & Agarwal 2022). As such, the current success of the 'plug-in' model that bridges the urban–rural divide in Dhenkanal offers exciting potential to replicate the approach for hundreds of other municipalities with functioning FSTPs and surrounding rural communities sustaining deficits in treatment options. The model also demonstrates that there are multiple variables involved in determining the feasibility and sustainability of this intervention.

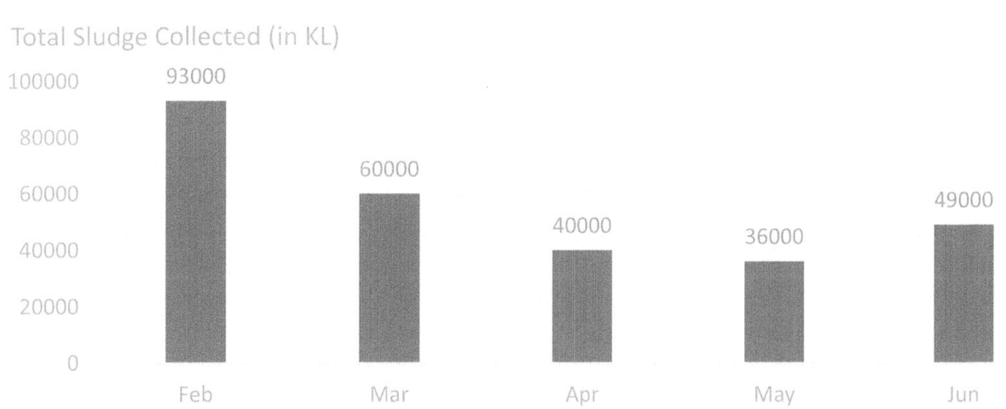

Figure 9 | Progress of the plug-in model.

First, it is imperative to understand the existing needs of the communities to be supported. This aligns with various best practices on sustainable community development (e.g. Narayan-Parker 1993, 1995; Chambers 1994). While precedence for the evaluation of community need is not surprising, the more nuanced characterization of salient community needs to be illuminated through this study that is useful and aligns with insights from other studies in the water and sanitation sector. For example, the pre-intervention survey assisted us in identifying not just the various types of OSS systems that existed – with the majority being single pits – but also the geographic spread and density of the households (Allen *et al.* 2006; OECD 2013), and income-sensitive willingness to pay for integrated treatment services (Whittington *et al.* 2009; Arouna & Dabbert 2012; Naiga & Penker 2014). Overall, the findings from the study survey provided qualitative feedback on the appetite for and awareness of FSM's importance.

Second, such approaches cannot function without agreed upon governance models (Caffyn & Dahlström 2005). This is especially true in India, where urban and rural services are segregated between ministries and departments serving each population separately. Given that this effort to connect urban services with rural populations is unprecedented, outlining strategies for stakeholder collaboration based on key roles and responsibilities served to generate consensus among policymakers and also ensure that there was shared accountability for the success of the model. Such forms of collective agreement and action are key activities in pursuit of sustainable water and sanitation development (Pugel *et al.* 2020, 2022a, 2022b). Previous studies in the area of inter-municipal cooperation have also highlighted the potential role played by urban–rural partnerships in efficiently delivering services and raising the capacity of local governments in Europe and the USA (Warner 2006; Hulst & van Montfort 2011; OECD 2013). In many developing countries, however, hybrid models of system delivery, consisting of a combination of state and non-state providers, remained prevalent in the sanitation sector albeit unregulated (Kelkar & Seetharam 2019). The current success of the 'plug-in' model that bridges the urban–rural divide in Dhenkanal offers exciting potential to replicate the approach for hundreds of other municipalities with functioning FSTPs and surrounding rural communities sustaining deficits in treatment options. In Dhenkanal specifically, the ULB signed MoAs between the rural local bodies to formalize the terms of the partnership and also ensured not just the horizontal alignment but also vertical integration, by restructuring the terms of labour for the service providers themselves, while also granting them the necessary autonomy for deliberating on tariffs and community engagement. Such urban–rural partnerships can help maintain distinct local governments, which may still be relevant due to persistent differences between these areas, while creating networks of governance that help coordinate the management (Warner 2006; OECD 2013) of urban–rural linkages. A final element that is continuing to play a key role in programme success is social and behaviour change communication (BCC). Developing the necessary information and communication materials has supported the uptake of the services offered to rural communities, which have previously not been familiar with the treatment option and thus risked having low demand for it. Sanitation literature point to BCC in general, through the thoughtful development of communication materials in particular, is an important tool in promoting sustainable service use and provision (Mwakitalima *et al.* 2018; Kalam *et al.* 2021). The proper communication on the impetus for and the proper use of FSM infrastructure is necessary, given that even though the tariff for collection is, on average, lower per household than previous offers, it is still a relatively large lump sum when compared to the average incomes earned by the rural households serviced. Therefore, information campaigns that advertised the importance of FSM, supported by government bodies, as well as making it easy to access the service, played a strong role in generating demand.

5. LIMITATIONS AND FUTURE WORK

Given that the study approach applied here has many uncontrolled components, there are limitations that would affect future replication and scale-up. Currently, while an MIS is being developed, it is not tracking all relevant indicators and thus would not be able to give the full set of data needed to analyse whether the model will continue to remain successful with various household densities over time and whether the tariff model is sustainable for the long term. What adjustments the partnership arrangement may need over the medium- to long term is still an open question that more experience could answer. Government cooperation and interest play a large role in the success of a social good – such as waste management – and the investment may vary depending on the day-to-day priorities of policymakers and leaders in various cities and rural habitations. Essentially, the levels of institutional strength and willingness would be a strong determinant of the approach's successful replicability in varying contexts. Finally, FSM is only one stream of waste management and while it is clear that tonnes of faecal waste treated safely are saving local resources from contamination, it cannot alone influence significant changes in public health. There will need to be more holistic investments in waste management across all streams – greywater, solids including plastics, animal waste, etc. – for there to be a sustained and positive impact on the health of communities and especially those of marginalized communities. Greywater management, in particular, will be critical as exploitation and water stress continues to burden households, and recycled water can have a large role to play in continuing FSM in stress-prone regions of the country (Cisneros 2011). Future research is needed that studies the implementation and formalization of such technologies for FSM service delivery and sustainability.

6. CONCLUSION

India has made unprecedented gains in improving access to sanitation among rural households in the last 5 years. In the run-up to 2030, leveraging these gains to institute safely managed sanitation will be critical to meet the WASH-related targets under the SDGs. The diverse characteristics of rural settlements and the differential preferences in toilet technologies necessitate a multi-pronged approach to ensuring that faecal waste is safely collected, contained, conveyed, and treated before it is discharged into the open environment. The focus on mainstreaming urban FSM in India presents a tremendous opportunity to bridge the urban–rural divide that has been typical of infrastructure and investment planning under the development programmes in the country.

The FSM model used here could serve as an exemplar of urban–rural convergence for sanitation as the first district in the country that has successfully leveraged urban infrastructure for the delivery of FSM services to the adjacent rural areas. It has demonstrated how to institutionalize the urban–rural convergence working within complex and multi-jurisdictional governance systems. It underscores the importance of a robust landscape assessment, capacity building, and regulation in achieving sustainable urban–rural cooperation towards regional environmental sanitation. Lastly, it maximizes the use of the investment made in sanitation and highlights the importance of evolving new governance models in transcending administrative silos to create more efficient pathways and rapidly scalable shifts to safely managed sanitation.

ACKNOWLEDGEMENTS

The paper has been prepared under the 'Dhenkanal Pilot Project for Solid and Liquid Waste Management (SLWM)' being implemented in the Dhenkanal district, Odisha. This project is in partnership with the Housing and Urban Development Department, Panchayati Raj and Drinking Water Department and supported by the UNICEF and the Centre for Policy Research.

DATA AVAILABILITY STATEMENT

Data cannot be made publicly available; readers should contact the corresponding author for details.

CONFLICT OF INTEREST

The authors declare there is no conflict.

REFERENCES

Agarwal, N., Mukherjee, A. & Dwivedi, A. 2020 *Towards a Sustha and Swachha Rural Dhenkanal: Institutionalizing Faecal Sludge Management for Achieving ODF Plus.* Centre for Policy Research, New Delhi. doi.10.13140/RG.2.2.14151.42404.

Allen, A., Davila, J. D., Hofman, P. & Aref, M. F. 2006 *Governance of Water and Sanitation Services for the Peri-Urban Poor: A Framework for Understanding and Action in Metropolitan Regions.* University College London, Development Planning Unit, London.

Arouna, A. & Dabbert, S. 2012 Estimating rural households' willingness to pay for water supply improvements: a Benin case study using a semi-nonparametric bivariate probit approach. *Water International* **37** (3), 293–304. https://doi.org/10.1080/02508060.2012.687507.

Bhol, A., Dasgupta, S. & Mukherjee, A. 2020 *Perceptions: Understanding On-site Sanitation System Choices in Large Dense Villages in India | SCI-FI.* Centre for Policy Research. Available from: https://scifi.cprindia.org/resources/perceptions-understanding-site-sanitation-system-choices-large-dense-villages-india.

Bureau of Indian Standards 1985 *Code of Practice for Installation of Septic Tanks – Part I.*

Caffyn, A. & Dahlström, M. 2005 Urban–rural interdependencies: joining up policy in practice. *Regional Studies* **39** (3), 283–296. https://doi.org/10.1080/0034340050086580.

Central Ground Water Board. Ground Water Year Book 2019-2020 Available from: http://cgwb.gov.in/Regions/SER/Reports/GROUND%20WATER%20YEAR%20BOOK%202019_2020.pdf.

Chambers, R. 1994 The origins and practice of participatory rural appraisal. *World Development* **22** (7), 953–969.

Cisneros, B. J. 2011 Safe sanitation in low economic development areas. In: *Treatise on Water Science*. Elsevier, Amsterdam, pp. 147–200. doi.0.1016/B978-0-444-53199-5.00082-8.

Dasgupta, S. & Agarwal, N. 2021 Managing non-sewered sanitation for achieving Sustainable Development Goal 6 in India. In: J. P. Tiefenbacher, ed. *Environmental Management: Pollution, Habitat, Ecology, and Sustainability*. IntechOpen, London, pp. 39-55.

Dasgupta, S. & Agarwal, N. 2022 Dichotomy to continuum: importance of faecal sludge management in shaping future urban–rural linkages in urbanizing India. *Environment and Urbanization ASIA*. https://doi.org/10.1177/09754253221083199.

Dasgupta, S., Roy, N. S., Bhol, A. & Raj, D. 2017 *Towards a New Research and Policy Paradigm: An Analysis of the Sanitation Situation in Large Dense Villages | SCI-FI*.

Centre for Policy Research. https://cprindia.org/briefsreports/towards-a-new-research-and-policy-paradigm-an-analysis-of-the-sanitation-situation-in-large-dense-villages/.

Dodane, P.-H., Mbéguéré, M., Sow, O. & Strande, L. 2012 Capital and operating costs of full-scale fecal sludge management and wastewater treatment systems in Dakar, Senegal. *Environmental Science & Technology* **46** (7), 3705–3711. https://doi.org/10.1021/es2045234.

FSSM Policy 2017 Amrut. Available from: http://www.amrut.gov.in/upload/newsrelease/5a5dc55188eb0FSSM_Policy_Report_23Feb.pdf. Accessed 22 November 2021.

Gaulke, L. 2006 On-site wastewater treatment and reuses in Japan. *Proceedings of the Institute of Civil Engineers – Water Management* **159** (2), 103–109.

Harada, H., Strande, L., Fujii, S., 2016 Challenges and opportunities of faecal sludge management for global sanitation. In: *Towards Future Earth: Challenges and Progress of Global Environmental Studies* (Katsumi, T. & Hashimoto, S., eds). pp. 81–100. Available from: https://www.researchgate.net/profile/Hidenori_Harada/publication/333041477_Challenges_and_Opportunities_of_Faecal_Sludge_Management_for_Global_Sanitation/links/5cd8d9a2a6fdccc9dda6a9c0/Challenges-and-Opportunities-of-Faecal-Sludge-Management-for-Global-Sa.

Hutton, G., Patil, S., Kumar, A., Osbert, N. & Odhiambo, F. 2020 Comparison of the costs and benefits of the Clean India Mission. *World Development* **134**, 105052. https://doi.org/10.1016/j.worlddev.2020.105052.

Hulst, J. R. & van Montfort, A. A. 2011 Institutional features of inter-municipal cooperation: Cooperative arrangements and their national contexts. *Public Policy and Administration* **27** (2), 121–144. https://doi.org/10.1177/0952076711403026.

Jadhav, A., Weitzman, A. & Smith-Greenaway, E. 2016 Household sanitation facilities and women's risk of non-partner sexual violence in India. *BMC Public Health* **16** (1), 1139. https://doi.org/10.1186/s12889-016-3797-z.

Kalam, M. A., Davis, T. P., Islam, M. A., Islam, S., Kittle, B. L. & Casas, M. P. 2021 Exploring behavioral determinants of handwashing with soap after defecation in an urban setting in Bangladesh: findings from a barrier analysis. *Journal of Water, Sanitation and Hygiene for Development* **11** (6), 1006–1015. https://doi.org/10.2166/washdev.2021.070.

Kelkar, V. & Seetharam, K. E. 2019 *A Literature Review Evaluating New Approaches to Resolving the Sanitation Challenge in Developing Asia*. Asian Development Bank Institute. http://hdl.handle.net/11540/10927.

Kwak, S. G. & Kim, J. H. 2017 Central limit theorem: the cornerstone of modern statistics. *Korean Journal of Anesthesiology* **70** (2), 144–156. https://doi.org/10.4097/kjae.2017.70.2.144.

Ministry of Jal Shakti, UNICEF India and SRI Delhi 2019 Study to ascertain the impact of the achievement of of Open Defecation Free (ODF) status, on the incidence of Environmental Contamination in Odisha, Bihar and West Bengal. Retrieved November 22, 2021, from https://www.jalshakti-ddws.gov.in/sites/default/files/ODF_Comprehensive_Report_Rev5.pdf.

Ministry of Jal Shakti 2022 Swachh Bharat Mission - Gramin (All-India). https://sbm.gov.in/sbmreport/home.aspx Retrieved August 18, 2022.

Ministry of Urban Development 2015 *AMRUT Mission Statement and Guidelines*. Available from: http://amrut.gov.in/upload/uploadfiles/files/AMRUT%20Guidelines%20(1).pdf.

Ministry of Housing and Urban Affairs (MoHUA) 2017 National Faecal Sludge and Septage Management (NFSSM) Policy. Available from: http://amrut.gov.in/upload/newsrelease/5a5dc55188eb0FSSM_Policy_Report_23Feb.pdf.

Ministry of Housing and Urban Affairs (MoHUA) 2022 Swacch Bharat Mission Urban 2.0 achievements: Sanitation. Retrieved August 18, 2022, from https://sbmurban.org/.

Mwakitalima, A., Massa, K., Seleman, A. & Kassile, T. 2018 Scaling up rural sanitation in Tanzania: evidence from the National Sanitation Campaign. *Journal of Water, Sanitation and Hygiene for Development* **8** (2), 290–306. https://doi.org/10.2166/washdev.2018.07.

Naiga, R. & Penker, M. 2014 Determinants of users' willingness to contribute to safe water provision in rural Uganda. *Lex localis – Journal of Local Self-Government* **12** (3), 695–714. https://doi.org/10.4335/12.3.695-714(2014).

Narayan-Parker, D. 1993 *Participatory Evaluation: Tools for Managing Change in Water and Sanitation*. World Bank, Washington, DC.

Narayan-Parker, D. 1995 *The Contribution of People's Participation: Evidence From 121 Rural Water Supply Projects*. World Bank, Washington, DC.

National FSM PMU 2020 Personal communication. https://guides.library.unr.edu/apacitation/cite-personalcommunications.

National Sample Survey Office 2018 Drinking Water, Sanitation, Hygiene and Housing condition in India (NSSO 76th Round, July–December 2018, Report No 484).

OECD 2013 *Rural-Urban Partnerships: An Integrated Approach to Economic Development. OECD Rural Policy Reviews.* OECD, Paris.

Prasad, C. S. S. & Ray, I. 2019 When the pits fill up: (in)visible flows of waste in urban India. *Journal of Water, Sanitation and Hygiene for Development* **9** (2), 338–347. https://doi.org/ 10.2166/washdev.2019.153.

Pugel, K., Javernick-Will, A., Koschmann, M., Peabody, S. & Linden, K. 2020 Adapting collaborative approaches for service provision to low-income countries: expert panel results. *Sustainability* **12** (7), 2612. https://doi.org/10.3390/su12072612.

Pugel, K., Javernick-Will, A., Mussa, M., Mekonta, L., Dimtse, D., Watsisi, M., Huston, A. & Linden, K. 2022a Pathways for securing government commitment for activities of collaborative approaches. *Journal of Water, Sanitation and Hygiene for Development.* https://doi.org/10.2166/washdev.2022.183.

Pugel, K., Javernick-Will, A., Peabody, S., Nyaga, C., Mussa, M., Mekonta, L., Dimtse, D., Watsisi, M., Buhungiro, E., Mulatu, T., Annis, J., Jordan, E., Sandifer, E. & Linden, K. 2022b Pathways for collaboratively strengthening water and sanitation systems. *Science of the Total Environment* **802**, 149854. https://doi.org/10.1016/j.scitotenv.2021.149854.

Strande, L., Ronteltap, M. & Brdjanovic, D. 2014 *Faecal Sludge Management: Systems Approach for Implementation and Operation.* IWA Publishing, London. doi.10.13140/2.1.2078.5285.

Strauss, M., Larmie, S. S., Heinss, U. & Montangero, A. 2000 Treating faecal sludges in ponds. *Water Science & Technology* **42** (10), 283–290.

Tilley, E., Ulrich, L., Luethi, C., Reymond, P. & Zurbruegg, C. 2014 *Compendium of Sanitation Systems and Technologies. 2nd Revised Edition | SSWM – Find Tools for Sustainable Sanitation and Water Management!* 2nd edn. Swiss Federal Institute of Aquatic Science and Technology (Eawag). Available from: https://sswm.info/node/5587.

UNICEF – Centre for Policy Research. 2020 *Solid and Liquid Waste Management in Dhenkanal District: Situation Assessment Report.* doi:10.13140/RG.2.2.14488.72967.

Warner, M. E. 2006 Market-based governance and the challenge for rural governments: US trends. *Social Policy & Administration* 40, 612–631. http://dx.doi.org/10.1111/j.1467-9515.2006.00523.x.

Water and Sanitation Program (WSP) 2014 Targeting the Urban Poor and Improving Services in Small Towns The Missing Link in Sanitation Service Delivery. A Review of Fecal Sludge Management in 12 Cities. https://www.wsp.org/sites/wsp/files/publications/WSP-Fecal-Sludge-12-City-Review-Research-Brief.pdf.

WaterAid India 2019 Strategy for Faecal Sludge Management in rural India. (n.d.). Retrieved 22 November 2021 from https://www.wateraidindia.in/sites/g/files/jkxoof336/files/strategy-for-faecal-sludge-management-in-rural-india-.pdf.

Whittington, D., Davis, J., Prokopy, L., Komives, K., Thorsten, R., Lukacs, H., Bakalian, A. & Wakeman, W. 2009 How well is the demand-driven, community management model for rural water supply systems doing? Evidence from Bolivia, Peru and Ghana. *Water Policy* **11** (6), 696. https://doi.org/10.2166/wp.2009.310.

WHO-UNICEF JMP 2018 *JMP Methodology – 2017 Update and SDG Baselines.* Available from: https://washdata.org/sites/default/files/documents/reports/2018-04/JMP-2017-update-methodology.pdf.

WHO and UNICEF 2021 Progress on household drinking water, sanitation and hygiene 2000-2020: Five years into the SDGs. World Health Organization (WHO) and the United Nations Children's Fund (UNICEF), Geneva. Licence: CC BY-NC-SA 3.0 IGO.

World Bank 2020 Overview. Retrieved 22 November 2021, from https://www.worldbank.org/en/country/india/overview#1.

First received 31 January 2021; accepted in revised form 1 August 2022. Available online 13 August 2022

© 2022 The Authors
doi: 10.2166/h2oj.2022.024

Increasing access to piped water for poor households: an analysis of water connection subsidy projects implemented by the GRET

Terra Michaels *, Chanmeakara Suong, Chanrith Chi, Lyheang Morm, Phanit Mab and Justine Denis

Department of Water, Sanitation, and Solid Waste Management, GRET, Phnom Penh, Cambodia
*Corresponding author. E-mail: michaels@gret.org

TM, 0000-0002-8731-6176; JD, 0000-0002-4706-6977

ABSTRACT

By 2030, Sustainable Development Goal 6 aims to ensure universal access to safe drinking water. The Royal Government of Cambodia has, thus, made a significant progress in improving the access to safe water supply in rural areas. However, results show only 26.9% of all poor households in a license area were connected when supported by the Groupe de Recherches et d'Echanges Technologique (GRET) programs compared to 45.6% of all households; nearly half of all poor households are located greater than 70 m or are of unknown location; and 56% of known households are within than 3 km. Logistic regression shows a negative relationship between household connections and increasing distance by 10 m, incremental water tariffs by 0.10 USD, and subsidized connection fees by 10 USD when compared to connections, with odds of connection decreasing in all cases (−23, −8.6, and −1%, respectively). Finally, 37% of unconnected households have never heard of the subsidy programs, indicating insufficient marketing programs. Recommendations include updating subsidies to cover all costs necessary for households to connect within 50 m of the distribution lines (an extra pipe, transaction fees, etc.); ensuring flat-rate connection fees paid by all poor households including all costs remaining 40 USD or less; subsidizing pipe extensions to reach unserved areas while maintaining fair tariffs for poor households and cost recoverability for piped water suppliers; shortening marketing and awareness programs to 3 months or less, with involvement from more local stakeholders; and adding poor household connection requirements to investment program requirements.

Key words: affordability, Cambodia, household water connections, piped water supply, poor households, subsidy

HIGHLIGHTS

- Sustainable development goals regarding the safe water supply.
- New solutions to reach poor households with clean water.
- Identification of innovative subsidies to reach poor households.

GRAPHICAL ABSTRACT

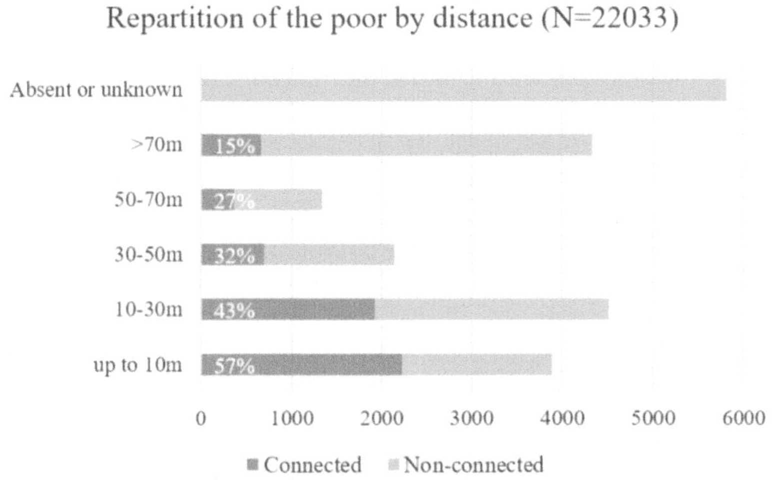

Independent Variables	Logistic Regression Coefficients
Increasing distance from the pipe network by 10 meter increments	-23%
Increasing the water tariff by 0.10 USD increments	-8.6%
Increasing the connection fees by 10 USD increments	-1%

INTRODUCTION

By 2030, Sustainable Development Goal 6 (SDG6) aims to 'ensure the availability and the sustainable management of water and sanitation for all' and includes goals for achieving universal, affordable access to safe drinking water (United Nations 2018) while the Human Rights to water emphasizes non-discrimination, sustainability, and accountability (United Nations 2014). The Royal Government of Cambodia and international donors have supported progress and an improved regulatory environment in the piped water sector through the Ministry of Industry, Science, Technology, and Handicraft (MISTI), resulting in services provided by 13 public and more than 400 licensed private water suppliers in all 25 provinces (Investing in Infrastructure, unpublished). Although the government aims to improve it (Cambodia, T. R. 2011), access to a piped water supply remains a challenge in rural areas, reaching 80% of urban households but only 17% of rural households in the country (WHO and UNICEF 2021) while the government aims to achieve 38 and 100% improved water supply access by 2025, respectively (Cambodia T. R. 2018).

For more than 20 years, the international NGO Groupe de Recherches et d'Echanges Technologique (GRET) in Cambodia has been supporting privately-owned, piped water suppliers in small- and medium-size towns (5,000–20,000 inhabitants) with improved planning, operations management, and infrastructure development. Many programs included pro-poor connection subsidies, where monitoring data anecdotally revealed that connecting the first ~30% of poor households in a license area is easy for operators, but connecting the last ~40% is difficult. Furthermore, one small study by Gret (2017) confirmed that subsidy programs do increase poor connection rates, but only reach just over a third of households; another similar study in Vietnam had similar findings, indicating that subsidies only marginally increase poor connection rates (Carrard *et al.* 2019).

The literature on poor households and subjects of water affordability is extensive, and seems to be largely focused on *consumption*, establishing that alternatives are more expensive (cited 10–32× more by Bakker 2007), and therefore piped water is the safest, most cost-effective, and sustainable resource for the poorest households. This also begs the evaluation of consumption subsidies and their inequality. While these subjects are not addressed in detail in this study, it is worth highlighting two multi-country studies regarding the ineffectiveness of subsidies: one compiled subsidy programs in 10 countries, finding that the wealthiest deciles, in fact, received the largest quantity of subsidy overall (Abramovsky *et al.* 2020), while a second evaluated the effectiveness of block tariffs, finding that they are not inherently pro-poor (Whittington & Nauges 2020).

On the other hand, the evaluation of *connection* subsidies appears to be more limited. Cook *et al.* (2020) provide an overview of non-tariff programs across the world, including 77 assistance programs in 45 low- or middle-income countries, where 27% included connection subsidies. The authors point out that there are few empirical studies specifically aiming to understand if these programs actually benefit their intended beneficiaries, despite the widespread policy recommendations. This is surprising since concerns have been highlighted since the early 2000s; Kayaga & Franceys (2007) is an early one of many studies which point out the wide-ranging barriers in connecting to the water supply.

Indeed, limited analysis has been undertaken in Cambodia to evaluate data on why households remain unconnected even when subsidies are available. With a growing need to 'reach the last mile' to achieve the 2030 SDGs, this study assesses some of the drivers and opportunities to reach poor households with improved water supply. It first includes a brief overview of the water regulatory context in Cambodia and then describes the materials and methods used including quantitative and qualitative analyses. Later, the results of the analysis are presented and discussed, including findings from experience during projects implemented by GRET and an assessment of the location of poor households and connections rates, followed by a regression analysis to understand interactions between connection fees, distance from the water supply, and tariffs. Finally, a spatial 'cluster' assessment is provided to understand the location of unconnected households and evaluate water supplier expansion possibilities, followed by recommendations and conclusions.

CAMBODIA REGULATORY BACKGROUND

There are several regulations for water suppliers in Cambodia which provide context on licensing, service provision, and expansion. First is the regulation on water tariff calculations (MISTI 2016, Article 15) which requires all water operators licensed by the MISTI to expand their pipe network to reach 100% of their license area within 5 years. A license area is typically determined by administrative commune boundaries and the 'reach' is defined by the presence of even a small pipe network within each village of the commune. Additionally, the same regulation (MISTI 2016, Article 4) requires that all suppliers connect all households within 50 m of a pipe network; these are called 'connectable households', and the area in which they lay is called the 'coverage area'.

In practice, these definitions do not account for key issues: (1) typically pipe networks are constructed along roads, although many households may be scattered at distances more than 50 m from the road and (2) defining the license area according to administrative boundaries does not account for technical limitations in the pipe network reach all villages (geographic barriers, lack of road access, distantly isolated villages, etc.). Furthermore, in practice, operators typically only provide a water meter and up to 10 m of pipe, requiring households to personally invest in the pipe needed to reach between the meter and their house.

In response, a newer regulation (MISTI 2020) specifically requires operators to provide all materials to reach households up to 50 m and defines a maximum connection fee of up to 75 USD, although the impacts of this regulation have not yet been observed. Furthermore, a pro-poor regulation (MISTI 2021) now recognizes and regulates the need for consistency among poor household subsidies, defining that they should pay no more than 20 USD for a piped water supply connection regardless of distance (within 50 m). However, monitoring and enforcement remain a challenge in practice, therefore poor household distance, connection fees, and other limitations may remain as barriers.

METHODS

This study used ID-Poor data produced by the Ministry of Planning (2020) of Cambodia and data from water operators supported by GRET's projects. Field assessments were carried out and supported with:

- A desk review of relevant studies and compiling of data from previous projects supporting 36 private water operators.
- Field assessments at five different water suppliers across five provinces (Kampong Cham, Kampong Speu, Prey Veng, Kratie, and Takeo), including:
  - Household surveys Table 1 to understand perceptions and challenges in accessing piped water ($n = 419$, using the Cochran formula and a 95% confidence interval with a margin of error of 10%, based on budget availability for the sample size), divided proportionally into connected ($n = 111$) and non-connected

Table 1 | The household survey sample sizes

District of water operator	No. of poor HH	No. of non-connected poor	No. of sample non-connected	No. of sample connected
Kampong Trabek	631	631	82	0
Steung Trang	661	528	61	0
Treang	1,832	1,205	65	0
Preak Prasab	527	151	59	77
Samraong Tong	189	123	41	34
Total	3,840	2,638	308	111

Note: HH, household.

($n = 308$) poor households in the five sites using previously collected GPS (Global Positioning System) data to randomly select households prior to data collection.

  - Five Focus Group Discussions with local authorities (primarily commune chiefs or councils) in the license area to understand (i) their perception of water supply development and the gap in their area, (ii) their participation and engagement with water service providers, and (iii) their plan to contribute and invest for water supply development in their area.
  - Five Water Operator interviews for understanding challenges in reaching the poor households including the capacity of water supply, water source issues, geography, market and demand, and connection promotion, especially for the poor households.
  - Field observation (distance to the existing pipe network, road conditions, etc.) to identify possible barriers or solutions to reach poor households.
- Data analysis to present results in various ways, including:
  - Compiling lessons learned from subsidy programs to compare subsidy amounts, connection rates, and other lessons learned from 36 operators.
  - Using 16,224 household GPS data points for households and pipe network plans for 28 operators to understand distances of connected and unconnected poor households.
  - Undertaking statistical analysis (logistic regression) of poor households located within 70 m of a pipe network ($n = 11,894$) in Microsoft Excel to assess the impacts of connection fees, water tariffs, and distance of households from the water supply pipe on poor household connections.
  - Using QGIS, satellite imagery, knowledge of the areas, and GPS locations of unconnected poor households in license areas of 13 water operators to identify clusters of households, aiming to understand the physical and spatial possibility of densifying and/or extending pipe networks.

LIMITATIONS OF THIS STUDY

The databases were developed under previously implemented programs to monitor connections of poor households and manage subsidy disbursements. Some were part of larger programs to support overall increased water supply and connections, whereas others only provided subsidies for poor households. A more ideal analysis would compare location and connection rates for non-poor and poor households; however, non-poor households were not mapped and budget limitations only allowed recent field data collection to only undertake short surveys as well as to observe physical limitations for extending pipe networks. The databases also only include poor households deemed reasonably connectable (i.e. close to a road, therefore an existing or potential pipe). Finally, if not already mapped, reasonable assumptions were made regarding the locations of pipes based on the location of connected households.

Potential biases include the perceptions of local authorities, courtesy bias on behalf of respondents to the household survey, and the nature of focus group discussions. There is also inherent bias with GRET analyzing its own data, although the data were viewed as objectively as possible.

Finally, regarding densification and extension, neither quality of operator services nor limitations in the quantity of water sources and treatment facilities were evaluated to determine if water availability and treatment capacity are sufficient to support additional supply.

FINDINGS

Projects supporting poor household subsidies

To begin to understand the limitations and successes of pro-poor programs, Table 2 presents the varying methods, achievements, and challenges after the implementation of GRET's pro-poor projects.

Program implementation was managed by locating and mapping ID-Poor households by GPS, allowing monitoring of connections and subsidy disbursement, followed by subsidy program marketing while operators were responsible for carrying out connections and claiming subsidies. Programs were also generally designed with varying flat-rate subsidies. Combined with varying connection fees (therefore unequal payment by poor households) and variation in tariffs across operators, this database creates an opportunity to evaluate the impacts of varying connection fees paid by households, distance from pipe networks, and tariffs on connection rates (regression analysis is given in the following).

Table 2 | A summary of past projects

Project	Description	Connection subsidy	Implementation period	Achievements	Challenges
Project 1 (2013–2017)	Kampong Cham and Kratie 3 operators	40 USD HHs pay: 0–25 USD	2–5 months	1,281 poor HHs identified 475 poor HHs connected Poor connection rate: 37%.	Short period of implementation of subsidy. Lack of interest from the water operators in the connection promotion since the poor must request the discount voucher from the commune.
Project 2 (2015–2019)	Countrywide 19 operators	40 USD HHs pay: varies according to the operator	1 year+	13,476 poor HHs identified 2,773 poor HHs connected Poor connection rate: 20%	Lack of interest from the water operators in the connection promotion. Late payment of subsidy disbursement due to the complex tool used for subsidy management.
Project 3 (2011–2017)	Kandal, Kampong Cham 10 operators	ID-Poor 1: 25 USD ID-Poor 2: 20 USD HHs pay: varies according to the operator	Around 6 months	2,331 poor HHs identified 806 poor HHs connected Poor connection rate: 35%	Delay for some poor households to save enough necessity for the fee during a short project period. The higher willingness of the operators to reach more poor households.
Project 4 (2018–2020)	Kampong Speu 2 operators	40+5 USD for 12 m additional pipe HHs pay: 15 or 25 USD	4 months	1,073 poor HHs identified 442 connected poor HHs. Poor connection rate: 41%	A short period of intervention that limits the connection installation and quickens the subsidy reimbursement process. Some poor could not afford the fee. One operator did not inform the poor about the subsidy; the other was very active.
Project 5 (2017–Present)	Kampong Speu and Kampong Chhnang 2 operators	Varies according to the operator fee: split between operator and project HHs pay: 10 USD	1 year+	558 poor HHs identified 64 poor HHs connected Poor connection rate: 11%	Water operators are not active in connection promotion; they rely on the support from the project and local authorities, but also must contribute themselves. The supply capacity of the water operator and network densification/expansion are key barriers for poor connecting to piped water.

Note: HH, household.

One thing to note is that poor households (and therefore connection rates) are quantified using publicly available databases of ID-Poor cardholders as identified by systematic processes through the Ministry of Planning (MoP). This is an important and useful tool for Cambodia, as the identification of the poor can be one barrier to distributing adequate services, such as piped water subsidies (Cook *et al.* 2020).

It should also be noted that Project 4 intended to test a type of *progressive subsidy* as defined in Abramovsky *et al.* (2020): those which progressively distribute larger subsidies through wealth categories, resulting in the poorest receiving the largest quantity. It provided additional subsidies for the extra pipe to reduce the investment cost barrier, aiming to enable farther households to connect. This program did indeed result in higher connection rates, but this could also be due to one particularly motivated operator.

Finally, two anecdotal findings from experience stand out. The first is active marketing by operators, which likely promoted the programs to more households. The second is the generally short, urgent implementation and payment timelines (less than 6 months, excluding preparation and monitoring phases, which require more extensive time for planning) resulted in higher connection rates. Both are evaluated further later in the paper.

Distance from the water supply

Data from 28 water operators across the country demonstrate the situation of poor households in relation to the piped water supply. Within the license area of these operators (134,675 households (NCDD 2016)), there are a total of 22,033 households (16.4%) which hold ID-Poor cards issued by MoP, equivalent to about 110,000 people. It is noted, however, that only present and relatively 'connectable' poor households – those lying relatively near roads and therefore potential pipe networks – were mapped during the project implementation period, identifying a total of 16,224 households or 73.6% of ID-Poor. The remaining 26.4% were presumed to be absent or primarily scattered between villages and located far from roads. There is little evidence to suggest which is most likely; however, it is noted that during the implementation, it was rare for unmapped households to request subsidies.

The distribution of these households classified by distance is shown in Table 3. It should be noted that GPS accuracy is well known to vary; therefore, while the MISTI policy requires households within 50 m to be connected, the analysis includes ranges up to 70 m to capture a fuller picture.

Just over half (53.9%) of poor households are within the coverage area (less than 70 m) and the majority (38.2%) are located within 30 m of the pipeline. The other half of the total poor population live more than 70 m away from the pipe network (19.7%) or are unknown (26.4%), the latter of which lie farther than any

Table 3 | Location of poor households by distance categories

Categories	Description	Location of the poor ($n = 22{,}033$ households)
Poor ≤10 m	Poor households located within 10 m from the pipe network have a *high potential* for connecting to piped water.	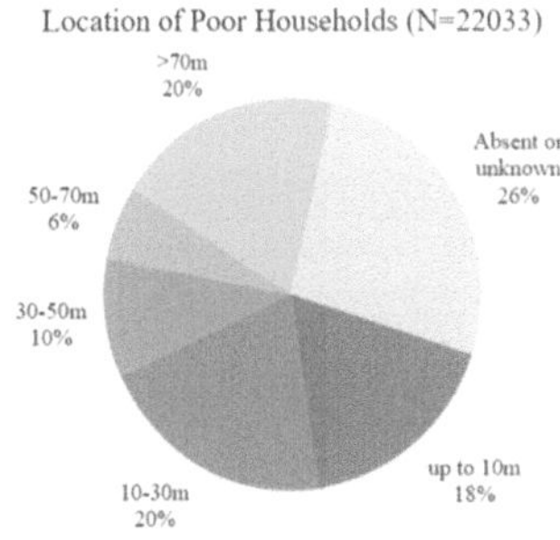
Poor 10–50 m	Poor households located between 10 and 50 m from the pipe network have a *lower potential* to connect, but operators are obligated to connect them according to their license.	
Poor 50–70 m	Poor households located 50–70 m from the pipe network in which operators are not legally obligated to connect, but *may still be connected.* This group is included in the assessment to (1) account for GPS inaccuracies and (2) understand where poor households are located in villages with piped water supply.	
Poor>70 m	Poor households located farther than 70 m from the pipe network which operators are *not legally obligated to connect* but were mapped because they were deemed connectable to *possible network extensions;* however, in many cases, whole villages remain without network coverage.	
Absent or unknown	Unmapped households, due to *absence or being physically scattered*, lying far from roads, villages, or pipe networks. Households may have been absent during mapping due to seasonal or other migratory activities, although this topic is not well understood.	

potential network. While few studies were found to have physical measurements of poor household proximity to water supply (likely because poor household databases similar to Cambodia's are rare), several note similar concerns: poor neighborhoods in Jakarta are more likely to have low network density (Bakker 2007) while poorer households in Maputo are more likely to be further than the mandated connection range of 25 m (Zuin & Nicholson 2021). Perhaps more similar data compare neighborhoods in Nicaragua (excluding the capital city) by wealth quintile, revealing that only 44% of the poorest while 85% of the richest quintiles had at least one piped water supply connection in their neighborhood (Angel-Urdinola & Wodon 2012). Although location data for non-poor households were not collected during the projects, these results of poor household pipe network reach are similar to the previous findings.

Two findings are notable from the previous data: most poor households are located either very close to or very far from the network. Connection rates of those close to the network are evaluated further subsequently. For those which are too far, however, it is necessary to consider innovative solutions to reach them; the final analysis section (cluster analysis) will map the locations of the known households to evaluate the possibility of connection via pipeline extension.

Connection rates of poor households

Endline data were collected for all households for 12 operators and reveal significant disparity in overall connection rates: 45.6% of total households were connected (including poor, by NCDD data), whereas the connection rate of poor households is only 26.9% (by MOP data). A study in Vietnam (Carrard *et al.* 2019) found similar results: overall, fewer poor households were connected than non-poor households in the study areas.

For further evaluation, the location of connected and unconnected poor households was repartitioned according to the distance in Figure 1. Connection rates decrease gradually for farther households, with 57% of poor households connected within 10 m of the network down to only 15% for those located more than 70 m. Another notable finding is that a significant portion of households located within 10 m has not connected. Both are addressed further subsequently.

Logistic regression comparing distance, connection fee, and water tariff

A logistic regression was undertaken to evaluate poor household connection data for households within 70 m of a pipe network ($n = 11{,}894$) *after* the implementation of subsidy projects. Notably, other studies included several socioeconomic indicators in a similar analysis, overall linking poor access to low wealth and education; Abubakar (2019) provides a review of several studies while Basani *et al.* (2008) showed the connection between average expenditures and probably of connection in Cambodia. However, because the data were collected for connection monitoring purposes, socioeconomic data were not collected. Therefore, this evaluation includes the evaluation of the impact of (1) water tariff (ranging from 0.4 to 0.75 USD), (2) distance (m), and (3)

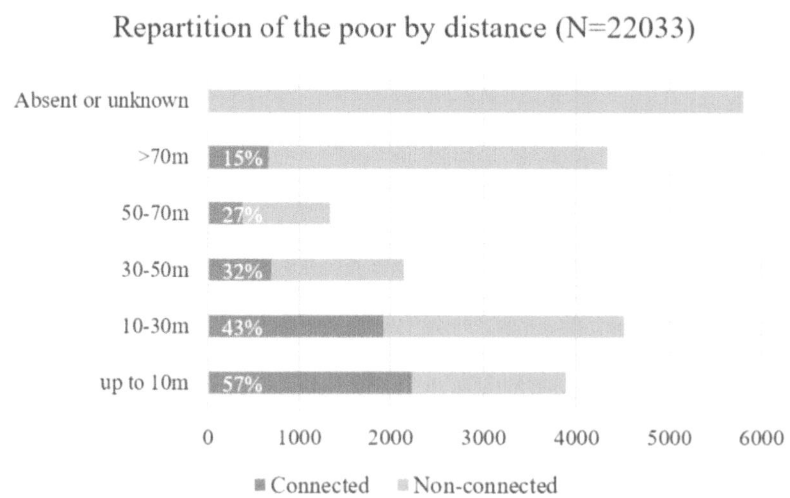

Figure 1 | A repartition of the poor by distance from their house to the pipe network ($n = 22{,}033$ households).

connection fee paid after subsidy (ranging from 3.75 to 42.55 USD) on connection rates (not connected: 0, connected: 1). The results are shown in Table 4.

Holding other values constant, the results show the following:

- For every 10 m farther that a poor household lies from a pipe network, the odds of connection decrease (−23%).
- Higher water tariffs have a negative impact on poor household connections: for every 0.10 USD increase in the water tariff, the odds of connection decrease (−8.6%).
- Increasing connection fees paid by poor households (after the subsidy) by 10 USD increments decreases (−1%) the odds of connection.

Accuracy, precision, and sensitivity are used to understand how well the data are connected and if the model is useful. Therefore, cutoff values were evaluated at 0.1 increments between to predict the rate of connection (value: 1) and non-connection (value: 2) and compared to the sample values (out-of-sample values were not compared).

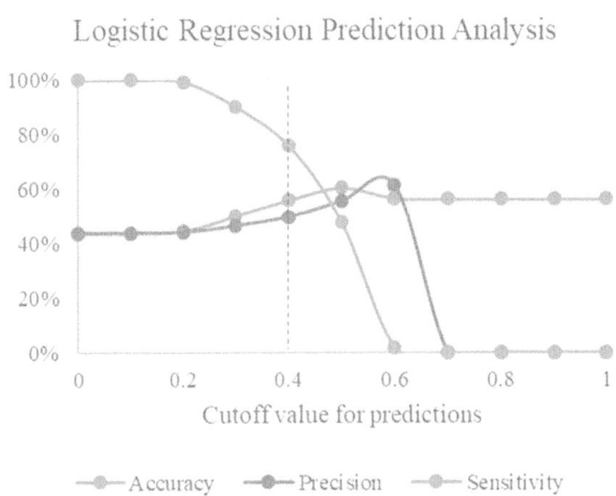

CUTOFF **0.4**

Connections		**Predicted 0**	**Predicted 1**	**Total**
Actual	0	2690	3988	6678
Actual	1	1264	3950	5214
	Total	3954	7939	11892

Cutoff value: 0.4
Accuracy: 56%
Precision: 50%
Sensitivity: 76%

For poor household connections, it is more interesting to correctly predict connections (precision), albeit at the expense of higher rates of false positives (sensitivity). Choosing a cutoff value of 0.4, the accuracy is 56% and in fact only reaches a maximum of 60%. However, correctly predicted connections (precision) are limited at 50% and the model correctly identifies positive connections (sensitivity) only 76% of the time. The relatively low accuracy and precision indicate that the model should only be used as a general guide to draw conclusions rather than as an analytical tool, while there are likely other influential factors that were not evaluated. Therefore, the results are discussed in this manner.

Subsidized connection fee affordability

Increasing connection fees paid by poor households from 10 to 40 USD decreases the odds of connection slightly and are indeed nearly zero. This is consistent with a study completed in 2017 of one GRET program and two programs run by other implementers in Cambodia (GRET 2017), which concluded that there is no correlation between the achieved connection results and the subsidy amount, the connection fee cost, the contribution from households, nor the contribution from operators. However, this is contrary to findings from other studies, which conclude unequal subsidies also impact the connection rates (Grant *et al.* 2016).

Table 4 | Results of logistic regression

Independent variables	Logistic regression coefficients (%)
Increasing distance from the pipe network by 10-m increments	−23
Increasing the water tariff by 0.10 USD increments	−8.6
Increasing the connection fees by 10 USD increments	−1

A study from Cambodia previously concluded that unsubsidized tariffs were indeed a barrier for connection and recommended that they be reduced for poor households (Basani *et al.* 2008). The results above, however, may indicate that affordability of poor connections may be up to 42 USD (the maximum paid by households in this study, resulting in 57% of households within 10 m to connect), but other issues are influencing their decision to connect. This is not dissimilar from the results in Maputo, where two-thirds of households connected when subsidized payments for the lowest quintile averaged 27.50 USD (Zuin & Nicholson 2021). Alternatively, for those who remained unconnected, it may indicate overall unaffordability for fees even of 10 USD. This is a complex issue with no clear answer. It is worth pointing out that while the overall result from the regression analysis indicates little impact from varying fees, this may be true for households located within 30 m, where the majority of connections lie. When the data are presented by distance and the subsidized connection fees in 10 USD increments (Figure 2), it appears that households beyond 30 m favor the lowest fees.

Distance from a pipe network as a barrier

The previously discussed results indicate that distance from the pipe network is a major barrier for connection. This is perhaps generally intuitive; Cook *et al.* (2020) also noted that households located farther than 12 m from the distribution lines were less likely to connect while Abubakar (2019) had similar findings, although the latter author consolidated studies mainly measuring the distance to public taps. This provides strong evidence for the theory that other investment costs, such as pipe to extend between the meter provided by the operator (only 10 m from the pipe network) and the household, are indeed barriers for connection. This notion would be consistent with other studies (Bakker 2007; Kayaga & Franceys 2007), which found that to actually achieve a connection, fees included pipe, road crossing costs, application fees, transportation, advance deposits, transactional costs for taking time to pay regular bills, and extra infrastructure needed for piped connections, adding up to 500 USD in one example in Uganda. It is, therefore, important to revisit the full costs for infrastructure and complexity of connection provided by the water supplier and then ensures subsidies and requirements are designed to cover these other invisible costs. This includes revisiting the issues regarding the densification of water supply in poor neighborhoods.

Water tariffs as a concern

Higher tariffs have a marginal, although less notable, impact on connection rates. Other studies have mixed results; Abubakar (2019) consolidated results from several urban studies which indicate that cost of water can also influence decisions about drinking water, while Basani *et al.* (2008) suggest that poor households in Cambodia may be able to afford tariffs after connecting. An important finding is that this is a more complex issue as highlighted in the introduction of this paper, particularly the issues about water alternatives and cost recovery.

Figure 3 shows the household distance from water supply pipes compared to the water tariff in 0.10 USD increments. It is worth noting that there are fewer of the highest tariffs included in this model (0.75 USD), as they are unusual in Cambodia, although the one suggests a larger percentage of poor households are located farther from

	# Water Operators	Counted Poor <=10m	Connected Poor <=10m	% Connected Poor <=10m	Counted Poor 10-30m	Connected Poor 10-30m	% Connected Poor 10-30m	Counted Poor 30-50m	Connected Poor 30-50m	% Connected Poor 30-50m	Counted Poor 50-70m	Connected Poor 50-70m	% Connected Poor 50-70m
Poor conection fee													
Up to 9.9 USD	2	70	49	27%	119	63	34%	88	38	21%	44	16	9%
10 - 19.9 USD	9	1530	857	37%	1708	740	32%	725	291	12%	556	183	8%
20 - 20.9 USD	8	1460	779	36%	1517	686	32%	670	230	11%	381	120	6%
30 - 30.9 USD	6	664	420	44%	1010	331	34%	604	104	11%	335	42	4%
>40 USD	3	169	127	47%	159	104	39%	62	30	11%	23	5	2%

Figure 2 | A comparison of connection fee and connection rate by distance of poor household.

	# Water Operators	Counted Poor <=10m	Connected Poor <=10m	% Connected Poor <=10m	Counted Poor 10-30m	Connected Poor 10-30m	% Connected Poor 10-30m	Counted Poor 30-50m	Connected Poor 30-50m	% Connected Poor 30-50m	Counted Poor 50-70m	Connected Poor 50-70m	% Connected Poor 50-70m	Counted Poor >70m	Connected Poor >70m	% Connected Poor >70m
Water tariff																
0.4 - 0.49 USD	4	503	287	12%	4513	1924	83%	188	77	3%	79	23	1%	294	14	1%
0.5 - 0.59 USD	12	2027	1076	34%	2364	983	31%	1074	363	12%	637	186	6%	3061	520	17%
0.6 - 0.69 USD	11	1287	861	42%	1308	691	34%	542	222	11%	440	143	7%	727	111	5%
0.7 - 0.79 USD	1	76	8	9%	369	17	19%	345	31	35%	183	14	16%	248	19	21%

Figure 3 | A comparison of water tariff and connection rate by distance of poor household.

the network, resulting in more, farther connections. This could suggest that networks in less dense and more rural areas require higher tariffs to achieve cost recovery for the private operator. Nevertheless, fair and affordable tariffs are even more important for the poor here.

Household perceptions and other reasons for not connecting

To assess more complex reasons for connection and non-connection, a household survey was undertaken in select sites. Households considered that one main reason to connect to piped water (Figure 4, $n = 111$) was the access to subsidy on the connection fee (82%) along with the ease of access to the tap (84%) and the water quality (68%). Similarly, the two main reasons for not connecting to piped water (non-connected households, Figure 5, $n = 308$) were having alternative water sources (47%) and the low affordability for the connection fee (32%). In addition, when asked the support needed to connect to piped water, a subsidy on the connection fee is the most-needed solution (Figure 6), with 42% asking for a free connection and 36% wanting a discount or subsidy for the connection fee.

Less than half of non-connected households were aware of the subsidy programs from the implemented projects (42%), but when segregated by distance from the piped network, farther households tend to have less knowledge about the subsidy: only 47% of households living farther than 50 m from the network were aware of the program compared to 63% for households located less than 50 m from the network (Figure 7). However, when asked if they would connect, 65% of the unconnected households also reported they would if they had been aware of it. Regarding the households that indicated they were not interested in connecting to the piped network (35%), the main reasons are due to their satisfaction towards their current water sources (35%) and the lack of

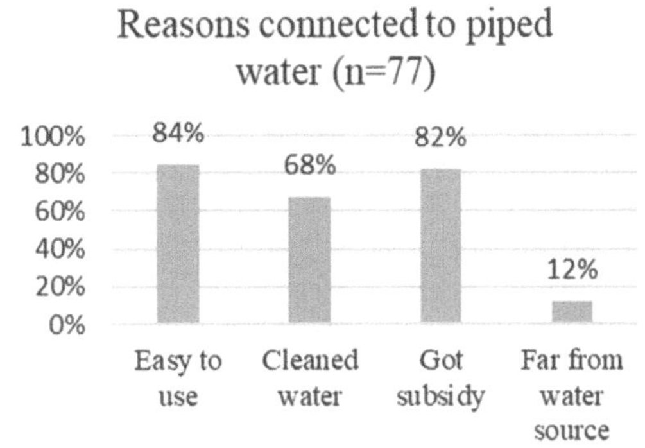

Figure 4 | Reasons for connecting to the piped network.

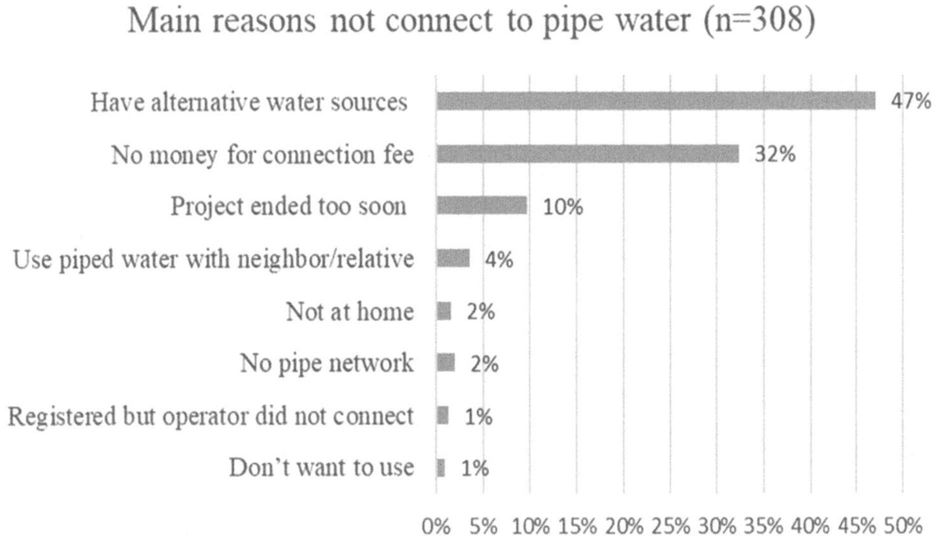

Figure 5 | Reasons for not connecting to the piped network.

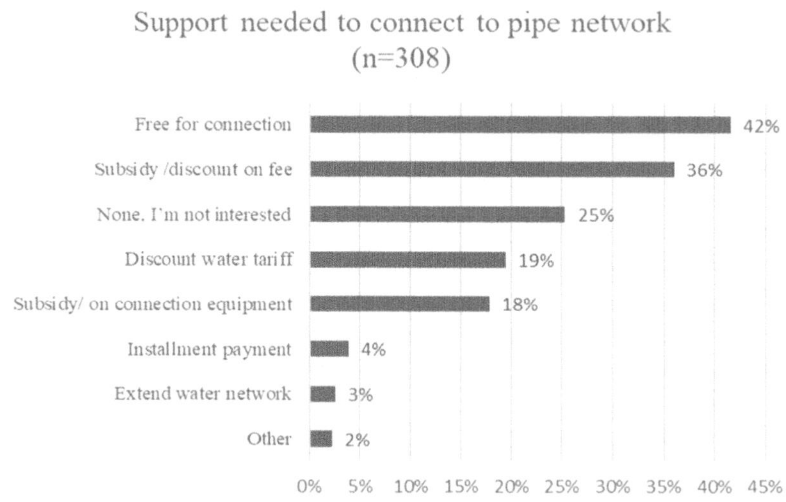

Figure 6 | The support needed to increase the connection rate.

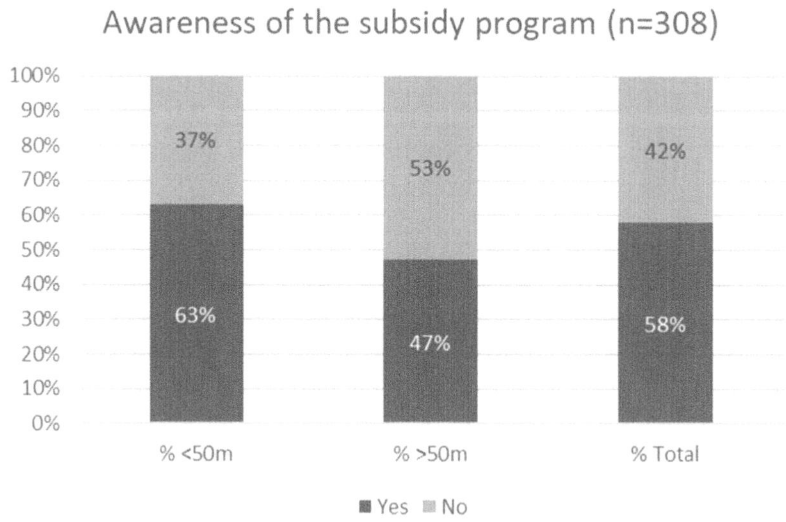

Figure 7 | Respondent's awareness of the subsidy program for connections.

capacity to pay for the connection fee even with a subsidy (37%). Similarly, local authorities in the focus group discussions raised concerns that poor households tend to lack knowledge about water and sanitation and many of them feel more comfortable with their water sources.

First is the concern of alternative water sources, although only a third of unconnected households cited their satisfaction with these as the reason for not connecting. As cited in the introduction, alternative sources can be significantly more expensive per volume. Another study from Jakarta suggested additional issues with alternatives: piped water quality can be perceived as lower than alternatives or competition from alternative vendors can be monopolistic and controlling (Bakker 2007). The use of alternatives is a complex behavior change issue not covered in this study; however, it remains an important consideration for future program implementation.

The second issue is about the perception of connection costs; households report affordability as a barrier, while the regression analysis showed that subsidized fees under 42 USD result in connecting a similar number of households, potentially indicating affordability. The household perceptions are relatively similar to those in Vietnam, where more than 50% of unconnected households reported unaffordability as the primary reason in service areas (Carrard *et al.* 2019), and in Maputo, where 36% perceived the connection cost to be too high (Zuin & Nicholson 2021). Interestingly, respondents in the latter study also reported that connection fees are an average of 20 USD *higher* than the actual fee. There are two possible reasons for this: other hidden fees are unaffordable or marketing of fees is insufficient. Furthermore, as discussed previously, subsidy programs must move away from flat-rate subsidies and instead focus on flat-rate fees paid by households for all within 50 m. This reaffirms the need for re-evaluating the full costs for connection, including extra pipe and transactional demands, to overcome the hidden fees, while promotional materials should make connection fees to be paid by households explicitly clear. It is also worth considering the use of 'connect now pay later' in installment campaigns (Zuin & Nicholson 2021), waiving down payments, and other flexibility in the delivery of payment (Bakker 2007), which have all proven to increase connection rates overall.

Finally, concerns about gaps in knowledge of subsidy programs are a consistent and notable issue across projects. In both Vietnam and Maputo, unconnected households reported being unaware of connection services (Carrard *et al.* 2019; Zuin & Nicholson 2021). There are several resources to suggest that this is likely connected to the anecdotal findings mentioned previously suggesting that inactive operators had lower overall connection rates and shorter implementation timelines which achieved higher connection rates. In Maputo, a few neighborhoods were targeted at a time in short targeted campaigns, which resulted in 80% of people receiving connection in a month or less and succeeded in connecting two-thirds of poor households (Zuin & Nicholson 2021). Results from Morocco showed that sending staff to the doorstep increased applicants by nearly 60% (Cook *et al.* 2020). In Vietnam, engaging with the local government and users early allowed empowerment and contributed to the overall success of the project, a women's union was engaged for some information and communication activities (World Bank 2013). Another study assessed overall bottom-up participation (by the community, local NGOs, etc.) across several studies, suggesting it can have an overall impact on improved access to individual taps, although the results are most significant in slums rather than general low-income areas (Narayanan *et al.* 2017).

As such, it is ideal to implement concentrated, short connection campaigns with a high level of marketing activity and involvement of more stakeholders in fewer villages or operators at one time in succession, rather than drawn-out programs supporting many operators simultaneously.

If too far to connect, where are the poor located?

The context of connecting poor households is not limited to connections alone; it also pertains to pipe network extensions. As per the law, all licensed water operators must expand their pipe network to cover 100% of their license area within 5 years, yet results show that less than 50% achieve this.

The aim of this 'cluster analysis' (Figure 8) was to understand the geographic distribution of groups of non-connected poor households which otherwise may have characteristics to be considered connectable and the feasibility of promoting pipeline expansion for 16 operators. The methodology is not an exact science: for example, a whole village that is not in the coverage area of a network may be considered a cluster, although groups of households set apart from the village centre may be as well. A total of 62 clusters identified as not having major barriers for connection (except perhaps distance, discussed subsequently).

The analysis shows that a significant number of poor households can be connected to the network with limited investment for the extension of the network (Table 5). Most clusters (53) range between 0 and 3 km from the existing piped network; this is a technically reasonable distance to plan extension lines and 56% of poor households lie within this range. Additionally, the larger clusters of households are farthest from the existing piped network. Beyond 3 km, however, while network extensions can be considered under certain conditions, it is ideal that alternative options be assessed. In particular, for the two clusters located more than 10 km away, the construction of independent piped water systems should certainly be explored.

Many authors agree that poor households are less likely to be located in service areas and to connect when they are in service areas, therefore both require subsidies (Kayaga & Franceys 2007; Angel-Urdinola Wodon 2012; Borja-Vega *et al.* 2019; Abramovsky *et al.* 2020). However, the suggestion to extend the water supply has no shortage of concerns. Complexity is one: multiple authors point out that the average cost per connection in poorer areas tends to be denser and disordered, leading to higher and more time for installation (Bakker 2007; Devkar *et al.* 2013; Grant *et al.* 2016). Others pointed out limited investment capacity by operators (Grant *et al.* 2016), lower tariffs for the poor reduced average revenue per volume of water, making cost recovery impossible (Bakker 2007), and operator reliance on connection fees for income (Grant *et al.* 2020). Adeoti & Fati (2020) specifically focus on the topic in their paper, citing in additional political interference, lack of organizational willingness, lack of government policy, and lack of budget funding.

While the analysis of this paper focused on poor household affordability, it is indeed important to balance the need to reach poor households with the viability of the water supplier: a World Bank study showed that the lack

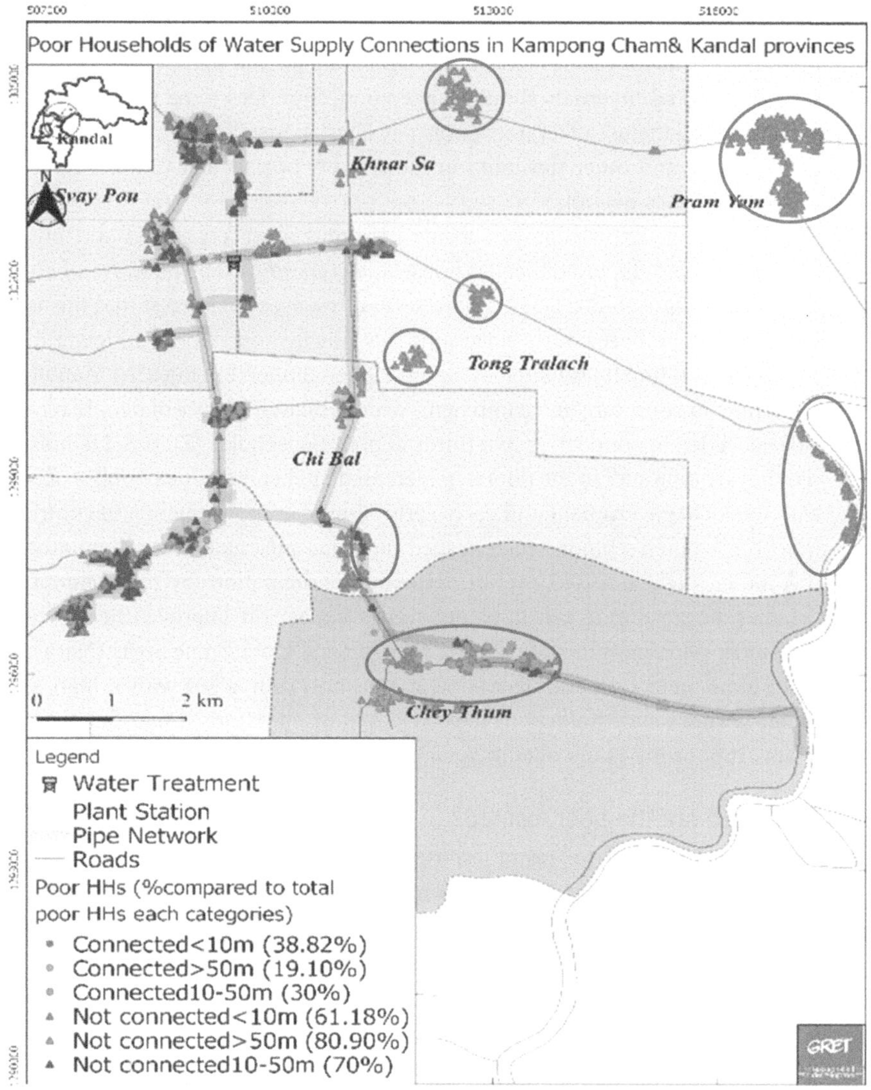

Figure 8 | The sample pipe network, location of poor households, and unconnected clusters.

Table 5 | Distance from the piped network and connectable households per cluster

		Estimated unconnected households (assuming 7–15% are poor)	
Distance along the road to the nearest pipeline (km)	**Total number of clusters**	**Average number of households per cluster**	**Estimated total number of households within the distance**
0–1	23	40–80	1,000–2,000
1–2	15	70–140	1,000–2,000
2–3	15	70–140	1,000–2,000
3–4	1	170–340	1,000–2,000
4–5	3	250–500	150–300
5–6	1	200–400	750–1,500
6–7	2	140–280	200–400
>10	2	490–980	300–500
Total	62	1,430–2,860	5,000–10,000

of staff capabilities, varying water service, and limited piped expansion were the key barriers to private water in Cambodia (World Bank 2015).

This brings the discussion back to cost recovery. The findings earlier on showed that higher tariff rates decrease the likelihood of poor household connection to piped water; however, block tariffs can be anti-poor if targets require cost recovery over connections (Bakker 2007). Furthermore, one study of two cities in Kenya showed that establishing pro-poor units in combination with pre-paid meters improved collection recovery of monthly bills, reducing the utilities' concerns over cost recovery concerns while reaching poor areas (Kemendi & Tutusaus 2018). An entire body of studies is dedicated to the equality of tariffs for the poor, which must be studied in detail before concluding tariff structures; however, one point is clear: connection of poor households must be emphasized, which also requires extension subsidies.

CONCLUSION AND RECOMMENDATIONS

The findings suggest that while pro-poor connection subsidy programs are necessary, the ones in Cambodia may be insufficient for their intended beneficiaries. The regression analysis showed that varying subsidized connection fees below 42 USD resulted in similar connection rates overall, but there is a significant decrease in connection rates for households that lie farther from a pipe network. A review of other studies shows that extra pipe and other hidden fees actually increase the overall cost of connection and are often overlooked. A new regulation by the MISTI attempts to correct this, now requiring operators to provide pipe up to 50 m for the set connection fee, but any impacts have not yet been observed in practice. It is necessary to fully review all costs related to connecting to piped water supply and ensure that a flat-rate cost borne by poor households is consistent for everyone.

Furthermore, nearly half of all poor households are at located more than 70 m from a pipe network and 56% of known households are located within 3 km. Although data for non-poor households were not available to compare services of wealthier inhabitants, other studies also suggest that inequalities in the pipe network reach poorer areas. The complexity for operators to reach poor neighborhoods is a valid concern, including water source, production, and cost recovery, yet the need for expansion remains. This is especially important, as the regression analysis also revealed that higher tariffs also reduce connection rates, although to a lesser degree than distance. Support is necessary for extension programs while maintaining fair tariffs and supporting operators to improve services.

Finally, perceptions and awareness of subsidy programs reveal flaws in marketing programs. Despite the regression analysis suggesting that subsidized connection fees are fair, 78% of unconnected poor households requested a free or subsidized connection, while nearly half had never heard of the subsidy, and 65% of these respondents said they would connect if they were aware of the subsidy. Findings from the programs in Cambodia and elsewhere suggest that short, highlight targeted programs (perhaps at the village level), with fast connection timelines (3 months or less) and significant involvement from the community, local authorities, and/or other local stakeholders are likely to improve overall connection rates.

While this summary provides an overview of key concerns, many other factors may influence connection (Cook *et al.* 2020). Therefore, the importance of improving subsidy programs as well as governance is also essential; Bakker (2007) suggests that government-mandated pro-poor policies and the inclusion of pro-poor targets in contracts could improve connections. Piped water supply licenses in Cambodia already include connection targets; however, it is suggested that plans and targets specifically aimed at connecting poor and vulnerable households to be included in all investment programs.

In summary, the following are recommended to improve programs:

1. *Update subsidies to cover all costs necessary for households to connect.* This will require a detailed review of associated costs, especially pipes needed to reach households within 50 m of the distribution lines, transaction fees, etc.
2. *Ensure flat-rate connection fees paid by all poor households* including all costs that remain ~40 USD or less.
3. *Subsidize pipe extensions to reach unserved areas while maintaining fair tariffs for poor households and cost recoverability for piped water suppliers*, including adequate support for water suppliers themselves to provide improved water supply.
4. *Shorten marketing and awareness programs to 3 months or less and include more local stakeholders*, as well as improve messages about processes and costs.
5. *Add poor household connection requirements* to investment programs and government requirements.

Recommendations for further analysis

To understand the true impact of subsidy programs, it would be useful to analyze poor household connection data compared to non-subsidized connection fees (ranging from 37.5 to 82.55 USD for this dataset, compared to 3.75 to 42.55 USD when subsidized) for older suppliers which have never received subsidies, which perhaps have been in operation for greater than 10 years. While the GRET collected poor household connection data before subsidies, it was not possible to account for old coverage areas versus newly expanded networks.

The issue of tariffs, expansion pipelines, and connection rates merits its own study, due to the complex interactions. It is ideal to consider investment programs' impacts on extension as well, to identify learned opportunities for future developments.

DATA AVAILABILITY STATEMENT

Data cannot be made publicly available; readers should contact the corresponding author for details.

CONFLICT OF INTEREST

The authors declare there is no conflict.

REFERENCES

Abramovsky, L., Luis, A., George, J., Rud, J. P., German, S. & Thibert, M. 2020 *Study of the Distributional Performance of Piped Water Consumption Subsidies in 10 Developing Countries.* World Bank Policy Research Working Paper 9245. Available from: https://ssrn.com/abstract=3599575

Abubakar, I. R. 2019 Factors influencing household access to drinking water in Nigeria. *Utilities Policy* **58**, 40–51. https://doi.org/10.1016/j.jup.2019.03.005.

Adeoti, O. & Fati, B. O. 2020 Barriers to extending piped water distribution networks: the case of Ekiti State, Nigeria. *Utilities Policy* **63**. https://doi.org/10.1016/j.jup.2019.100983.

Angel-Urdinola, D. & Wodon, Q. 2012 Does increasing access to infrastructure services improve the targeting performance of water subsidies? *Journal of International Development* **24** (1), 88–101.

Bakker, K. 2007 Trickle down? Private sector participation and the pro-poor water supply debate in Jakarta, Indonesia. *Geoforum* **38**, 855–868. https://doi.org/10.1016/j.geoforum.2005.11.011.

Basani, M., Isham, J. & Reilly, B. 2008 The determinants of water connection and water consumption: empirical evidence from a Cambodian household survey. *World Development* **36** (5), 953–968. https://doi.org/10.1016/j.worlddev.2007.04.021.

Borja-Vega, C., Garcia Morales, E. E. & Gonzalez, J. A. 2019 Incidence of subsidies in residential public services in Mexico: the case of the water sector. *Water* **11** (10), 2078. https://doi.org/10.3390/w11102078.

Cambodia, T. R. 2011 *National Strategy for Water Supply, Sanitation and Hygiene 2011–2025.*

Cambodia, T. R. 2018 *Cambodian Sustainable Development Goals (CSDGs) Framework (2016–2030).*

Carrard, N., Madden, B., Chong, J., Grant, M., Nghiêm, T. P., Bùi, L. H., Hà, H. T. T. & Willetts, J. 2019 Are piped water services reaching poor households? Empirical evidence from rural Viet Nam. *Water Research* **153**, 239–250. https://doi.org/10.1016/j.watres.2019.01.026.

Cook, J., Fuente, D., Matichich, M. & Whittington, D. 2020 A global assessment of nontariff customer assistance programs in water supply and sanitation. In: *Development Studies in Regional Science. New Frontiers in Regional Science: Asian Perspectives*. Springer, Singapore, p. 42. https://doi.org/10.1007/978-981-15-1435-7_17.

Devkar, G. A., Mahalingam, A., Deep, A. & Thillairajan, A. 2013 Impact of private sector participation on access and quality in provision of electricity, telecom and water services in developing countries: a systematic review. *Utilities Policy* **27**, 65–81. http://dx.doi.org/10.1016/j.jup.2013.08.002.

Grant, M. L., Dominish, E., Carrard, N., Bui, L., Ha, H., Nghiem, T. & Willetts, J. 2016 Reducing or increasing inequalities? The role of private water enterprises in rural Viet Nam. *Development Bulletin* **77**, 31–36.

Grant, M., Foster, T., Dinh, D. V., Willetts, J. & Davis, G. 2020 Life-cycle costs approach for private piped water service delivery: a study in rural Viet Nam. *Journal of Water, Sanitation and Hygiene* **10** (4), 659–669. https://doi.org/10.2166/washdev.2020.037.

GRET 2017 *Internal Report – Documentation of Experience on the Use of Pro-Poor Subsidies for Connection to Piped Water Supply. Unpublished.* Phnom Penh.

Investing in Infrastructure Program (3i), unpublished. *Provincial Investment Plan for Piped Water Supply in Cambodia.* (https://www.springfieldcentre.com/wp-content/uploads/2018/12/3i-Viability-Gap-Funding-for-Private-Water-Operators-in-Rural-Cambodia.pdf).

Kayaga, S. & Franceys, R. 2007 Costs of urban utility water connections: excessive burden to the poor. *Utilities Policy* **15**, 270–277. https://doi.org/10.1016/j.jup.2007.06.002.

Kemendi, T. J. & Tutusaus, M. 2018 The impact of pro-poor interventions on the performance indicators of a water utility: case studies of Nakuru and Kisumu. *Journal of Water, Sanitation and Hygiene for Development* **8** (2), 208–216. https://doi.org/10.2166/washdev.2018.088.

Ministry of Planning 2020. Available from: https://www.idpoor.gov.kh/ (Accessed 13 August).
MISTI 2016 *Prakas 069 MIH/2016 Procedure for Water Tariff Calculation.*
MISTI 2020 *Prakas 320 MISTI/2020 Determination of the Connection Fee.*
MISTI 2021 *Prakas 050 MISTI/2021 Guidelines for Pro-Poor Household Subsidies.*
Narayanan, S., Rajan, A. T., Jebaraj, P. & Elayaraja, M. S. 2017 Delivering basic infrastructure services to the urban poor: a meta-analysis of the effectiveness of bottom-up approaches. *Utilities Policy* **44**, 50–62. https://doi.org/10.1016/j.jup.2017.01.002.
NCDD 2016 *2016 Commune Database.* National Committee for Sub-National Democratic Development.
United Nations 2014 *Special Rapporteur on the Human Right to Safe Drinking Water and Sanitation. Realising the Human Rights to Water and Sanitation: A Handbook by the UN Special Rapporteur Catarina De Albuquerque.* First published in Portugal. https://www.ohchr.org/sites/default/files/Documents/Issues/Water/Handbook/Book1_intro_.pdf.
United Nations 2018 *Sustainable Development Goal 6, Synthesis Report on Water and Sanitation.* New York.
Whittington, D. & Nauges, C. 2020 An assessment of the widespread use of increasing block tariffs in the municipal water supply sector. In: *Oxford Research Encyclopedia of Global Public Health.* https://doi.org/10.1093/acrefore/9780190632366.013.243.
World Bank 2013 *Implementation Completion and Results Report (IDA-41150 IDA-47100) (Red River Delta Rural Water Supply and Sanitation Project).*
World Bank 2015 *Water Supply and Sanitation in Cambodia, Turning Finance into Services for the Future Service Delivery Assessment.*
World Health Organization (WHO) and the United Nations Children's Fund (UNICEF) 2021 *Progress on Household Drinking Water, Sanitation and Hygiene 2000–2020: Five Years Into the SDGs.* Geneva.
Zuin, V. & Nicholson, M. 2021 The impact of pro-poor reforms on consumers and the water utility in Maputo, Mozambique. *Water Alternatives* **14** (1), 158–185.

First received 29 January 2021; accepted in revised form 21 August 2022. Available online 6 September 2022

IWA Publishing's authorised EU representative for General Product Safety Regulations is Diane D'Arras, 15 rue Duret, 75116 Paris, France, e-mail: safety@iwap.co.uk.

Printed and bound by CPI Group (UK) Ltd, Croydon, CR0 4YY
12/05/2026
02108881-0016